Prealgebra

Prealgebra
Fourth Edition

Charles P. McKeague
CUESTA COLLEGE

BROOKS/COLE

™

THOMSON LEARNING

Australia • Canada • Mexico • Singapore • Spain • United Kingdom • United States

BROOKS/COLE

THOMSON LEARNING

Publisher: *Bob Pirtle*
Marketing Manager: *Leah Thomson*
Advertising: *Samantha Cabaluna*
Editorial Assistant: *Erin Mettee-McCutcheon*
Production Editor: *Mary Vezilich*
Production Service: *Heckman & Pinette*
Manuscript Editor: *Margaret Pinette*
Permissions Editor: *Sue Ewing*
Interior Design: *Design Associates*

Cover Design: *Roy Neuhaus*
Cover Photo: *PhotoDisc*
Art Editor: *Lisa Torri*
Interior Illustration: *Lori Heckelman*
Photo Permissions Editor: *Jennifer Mackres*
Print Buyer: *Vena Dyer*
Typesetting: *The Beacon Group*
Printing and Binding: *Transcontinental Printing, Inc.*

Printed in Canada.

10   9   8   7   6   5   4   3   2   1

Library of Congress Cataloging-in-Publication Data

McKeague, Charles P.
  Prealgebra/Charles P. McKeague.--4th ed.
    p. cm.
  ISBN 0-534-37893-5
    1. Mathematics. I. Title.

QA39.2.M418 2001   00-042901
513'.14--dc21

Brief Contents

Contents

Preface to the Instructor

Prealgebra was originally written to bridge the gap between arithmetic and introductory algebra. The basic concepts of algebra are introduced early in the book and then applied to new topics as they are encountered.

General Overview of the Text

The commutative, associative, and distributive properties are covered in Chapter 1, along with the arithmetic of whole numbers. Chapter 2 gives a thorough coverage of negative numbers, along with an introduction to exponents and polynomials. Fractions (both positive and negative), mixed numbers, and simple algebraic fractions are covered in Chapter 3. Linear equations in one variable are covered in Chapter 4, along with graphing in two dimensions. All the material on positive and negative numbers and linear equations is carried into Chapter 5 on decimals. Chapters 6 and 7 cover ratio and proportion, measurement, and percent.

Features of the Text

Flexibility In a lecture-format class, each section of the book can be discussed in a 45- or 50-minute class session. A technique that I have used successfully in lecture classes is to have the students work some of the Practice Problems in the margins of the textbook after I have done an example on the board or overhead projector.

In a self-paced class, the Practice Problems in the margins allow the student to become actively involved with the material before starting the Problem Set that follows that section

Emphasis on Visualization of Topics This edition contains many more diagrams, charts, and graphs than the previous edition. The purpose of this is to give students additional information, in visual form, to help them understand the topics covered.

Practice Problems The Practice Problems, with their answers and solutions, are the key to moving students through the material in this book. Our method of getting students through the material is unique and successful. We call it the EPAS system for Example, Practice Problem, Answer, Solution. Here is how it works: In the margin, next to each example, is a Practice Problem. Students begin by reading through the text, stopping after they have read through an example. Then they work the Practice Problem in the margin next to the example. When they are finished, they check the answer to the Practice Problem at the bottom of the page. If they have made a mistake, they try the problem again. If they still cannot get it right, they look at the solution to the Practice Problem in the back of the book. After working their way through the section in this manner, they are ready to start on the problems in the Problem Set. As an added bonus, there is a video lesson to accompany each section of the text. Each video lesson is 10 to 20 minutes in length. On the video I work some of the odd-numbered problems in the Problem Set.

Descriptive Statistics Beginning in Chapter 1 and then continuing through the rest of the book, students are introduced to descriptive statistics. In Chapter 1 we cover tables and bar charts, as well as mean, median, and mode. These topics are carried through the rest of the book. Along the way we add to the list of descriptive statistics by including scatter diagrams and line graphs. In Chapter 4, we move on to graph ordered pairs and linear equations on a rectangular coordinate system.

Using Technology Scattered throughout the book is new material that shows how scientific calculators, graphing calculators, spreadsheet programs, and the Internet can be used to enhance the topics being covered. This material is easy to find because it appears in the box *Using Technology*.

Facts from Geometry The material on perimeter, area, and volume is introduced in Chapter 1, and then integrated throughout the rest of the book. Each new idea from geometry is introduced with the heading *Facts from Geometry*. In most cases, an example or two accompanies each of the facts to give students a chance to see how topics from geometry are related to the algebra they are learning.

Chapter Openings Each chapter opens with an introduction in which a real-world application, many of which are new to this edition, is used to stimulate interest in the chapter. These opening applications are expanded on later in the chapter. Most of them are explained using the rule of four: in words, numerically, graphically, and algebraically.

Section Openings Many sections begin with a real-world application that leads into the topic developed in the section. This way, we let the application motivate the mathematics, rather than having the mathematics motivate the application.

Blueprint for Problem Solving The Blueprint for Problem Solving is a detailed outline of the steps needed to successfully attempt application problems involving equations. See Section 4.5 for more details.

Study Skills The first five chapter introductions contain a list of study skills to help students organize their time efficiently. More detailed than the general study skills listed in the Preface to the Student, these study skills point students in the direction of success. They are intended to benefit students not only in this course but also throughout their college careers.

Problem Set Organization Following each section is a Problem Set. For this edition, Problem Sets have been expanded to include more work on estimation and reading and using graphs. Problem Sets contain the following items:

1. *Drill:* There are enough drill problems in each Problem Set to ensure that students working the odd-numbered problems become proficient with the material.

2. *Progressive Difficulty:* The problems increase in difficulty as the Problem Set progresses.

3. *Odd-Even Similarities:* Whenever possible, each even-numbered problem is similar to the odd-numbered problem that precedes it. The answers to the odd-numbered problems are listed in the back of the book.

4. *Applying the Concepts:* Application problems are emphasized throughout the book. The placement of the word problems is such that the student must do more than just read the section title to decide how to set up and solve a word problem. For example, Problem Set 1.7 covers multiplication with whole numbers. The solutions to some of the application problems in the Problem Set require multiplication, but others require addition and subtraction as well.

5. *Calculator Problems:* Most of the Problem Sets contain a set of problems to be worked on a calculator.

6. *Review Problems:* Beginning in Chapter 2, each Problem Set contains a few review problems. Whenever possible, these problems review concepts from preceding sections that are needed for the development of the following section.

7. *Extending the Concepts:* These problems, most of which can be categorized as critical thinking problems, extend the concepts covered in the section.

8. *Answers:* The answers to all odd-numbered problems in the Problem Sets and Chapter Reviews, and the answers to all problems in the Chapter Tests, are given in the back of the book.

End-of-Chapter Retrospective Each chapter ends with the following items that together give a comprehensive reexamination of the chapter and some of the important problems from previous chapters:

1. *Chapter Summary:* These list all main points from the chapter. In the margin, next to each topic being reviewed, is an example that illustrates the type of problem associated with the topic being reviewed.

2. *Chapter Review:* An extensive set of problems that review all the main topics in the chapter. The chapter reviews contain more problems than the chapter tests.

3. *Chapter Test:* A set of problems representative of all the main points of the chapter.

4. *Cumulative Review:* Starting in Chapter 2, each chapter ends with a set of problems that reviews material from all preceding chapters. The cumulative review keeps students current with past topics and helps them retain the information they study.

A Sample of the Features of *Prealgebra*

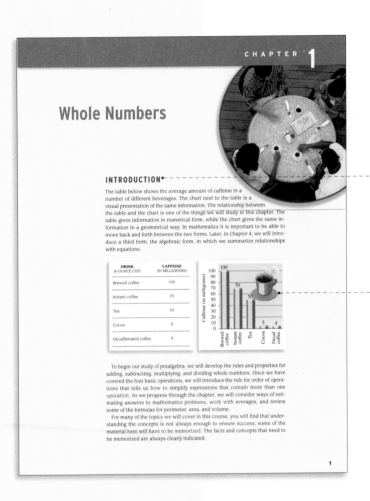

Whole Numbers

CHAPTER 1

INTRODUCTION

The table below shows the average amount of caffeine in a number of different beverages. The chart next to the table is a visual presentation of the same information. The relationship between the table and the chart is one of the things we will study in this chapter. The table gives information in numerical form, while the chart gives the same information in a geometrical way. In mathematics it is important to be able to move back and forth between the two forms. Later, in Chapter 4, we will introduce a third form, the algebraic form, in which we summarize relationships with equations.

DRINK (6-OUNCE CUP)	CAFFEINE (IN MILLIGRAMS)
Brewed coffee	100
Instant coffee	70
Tea	50
Cocoa	5
Decaffeinated coffee	4

To begin our study of prealgebra, we will develop the rules and properties for adding, subtracting, multiplying, and dividing whole numbers. Once we have covered the four basic operations, we will introduce the rule for order of operations that tells us how to simplify expressions that contain more than one operation. As we progress through the chapter, we will consider ways of estimating answers to mathematics problems, work with averages, and review some of the formulas for perimeter, area, and volume.

For many of the topics we will cover in this course, you will find that understanding the concepts is not always enough to ensure success; some of the material here will have to be memorized. The facts and concepts that need to be memorized are always clearly indicated.

1

Chapter Introduction

A real-world application now opens each chapter, and is often expanded and reiterated in the book.

Emphasis on Visualization of Topics

Clear, informative diagrams, charts, and graphs enliven the new edition and help students visualize the topics they are studying.

New! *A Digital Video Companion to Prealgebra (DVC)* Packaged with each book is a single CD-ROM containing over eight hours of video instruction. The foundation of the DVC is video instruction presented by author Charles P. McKeague. There is one video lesson for each section of the book. Each video lesson is eight to twelve minutes long. The problems worked during each video lesson are listed next to the viewing screen, so that students can work them ahead of time, if they choose. In addition, each section of the DVC contains a study guide to help students work their way through the book and prepare for the video lesson. In order to help students evaluate their progress, each chapter of the DVC contains a chapter test, with answers.

More Application Problems

Each section of the text contains a number of application problems under the heading *Applying the Concepts*. There are many new application problems in this edition. Each application problem has a title for easy classification.

Increased Use of Charts and Graphs

This edition contains many pie charts, bar charts, and line graphs to give a visual component to the topics being studied.

Increased Use of Tables

Tables are used throughout the text, beginning in Chapter 1. Students learn to use tables to help organize data and to assist them in recognizing patterns that occur regularly in their study of mathematics.

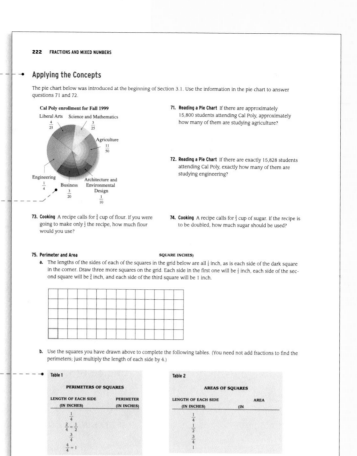

Practice Problems

A hallmark of the McKeague texts, Practice Problems in the margin are the key to moving students through the material. Answers are at the bottom of the page. Solutions are at the back of the book

Using Technology

This boxed material is spread throughout the text. It shows how calculators, both scientific and graphing, can be applied to the topics being studied. Spreadsheet programs and the Internet are also considered in *Using Technology* boxes.

Descriptive Statistics

Also spread throughout the text, this material is set off from the surrounding material. Among the topics covered under this heading are bar charts, line graphs, mean, median, and mode.

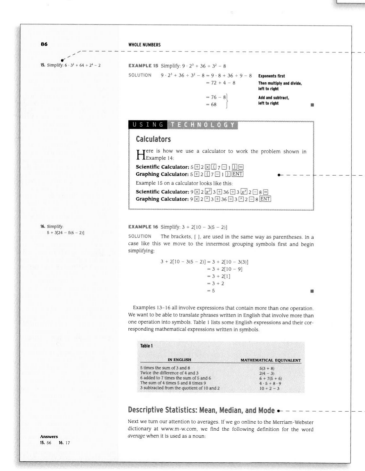

Supplements to the Text

For the Instructor

Annotated Instructor's Edition This is a specially bound version of the complete student text with exercise answers printed adjacent to each exercise and teaching notes written for the instructor by the author.

Test Bank The test bank includes three tests per chapter as well as three final exams. The tests are made up of a combination of multiple-choice, free-response, true/false, and fill-in-the-blank questions.

BCA Testing With a balance of efficiency and high performance, simplicity and versatility, *Brooks/Cole Assessment* gives you the power to transform the learning and teaching experience. This revolutionary, Internet-ready testing suite is text-specific and allows instructors to customize exams and track student progress in an accessible, browser-based format. BCA offers full algorithmic generation of problems and free-response mathematics. No longer are you limited to multiple-choice or true/false test questions. The complete integration of the testing and course management components simplifies your routine tasks. Test results flow automatically to your gradebook and you can easily communicate to individuals, sections, or entire courses.

Videotapes Created and hosted by the author, this set of videotapes is available free upon adoption of the text. Each tape offers one chapter of the text and is broken down into 10- to 20-minute problem-solving lessons that cover each section of the chapter.

For the Student

Student Solutions Manual The student solutions manual provides worked-out solutions to the odd-numbered problems.

BCA Tutorial This text-specific, interactive tutorial software is delivered via the Web (at http://bca.brookscole.com) and is offered in both student and instructor versions. Like *BCA Testing*, it is browser-based, making it an intuitive mathematical guide even for students with little technological proficiency. So sophisticated it's simple, *BCA Tutorial* allows students to work with real math notation in real time, providing instant analysis and feedback. The tracking program built into the instructor version of the software enables instructors to carefully monitor student progress.

A Digital Video Companion (DVC) This single CD-ROM containing over eight hours of video instruction is packaged with each book. The foundation of the DVC is the video lessons. There is one video lesson for each section of the book. Each video lesson is eight to twelve minutes long. The problems worked during each video lesson are listed next to the viewing screen, so that students can work them ahead of time, if they choose.

Video Notebook This workbook contains all of the problems that are worked out on the videotapes in a format that allows the student to work along with the video.

Acknowledgments

What a wonderful team of people we had working on this book! To begin, Bob Pirtle, my editor at Brooks/Cole, was very helpful and encouraging in all parts of the revision process. Mary Vezilich and Margaret Pinette handled the production part of this revision, and I am delighted to have been able to work with such efficient, professional, pleasant people. Because of them, the production process on this book has been one of the best experiences I have had in publishing. Assisting Mary and Margaret was the team at Brooks/Cole: Vernon Boes, Art Director; Roy Neuhaus, Senior Design Editor; Jennifer Mackres, Photo Permissions Editor; Lisa Torri, Art Editor; and Rachael Bruckman, Assistant Editor. They all worked together to integrate the different parts of this project, and I think you will agree that they did an outstanding job. Dudley Brooks, and my son, Patrick, suggested many new applications and exercises for the problem sets. Their suggestions have strengthened this revision. Mark Turner and John Banks did an excellent job of accuracy checking and proofreading. Mark's input and suggestions for the Digital Video Companion that accompanies the text were also very helpful. Finally, Sue Ewing worked very hard to get the many permissions and sources straightened out for the new real-data problems. My thanks to all these people; this book would not have been possible without them.

Thanks also to Diane McKeague, for her encouragement with all my writing endeavors, and to Amy Jacobs for all her advice and encouragement.

Finally, I am grateful to the following instructors for their suggestions and comments on this revision. Some reviewed the entire manuscript, while others were asked to evaluate the development of specific topics, or the overall sequence of topics. My thanks go to the people listed below:

Renee Starr, *Beaver College*
Stan VerNooy, *Community College of Southern Nevada*
Shirley Daniels, *Cuesta College*
Mary Ellen Gallegos, *Santa Fe Community College*

—Pat McKeague

Preface to the Student

Many of the people who enroll in my introductory math classes are apprehensive at first. They think that since they have had a difficult time with mathematics in the past, they are in for a difficult time again. Some of them feel that they have never really understood mathematics and probably never will.

Do you feel like that?

Most people who have a difficult time with mathematics expect something from it that is not there. They believe that since mathematics is a logical subject, they should be able to understand it without any trouble. Mathematics is probably the most logical subject there is, but that doesn't mean it is always easy to understand. Some topics in mathematics must be read and thought about many times before they become understandable.

If you are interested in having a positive experience in this course and have had difficulty with mathematics in the past, then the following list will be of interest to you. The list explains the things you can do to make sure that you are successful in mathematics.

How to Be Successful in Mathematics

1. **If you are in a lecture class, be sure to attend all class sessions on time.** You cannot know exactly what goes on in class unless you are there. Missing class and then expecting to find out what went on from someone else is not the same as being there yourself.

2. **Read the book and work the Practice Problems.** It is best to read the section that will be covered in class beforehand. Reading in advance, even if you do not understand everything you read, is still better than going to class with no idea of what will be discussed. As you read through each section, be sure to work the Practice Problems in the margin of the text. Each Practice Problem is similar to the example with the same number. Look over the example and then try the corresponding Practice Problem. The answers to most of the Practice Problems are given on the same page as the problems. If you don't get the correct answer, see if you can rework the problem correctly. If you miss it a second time, check your solution with the solution to the Practice Problem in the back of the book.

3. **Work problems every day and check your answers.** This is especially important if you are studying in a self-paced environment. The key to success in mathematics is working problems. The more problems you work, the better you will become at working them. The answers to the odd-numbered problems are given in the back of the book. When you have finished an assignment, be sure to compare your answers with those in the book. If you have made a mistake, find out what it is, and correct it.

4. **Do it on your own.** Don't be mislead into thinking someone else's work is your own. Watching someone else work though a problem is not the same as working the same problem yourself. It is okay to get help when you are stuck; as a matter of fact, it is a good idea. Just be sure that you do the *work* yourself.

5. **Don't expect to understand every new topic the first time you see it.** Sometimes you will understand everything you are doing, and sometimes you won't. That's just the way things are in mathematics. Expecting to understand each topic the first time you see it will only lead to disappointment and frustration. The process of understanding mathematics takes time. It requires that you read the book, work the problems, and get your questions answered.

6. **Review every day.** After you have finished the assigned problems on a certain day, take at least another 15 minutes to go back and review a section you did previously. You can review by working the Practice Problems in the margin or by doing some of the problems in the Problem Set. The more you review, the longer you will retain the material. Also, there are times when material that at first seemed unclear will become understandable when you review it.

7. **Spend as much time as it takes for you to master the material.** There is no set formula for the exact amount of time you will need to spend on the material in this course in order to master it. You will find out quickly—probably on the first test—if you are spending enough time studying. Even if it turns out that you have to spend 2 or 3 hours on each section to master it, then that's how much time you should take. Trying to get by with less time will not work.

8. **Relax.** It's probably not as difficult as you think.

Prealgebra

Whole Numbers

INTRODUCTION

The table below shows the average amount of caffeine in a number of different beverages. The chart next to the table is a visual presentation of the same information. The relationship between the table and the chart is one of the things we will study in this chapter. The table gives information in numerical form, while the chart gives the same information in a geometrical way. In mathematics it is important to be able to move back and forth between the two forms. Later, in Chapter 4, we will introduce a third form, the algebraic form, in which we summarize relationships with equations.

DRINK (6-OUNCE CUP)	CAFFEINE (IN MILLIGRAMS)
Brewed coffee	100
Instant coffee	70
Tea	50
Cocoa	5
Decaffeinated coffee	4

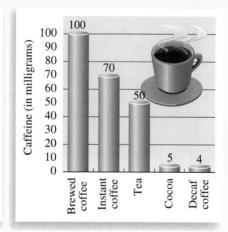

To begin our study of prealgebra, we will develop the rules and properties for adding, subtracting, multiplying, and dividing whole numbers. Once we have covered the four basic operations, we will introduce the rule for order of operations that tells us how to simplify expressions that contain more than one operation. As we progress through the chapter, we will consider ways of estimating answers to mathematics problems, work with averages, and review some of the formulas for perimeter, area, and volume.

For many of the topics we will cover in this course, you will find that understanding the concepts is not always enough to ensure success; some of the material here will have to be memorized. The facts and concepts that need to be memorized are always clearly indicated.

STUDY SKILLS

Some of the students enrolled in my mathematics classes develop difficulties early in the course. Their difficulties are not associated with their ability to learn mathematics; they all have the potential to pass the course. Students who get off to a poor start do so because they have not developed the study skills necessary to be successful in mathematics; they do not put themselves on an effective homework schedule, and when they work problems, they do it their way, not my way. Here is a list of things you can do to begin to develop effective study skills.

1. Put Yourself on a Schedule. The general rule is that you spend two hours on homework for every hour you are in class. Make a schedule for yourself in which you set aside two hours each day to work on this course. Once you make the schedule, stick to it. Don't just complete your assignments and then stop. Use all the time you have set aside. If you complete the assignment and have time left over, read the next section in the book, and then work more problems. As the course progresses you may find that two hours a day is not enough time to master the material in this course. If it takes you longer than two hours a day to reach your goals for this course, then that's how much time it takes. Trying to get by with less will not work.

2. Find Your Mistakes and Correct Them. There is more to studying mathematics than just working problems. You must always check your answers with the answers in the back of the book. When you have made a mistake, find out what it is, and then correct it. Making mistakes is part of the process of learning mathematics. I have never had a successful student who didn't make mistakes—lots of them. Your mistakes are your guides to understanding; look forward to them.

3. Imitate Success. Your work should look like the work you see in this book and the work your instructor shows. The steps shown in solving problems in this book were written by someone who has been successful in mathematics. The same is true of your instructor. Your work should imitate the work of people who have been successful in mathematics.

If you are determined to pass this course, make it a point to become organized and effective right from the start and to show your work in the same way as it is shown in this book.

1.1 ## Place Value and Names for Numbers

Introduction . . .

The two diagrams below are known as Pascal's triangle, after the French mathematician and philosopher Blaise Pascal (1623–1662). Both diagrams contain the same information. The one on the left contains numbers in our number system; the one on the right uses numbers from Japan in 1781. If you study the two diagrams, you will find that you can count to 20 in Japanese.

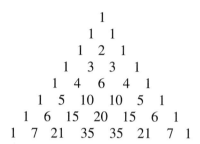

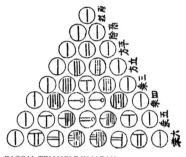

PASCAL TRIANGLE IN JAPAN
From Murai Chūzen's *Sampō Dōshi-mon* (1781)

Our number system is based on the number 10 and is therefore called a "base 10" number system. We write all numbers in our number system using the *digits* 0, 1, 2, 3, 4, 5, 6, 7, 8, and 9. The positions of the digits in a number determine the values of the digits. For example, the 5 in the number 251 has a different value from the 5 in the number 542.

The *place values* in our number system are as follows: The first digit on the right is in the *ones column*. The next digit to the left of the ones column is in the *tens column*. The next digit to the left is in the *hundreds column*. For a number like 542, the digit 5 is in the hundreds column, the 4 is in the tens column, and the 2 is in the ones column.

If we keep moving to the left, the columns increase in value. The following diagram shows the name and value of each of the first seven columns in our number system:

MILLIONS COLUMN	HUNDRED THOUSANDS COLUMN	TEN THOUSANDS COLUMN	THOUSANDS COLUMN	HUNDREDS COLUMN	TENS COLUMN	ONES COLUMN
1,000,000	100,000	10,000	1,000	100	10	1

EXAMPLE 1 Give the place value of each digit in the number 305,964.

SOLUTION Starting with the digit at the right, we have:

4 in the ones column, 6 in the tens column, 9 in the hundreds column, 5 in the thousands column, 0 in the ten thousands column, and 3 in the hundred thousands column. ■

To find the place values of digits in larger numbers, we can refer to Table 1.

Practice Problems
1. Give the place value of each digit in the number 46,095.

Answer
1. 5 ones, 9 tens, 0 hundreds, 6 thousands, 4 ten thousands

Table 1

Hundred Billions 100,000,000,000	Ten Billions 10,000,000,000	Billions 1,000,000,000	Hundred Millions 100,000,000	Ten Millions 10,000,000	Millions 1,000,000	Hundred Thousands 100,000	Ten Thousands 10,000	Thousands 1,000	Hundreds 100	Tens 10	Ones 1

2. Give the place value of each digit in the number 21,705,328,456.

EXAMPLE 2   Give the place value of each digit in the number 73,890,672,540.

SOLUTION   The following diagram shows the place value of each digit.

Ten Billions	Billions	Hundred Millions	Ten Millions	Millions	Hundred Thousands	Ten Thousands	Thousands	Hundreds	Tens	Ones
7	3,	8	9	0,	6	7	2,	5	4	0

We can use the idea of place value to write numbers in *expanded form*. For example, the number 542 can be written in expanded form as

$$542 = 500 + 40 + 2$$

because the 5 is in the hundreds column, the 4 is in the tens column, and the 2 is in the ones column.

Here are more examples of numbers written in expanded form.

3. Write 3,972 in expanded form.

EXAMPLE 3   Write 5,478 in expanded form.

SOLUTION $5,478 = 5,000 + 400 + 70 + 8$

We can use money to make the results from Example 3 more intuitive. Suppose you have $5,478 in cash as follows:

$5,000 $400 $70 $8

Using this diagram as a guide, we can write

$$\$5,478 = \$5,000 + \$400 + \$70 + \$8$$

which shows us that our work writing numbers in expanded form is consistent with our intuitive understanding of the different denominations of money.

4. Write 271,346 in expanded form.

EXAMPLE 4   Write 354,798 in expanded form.

SOLUTION $354,798 = 300,000 + 50,000 + 4,000 + 700 + 90 + 8$

Answers
2. 6 ones, 5 tens, 4 hundreds,
   8 thousands, 2 ten thousands,
   3 hundred thousands, 5 millions,
   0 ten millions, 7 hundred millions,
   1 billion, 2 ten billions
3. 3,000 + 900 + 70 + 2
4. 200,000 + 70,000 + 1,000 +
   300 + 40 + 6

EXAMPLE 5 Write 56,094 in expanded form.

SOLUTION Notice that there is a 0 in the hundreds column. This means we have 0 hundreds. In expanded form we have

$$56{,}094 = 50{,}000 + 6{,}000 + 90 + 4$$

**Note that we don't have
to include the 0 hundreds** ■

EXAMPLE 6 Write 5,070,603 in expanded form.

SOLUTION The columns with 0 in them will not appear in the expanded form.

$$5{,}070{,}603 = 5{,}000{,}000 + 70{,}000 + 600 + 3$$ ■

The idea of place value and expanded form can be used to help write the names for numbers. Naming numbers and writing them in words takes some practice. Let's begin by looking at the names of some two-digit numbers. Table 2 lists a few. Notice that the two-digit numbers that do not end in 0 have two parts. These parts are separated by a hyphen.

Table 2

NUMBER	IN ENGLISH	NUMBER	IN ENGLISH
25	*Twenty-five*	30	*Thirty*
47	*Forty-seven*	62	*Sixty-two*
93	*Ninety-three*	77	*Seventy-seven*
		50	*Fifty*

The following examples give the names for some larger numbers. In each case the names are written according to the place values given in Table 1.

EXAMPLE 7 Write each number in words.
    a. 452   **b.** 397   **c.** 608

SOLUTION   **a.** Four hundred fifty-two
       b. Three hundred ninety-seven
       c. Six hundred eight ■

EXAMPLE 8 Write each number in words.
    a. 3,561   **b.** 53,662   **c.** 547,801

SOLUTION   **a.** Three thousand, five hundred sixty-one

**Notice how the comma separates
the thousands from the hundreds**

      b. Fifty-three thousand, six hundred sixty-two
      c. Five hundred forty-seven thousand, eight hundred one ■

---

5. Write 71,306 in expanded form.

6. Write 4,003,560 in expanded form.

7. Write each number in words.
  a. 724
  b. 595
  c. 307

8. Write each number in words.
  a. 4,758
  b. 62,779
  c. 305,440

Answers
5. 70,000 + 1,000 + 300 + 6
6. 4,000,000 + 3,000 + 500 + 60
7. a. Seven hundred twenty-four
  b. Five hundred ninety-five
  c. Three hundred seven
8. a. Four thousand, seven hundred fifty-eight
  b. Sixty-two thousand, seven hundred seventy-nine
  c. Three hundred five thousand, four hundred forty

9. Write each number in words.
  a. 707,044,002
  b. 452,900,008
  c. 4,008,002,001

EXAMPLE 9 Write each number in words
  a. 507,034,005
  b. 739,600,075
  c. 5,003,007,006

SOLUTION **a.** Five hundred seven million, thirty-four thousand, five
  b. Seven hundred thirty-nine million, six hundred thousand, seventy-five
  c. Five billion, three million, seven thousand, six ∎

The next examples show how we write a number given in words as a number written with digits.

10. Write six thousand, two hundred twenty-one using digits instead of words.

EXAMPLE 10 Write five thousand, six hundred forty-two, using digits instead of words.

SOLUTION *Five Thousand,*   *Six Hundred*   *Forty-two*
              5,   6   4 2 ∎

11. Write each number with digits instead of words.
  a. Eight million, four thousand, two hundred
  b. Twenty-five million, forty
  c. Nine million, four hundred thirty-one

EXAMPLE 11 Write each number with digits instead of words.
  a. Three million, fifty-one thousand, seven hundred
  b. Two billion, five
  c. Seven million, seven hundred seven

SOLUTION **a.** 3,051,700
  b. 2,000,000,005
  c. 7,000,707 ∎

In mathematics a collection of numbers is called a *set*. In this chapter we will be working with the set of counting numbers and the set of *whole numbers*, which are defined as follows:

Counting numbers = {1, 2, 3, . . .}
Whole numbers = {0, 1, 2, 3, . . .}

The dots mean "and so on," and the braces { } are used to group the numbers in the set together.

Another way to visualize the whole numbers is with a *number line*. To draw a number line, we simply draw a straight line and mark off equally spaced points along the line, as shown in Figure 1. We label the point at the left with 0 and the rest of the points, in order, with the numbers 1, 2, 3, 4, 5, and so on.

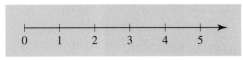

FIGURE 1

The arrow on the right indicates that the number line can continue in that direction forever. When we refer to numbers in this chapter, we will always be referring to the whole numbers.

Answers
9. a. Seven hundred seven million, forty-four thousand, two
  b. Four hundred fifty-two million, nine hundred thousand, eight
  c. Four billion, eight million, two thousand, one
10. 6,221
11. a. 8,004,200
  b. 25,000,040
  c. 9,000,431

Problem Set 1.1

Give the place value of each digit in the following numbers.

1. 78   **2.** 93   **3.** 45   **4.** 79   **5.** 348   **6.** 789

7. 608   **8.** 450   **9.** 2,378   **10.** 6,481   **11.** 273,569   **12.** 768,253

Give the place value of the 5 in each of the following numbers.

13. 458,992   **14.** 75,003,782   **15.** 507,994,787   **16.** 320,906,050

17. 267,894,335   **18.** 234,345,678,789   **19.** 4,569,000   **20.** 50,000

Write each of the following numbers in expanded form.

21. 658   **22.** 479

23. 68   **24.** 71

25. 4,587   **26.** 3,762

27. 32,674   **28.** 54,883

29. 3,462,577   **30.** 5,673,524

31. 407   **32.** 508

33. 30,068   **34.** 50,905

35. 3,004,008   **36.** 20,088,060

Write each of the following numbers in words.

37. 29

38. 75

39. 40

40. 90

41. 573

42. 895

43. 707

44. 405

45. 770

46. 450

47. 23,540

48. 56,708

49. 3,004

50. 5,008

51. 3,040

52. 5,080

53. 104,065,780

54. 637,008,500

55. 5,003,040,008

56. 7,050,800,001

57. 2,546,731

58. 6,998,454

Write each of the following numbers with digits instead of words.

59. Three hundred twenty-five

60. Forty-eight

61. Five thousand, four hundred thirty-two

62. One hundred twenty-three thousand, sixty-one

63. Eighty-six thousand, seven hundred sixty-two

64. One hundred million, two hundred thousand, three hundred

65. Two million, two hundred

66. Two million, two

67. Two million, two thousand, two hundred

68. Two billion, two hundred thousand, two hundred two

Applying the Concepts

69. Population The population of San Luis Obispo, California, was 43,704 in 1999. What is the place value of the 3 in this number?

Welcome to
San Luis Obispo
Population 43,704
Founded 1772

70. Seating Arrangements The number of different ways in which 10 people can be seated at a table with 10 places is 3,628,800. What is the place value of the 3 in this number?

71. Seating Capacity The Rose Bowl has a seating capacity of 106,721. Write this number in expanded form.

72. Astronomy The distance from the sun to the earth is 92,897,416 miles. Write this number in expanded form.

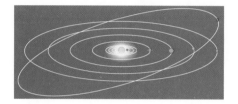

73. Baseball Salaries The November 10, 1999, edition of *USA Today* reported that major league baseball's 1999 average player salary was $1,384,530, representing an increase of 13.24% from the previous season's average. Write 1,384,530 in words.

74. Tallest Mountain The world's tallest mountain is Mount Everest. On May 5, 1999, it was found to be 7 feet taller than it had previously been thought to be. Before this date, Everest was thought to be 29,028 feet high. That height was determined by B. L. Gulatee in 1954. The first measurement of Everest was in 1847. At that time the height was given as 29,002 feet. Write 29,002 in words.

Populations of Countries The table below gives estimates of the populations of some countries for the year 2000. The first column under *Population* gives the population in digits. The second column gives the population in words. Fill in the blanks.

COUNTRY	POPULATION DIGITS	POPULATION WORDS
75. United States	_____	Two hundred seventy-five million
76. Peoples Republic of China	_____	One billion, two hundred fifty-six million
77. Japan	127,000,000	_____
78. United Kingdom	59,000,000	_____

(From U.S. Census Bureau, International Data Base)

Populations of Cities The table below gives estimates of the populations of some cities for the year 2000. The first column under *Population* gives the population in digits. The second column gives the population in words. Fill in the blanks.

CITY	POPULATION DIGITS	POPULATION WORDS
79. Tokyo	_____	Thirty million
80. Los Angeles	_____	Eleven million
81. Paris	8,800,000	_____
82. London	8,600,000	_____

Extending the Concepts

Many of the Problem Sets in this book will end with a few problems like the ones below. These problems challenge you to extend your knowledge of the material in the Problem Set. In most cases, there are no examples in the text similar to these problems. You should approach these problems with a positive point of view; even though you may not always work them correctly, just the process of attempting them will increase your knowledge and ability in mathematics.

The numbers on the number line below are each 1 inch apart. As with all number lines, the arrow indicates that the number line continues to the right indefinitely. Use this number line to answer Problems 83–92.

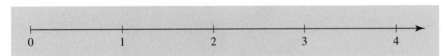

83. How far apart are the numbers 1 and 4?

84. How far apart are the numbers 2 and 5?

85. How far apart are the numbers 0 and 5?

86. How far apart are the numbers 0 and 10?

87. What number is 4 inches to the right of 2?

88. What number is 4 inches to the right of 3?

89. What number is 4 inches to the left of 7?

90. What number is 4 inches to the left of 12?

91. If 1 foot is 12 inches in length, what number is 1 foot from 0?

92. What number is 2 feet from 0?

1.2 Properties and Facts of Addition

Introduction . . .

Suppose we put two pieces of wood end to end and measure their combined length with a tape measure. If one of the pieces is 3 feet long, and the other is 5 feet long, we could draw a diagram like this:

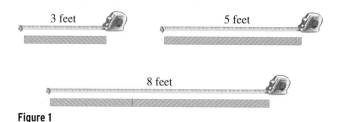

3 feet   5 feet

8 feet

Figure 1

We can summarize all the information above with the following addition problem:

3 feet + 5 feet = 8 feet

Using lengths to visualize addition can be very helpful. In mathematics we generally do so by using the number line. For example, we add 3 and 5 on the number line like this: Start at 0 and move to 3, as shown in Figure 2. From 3, move 5 more units to the right. This brings us to 8. Therefore, $3 + 5 = 8$.

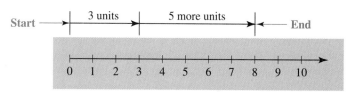

Start ⟶ 3 units | 5 more units ⟶ **End**

0   1   2   3   4   5   6   7   8   9   10

Figure 2

If we do this kind of addition on the number line with all combinations of the numbers 0 through 9, we get the results summarized in Table 1.

Table 1

ADDITION TABLE

+	0	1	2	3	4	5	6	7	8	9
0	0	1	2	3	4	5	6	7	8	9
1	1	2	3	4	5	6	7	8	9	10
2	2	3	4	5	6	7	8	9	10	11
3	3	4	5	6	7	8	9	10	11	12
4	4	5	6	7	8	9	10	11	12	13
5	5	6	7	8	9	10	11	12	13	14
6	6	7	8	9	10	11	12	13	14	15
7	7	8	9	10	11	12	13	14	15	16
8	8	9	10	11	12	13	14	15	16	17
9	9	10	11	12	13	14	15	16	17	18

Note
Table 1 is a summary of the addition facts that you *must* know in order to make a successful start in your study of prealgebra. You *must* know how to add any pair of numbers that come from the list. You *must* be fast and accurate. You don't want to have to think about the answer to 7 + 9. You should know it's 16. Memorize these facts now. Don't put it off until later.

We read Table 1 in the following manner: Suppose we want to use the table to find the answer to 3 + 5. We locate the 3 in the column on the left and the 5 in the row at the top. We read *across* from the 3 and *down* from the 5. The entry in the table that is across from 3 and below 5 is 8. To see that 3 + 5 = 8, we read across from 3 and down from 5.

Variables: An Intuitive Look

When you filled out the application for the school you are attending, there was a space to fill in your first name. "First Name" is a variable quantity, because the value it takes on depends on who is filling out the application. For example, if your first name is Raul, then the value of "First Name" is Raul. On the other hand, if your first name is Christa, then the value of "First Name" is Christa.

If we abbreviate "First Name" with *FN*, "Last Name" with *LN*, and "Whole Name" with *WN*, then we take the concept of a variable further and write the relationship between the names this way:

$$FN + LN = WN$$

(We are using the + symbol loosely here to represent writing the names together with a space between them.) The relationship that we have written above holds for all people who have only a first name and a last name. For those people who have a middle name, the relationship between the names is:

$$FN + MN + LN = WN$$

A similar situation exists in mathematics when we let a letter stand for a number or a group of numbers. For instance, if we say "let *a* and *b* represent numbers," then *a* and *b* are called *variables*, because the values they take on vary. We use the variables *a* and *b* in the definition below because we want you to know that the definition is true for all numbers that you will encounter in this book.

Vocabulary

The word we use to indicate addition is the word *sum*. If we say "the sum of 3 and 5 is 8," what we mean is 3 + 5 = 8. The word *sum* always indicates addition. We can state this fact in symbols by using the letters *a* and *b* to represent numbers.

> **DEFINITION**
>
> If *a* and *b* are any two numbers, then the **sum** of *a* and *b* is $a + b$. To find the sum of two numbers, we add them.

Table 2 gives some word statements and their mathematical equivalents written in symbols.

Application

Type or print in ink.

First Name_____

Middle Name_____

Last Name_____

Note

When mathematics is used to solve everyday problems, the problems are almost always stated in words. The translation of English to symbols is a very important part of mathematics.

Table 2

IN ENGLISH	IN SYMBOLS
The sum of 4 and 1	$4 + 1$
4 added to 1	$1 + 4$
8 more than m	$m + 8$
x increased by 5	$x + 5$
The sum of x and y	$x + y$
The sum of 2 and 4 is 6	$2 + 4 = 6$

EXAMPLE 1 Find the sum of 6 and 9.

SOLUTION   Because the word *sum* indicates addition, we have $6 + 9 = 15$. We say that the sum of 6 and 9 is 15. If you don't know where we got 15, go back to Table 1, and work on memorizing the addition facts. Both the expressions $6 + 9$ and 15 are the sum of 6 and 9. We could also write this sum in column form as

$$\begin{array}{r} 6 \\ + 9 \\ \hline 15 \end{array}$$

∎

Properties of Addition

Once we become familiar with the addition table, we may notice some facts about addition that are true regardless of the numbers involved. The first of these facts involves the number 0 (zero).

Notice in Table 1 that whenever we add 0 to a number, the result is the original number. That is:

$0 + 0 = 0$	$5 + 0 = 5$	$0 + 0 = 0$	$0 + 5 = 5$
$1 + 0 = 1$	$6 + 0 = 6$	$0 + 1 = 1$	$0 + 6 = 6$
$2 + 0 = 2$	$7 + 0 = 7$	$0 + 2 = 2$	$0 + 7 = 7$
$3 + 0 = 3$	$8 + 0 = 8$	$0 + 3 = 3$	$0 + 8 = 8$
$4 + 0 = 4$	$9 + 0 = 9$	$0 + 4 = 4$	$0 + 9 = 9$

Because this fact is true no matter what number we add to 0, we call it a property of 0.

Addition Property of 0

If we let *a* represent any number, then it is always true that

$$a + 0 = a \quad \text{and} \quad 0 + a = a$$

In words: Adding 0 to any number leaves that number unchanged.

A second property we notice by becoming familiar with the addition table is that the order of two numbers in a sum can be changed without changing the result.

$$3 + 5 = 8 \quad \text{and} \quad 5 + 3 = 8$$
$$4 + 9 = 13 \quad \text{and} \quad 9 + 4 = 13$$

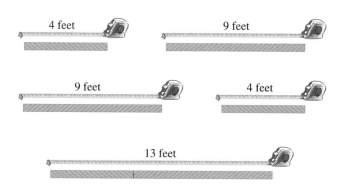

This fact about addition is true for *all* numbers—not just the numbers listed in the addition table. The order in which you add two numbers doesn't affect the result. We call this fact the *commutative property* of addition, and we write it in symbols as follows.

Commutative Property of Addition

If *a* and *b* are any two numbers, then it is always true that

$$a + b = b + a$$

In words: Changing the order of two numbers in a sum doesn't change the result.

EXAMPLE 2 Use the commutative property of addition to rewrite each sum.
  a. $4 + 6$ **b.** $5 + 9$ **c.** $3 + 0$ **d.** $7 + n$

SOLUTION   The commutative property of addition indicates that we can change the order of the numbers in a sum without changing the result. Applying this property we have:

  a. $4 + 6 = 6 + 4$
  b. $5 + 9 = 9 + 5$
  c. $3 + 0 = 0 + 3$
  d. $7 + n = n + 7$

Notice that we did not actually add any of the numbers. The instructions were to use the commutative property, and the commutative property involves only the order of the numbers in a sum. ■

The last property of addition we will consider here has to do with sums of more than two numbers. Suppose we want to find the sum of 2 and 3 and then add 4 to it. Because we want to find the sum of 2 and 3 first, and then add 4 to what we get, we use parentheses:

$$(2 + 3) + 4$$

The parentheses are grouping symbols. They indicate that we are adding the 2 and 3 first.

$$(2 + 3) + 4 = 5 + 4 \quad \textbf{Add 2 and 3 first}$$
$$= 9 \quad \textbf{Then add 4 to the result}$$

Now let's find the sum of the same three numbers, but this time we will group the 3 and 4 together.

$$2 + (3 + 4) = 2 + 7 \quad \textbf{Add 3 and 4 first}$$
$$= 9 \quad \textbf{Then add 2 to the result}$$

The result in both cases is the same. If we try this with any other numbers, the same thing happens. We call this fact about addition the *associative property* of addition, and we write it in symbols as follows.

Associative Property of Addition

If *a*, *b*, and *c* represent any three numbers, then

$$(a + b) + c = a + (b + c)$$

In words: Changing the grouping of three numbers in a sum doesn't change the result.

---

2. Use the commutative property of addition to rewrite each sum.
 a. $7 + 9$
 b. $6 + 3$
 c. $4 + 0$
 d. $5 + n$

Note
This discussion is here to show why we write the next property the way we do. Sometimes it is helpful to look ahead to the property itself (in this case, the associative property of addition) to see what it is that is being justified.

Answers
2. a. $9 + 7$ **b.** $3 + 6$ **c.** $0 + 4$
  d. $n + 5$

EXAMPLE 3 Use the associative property of addition to rewrite each sum.

    a. $(5 + 6) + 7$ **b.** $(3 + 9) + 1$ **c.** $6 + (8 + 2)$ **d.** $4 + (9 + n)$

SOLUTION   The associative property of addition indicates that we are free to regroup the numbers in a sum without changing the result.

    a. $(5 + 6) + 7 = 5 + (6 + 7)$

    b. $(3 + 9) + 1 = 3 + (9 + 1)$

    c. $6 + (8 + 2) = (6 + 8) + 2$

    d. $4 + (9 + n) = (4 + 9) + n$ ■

The following examples show how we can use the associative property of addition to simplify expressions that contain both numbers and variables.

In Examples 4 and 5, use the associative property of addition to regroup the numbers and variables so that the expressions can be simplified.

EXAMPLE 4 $(x + 3) + 7 = x + (3 + 7)$ **Associative property**

                  $= x + 10$ **Addition**   ■

EXAMPLE 5 $2 + (4 + y) = (2 + 4) + y$ **Associative property**

                  $= 6 + y$ **Addition**   ■

The next examples show that it is sometimes necessary to use the commutative property of addition as well as the associative property to simplify expressions.

EXAMPLE 6 $(7 + x) + 2 = (x + 7) + 2$ **Commutative property**

                  $= x + (7 + 2)$ **Associative property**

                  $= x + 9$ **Addition**   ■

EXAMPLE 7 $5 + (a + 9) = 5 + (9 + a)$ **Commutative property**

                  $= (5 + 9) + a$ **Associative property**

                  $= 14 + a$ **Addition**   ■

Solving Equations

We can use the addition table to help solve some simple equations. If n is used to represent a number, then the *equation*

$$n + 3 = 5$$

will be true if n is 2. The number 2 is therefore called a *solution* to the equation, because, when we replace n with 2, the equation becomes a true statement:

$$2 + 3 = 5$$

Equations like this are really just puzzles, or questions. When we say, "Solve the equation $n + 3 = 5$," we are asking the question, "What number do we add to 3 to get 5?"

---

3. Use the associative property of addition to rewrite each sum.

    a. $(3 + 2) + 9$

    b. $(4 + 10) + 1$

    c. $5 + (9 + 1)$

    d. $3 + (8 + n)$

Use the associative property to simplify.

4. $(x + 5) + 9$

5. $6 + (8 + y)$

Use the commutative property and the associative property to simplify each expression.

6. $(1 + x) + 4$

7. $8 + (a + 4)$

Answers

3. **a.** $3 + (2 + 9)$ **b.** $4 + (10 + 1)$

    c. $(5 + 9) + 1$ **d.** $(3 + 8) + n$

4. $x + 14$

5. $14 + y$

6. $x + 5$

7. $12 + a$

8. Use the addition table to find the solution to each equation.
  a. $n + 9 = 17$
  b. $n + 2 = 10$
  c. $8 + n = 9$
  d. $16 = n + 10$

EXAMPLE 8 Use the addition table to find the solution to each equation.
  a. $n + 5 = 9$ **b.** $n + 6 = 12$ **c.** $4 + n = 5$ **d.** $13 = n + 8$

SOLUTION   We find the solution to each equation by using the addition facts given in Table 1.

  a. The solution to $n + 5 = 9$ is 4, because $4 + 5 = 9$.
  b. The solution to $n + 6 = 12$ is 6, because $6 + 6 = 12$.
  c. The solution to $4 + n = 5$ is 1, because $4 + 1 = 5$.
  d. The solution to $13 = n + 8$ is 5, because $13 = 5 + 8$. ■

When we solve equations by reading the equation to ourselves and then stating the solution, as we did with the equation above, we are solving the equation by inspection. The next equations we solve require some simplification before the solution can be found.

9. Solve: $(x + 5) + 4 = 10$

EXAMPLE 9 Solve: $(x + 2) + 3 = 5$

SOLUTION   We begin by applying the associative property to the left side of the equation to simplify it. Once the left side has been simplified, we solve the equation by inspection.

$$(x + 2) + 3 = 5 \quad \text{Original equation}$$
$$x + (2 + 3) = 5 \quad \text{Associative property}$$
$$x + 5 = 5 \quad \text{Add 2 and 3}$$
$$x = 0 \quad \text{Solve by inspection} \quad ■$$

10. Solve: $(a + 4) + 2 = 7 + 9$

EXAMPLE 10 Solve: $(a + 4) + 3 = 2 + 9$

SOLUTION   We simplify each side of the equation separately, and then we solve by inspection. Here are the steps:

$$(a + 4) + 3 = 2 + 9$$
$$a + (4 + 3) = 11 \quad \text{Simplify each side}$$
$$a + 7 = 11$$
$$a = 4 \quad \text{Solve by inspection} \quad ■$$

Writing about Mathematics

In the "Extending the Concepts" section at the end of the next problem set, you will be asked to give written answers to some questions. Writing about mathematics is a valuable exercise because it requires you to organize your thoughts so that you can communicate what you know. You may think that you have mastered a topic because you can work problems involving that topic. If you write with the intention of explaining and communicating what you know to someone else, you will find that you understand the topic you are writing about even better than you did when you were just working problems.

Here is an example of what one of my students wrote when asked to explain the commutative property of addition: "The commutative property of addition states that changing the order of the numbers in an addition problem will not change the answer. For example, the commutative property of addition tells us that $3 + 5$ is the same as $5 + 3$, $9 + 7$ is the same as $7 + 9$, and $3 + x$ is the same as $x + 3$."

Answers
8. a. 8   **b.** 8   **c.** 1   **d.** 6
9. 1
10. 10

Problem Set **1.2**

Before you try to work these problems, turn to Appendix 1 in the back of the book. It contains 100 addition problems that correspond to the addition facts given in Table 1. You should be able to do all 100 problems quickly and accurately before you go any further.

Find each of the following sums. (Add.)

1. $(3 + 1) + 8$

2. $3 + (1 + 8)$

3. $(5 + 2) + 7$

4. $5 + (2 + 7)$

5. $(6 + 1) + 4$

6. $6 + (8 + 1)$

7. $9 + (4 + 0)$

8. $0 + (3 + 6)$

9. $(1 + 8) + 6$

10. $(3 + 2) + 6$

11. $(6 + 3) + 7$

12. $(4 + 4) + 4$

13. $(7 + 1) + 5$

14. $(2 + 5) + 7$

15. $(8 + 1) + (7 + 2)$

16. $(6 + 3) + (5 + 1)$

17. $(4 + 3) + (3 + 4)$

18. $(5 + 1) + (1 + 5)$

19. $(2 + 1) + (6 + 3)$

20. $(1 + 2) + (3 + 6)$

Rewrite each of the following using the commutative property of addition.

21. $5 + 9$

22. $2 + 1$

23. $3 + 8$

24. $9 + 2$

25. $6 + 4$

26. $1 + 7$

27. $n + 5$

28. $n + 8$

29. $x + 1$

30. $x + 2$

Rewrite each of the following using the associative property of addition.

31. $(1 + 2) + 3$

32. $(4 + 5) + 9$

33. $(2 + 1) + 6$

34. $(2 + 3) + 8$

35. $1 + (9 + 1)$

36. $2 + (8 + 2)$

37. $(4 + n) + 1$

38. $(n + 8) + 1$

39. $(a + b) + c$

40. $(x + y) + z$

Use the associative property of addition to simplify each of the following expressions.

41. $(x + 5) + 6$

42. $(x + 4) + 2$

43. $(a + 7) + 9$

44. $(a + 9) + 7$

45. $3 + (2 + y)$

46. $5 + (7 + y)$

47. $8 + (8 + t)$

48. $7 + (7 + t)$

Find a solution for each equation.

49. $n + 1 = 6$ **50.** $n + 2 = 3$ **51.** $n + 6 = 10$ **52.** $n + 4 = 7$

53. $n + 8 = 13$ **54.** $n + 6 = 15$ **55.** $4 + n = 12$ **56.** $5 + n = 7$

57. $7 + n = 16$ **58.** $8 + n = 14$ **59.** $17 = n + 9$ **60.** $13 = n + 5$

Complete the following tables.

61.

FIRST NUMBER	SECOND NUMBER	THEIR SUM
a	b	$a + b$
1	4	
2	3	
3	2	
4	1	

62.

FIRST NUMBER	SECOND NUMBER	THEIR SUM
a	b	$a + b$
2	8	
4	6	
6	4	
8	2	

63.

FIRST NUMBER	SECOND NUMBER	THEIR SUM
a	b	$a + b$
1	7	
3	5	
5	3	
7	1	

64.

FIRST NUMBER	SECOND NUMBER	THEIR SUM
a	b	$a + b$
1	8	
3	6	
5	4	
7	2	

Identify each statement in Problems 65–76 as an example of one of the following properties:

a. Addition property of 0
b. Commutative property of addition
c. Associative property of addition

65. $4 + 7 = 7 + 4$ **66.** $6 + (5 + 4) = (6 + 5) + 4$

67. $(1 + 2) + 3 = 1 + (2 + 3)$ **68.** $3 + 0 = 3$

69. $(4 + 1) + 5 = 5 + (4 + 1)$ **70.** $(6 + 8) + 2 = 6 + (8 + 2)$

71. $0 + 9 = 9$ **72.** $(6 + 2) + 3 = (2 + 6) + 3$

73. $(x + 5) + 6 = x + (5 + 6)$ **74.** $(x + 4) + 2 = x + (4 + 2)$

75. $2(a + 3) = 2(3 + a)$ **76.** $5(a + 4) = 5(4 + a)$

Solve each equation below by first simplifying each side of the equation as much as possible and then finding the solution by inspection.

77. $(x + 1) + 3 = 9$

78. $(x + 3) + 1 = 9$

79. $(a + 4) + 5 = 16$

80. $(a + 6) + 2 = 20$

81. $2 + (5 + y) = 17$

82. $3 + (4 + y) = 15$

83. $6 + (6 + x) = 12$

84. $7 + (7 + x) = 14$

85. $(a + 3) + 5 = 4 + 7$

86. $(a + 2) + 4 = 3 + 9$

87. $4 + (3 + t) = 5 + 6$

88. $5 + (2 + t) = 7 + 8$

Write each of the following expressions in words. Use the word *sum* in each case.

89. $4 + 9$

90. $9 + 4$

91. $8 + 1$

92. $9 + 9$

93. $2 + 3 = 5$

94. $8 + 2 = 10$

95. $a + b = c$

96. $x + y = z$

Write each of the following phrases in symbols.

97. The sum of 5 and 2

98. The sum of 3 and 1

99. The sum of 5 and a

100. The sum of a and 5

101. The sum of 6 and 8 is 14

102. The sum of 5 and 1 is 6

103. 8 more than 2

104. 3 more than 5

105. x increased by 4

106. y increased by 8

107. 6 added to t

108. 7 added to r

Extending the Concepts: Writing about Mathematics

Some of the "Extending the Concepts" problems in this book will ask you to give written explanations and answers to questions about the topics we have covered. When a written answer is required, always write in complete sentences. As with all the "Extending the Concepts" problems, you should approach them with a positive point of view. You will get better at giving written responses to questions as you progress through the course. Even if you don't feel comfortable writing about mathematics, just the process of attempting to do so will increase your understanding and ability in mathematics.

109. In your own words and in complete sentences, explain the commutative property of addition.

110. In your own words and in complete sentences, explain the associative property of addition.

111. In the next section of the text, we will cover subtraction with real numbers. Do you think subtraction will be a commutative operation? Write your answer using complete sentences, and include examples that justify your answer.

112. Do you think multiplication is a commutative operation? Explain your answer using complete sentences and include examples that justify your answer.

1.3 Addition with Whole Numbers and Perimeter

Introduction . . .

The engraving below is by Albrecht Dürer (1471–1528). It is called *Melancholia* and was finished in 1514. Dürer was interested in the applications of mathematics to the arts. The small square in the upper-right-hand corner has been enlarged. Notice how Dürer included the year *Melancholia* was completed in the bottom row of the square. This square is an example of what we call a "magic square."

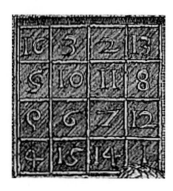

A magic square is a square array of numbers for which the sum of the numbers in any row, any column, and each diagonal are equal. To check that the Dürer square is a magic square, you must be able to add whole numbers. And that is what you will learn in this section.

EXAMPLE 1   Add: 43 + 52

SOLUTION   This type of addition is best done vertically. If we write each number showing the place value of the digits, we have

$$
\begin{array}{r}
43 = 4 \text{ tens} + 3 \text{ ones} \\
+52 = 5 \text{ tens} + 2 \text{ ones} \\
\hline
9 \text{ tens} + 5 \text{ ones}
\end{array}
$$

Write each number showing the place value of each digit

Add digits with the same place value to get these numbers ■

We can write the sum 9 tens + 5 ones in standard form as 95. So, to perform addition we add digits with the same place value.

We can visualize the problem from Example 1 using money in $10 bills and $1 bills.

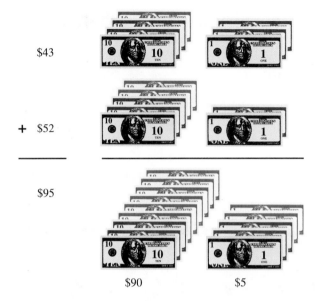

$43

+ $52
——————————
$95

              $90 $5

2. Add: 342 + 605

EXAMPLE 2   Add: 165 + 801

SOLUTION   Writing the sum vertically and showing the place values of the digits, we have

$$
\begin{array}{l}
165 = 1 \text{ hundred } + 6 \text{ tens} + 5 \text{ ones} \\
\underline{+801 = 8 \text{ hundreds} + 0 \text{ tens} + 1 \text{ one}} \\
 9 \text{ hundreds} + 6 \text{ tens} + 6 \text{ ones}
\end{array}
$$

This last number written in standard form is 966.   ■

Note

We are showing the place value of the digits in each number only for the purpose of explanation. The numbers are written like this only to show *why* we add digits with the same place value.

Addition with Carrying

In Examples 1 and 2, the sums of the digits with the same place value were always 9 or less. There are many times when the sum of the digits with the same place value will be a number larger than 9. In these cases we have to do what is called *carrying* in addition. The following examples illustrate this process.

3. Add: 375 + 121 + 473

EXAMPLE 3   Add: 173 + 224 + 342

SOLUTION   We write each number showing the place value of each digit and then add digits with the same place value:

$$
\begin{array}{l}
173 = 1 \text{ hundred } + 7 \text{ tens} + 3 \text{ ones} \\
224 = 2 \text{ hundreds} + 2 \text{ tens} + 4 \text{ ones} \\
\underline{+342 = 3 \text{ hundreds} + 4 \text{ tens} + 2 \text{ ones}} \\
 6 \text{ hundreds} + 13 \text{ tens} + 9 \text{ ones}
\end{array}
$$

Look at the tens column. We have 13 tens, which is

$$
\begin{aligned}
13 \text{ tens} &= 13(10) \\
&= 130 \\
&= 100 + 30 \\
&= 1 \text{ hundred} + 3 \text{ tens}
\end{aligned}
$$

Note

Look over Example 3 carefully before trying Practice Problem 3. The explanation of why we do carrying in addition is contained in Example 3.

Answers
2. 947
3. 969

Because 13 tens is the same as 1 hundred + 3 tens, we carry the 1 hundred to the hundreds column to get

6 hundreds + 1 hundred = 7 hundreds

Our result is

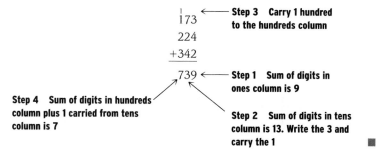

$$6 \text{ hundreds} + 13 \text{ tens} + 9 \text{ ones}$$
$$= 6 \text{ hundreds} + 1 \text{ hundred} + 3 \text{ tens} + 9 \text{ ones}$$
$$= 7 \text{ hundreds} + 3 \text{ tens} + 9 \text{ ones}$$

We can summarize this whole process by using the shorthand notation for carrying:

```
  1
 173  ←——— Step 3   Carry 1 hundred
              to the hundreds column
 224
+342
─────
 739  ←——— Step 1   Sum of digits in
              ones column is 9
```

Step 4   Sum of digits in hundreds column plus 1 carried from tens column is 7

Step 2   Sum of digits in tens column is 13. Write the 3 and carry the 1 ■

This shorthand notation for addition is much easier to use than writing out the numbers showing the place values of the digits. We will use the shorthand notation for the rest of the examples in this section. If you have any problems understanding the concept of carrying with addition, you may want to try a few problems the long way. However, the final goal is to be able to find sums quickly and accurately.

We can visualize the carrying done in the tens column by using money. We can think of each number in the tens column as representing a $10 bill.

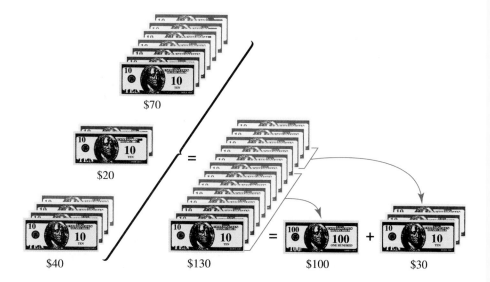

$70

$20

$40

$130

=

$100 + $30

4. Add: 57,904 + 7,193 + 655

EXAMPLE 4   Add: 46,789 + 2,490 + 864

SOLUTION   We write the sum vertically—with the digits with the same place value aligned—and then use the shorthand form of addition.

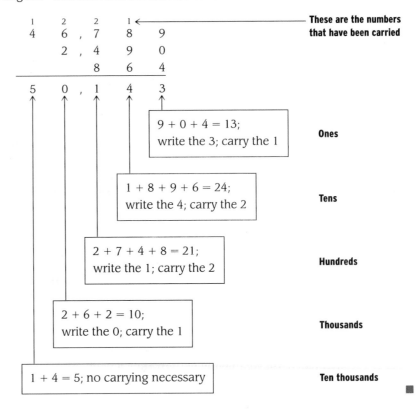

Adding numbers as we are doing here takes some practice. Most people don't make mistakes in carrying. Most mistakes in addition are made in adding the numbers in the columns. That is why it is so important that you are accurate with the basic addition facts given in this chapter.

Facts from Geometry: Perimeter

We end this section with an introduction to perimeter. Let's start with the definition of a *polygon*:

> **DEFINITION**
>
> A **polygon** is a closed geometric figure, with at least three sides, in which each side is a straight line segment.

The most common polygons are squares, rectangles, and triangles. Examples of these are shown in Figure 1.

Note
In the triangle the small square where the broken line meets the base is the notation we use to show that the two line segments meet at right angles. That is, the height *h* and the base *b* are perpendicular to each other; the angle between them is 90°.

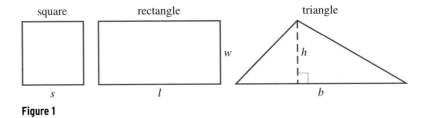

Figure 1

Answer
4. 65,752

In the square, *s* is the length of the side, and each side has the same length. In the rectangle, *l* stands for the length, and *w* stands for the width. The width is usually the lesser of the two. The *b* and *h* in the triangle are the base and height, respectively. The height is always perpendicular to the base. That is, the height and base form a 90°, or right, angle where they meet.

DEFINITION

The **perimeter** of any polygon is the sum of the lengths of the sides, and it is denoted with the letter *P*.

To find the perimeter of a polygon we add all the lengths of the sides together.

EXAMPLE 5 Find the perimeter of each geometric figure.

a.

15 inches

b.

24 feet
37 feet

c.
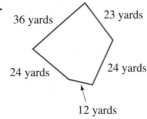
36 yards
23 yards
24 yards
24 yards
12 yards

SOLUTION In each case we find the perimeter by adding the lengths of all the sides.

a. The figure is a square. Because the length of each side in the square is the same, the perimeter is

$P = 15 + 15 + 15 + 15 = 60$ inches

b. In the rectangle, two of the sides are 24 feet long, and the other two are 37 feet long. The perimeter is the sum of the lengths of the sides.

$P = 24 + 24 + 37 + 37 = 122$ feet

c. For this polygon, we add the lengths of the sides together. The result is the perimeter.

$P = 36 + 23 + 24 + 12 + 24 = 119$ yards ∎

5. Find the perimeter of each geometric figure.

a.

7 feet

b.

33 inches
88 inches

c.

44 yards
66 yards
77 yards

Answers
5. a. 28 feet **b.** 242 inches
c. 187 yards

Calculators

From time to time we will include some notes like this one, which show how a calculator can be used to assist us with some of the calculations in the book. Most calculators on the market today fall into one of two categories: those with algebraic logic and those with function logic. Calculators with algebraic logic have a key with an equals sign on it. Calculators with function logic do not have an equals key. Instead they have a key labeled ENTER or EXE (for execute). Scientific calculators use algebraic logic, and graphing calculators, such as the TI-82 and CASIO 7700, use function logic.

Here are the sequences of keystrokes to use to work the problem shown in Part c of Example 6.

Scientific Calculator: 36 $\boxed{+}$ 23 $\boxed{+}$ 24 $\boxed{+}$ 12 $\boxed{+}$ 24 $\boxed{=}$

Graphing Calculator: 36 $\boxed{+}$ 23 $\boxed{+}$ 24 $\boxed{+}$ 12 $\boxed{+}$ 24 $\boxed{\text{ENT}}$

Problem Set 1.3

Find each of the following sums. (Add.)

1. 3 + 5 + 7   **2.** 2 + 8 + 6   **3.** 1 + 4 + 9   **4.** 2 + 8 + 3

5. 5 + 9 + 4 + 6   **6.** 8 + 1 + 6 + 2   **7.** 1 + 2 + 3 + 4 + 5   **8.** 5 + 6 + 7 + 8 + 9

9. 9 + 1 + 8 + 2   **10.** 7 + 3 + 6 + 4

Add each of the following. (There is no carrying involved in these problems.)

11. 43	**12.** 56	**13.** 81	**14.** 37	**15.** 4,281	**16.** 2,749
25	23	17	22	3,016	1,250

17. 3,482	**18.** 2,496	**19.** 32	**20.** 521	**21.** 6,245	**22.** 27
3,005	7,503	21	340	203	4,510
		43	135	1,001	342

Add each of the following. (All problems involve carrying in at least one column.)

23. 49	**24.** 85	**25.** 74	**26.** 36	**27.** 682	**28.** 439
16	29	28	46	193	270

29. 638	**30.** 444	**31.** 4,963	**32.** 8,291	**33.** 6,205	**34.** 8,888
191	595	5,428	7,489	9,999	9,999

35. 56,789	**36.** 45,678	**37.** 52,468	**38.** 13,579	**39.** 4,296	**40.** 5,637
98,765	87,654	58,642	97,531	8,720	481
				4,375	7,899

41. 4,994	**42.** 6,824	**43.** 12	**44.** 21	**45.** 999	**46.** 646
449	371	34	43	444	464
9,449	4,857	56	65	555	525
		78	87	222	252

47. 9,245	**48.** 45
672	9,876
8,341	54
27	6,789

Complete the following tables.

49.

FIRST NUMBER	SECOND NUMBER	THEIR SUM
a	b	a + b
61	38	
63	36	
65	34	
67	32	

50.

FIRST NUMBER	SECOND NUMBER	THEIR SUM
a	b	a + b
10	45	
20	35	
30	25	
40	15	

51.

FIRST NUMBER	SECOND NUMBER	THEIR SUM
a	b	a + b
9	16	
36	64	
81	144	
144	256	

52.

FIRST NUMBER	SECOND NUMBER	THEIR SUM
a	b	a + b
25	75	
24	76	
23	77	
22	78	

Applying the Concepts

The application problems that follow are related to addition of whole numbers. Read each problem carefully to determine exactly what you are being asked to find. Don't assume that just because a number appears in a problem you have to use it to solve the problem. Sometimes you do, and sometimes you don't.

53. Gallons of Gasoline Tim bought gas for his economy car twice last month. The first time he bought 18 gallons and the second time he bought 16 gallons. What was the total amount of gasoline Tim bought last month?

54. Tallest Mountain The world's tallest mountain is Mount Everest. On May 5, 1999, it was found to be 7 feet taller than it was previously thought to be. Before this date, Everest was thought to be 29,028 feet high. That height was determined by B. L. Gulatee in 1954. What is the current height of Mount Everest?

55. Checkbook Balance On Monday Bob had a balance of $241 in his checkbook. On Tuesday he made a deposit of $108, and on Thursday he wrote a check for $24. What was the balance in his checkbook on Wednesday?

56. Number of Passengers A plane flying from Los Angeles to New York left Los Angeles with 67 passengers on board. The plane stopped in Bakersfield and picked up 28 passengers, and then it stopped again in Dallas where 57 more passengers came on board. How many passengers were on the plane when it landed in New York?

		RECORD ALL CHARGES OR CREDITS THAT AFFECT YOUR ACCOUNT					BALANCE	
NUMBER	DATE	DESCRIPTION OF TRANSACTION	PAYMENT/DEBIT (−)	√ T	FEE (IF ANY) (+)	DEPOSIT/CREDIT (+)	$ 241	00
	11/16	Deposit	$		$	$ 108 00	?	
1401	11/18	Postage Stamps	24 00				?	

57. Living Expenses A couple estimates they will have the following monthly expenses:

ITEM	EXPENSE
Rent	$600
Food	145
Utilities	56
Gasoline	97
Miscellaneous	250

What are the couple's total monthly living expenses?

58. Number of Calories Tim is on a diet and decides to keep track of the number of calories he consumes at each meal. At the end of the day on Monday he has the following totals:

MEAL	CALORIES
Breakfast	342
Lunch	480
Dinner	217
Snacks	125

What was Tim's total caloric intake on this particular day?

59. Magic Square The magic square shown below is from the introduction to this section. The sum of the numbers in each row, each column, and each diagonal of the magic square is same.

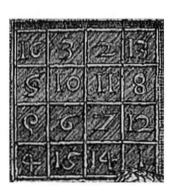

a. Add the numbers in the top row.

b. Add the numbers in the bottom row.

c. Add the numbers in the first column.

d. Add the numbers in either diagonal.

60. Travel Expenses On a recent business trip, Nancy spent $29 for gas, $68 for lodging, and $49 for food. What was her total cost for food and lodging?

Find the perimeter of each figure. (Note that we have abbreviated the units on each figure to save space. The abbreviation for feet is ft, inches is in., and yards is yd.) The first four figures are squares.

61.

3 in.

62.

9 in.

63.

4 ft

64.

2 ft

65.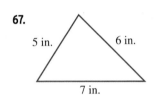

3 yd

10 yd

66.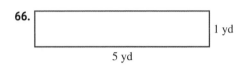

1 yd

5 yd

67.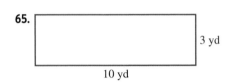

5 in.   6 in.

7 in.

68.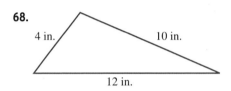

4 in.   10 in.

12 in.

69.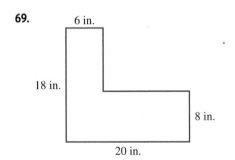

6 in.

18 in.

8 in.

20 in.

70.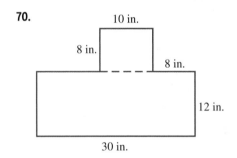

10 in.

8 in.   8 in.

12 in.

30 in.

71.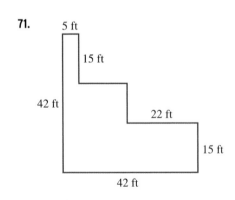

5 ft

15 ft

42 ft

22 ft

15 ft

42 ft

72.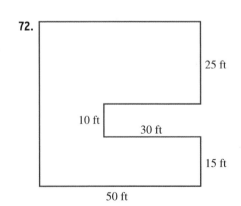

25 ft

10 ft   30 ft

15 ft

50 ft

Calculator Problems

From now on most Problem Sets will contain a number of problems to be worked on a calculator. These problems will always be marked by the heading "Calculator Problems."

Find each of the following sums. (Add.)

73. 4,793,420
    6,825,781

74. 8,592,486
    9,876,987

75. 654,321
    123,456
    973,654
    842,183

76. 7,913,428
    6,594,721
    8,873,255

State Populations The populations of the four largest states in the United States in 1999 are given in the table below. Use these numbers to answer Problems 77–80.

STATE	POPULATION
California	33,145,121
Texas	20,044,141
New York	18,196,601
Florida	15,111,244

(From U.S. Census Bureau, International Data Base)

77. What was the combined population of New York and California in 1999?

78. What was the total number of people living in Texas and Florida?

79. Give the combined population of the three largest states.

80. Give the combined population of the four largest states.

81. Geometry Suppose a rectangle has a perimeter of 12 inches. If the length and the width are whole numbers, give all the possible values for the width.

82. Geometry Suppose a rectangle has a perimeter of 10 inches. If the length and the width are whole numbers, give all the possible values for the width.

83. Geometry If a rectangle has a perimeter of 20 feet, is it possible for the rectangle to be a square? Explain your answer.

84. Geometry If a rectangle has a perimeter of 10 feet, is it possible for the rectangle to be a square? Explain your answer.

Extending the Concepts: Number Sequences and Inductive Reasoning

Suppose you were asked to give the next number in the sequence of numbers below. (The dots indicate that the sequence continues in the same pattern forever.)

$$3, 7, 11, 15, \ldots$$

Most people would give the next number as 19, by observing that each number after the first number is 4 more than the number before it. Therefore, adding 4 to 15 gives the next number: $15 + 4 = 19$. Reasoning in this manner is called *inductive reasoning.* In mathematics we use inductive reasoning when we notice a pattern to a sequence of numbers and then extend the sequence using that pattern.

For each sequence below, give the next number in the sequence.

85. 1, 2, 3, 4, ... (The sequence of counting numbers)

86. 0, 1, 2, 3, ... (The sequence of whole numbers)

87. 2, 4, 6, 8, ... (The sequence of even numbers)

88. 1, 3, 5, 7, ... (The sequence of odd numbers)

An *arithmetic* (pronounced air ith meh′ tic) *sequence* is a sequence of numbers in which each number is obtained from the previous number by adding the same number each time. Each sequence below is an arithmetic sequence. Find the next number in each sequence, and then give a written description of the sequence. (For example, one way to give a written description of the sequence 3, 7, 11, 15, ... would be to write "The sequence starts with 3. Then each number comes from the number before it by adding 4 each time.")

89. 5, 8, 11, ...

90. 3, 8, 13, ...

91. 10, 35, 60, ...

92. 35, 45, 55, ...

1.4 Rounding Numbers, Estimating Answers, and Displaying Information

Introduction . . .

Many times when we talk about numbers, it is helpful to use numbers that have been *rounded off*, rather than exact numbers. For example, the city where I live has a population of 43,704. But when I tell people how large the city is, I usually say, "The population is about 44,000." The number 44,000 is the original number rounded to the nearest thousand. The number 43,704 is closer to 44,000 than it is to 43,000, so it is rounded to 44,000. We can visualize this situation on the number line.

Rounding

The steps used in rounding numbers are given below.

Steps for Rounding Whole Numbers

1. Locate the digit just to the right of the place you are to round to.
2. If that digit is less than 5, replace it and all digits to its right with zeros.
3. If that digit is 5 or more, replace it and all digits to its right with zeros, and add 1 to the digit to its left.

You can see from these rules that in order to round a number you must be told what column (or place value) to round to.

EXAMPLE 1   Round 5,382 to the nearest hundred.

SOLUTION   The 3 is in the hundreds column. We look at the digit just to its right, which is 8. Because 8 is greater than 5, we add 1 to the 3, and we replace the 8 and 2 with zeros:

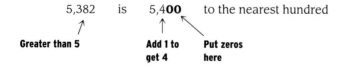

      5,382   is   5,4**00**   to the nearest hundred

Greater than 5   **Add 1 to**   **Put zeros**
                         get 4   **here**

EXAMPLE 2   Round 94 to the nearest ten.

SOLUTION   The 9 is in the tens column. To its right is 4. Because 4 is less than 5, we simply replace it with 0:

      94   is   9**0**   to the nearest ten

Less than 5   **Replaced with zero**

Welcome to
San Luis Obispo
Population 43,704
Founded 1772

Note
After you have used the steps listed here to work a few problems, you will find that the procedure becomes almost automatic.

Practice Problems
1. Round 5,742 to the nearest hundred.

2. Round 87 to the nearest ten.

Answers
1. 5,700   **2.** 90

3. Round 980 to the nearest hundred.

EXAMPLE 3 Round 973 to the nearest hundred.

SOLUTION   We have a 9 in the hundreds column. To its right is 7, which is greater than 5. We add 1 to 9 to get 10, and then replace the 7 and 3 with zeros:

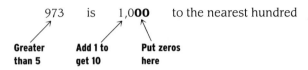

973   is   1,0**00**   to the nearest hundred

Greater   **Add 1 to**   **Put zeros**
than 5   **get 10**   **here**

4. Round 376,804,909 to the nearest million.

EXAMPLE 4 Round 47,256,344 to the nearest million.

SOLUTION   We have 7 in the millions column. To its right is 2, which is less than 5. We simply replace all the digits to the right of 7 with zeros to get:

47,256,344   is   47,**000,000**   to the nearest million

Less than 5   **Leave as is**   **Replaced with zeros**

Table 1 gives more examples of rounding.

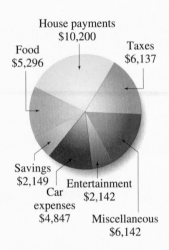

House payments
$10,200
Food
$5,296
Taxes
$6,137
Savings
$2,149
Entertainment
$2,142
Car
expenses
$4,847
Miscellaneous
$6,142

Table 1

ORIGINAL NUMBER	ROUNDED TO THE NEAREST		
	TEN	HUNDRED	THOUSAND
6,914	6,910	6,900	7,000
8,485	8,490	8,500	8,000
5,555	5,560	5,600	6,000
1,234	1,230	1,200	1,000

RULE: Calculating and Rounding

If we are doing calculations and are asked to round our answer, we do all our arithmetic first and then round the result. That is, the last step is to round the answer; we don't round the numbers first and then do the arithmetic.

5. Use the pie chart above to answer these questions.
   a. To the nearest ten dollars, what is the total amount spent on food and car expenses?
   b. To the nearest hundred dollars, how much is spent on savings and taxes?
   c. To the nearest thousand dollars, how much is spent on items other than food and entertainment?

EXAMPLE 5 The pie chart in the margin shows how a family earning $36,913 a year spends their money.
   a. To the nearest hundred dollars, what is the total amount spent on food and entertainment?
   b. To the nearest thousand dollars, how much of their income is spent on items other than taxes and savings?

SOLUTION   In each case we add the numbers in question and then round the sum to the indicated place.
   a. We add the amounts spent on food and entertainment and then round that result to the nearest hundred dollars.

Food $5,296
Entertainment   2,142

Total $7,438 = $7,400 to the nearest hundred dollars

Answers
3. 1,000
4. 377,000,000
5. a. $10,140   **b.** $8,300
    c. $29,000

b. We add the numbers for all items except taxes and savings.

House payments	$10,200
Food	5,296
Car expenses	4,847
Entertainment	2,142
Miscellaneous	6,142
Total	$28,627 = $29,000 to the nearest thousand dollars ∎

Estimating

When we *estimate* the answer to a problem, we simplify the problem so that an approximate answer can be found quickly. There are a number of ways of doing this. One common method is to use rounded numbers to simplify the arithmetic necessary to arrive at an approximate answer, as our next example shows.

EXAMPLE 6   Estimate the answer to the following problem by rounding each number to the nearest thousand.

$$
\begin{array}{r}
4,872 \\
1,691 \\
777 \\
+6,124 \\
\end{array}
$$

SOLUTION   We round each of the four numbers in the sum to the nearest thousand. Then we add the rounded numbers.

4,872	rounds to	5,000
1,691	rounds to	2,000
777	rounds to	1,000
+6,124	rounds to	+ 6,000
		14,000

We estimate the answer to this problem to be approximately 14,000. The actual answer, found by adding the original unrounded numbers, is 13,464.   ∎

Note The method used in Example 6 above does not conflict with the rule we stated before Example 5. In Example 6 we are asked to *estimate* an answer, so it is okay to round the numbers in the problem before adding them. In Example 5 we are asked for a rounded answer, meaning that we are to find the exact answer to the problem and then round to the indicated place. In this case we must not round the numbers in the problem before adding. Look over the instructions, solutions, and answers to Examples 5 and 6 until you understand the difference between the problems shown there.

6. Estimate the answer by first rounding each number to the nearest thousand.

$$
\begin{array}{r}
5,287 \\
2,561 \\
888 \\
+4,898 \\
\end{array}
$$

Descriptive Statistics: Bar Charts

In the introduction to this chapter, we gave two representations for the amount of caffeine in five different drinks, one numeric and the other visual. Those two representations are shown below in Table 2 and Figure 1.

Table 2

CAFFEINE CONTENT OF HOT DRINKS

DRINK (6-OUNCE CUP)	CAFFEINE (IN MILLIGRAMS)
Brewed coffee	100
Instant coffee	70
Tea	50
Cocoa	5
Decaffeinated coffee	4

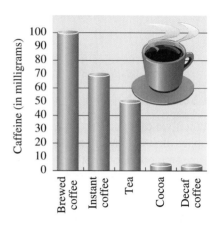

Figure 1

The diagram in Figure 1 is called a *bar chart*, or *histogram*. The horizontal line below which the drinks are listed is called the *horizontal axis*, while the vertical line that is labeled from 0 to 100 is called the *vertical axis*. The key to constructing a readable bar chart is in labeling the axes in a simple and straightforward manner, so that the information shown in the bar chart is easy to read.

7. The results of crash tests for two other cars are given below. Extend the horizontal axis in Figure 3 past the Nissan Maxima, so you can add labels for these two cars. Then, for each car below, draw in bars to represent the amount of damage done to each of the two cars.

Honda Accord $1,433
Chevrolet Lumina $2,629

EXAMPLE 7   The information in Table 3 was published in 1995 by the Insurance Institute for Highway Safety. It gives the total repair bill from damage done to a car from four separate crashes at 5 miles per hour. Construct a bar chart that gives a clear visual summary of the information in the table.

Table 3

DAMAGE AT 5 MILES PER HOUR

CAR	DAMAGE
Chevrolet Cavalier	$1,795
Mazda Millenia	$2,031
Toyota Camry LE	$2,328
Ford Taurus GL	$2,814
Ford Contour GL	$3,188
Nissan Maxima	$3,605

SOLUTION   Although there are many ways to construct a bar chart of this information, we will construct one in which the information in the first column of the table is given along the horizontal axis and the information in the second column is associated with the vertical axis, as shown in Figure 2 on the facing page.

Answer

7. See solutions section.

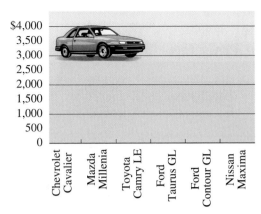

Figure 2

To complete the bar chart, we mentally round the numbers in the table to the nearest hundred to better estimate how tall we want the bars. Using the rounded numbers to draw the bars, we have the bar chart shown in Figure 3.

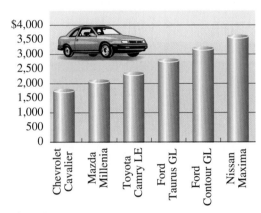

Figure 3

Spreadsheet Programs

When I put together the manuscript for this book, I used a spreadsheet program to draw the bar charts, pie charts, and some of the other diagrams you will see as you progress through the book.

Figure 2 shows how the screen on my computer looked when I was preparing the bar chart for Figure 1 in this section. Notice that I also used the computer to create a pie chart from the same data.

If you have a computer with a spreadsheet program, you may want to use it to create some of the charts you will be asked to create in the problem sets throughout the book.

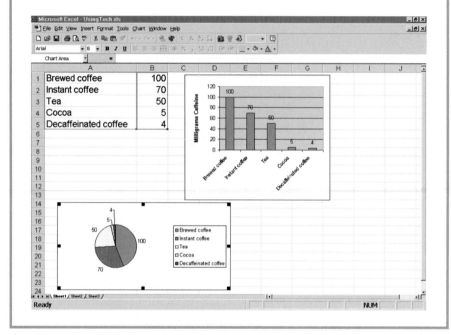

Problem Set 1.4

Round each of the numbers to the nearest ten.

1. 42	**2.** 44	**3.** 46	**4.** 48	**5.** 45	**6.** 73
7. 77	**8.** 75	**9.** 458	**10.** 455	**11.** 471	**12.** 680
13. 56,782	**14.** 32,807	**15.** 4,504	**16.** 3,897		

Round each of the numbers to the nearest hundred.

17. 549	**18.** 954	**19.** 833	**20.** 604	**21.** 899	**22.** 988
23. 1090	**24.** 6,778	**25.** 5,044	**26.** 56,990	**27.** 39,603	**28.** 31,999

Round each of the numbers to the nearest thousand.

29. 4,670	**30.** 9,054	**31.** 9,760	**32.** 4,444	**33.** 978	**34.** 567
35. 657,892	**36.** 688,909	**37.** 509,905	**38.** 608,433	**39.** 3,789,345	**40.** 5,744,500

Complete the following table by rounding the numbers on the left as indicated by the headings in the table.

ORIGINAL NUMBER	ROUNDED TO THE NEAREST		
	TEN	HUNDRED	THOUSAND
41. 7,821			
42. 5,945			
43. 5,999			
44. 4,353			
45. 10,985			
46. 11,108			
47. 99,999			
48. 95,505			

Applying the Concepts

49. Average Salary The November 10, 1999, edition of *USA Today* reported that major league baseball's average player salary for the 1999 season was $1,384,530, representing an increase of 13.24% over the previous season's average. Round the 1999 average player salary to the nearest hundred thousand.

50. Tallest Mountain The world's tallest mountain is Mount Everest. On May 5, 1999, it was found to be 7 feet taller than it was previously thought to be. Before this date, Everest was thought to be 29,028 feet high. That height was determined by B. L. Gulatee in 1954. The first measurement of Everest was in 1847. At that time the height was given as 29,002 feet. Round the current height, the 1954 height, and the 1847 height of Mount Everest to the nearest thousand.

51. School Enrollment In 1900, the enrollment in public schools in the United States was 15,503,711. Round this number to the nearest hundred thousand.

52. School Enrollment In 1987, the enrollment of students in public schools in the United States was 39,837,241. Round this number to the nearest hundred thousand.

53. Federal Grants In 1990, the federal government spent $4,680,368,000 on Pell grants to students pursuing higher education. Round this number to the nearest ten million.

54. Library Content The libraries at Harvard University have over 11,496,906 bound volumes. Round this number to the nearest million.

Business Expenses The pie chart shows one year's worth of expenses for a small business. Use the chart to answer Problems 55–58.

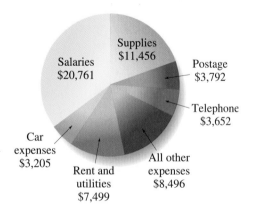

55. To the nearest hundred dollars, how much was spent on postage and supplies?

56. Find the total amount spent, to the nearest hundred dollars, on rent and utilities and car expenses.

57. To the nearest thousand dollars, how much was spent on items other than salaries and rent and utilities?

58. To the nearest thousand dollars, how much was spent on items other than postage, supplies, and car expenses?

Estimating Estimate the answer to each of the following problems by rounding each number to the indicated place value and then adding.

59. hundred	**60.** thousand	**61.** hundred	**62.** hundred	**63.** thousand	**64.** thousand
750	1,891	472	399	25,399	9,999
275	765	422	601	7,601	8,888
+120	+3,223	536	744	18,744	7,777
		+511	+298	+ 6,298	+6,666

The bar chart below is similar to the ones we studied in this section. It was given to me by a friend who owns and operates an alcohol dragster. The dragster contains a computer that gives information about each of his races. This particular race was run during the 1993 Winternationals. The bar chart gives the speed of a race car in a quarter-mile drag race every second during the race. The horizontal lines have been added to assist you with Problems 65–68.

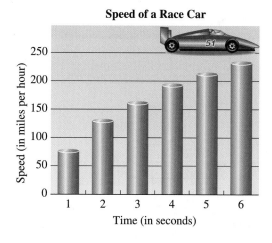

Speed of a Race Car

65. Is the speed of the race car after 3 seconds closer to 160 miles per hour or 190 miles per hour?

66. After 4 seconds, is the speed of the race car closer to 150 miles per hour or 190 miles per hour?

67. Estimate the speed of the car after 1 second.

68. Estimate the speed of the car after 6 seconds.

69. Caffeine Content The following table lists the amount of caffeine in five different soft drinks. Use the axes in the figure below to construct a bar chart from the information in the table.

CAFFEINE CONTENT IN SOFT DRINKS	
DRINK	**CAFFEINE** (IN MILLIGRAMS)
Jolt	100
Mountain Dew	55
Coca-Cola	45
Diet Pepsi	36
7 Up	0

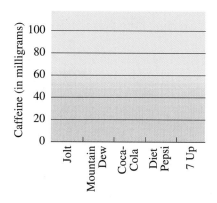

70. Caffeine Content The following table lists the amount of caffeine in five different nonprescription drugs. Use the axes in the figure below to construct a bar chart from the information in the table.

NONPRESCRIPTION DRUG	CAFFEINE (IN MILLIGRAMS)
CAFFEINE CONTENT IN NONPRESCRIPTION DRUGS	
Dexatrim	200
No Doz	100
Excedrin	65
Triaminicin tablets	30
Dristan tablets	16

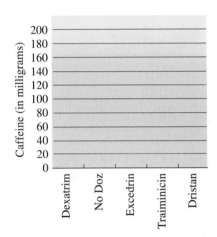

71. Exercise The following table lists the number of calories burned in 1 hour of exercise by a person who weighs 150 pounds. Use the axes in the figure below to construct a bar chart from the information in the table.

ACTIVITY	CALORIES
CALORIES BURNED BY A 150-POUND PERSON IN ONE HOUR	
Bicycling	374
Bowling	265
Handball	680
Jazzercize	340
Jogging	680
Skiing	544

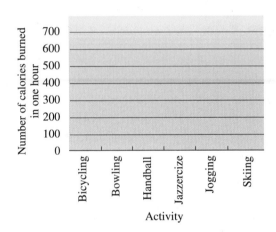

72. Fast Food The following table lists the number of calories consumed by eating some popular fast foods. Use the axes in the figure below to construct a bar chart from the information in the table.

FOOD	CALORIES
CALORIES IN FAST FOOD	
McDonald's hamburger	270
Burger King hamburger	260
Jack in the Box hamburger	280
McDonald's Big Mac	510
Burger King Whopper	630
Jack in the Box Colossus burger	940

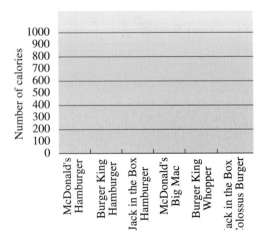

1.5 Subtraction with Whole Numbers

Introduction . . .

Suppose your weekly salary is $324, and you have $109 withheld from your check for taxes and retirement. If your paycheck is for $215, is it the correct amount? You can find out by adding your take-home pay and the amount withheld for taxes and retirement: $215 + $109 = $324. Your take-home pay is the correct amount.

A similar problem, and the one your employer would use to find the amount of your take-home pay, is a subtraction problem: $324 − $109 = $215.

Subtraction is the opposite operation of addition. If you understand addition and can work simple addition problems quickly and accurately, then subtraction shouldn't be difficult for you.

Vocabulary

The word **difference** always indicates subtraction. We can state this in symbols by letting the letters a and b represent numbers.

> **DEFINITION**
>
> The **difference** of two numbers a and b is
>
> $$a - b$$

Table 1 gives some word statements involving subtraction and their mathematical equivalents written in symbols.

Table 1

IN ENGLISH	IN SYMBOLS
The difference of 9 and 1	$9 - 1$
The difference of 1 and 9	$1 - 9$
The difference of m and 4	$m - 4$
The difference of x and y	$x - y$
3 subtracted from 8	$8 - 3$
2 subtracted from t	$t - 2$
The difference of 7 and 4 is 3	$7 - 4 = 3$
The difference of 9 and 3 is 6	$9 - 3 = 6$

The Meaning of Subtraction

When we want to subtract 3 from 8, we write

$$8 - 3, \quad 8 \text{ subtract } 3, \quad \text{or} \quad 8 \text{ minus } 3$$

The number we are looking for here is the difference between 8 and 3, or the number we add to 3 to get 8. That is:

$$8 - 3 = ? \quad \text{is the same as} \quad ? + 3 = 8$$

In both cases we are looking for the number we add to 3 to get 8. The number we are looking for is 5. We have two ways to write the same statement.

Subtraction		**Addition**
$8 - 3 = 5$	or	$5 + 3 = 8$

For every subtraction problem, there is an equivalent addition problem. Table 2 lists some examples. The diagram next to the table gives a visual interpretation of the first subtraction problem in the table.

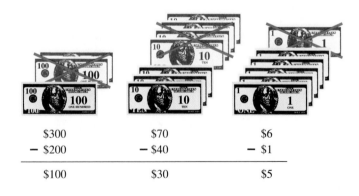

Table 2		
SUBTRACTION		**ADDITION**
7 − 3 = 4	because	4 + 3 = 7
9 − 7 = 2	because	2 + 7 = 9
10 − 4 = 6	because	6 + 4 = 10
15 − 8 = 7	because	7 + 8 = 15

To subtract numbers with two or more digits, we align the numbers vertically and subtract in columns.

EXAMPLE 1   Subtract: 376 − 241

SOLUTION   We write the problem vertically with the place values of the digits showing, and then subtract in columns:

$$\begin{array}{rll} 376 = & 3 \text{ hundreds} + 7 \text{ tens} + 6 \text{ ones} \\ -241 = & 2 \text{ hundreds} \quad 4 \text{ tens} \quad 1 \text{ one} \\ \hline & 1 \text{ hundred} + 3 \text{ tens} + 5 \text{ ones} \end{array}$$

Subtract the bottom number in each column from the number above it.

The difference is 1 hundred + 3 tens + 5 ones, which we write in standard form as 135.   ■

We can visualize Example 1 using money.

$$\begin{array}{ccc} \$300 & \$70 & \$6 \\ -\ \$200 & -\ \$40 & -\ \$1 \\ \hline \$100 & \$30 & \$5 \end{array}$$

EXAMPLE 2   Subtract 503 from 7,835.

SOLUTION   In symbols this statement is equivalent to

7,835 − 503

To subtract we write 503 below 7,835 and then subtract in columns. This time we will not show the place value of each digit but will simply subtract in columns.

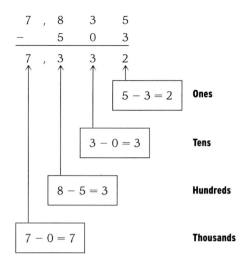

As you can see, subtraction problems like the ones in Examples 1 and 2 are fairly simple. We write the problem vertically, linking up the digits with the same place value, and subtract in columns. We always subtract the bottom number from the top number.

Subtraction with Borrowing

Subtraction must involve *borrowing* when the bottom digit in any column is larger than the digit above it. In one sense borrowing is the reverse of the carrying we did in addition.

EXAMPLE 3   Subtract: $92 - 45$

SOLUTION   We write the problem vertically with the place values of the digits showing:

$$92 = 9 \text{ tens} + 2 \text{ ones}$$
$$-45 = 4 \text{ tens} \quad 5 \text{ ones}$$

Look at the ones column. We cannot subtract immediately, because 5 is larger than 2. Instead, we borrow 1 ten from the 9 tens in the tens column. We can rewrite the number 92 as

$$9 \text{ tens} + \quad 2 \text{ ones}$$
$$= 8 \text{ tens} + 1 \text{ ten} + 2 \text{ ones}$$
$$= 8 \text{ tens} + 12 \text{ ones}$$

Now we are in a position to subtract.

$$92 = 9 \text{ tens} + 2 \text{ ones} = 8 \text{ tens} + 12 \text{ ones}$$
$$-45 = 4 \text{ tens} \quad 5 \text{ ones} = 4 \text{ tens} \quad 5 \text{ ones}$$
$$4 \text{ tens} + \quad 7 \text{ ones}$$

The result is 4 tens + 7 ones, which can be written in standard form as 47.

Writing the problem out in this way is more trouble than is actually necessary. The shorthand form of the same problem looks like this:

3. Subtract: $63 - 47$

Note
The discussion here shows why borrowing is necessary and how we go about it. To understand borrowing you should pay close attention to this discussion.

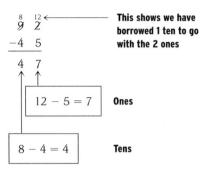

This shortcut form shows all the necessary work involved in subtraction with borrowing. We will use it from now on. ■

The borrowing that changed 9 tens + 2 ones into 8 tens + 12 ones can be visualized with money.

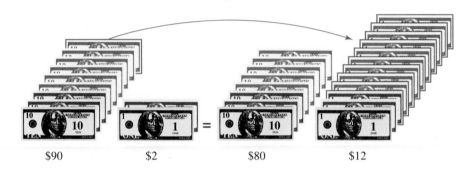

$90 $2 $80 $12

4. Find the difference of 656 and 283.

EXAMPLE 4   Find the difference of 549 and 187.

SOLUTION   In symbols the difference of 549 and 187 is written

$$549 - 187$$

Writing the problem vertically so that the digits with the same place value are aligned, we have

$$\begin{array}{r} 549 \\ -187 \end{array}$$

The top number in the tens column is smaller than the number below it. This means that we will have to borrow from the next larger column.

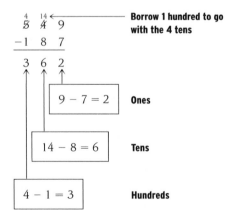

The actual work we did in borrowing looks like this:

$$5 \text{ hundreds} + 4 \text{ tens} + 9 \text{ ones}$$
$$= 4 \text{ hundreds} + 1 \text{ hundred} + 4 \text{ tens} + 9 \text{ ones}$$
$$= 4 \text{ hundreds} + 14 \text{ tens} + 9 \text{ ones}$$

EXAMPLE 5   Jo Ann has $742 in her checking account. If she writes a check for $615 to pay the rent, how much is left in her checking account?

SOLUTION   To find the amount left in the account after she has written the rent check, we subtract:

$$\begin{array}{r} \$7\,\overset{3}{\cancel{4}}\,\overset{12}{\cancel{2}} \\ -\ 6\ 1\ 5 \\ \hline \$1\ 2\ 7 \end{array}$$

NUMBER	DATE	DESCRIPTION OF TRANSACTION	PAYMENT/DEBIT (−)	√ T	FEE (IF ANY) (+)	DEPOSIT/CREDIT (+)	BALANCE $ 742 00
		RECORD ALL CHARGES OR CREDITS THAT AFFECT YOUR ACCOUNT					
1402	12/1	Rent	$ 615 00		$	$	?

She has $127 left in her account after writing a check for the rent.

USING TECHNOLOGY

Calculators

Here is how we would work the problem shown in Example 5 on a calculator:

Scientific Calculator: 742 ⊟ 615 ▭
Graphing Calculator: 742 ⊟ 615 |ENT|

Estimating

One way to estimate the answer to the problem shown in Example 5 is to round 742 to 700 and 615 to 600 and then subtract 600 from 700 to obtain 100, which is an estimate of the difference. Making a mental estimate in this manner will help you catch some of the errors that will occur if you press the wrong buttons on your calculator.

5. Suppose Jo Ann writes another check for $52. How much does she then have left in her checking account?

Problem Set 1.5

1. Subtract 24 from 56.

2. Subtract 71 from 89.

3. Subtract 23 from 45.

4. Subtract 97 from 98.

5. Find the difference of 29 and 19.

6. Find the difference of 37 and 27.

7. Find the difference of 126 and 15.

8. Find the difference of 348 and 32.

Work each of the following subtraction problems.

9.
$$975 \\ -663$$

10.
$$480 \\ -260$$

11.
$$904 \\ -501$$

12.
$$657 \\ -507$$

13.
$$9{,}876 \\ -8{,}765$$

14.
$$5{,}008 \\ -3{,}002$$

15.
$$7{,}976 \\ -3{,}432$$

16.
$$6{,}980 \\ -\ \ 470$$

Find the difference in each case. (These problems all involve borrowing.)

17. $52 - 37$

18. $65 - 48$

19. $70 - 37$

20. $90 - 21$

21. $74 - 69$

22. $31 - 28$

23. $51 - 18$

24. $64 - 58$

25. $329 - 234$

26. $518 - 492$

27. $348 - 196$

28. $759 - 661$

29.
$$932 \\ -658$$

30.
$$895 \\ -597$$

31.
$$647 \\ -159$$

32.
$$842 \\ -199$$

33.
$$905 \\ -367$$

34.
$$804 \\ -238$$

35.
$$600 \\ -437$$

36.
$$800 \\ -342$$

37.
$$4{,}583 \\ -2{,}973$$

38.
$$7{,}849 \\ -2{,}957$$

39.
$$6{,}090 \\ -3{,}947$$

40.
$$4{,}218 \\ -1{,}039$$

Complete the following tables.

41.

FIRST NUMBER a	SECOND NUMBER b	THE DIFFERENCE OF a AND b $a - b$
25	15	
24	16	
23	17	
22	18	

42.

FIRST NUMBER a	SECOND NUMBER b	THE DIFFERENCE OF a AND b $a - b$
90	79	
80	69	
70	59	
60	49	

43.

FIRST NUMBER a	SECOND NUMBER b	THE DIFFERENCE OF a AND b $a - b$
400	256	
400	144	
225	144	
225	81	

44.

FIRST NUMBER a	SECOND NUMBER b	THE DIFFERENCE OF a AND b $a - b$
100	36	
100	64	
25	16	
25	9	

Applying the Concepts

Not all of the following application problems involve only subtraction. Some involve addition as well. Be sure to read each problem carefully.

45. Checkbook Balance Diane has $504 in her checking account. If she writes five checks for a total of $249, how much does she have left in her account?

46. Checkbook Balance Larry has $763 in his checking account. If he writes a check for each of the three bills listed below, how much will he have left in his account?

ITEM	AMOUNT
Rent	$418
Phone	25
Car repair	117

47. Tallest Mountain The world's tallest mountain is Mount Everest. On May 5, 1999, it was found to be 7 feet taller than it was previously thought to be. Before this date, Everest was thought to be 29,028 feet high. That height was determined by B. L. Gulatee in 1954. The first measurement of Everest was in 1847. At that time the height was thought to be 29,002 feet. What is the difference between the current height of Everest and the height measured in 1847?

48. Home Prices In 1985, Mr. Hicks paid $137,500 for his home. He sold it in 2000 for $260,600. What is the difference between what he sold it for and what he bought it for?

49. Enrollment Six years ago, there were 567 students attending Smith Elementary School. Today the same school has an enrollment of 399 students. How much of a decrease in enrollment has there been in the last six years at Smith School?

50. Oil Spills In March 1977, an oil tanker hit a reef off Taiwan and spilled 3,134,500 gallons of oil. In March 1989, an oil tanker hit a reef off Alaska and spilled 10,080,000 gallons of oil. How much more oil was spilled in the 1989 disaster?

Checkbook Balance On Monday Gil has a balance of $425 in his checkbook. On Tuesday he deposits $149 into the account. On Wednesday he writes a check for $37, and on Friday he writes a check for $188. Use this information to answer Problems 51–54.

RECORD ALL CHARGES OR CREDITS THAT AFFECT YOUR ACCOUNT

NUMBER	DATE	DESCRIPTION OF TRANSACTION	PAYMENT/DEBIT (−)	√T	FEE (IF ANY) (+)	DEPOSIT/CREDIT (+)	BALANCE $ 425 00
	10/10	Deposit	$		$	$ 149 00	?
1405	10/11	Market	37 00				?
1406	10/13	Credit Card	188 00				?

51. Find Gil's balance after he makes the deposit on Tuesday.

52. What is his balance after he writes the check on Wednesday?

53. To the nearest ten dollars, what is his balance at the end of the week?

54. To the nearest ten dollars, what is his balance before he writes the check on Friday?

55. Digital Camera Sales The bar chart below shows the sales of digital cameras from 1996–1999.

Digital Camera Sales

a. Use the information in the bar chart to fill in the missing entries in the table.

YEAR	SALES ($ millions)
1996	
1997	
1998	
1999	819

b. What is the difference in camera sales between 1999 and 1997?

56. Wireless Phone Costs The bar chart below shows the projected costs of wireless phone use through 2003.

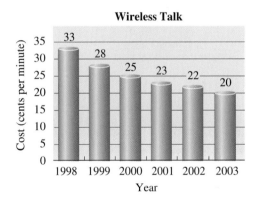

Wireless Talk

a. Use the chart to fill in the missing entries in the table.

YEAR	Cents/Minute
1998	33
1999	
2000	
2001	23
2002	
2003	20

b. What is the difference in cost between 1998 and 1999?

Estimating

Mentally estimate the answer to each of the following problems by rounding each number to the indicated place and then subtracting.

57.   750   hundred
   −120   hundred

58.   5,891   thousand
   −3,223   thousand

59.   3,472   hundred
   −   511   hundred

60.   399   hundred
   −298   hundred

61.   26,399   thousand
   − 6,298   thousand

62.   9,999   thousand
   +6,666   thousand

Speed of a Race Car The bar chart below was also shown in Problem Set 1.4. It gives the speed of a race car during a quarter-mile drag race every second during the race. Use the bar chart to answer Problems 63–66.

Speed of a Race Car

Time (in seconds)

63. Is the difference in the speed of the car after 3 seconds and after 2 seconds closer to 50 miles per hour or 75 miles per hour?

64. Is the difference in speed after 6 seconds and after 5 seconds closer to 25 miles per hour or 50 miles per hour?

65. Estimate the difference in speed between 2 seconds and 1 second.

66. Estimate the difference in speed between 6 seconds and 2 seconds.

Calculator Problems

Use a calculator to find each of the following differences.

67. 9,432,087 − 7,659,989

68. 37,892,401 − 6,471,975

69. Number of Marriages What is the difference in the number of marriages in 1935 and 1988 in the United States, if there were 1,327,000 in 1935 and 2,389,000 in 1988?

70. Number of Births In 1988, there was a total of 3,913,000 births in the United States. That same year there was a total of 2,171,000 deaths. How many more births than deaths were there in the United States in 1988?

71. Number of Births In 1955, there were 4,097,000 births in the United States. Use this information, along with the information in Problem 70, to find the decrease in the number of births from 1955 to 1988.

72. Births and Deaths As you may have guessed, the large numbers in Problems 69–71 have been rounded to the nearest thousand. Given this fact, what is the maximum amount that your answer to Problem 70 can differ from the actual difference between births and deaths in 1988?

1.6 Properties and Facts of Multiplication

Introduction . . .

If Diane makes $6 an hour for the first 40 hours she works each week and $9 for every hour after that, how much will she make if she works 45 hours in one week? One way to find the answer to this question is with *multiplication*. We multiply 6 by 40 and 9 by 5 and then add the results. In symbols, the problem looks like this:

$$6(40) + 9(5) = 240 + 45$$
$$= 285$$

She would make $285 for the week.

The type of multiplication we used in this problem is what we will learn in this section. To begin we can think of multiplication as shorthand for repeated addition. That is, multiplying 3 times 4 can be thought of this way:

$$3 \text{ times } 4 = 4 + 4 + 4 = 12$$

Multiplying 3 times 4 means to add three 4s. In symbols we write 3 times 4 as 3×4, or $3 \cdot 4$.

An instructor is teaching a seminar to 12 students. We can visualize $12 = 3 \times 4$ by thinking of the instructor arranging chairs for the 12 students into 3 rows with 4 chairs in each row.

3 rows

4 chairs in
each row

We can construct a multiplication table using the fact that multiplication can be thought of as repeated addition.

Table 1

MULTIPLICATION FACTS

×	0	1	2	3	4	5	6	7	8	9
0	0	0	0	0	0	0	0	0	0	0
1	0	1	2	3	4	5	6	7	8	9
2	0	2	4	6	8	10	12	14	16	18
3	0	3	6	9	12	15	18	21	24	27
4	0	4	8	12	16	20	24	28	32	36
5	0	5	10	15	20	25	30	35	40	45
6	0	6	12	18	24	30	36	42	48	54
7	0	7	14	21	28	35	42	49	56	63
8	0	8	16	24	32	40	48	56	64	72
9	0	9	18	27	36	45	54	63	72	81

Note
As we have indicated, the numbers in this table were found by repeated addition. To understand multiplication, you must know that 3 times $5 = 5 + 5 + 5$. Once you understand multiplication, the next step is to memorize the multiplication facts contained in Table 1. It is absolutely essential that you become fast and accurate at this kind of multiplication.

Note
The kind of notation we will use to indicate multiplication will depend on the situation. For example, when we are solving equations that involve letters, it is not a good idea to indicate multiplication with the symbol ×, since it could be confused with the letter *x*. The symbol we will use to indicate multiplication most often in this book is the multiplication dot.

Notation

There are many ways to indicate multiplication. All the following statements are equivalent. They all indicate multiplication with the numbers 3 and 4.

$$3 \cdot 4, \qquad 3 \times 4, \qquad 3(4), \qquad (3)4, \qquad (3)(4), \qquad \begin{array}{r} 4 \\ \times\,3 \\ \hline \end{array}$$

If one or both of the numbers we are multiplying are represented by letters, we may also use the following notation:

$5n$	means	5 times n
ab	means	a times b

Vocabulary

We use the word *product* to indicate multiplication. If we say "The product of 3 and 4 is 12," then we mean

$$3 \cdot 4 = 12$$

Both $3 \cdot 4$ and 12 are called the product of 3 and 4. The 3 and 4 are called *factors*. That is, we say 3 and 4 are factors of 12 because their product is 12.

Table 2 gives some additional translations of statements involving multiplication.

Table 2

IN ENGLISH	IN SYMBOLS
The product of 2 and 5	$2 \cdot 5$
The product of 5 and 2	$5 \cdot 2$
The product of 4 and n	$4n$
The product of x and y	xy
The product of 9 and 6 is 54	$9 \cdot 6 = 54$
The product of 2 and 8 is 16	$2 \cdot 8 = 16$

The examples below illustrate the vocabulary associated with multiplication.

EXAMPLE 1 Identify the products and factors in the statement

$$9 \cdot 8 = 72$$

SOLUTION The factors are 9 and 8, and the products are $9 \cdot 8$ and 72. ∎

EXAMPLE 2 Identify the products and factors in the statement

$$30 = 2 \cdot 3 \cdot 5$$

SOLUTION The factors are 2, 3, and 5. The products are $2 \cdot 3 \cdot 5$ and 30. ∎

Properties of Multiplication

You can see from Table 1 that the product of any number and 0 is 0. We state this as a property of 0.

Practice Problems

1. Identify the products and factors in the statement
$$6 \cdot 7 = 42$$

2. Identify the products and factors in the statement
$$20 = 2 \cdot 2 \cdot 5$$

Answers

1. Factors: 6, 7;
products: $6 \cdot 7$ and 42
2. Factors: 2, 2, 5;
products: $2 \cdot 2 \cdot 5$ and 20

Multiplication Property of 0

If *a* represents any number, then

$$a \cdot 0 = 0 \quad \text{and} \quad 0 \cdot a = 0$$

In words: Multiplication by 0 always results in 0.

We also notice by looking through Table 1 that whenever we multiply a number by 1, the result is the original number.

Multiplication Property of 1

If *a* represents any number, then

$$a \cdot 1 = a \quad \text{and} \quad 1 \cdot a = a$$

In words: Multiplying any number by 1 leaves that number unchanged.

It is also apparent from Table 1 that multiplication is a commutative operation. For example, the results of multiplying $4 \cdot 5$ and $5 \cdot 4$ are both 20. We can change the order of the factors in a product without changing the result.

Commutative Property of Multiplication

If *a* and *b* are any two numbers, then

$$ab = ba$$

In words: The order of the numbers in a product doesn't affect the result.

To visualize the commutative property, we can think of the seminar instructor mentioned earlier. He can arrange the chairs for his 12 students into 3 rows with 4 chairs in a row, or he can arrange them into 4 rows with 3 chairs in a row. Either way, he will have 12 chairs for his 12 students.

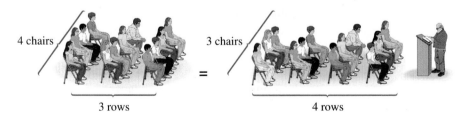

4 chairs   3 chairs

=

3 rows   4 rows

3. Use the commutative property of multiplication to rewrite each of the following products.
   a. $5 \cdot 8$
   b. $(4)(6)$
   c. $7(2)$
   d. $9n$

EXAMPLE 3   Use the commutative property of multiplication to rewrite each of the following products:

   a. $7 \cdot 9$ **b.** $(3)(2)$ **c.** $4(6)$ **d.** $8 \cdot n$

SOLUTION   Applying the commutative property to each expression, we have:

   a. $7 \cdot 9 = 9 \cdot 7$ **b.** $(3)(2) = (2)(3)$

   c. $4(6) = 6(4)$ **d.** $8 \cdot n = n \cdot 8$

Answers
3. a. $8 \cdot 5$ **b.** $(6)(4)$
   c. $2(7)$ **d.** $n9$

And finally, multiplication is also an associative operation.

Associative Property of Multiplication

If a, b, and c represent any three numbers, then

$$(ab)c = a(bc)$$

In words: We can change the grouping of the numbers in a product without changing the result.

---

4. Use the associative property of multiplication to rewrite each of the following products.
  a. $(5 \cdot 7) \cdot 4$
  b. $(3 \times 9) \times 8$
  c. $4 \cdot (6 \cdot 4)$
  d. $6 \cdot (9 \cdot n)$

EXAMPLE 4 Use the associative property of multiplication to rewrite each of the following products:
  a. $(2 \cdot 7) \cdot 9$ **b.** $(4 \times 5) \times 6$ **c.** $3 \cdot (8 \cdot 2)$ **d.** $(5 \cdot 7) \cdot n$

SOLUTION Applying the associative property of multiplication, we regroup as follows:

  a. $(2 \cdot 7) \cdot 9 = 2 \cdot (7 \cdot 9)$ **b.** $(4 \times 5) \times 6 = 4 \times (5 \times 6)$
  c. $3 \cdot (8 \cdot 2) = (3 \cdot 8) \cdot 2$ **d.** $(5 \cdot 7) \cdot n = 5 \cdot (7 \cdot n)$ ∎

We can use the associative property of multiplication to multiply expressions that contain numbers and variables. The next examples illustrate.

Apply the associative property of multiplication and then multiply.
5. $4(7x)$

EXAMPLE 5 Apply the associative property of multiplication and then multiply.

$$3(5x) = (3 \cdot 5)x \quad \textbf{Associative property of multiplication}$$
$$= 15x \quad \textbf{Multiply 3 and 5}$$ ∎

6. $7(9y)$

EXAMPLE 6 $6(8y) = (6 \cdot 8)y \quad \textbf{Associative property of multiplication}$
$$= 48y \quad \textbf{Multiply 6 and 8}$$ ∎

Solving Equations

We can use the facts contained in the multiplication table to help solve some equations. If n is used to represent a number, then the equation

$$4 \cdot n = 12$$

is read "4 times n is 12," or "The product of 4 and n is 12." This means that we are looking for the number we multiply by 4 to get 12. The number is 3. Because the equation becomes a true statement if n is 3, we say that 3 is the solution to the equation.

7. Use the multiplication table to find the solution to each of the following equations.
  a. $5 \cdot n = 35$
  b. $8 \cdot n = 72$
  c. $49 = 7 \cdot n$
  d. $27 = 9 \cdot n$

EXAMPLE 7 Use the multiplication table to find the solution to each of the following equations:
  a. $6 \cdot n = 24$ **b.** $4 \cdot n = 36$ **c.** $15 = 3 \cdot n$ **d.** $21 = 3 \cdot n$

SOLUTION **a.** The solution to $6 \cdot n = 24$ is 4, because $6 \cdot 4 = 24$.
  b. The solution to $4 \cdot n = 36$ is 9, because $4 \cdot 9 = 36$.
  c. The solution to $15 = 3 \cdot n$ is 5, because $15 = 3 \cdot 5$.
  d. The solution to $21 = 3 \cdot n$ is 7, because $21 = 3 \cdot 7$. ∎

Answers
4. a. $5 \cdot (7 \cdot 4)$ **b.** $3 \times (9 \times 8)$
  c. $(4 \cdot 6) \cdot 4$ **d.** $(6 \cdot 9) \cdot n$
5. $28x$ **6.** $63y$
7. a. 7   **b.** 9   **c.** 7   **d.** 3

Now let's see what happens when we multiply some familiar whole numbers with multiples of 10.

EXAMPLE 8 Multiply: 3 · 100

SOLUTION Using the definition of multiplication as repeated addition, we have

$$3 \cdot 100 = 100 + 100 + 100$$
$$= 300$$

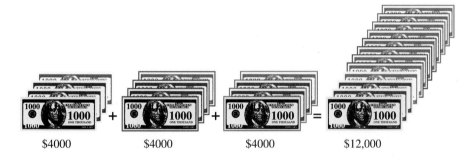

Notice that this is the same result we would get if we just multiplied 3 · 1 and attached two zeros to the result. ■

EXAMPLE 9 Multiply: 3 · 4,000

SOLUTION Proceeding as we did in Example 8, we have

$$3 \cdot 4,000 = 4,000 + 4,000 + 4,000$$
$$= 12,000$$

$4000 \qquad $4000 \qquad $4000 \qquad $12,000

Again, if we had multiplied 3 and 4 to get 12 and then attached three zeros on the right, the result would have been the same. ■

In problems like the ones given in Examples 8 and 9, we can normally leave out the intermediate step that requires showing the repeated addition and just write the result.

Problem Set 1.6

Before you try to work these problems, turn to Appendix 2 in the back of the book. Appendix 2 contains 100 multiplication problems that correspond to the multiplication facts listed in Table 1. You should be able to do all 100 problems quickly and accurately before you go any further.

Write each of the following expressions in words, using the word *product*.

1. $6 \cdot 7$

2. 7×6

3. $4(9)$

4. $9(4)$

5. $2 \cdot n$

6. $5 \cdot x$

7. $9 \cdot 7 = 63$

8. $(5)(6) = 30$

9. $(2)(9) = 18$

10. $9 \cdot 9 = 81$

Write each of the following phrases in symbols.

11. The product of 8 and 3

12. The product of 2 and 3

13. The product of 6 and 6

14. The product of 0 and 1

15. The product of 7 and n

16. The product of 9 and x

17. The product of 6 and 7 is 42.

18. The product of 8 and 9 is 72.

19. The product of 0 and 6 is 0.

20. The product of 1 and 6 is 6.

Identify the products in each statement.

21. $9 \cdot 7 = 63$

22. $2(6) = 12$

23. $4(4) = 16$

24. $5 \cdot 5 = 25$

Identify the factors in each statement.

25. $2 \cdot 3 \cdot 4 = 24$

26. $6 \cdot 1 \cdot 5 = 30$

27. $12 = 2 \cdot 2 \cdot 3$

28. $42 = 2 \cdot 3 \cdot 7$

Rewrite each of the following using the commutative property of multiplication.

29. 5(9)
30. 4(3)
31. $6 \cdot 7$
32. $8 \cdot 3$

33. (1)0
34. (6)(9)
35. $3 \cdot n$
36. $5 \cdot x$

Rewrite each of the following using the associative property of multiplication.

37. $2 \cdot (7 \cdot 6)$
38. $4 \cdot (8 \cdot 5)$
39. $3 \times (9 \times 1)$
40. $5 \times (8 \times 2)$

41. $2 \cdot (6 \cdot 5)$
42. $8 \cdot (7 \cdot 6)$
43. $3 \cdot (4 \cdot n)$
44. $(7 \cdot 2) \cdot x$

Give the letter of the property or properties illustrated in Problems 45–54.

a. Multiplication property of 0
b. Multiplication property of 1
c. Commutative property of multiplication
d. Associative property of multiplication

45. $(5 \cdot 2) \cdot 9 = 5 \cdot (2 \cdot 9)$
46. $3 \cdot 4 = 4 \cdot 3$
47. $4(0) = 0$
48. $4 \cdot (1 \cdot 8) = (4 \cdot 1) \cdot 8$

49. $7(2) = 2(7)$
50. $2(1) = 2$
51. $7(6) = 6(7)$
52. $8 \cdot (6 \cdot 4) = 8 \cdot (4 \cdot 6)$

53. $9 \cdot (4 \cdot 1) = 9 \cdot (1 \cdot 4)$
54. $0(6) = 0$

Apply the associative property, and then multiply.

55. $6(7x)$
56. $8(4x)$
57. $9(3x)$
58. $3(6x)$

59. $4(7a)$
60. $7(8a)$
61. $2(9y)$
62. $3(2y)$

Find a solution for each equation.

63. $4 \cdot n = 12$
64. $3 \cdot n = 12$
65. $9 \cdot n = 81$
66. $6 \cdot n = 36$

67. $0 = n \cdot 5$
68. $6 = 1 \cdot n$
69. $9 \cdot n = 72$
70. $7 \cdot n = 49$

71. $27 = 9 \cdot n$
72. $48 = 6 \cdot n$
73. $42 = 6 \cdot n$
74. $0 = 9 \cdot n$

Multiply each of the following without showing any work.

75. 6 · 10   **76.** 4 · 10   **77.** 15 · 10   **78.** 36 · 10   **79.** 9 · 100   **80.** 2 · 100

81. 3 · 100   **82.** 7 · 100   **83.** 3 · 200   **84.** 4 · 200   **85.** 6 · 500   **86.** 8 · 400

87. 5 · 1,000   **88.** 8 · 1,000   **89.** 3 · 7,000   **90.** 6 · 7,000   **91.** 9 · 9,000   **92.** 7 · 7,000

93. 4 · 600,000   **94.** 2 · 800,000   **95.** 5 · 60   **96.** 6 · 50   **97.** 5 · 6,000   **98.** 6 · 5,000

99. 9 · 80   **100.** 8 · 9,000   **101.** 9 · 80,000   **102.** 8 · 90   **103.** 2 · 200   **104.** 4 · 400

Applying the Concepts

Most, but not all, of the application problems that follow require multiplication. Read the problems carefully before trying to solve them.

105. Hourly Wages A person earning $9 an hour works 30 hours one week. How much will the person be paid?

106. Hourly Wages A person making $12 an hour works 10 hours one week. How much will the person be paid?

107. Number of Passengers A plane is flying from Los Angeles to New York by way of Dallas. There are 207 passengers on the plane when it leaves Los Angeles. When the plane stops in Dallas, 55 of the passengers leave the plane and do not get back on, while 73 new passengers board there. How many passengers are on the plane when it arrives in New York?

108. Gallons of Gasoline Bob leaves on a trip to Las Vegas with 12 gallons of gas in his car. By the time he gets to Barstow, he has used 8 gallons of gas. In Barstow he stops at a gas station and puts 10 gallons in the tank. If the trip from Barstow to Las Vegas takes 11 gallons of gas, how much gas is in the tank when he arrives in Las Vegas?

109. Number of Desks A classroom has 5 rows of desks. In each row there are 9 desks. How many desks are there in the classroom?

110. Number of Parking Places A parking lot has 3 rows of spaces. If each row has 40 spaces in it, how many cars can park in the lot?

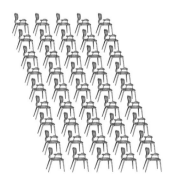

Concert Tickets Tickets for a concert cost $6 for adults and $3 for children. One of the performances is attended by 300 adults and 80 children. Use the information given here to answer Problems 111–114.

CONCERT
Adult's ticket $6
Children's ticket $3

111. How many people attended the concert?

112. How many more adults were in attendance than children?

113. How much money was taken in from adults' tickets?

114. How much money was taken in from children's tickets?

115. Weekly Wages Jeff makes $6 an hour for the first 40 hours he works in a week and $9 an hour for every hour after that. How much does he make if he works 48 hours in one week?

116. Weekly Wages Barbara earns $8 an hour for the first 40 hours she works in a week and $12 an hour for every hour after that. How much does she make if she works 50 hours in one week?

117. Possibilities The product of two whole numbers is 12. List all the possibilities for the two numbers.

118. Possibilities The product of two whole numbers is 15. List all the possibilities for the two numbers.

119. Comparing Products and Sums The product of 8 and 5 is how much larger than the sum of 8 and 5?

120. Comparing Products and Sums The sum of 7 and 9 is how much smaller than the product of 7 and 9?

Extending the Concepts: Writing about Mathematics

In your own words, using complete sentences, give a written description for each of the phrases in Problems 121–124.

121. The commutative property of multiplication

122. The associative property of multiplication

123. A factor of a number

124. The product of two numbers

1.7 The Distributive Property and Multiplication with Whole Numbers

If a supermarket bought 35 cases of soft drinks, and each case contained 24 bottles, then the number of bottles bought can be found by multiplying 35 by 24. The methods we used to multiply numbers in the previous section will not work well for products like 35(24). To develop a more efficient method of multiplication, we need to use what is called the *distributive property*. To begin consider the following two problems:

Problem 1	**Problem 2**
3(4 + 5)	3(4) + 3(5)
= 3(9)	= 12 + 15
= 27	= 27

The result in both cases is the same number, 27. This indicates that the original two expressions must have been equal also. That is,

$$3(4 + 5) = 3(4) + 3(5)$$

This is an example of the distributive property. We say that multiplication *distributes* over addition.

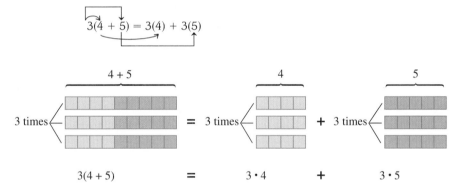

We can write this property in symbols using the letters, a, b, and c to represent any three whole numbers.

Distributive Property

If a, b, and c represent any three whole numbers, then

$$a(b + c) = a(b) + a(c)$$

The following examples illustrate how we apply the distributive property.

EXAMPLE 1
$$4(5 + 9) = 4(5) + 4(9) \quad \textbf{Distributive property}$$
$$= 20 + 36 \quad \textbf{Multiplication}$$
$$= 56 \quad \textbf{Addition}$$ ◼

EXAMPLE 2
$$7(8 + 6) = 7(8) + 7(6) \quad \textbf{Distributive property}$$
$$= 56 + 42 \quad \textbf{Multiplication}$$
$$= 98 \quad \textbf{Addition}$$ ◼

Practice Problems

Use the distributive property to rewrite each of the following expressions, and then simplify.

1. 3(5 + 7)

2. 5(6 + 8)

Answers

1. 36 **2.** 70

Apply the distributive property.

3. $4(x + 2)$

4. $7(a + 9)$

Multiply.

5. $4(7x + 2)$

6. $7(9y + 6)$

The distributive property can also be applied to expressions that contain variables as well as numbers, as shown in Examples 3 and 4.

EXAMPLE 3 $\begin{aligned} 3(x + 5) &= 3 \cdot x + 3 \cdot 5 && \textbf{Distributive property} \\ &= 3x + 15 && \textbf{Multiply 3 and 5} \end{aligned}$ ■

EXAMPLE 4 $\begin{aligned} 9(a + 8) &= 9 \cdot a + 9 \cdot 8 && \textbf{Distributive property} \\ &= 9a + 72 && \textbf{Multiply 9 and 8} \end{aligned}$ ■

The next examples show how we use the distributive property and then the associative property of multiplication to multiply.

EXAMPLE 5 $\begin{aligned} 5(3x + 2) &= 5(3x) + 5 \cdot 2 && \textbf{Distributive property} \\ &= (5 \cdot 3)x + 5 \cdot 2 && \textbf{Associative property} \\ &= 15x + 10 && \textbf{Multiply} \end{aligned}$ ■

EXAMPLE 6 $\begin{aligned} 6(8y + 3) &= 6(8y) + 6 \cdot 3 && \textbf{Distributive property} \\ &= (6 \cdot 8)y + 6 \cdot 3 && \textbf{Associative property} \\ &= 48y + 18 && \textbf{Multiply} \end{aligned}$ ■

Multiplication with Whole Numbers

We can use the distributive property and information from our discussion on multiplication in Section 1.6 to explain how we do multiplication with whole numbers that have two or more digits.

Suppose we want to find the product 7(65). By writing 65 as 60 + 5 and applying the distributive property, we have:

$$\begin{aligned} 7(65) &= 7(60 + 5) && \textbf{65 = 60 + 5} \\ &= 7(60) + 7(5) && \textbf{Distributive property} \\ &= 420 + 35 && \textbf{Multiplication} \\ &= 455 && \textbf{Addition} \end{aligned}$$

We can write the same problem vertically like this:

$$\begin{array}{r} 60 + 5 \\ \times \qquad 7 \\ \hline 35 \quad \leftarrow \quad \textbf{7(5) = 35} \\ + \quad\; 420 \quad \leftarrow \quad \textbf{7(60) = 420} \\ \hline 455 \end{array}$$

This saves some space in writing. But notice that we can cut down on the amount of writing even more if we write the problem this way:

Step 2   **7(6) = 42; add the** $\longrightarrow$ $\overset{3}{6}5$ **Step 1**   **7(5) = 35; carry the 3 to**
3 we carried to $\times\; 7$ **the tens column, and then write**
42 to get 45 $\longrightarrow$ $455 \leftarrow$ **the 5 in the ones column**

This shortcut notation takes some practice. We will do two more problems showing both the long method and the shortcut method.

Answers

3. $4x + 8$ **4.** $7a + 63$
5. $28x + 8$ **6.** $63y + 42$

EXAMPLE 7   Multiply: 9(43)

SOLUTION   We show both methods. The shortcut method is the one you want to become proficient with. The long method is shown so you can see the correlation between the two.

Long Method

$43 = 40 + 3 \longrightarrow$ $40 + 3$

$$\times \quad 9$$

$27 \longleftarrow$ $9(3) = 27$

$+360 \longleftarrow$ $9(40) = 360$

$387 \longleftarrow$ $360 + 27 = 387$

Shortcut Method

Step 2   9(4) = 36; add the $\longrightarrow$ $\overset{2}{4}3$ **Step 1**   9(3) = 27; carry the 2 to
2 we carried to   the tens column, and then write
36 to get 38 ————————→ $\times\ 9$ the 7 in the ones column
$\qquad\qquad\qquad\qquad\qquad 387 \longleftarrow$

■

EXAMPLE 8   Multiply: 8(487)

SOLUTION   Again we show both methods. This time we will leave out the explanations. Study these solutions by comparing them with the solutions in Example 7 until you understand the relationship between the two methods.

Long Method   **Shortcut Method**

$400 + 80 + 7$ $\overset{65}{487}$

$$\times \qquad\qquad 8$$ $\times\ \ 8$

56 $3{,}896$

640

$+3{,}200$

$3{,}896$

■

EXAMPLE 9   Multiply: 52(37)

SOLUTION   This is the same as 52(30 + 7) or

$$52(30) + 52(7)$$

We can find each of these products by using the shortcut method:

$$52 \qquad\qquad \overset{1}{5}2$$
$$\times\ 30 \qquad\qquad \times\ 7$$
$$1{,}560 \qquad\qquad 364$$

The sum of these two numbers is 1,560 + 364 = 1,924. Here is a summary of what we have so far:

$52(37) = 52(30 + 7)$ **37 = 30 + 7**

$\quad\quad\ = 52(30) + 52(7)$ **Distributive property**

$\quad\quad\ = 1{,}560 + 364$ **Multiplication**

$\quad\quad\ = 1{,}924$ **Addition**

The shortcut form for this problem is

$$52$$
$$\times 37$$
$$364 \longleftarrow \ \ 7(52) = 364$$
$$+1{,}560 \longleftarrow \ \ 30(52) = 1{,}560$$
$$1{,}924$$

7. Multiply: 8(57)

8. Multiply: 6(572)

9. Multiply: 45(62)

Note
This discussion is to show why we multiply the way we do. You should go over it in detail, so you will understand the reasons behind the process of multiplication. Besides being able to do multiplication, you should understand it.

Answers
7. 456   **8.** 3,432   **9.** 2,790

In this case we have not shown any of the numbers we carried, simply because it becomes very messy. Notice that because multiplication is a commutative operation, we can get the same result by reversing the order of the factors.

$$
\begin{array}{r}
37 \\
\times 52 \\
\hline
74 \\
+1,850 \\
\hline
1,924
\end{array}
$$

74 ⟵ **2(37) = 74**

+1,850 ⟵ **50(37) = 1,850**

■

10. Multiply: 356(641)

EXAMPLE 10 Multiply: 279(428)

SOLUTION

$$
\begin{array}{r}
279 \\
\times 428 \\
\hline
2,232 \\
5,580 \\
+111,600 \\
\hline
119,412
\end{array}
$$

2,232 ⟵ **8(279) = 2,232**

5,580 ⟵ **20(279) = 5,580**

+111,600 ⟵ **400(279) = 111,600**

If we reverse the order of the factors, we again arrive at the same result. (This just demonstrates that multiplication is commutative and gives us another example to look at.)

$$
\begin{array}{r}
428 \\
\times 279 \\
\hline
3,852 \\
29,960 \\
+85,600 \\
\hline
119,412
\end{array}
$$

3,852 ⟵ **9(428) = 3,852**

29,960 ⟵ **70(428) = 29,960**

+85,600 ⟵ **200(428) = 85,600**

■

USING TECHNOLOGY

Calculators

Here is how we would work the problem shown in Example 10 on a calculator:

Scientific Calculator: 279 ⊠ 428 ▣

Graphing Calculator: 279 ⊠ 428 ENT

Estimating

One way to estimate the answer to the problem shown in Example 10 is to round each number to the nearest hundred and then multiply the rounded numbers. Doing so would give us this:

$$300(400) = 120,000$$

Our estimate of the answer is 120,000, which is close to the actual answer, 119,412. Making estimates is important when we are using calculators; having an estimate of the answer will keep us from making major errors in multiplication.

Answer

10. 228,196

Applications

EXAMPLE 11 A supermarket orders 35 cases of a certain soft drink. If each case contains 24 bottles of the drink, how many bottles were ordered?

SOLUTION   We have 35 cases and each case has 24 bottles. The total number of bottles is the product of 35 and 24, which is 35(24):

$$
\begin{array}{r}
24 \\
\times 35 \\
\hline
120 \\
+720 \\
\hline
840
\end{array}
$$

120 ⟵ **5(24) = 120**
+720 ⟵ **30(24) = 720**

There is a total of 840 bottles of the soft drink. ∎

EXAMPLE 12 Shirley earns $12 an hour for the first 40 hours she works each week. If she has $109 deducted from her weekly check for taxes and retirement, how much money will she take home if she works 38 hours this week?

SOLUTION   To find the amount of money she earned for the week, we multiply 12 and 38. From that total we subtract 109. The result is her take-home pay. Without showing all the work involved in the calculations, here is the solution:

$$38(\$12) = \$456 \quad \textbf{Her total weekly earnings}$$
$$\$456 - \$109 = \$347 \quad \textbf{Her take-home pay}$$ ∎

EXAMPLE 13 In 1993, the government standardized the way in which nutrition information is presented on the labels of most packaged food products. Figure 1 on the following page shows one of these standardized food labels. It is from a package of Fritos Corn Chips that I ate the day I was writing this example. Approximately how many chips are in the bag, and what is the total number of calories consumed if all the chips in the bag are eaten?

SOLUTION   Reading toward the top of the label, we see that there are about 32 chips in one serving, and approximately 3 servings in the bag. Therefore, the total number of chips in the bag is

$$3(32) = 96 \text{ chips}$$

This is an approximate number, because each serving is approximately 32 chips. Reading further we find that each serving contains 160 calories. Therefore, the total number of calories consumed by eating all the chips in the bag is

$$3(160) = 480 \text{ calories}$$

As we progress through the book, we will study more of the information in nutrition labels. ∎

11. If each tablet of vitamin C contains 550 milligrams of vitamin C, what is the total number of milligrams of vitamin C in a bottle that contains 365 tablets?

12. If Shirley works 36 hours the next week and has the same amount deducted from her check for taxes and retirement, how much will she take home?

13. The amounts given in the middle of the nutrition label in Figure 1 are for one serving of chips. If all the chips in the bag are eaten, how much fat has been consumed? How much sodium?

Answers
11. 200,750 milligrams
12. $323
13. 30 g of fat, 480 mg of sodium

Nutrition Facts
Serving Size 1 oz. (28g/About 32 chips)
Servings Per Container: 3

Amount Per Serving

Calories 160   Calories from fat 90

% Daily Value*

Total Fat 10 g	16%
Saturated Fat 1.5g	8%
Cholesterol 0mg	0%
Sodium 160mg	7%
Total Carbohydrate 15g	5%
Dietary Fiber 1g	4%
Sugars 0g	
Protein 2g	

Vitamin A 0%	•	Vitamin C 0%
Calcium 2%	•	Iron 0%

*Percent Daily Values are based on a 2,000 calorie diet

Figure 1

14. If a 150-pound person bowls for 3 hours, will he or she burn all the calories consumed by eating two bags of the chips mentioned in Example 13?

EXAMPLE 14 The table below was first introduced in Section 1.4. Suppose a 150-pound person goes bowling for 2 hours after having eaten the bag of chips mentioned in Example 13. Will he or she burn all the calories consumed from the chips?

ACTIVITY	CALORIES BURNED IN 1 HOUR BY A 150-POUND PERSON
Bicycling	374
Bowling	265
Handball	680
Jazzercize	340
Jogging	680
Skiing	544

SOLUTION   Each hour of bowling burns 265 calories. If the person bowls for 2 hours, a total of

$$2(265) = 530 \text{ calories}$$

will have been burned. Because the bag of chips contained only 480 calories, all of them have been burned with 2 hours of bowling. ∎

Answer
14. No

Problem Set 1.7

Use the distributive property to rewrite each expression.

1. $7(2 + 3)$ **2.** $4(5 + 8)$ **3.** $9(4 + 7)$ **4.** $6(9 + 5)$ **5.** $3(x + 1)$ **6.** $5(x + 8)$

7. $4(a + 9)$ **8.** $3(a + 10)$ **9.** $5(3x + 4)$ **10.** $7(2x - 5)$ **11.** $6(8y + 2)$ **12.** $9(6y + 3)$

Find each of the following products. (Multiply.) In each case use the shortcut method.

13. $\begin{array}{r} 25 \\ \times\ 4 \\ \hline \end{array}$
14. $\begin{array}{r} 87 \\ \times\ 9 \\ \hline \end{array}$
15. $\begin{array}{r} 38 \\ \times\ 6 \\ \hline \end{array}$
16. $\begin{array}{r} 45 \\ \times\ 7 \\ \hline \end{array}$
17. $\begin{array}{r} 18 \\ \times\ 2 \\ \hline \end{array}$
18. $\begin{array}{r} 29 \\ \times\ 3 \\ \hline \end{array}$

19. $\begin{array}{r} 72 \\ \times 20 \\ \hline \end{array}$
20. $\begin{array}{r} 68 \\ \times 30 \\ \hline \end{array}$
21. $\begin{array}{r} 19 \\ \times 50 \\ \hline \end{array}$
22. $\begin{array}{r} 24 \\ \times 40 \\ \hline \end{array}$
23. $\begin{array}{r} 69 \\ \times 25 \\ \hline \end{array}$
24. $\begin{array}{r} 27 \\ \times 36 \\ \hline \end{array}$

25. $\begin{array}{r} 11 \\ \times 11 \\ \hline \end{array}$
26. $\begin{array}{r} 12 \\ \times 21 \\ \hline \end{array}$
27. $\begin{array}{r} 97 \\ \times 16 \\ \hline \end{array}$
28. $\begin{array}{r} 24 \\ \times 39 \\ \hline \end{array}$
29. $\begin{array}{r} 168 \\ \times\ 25 \\ \hline \end{array}$
30. $\begin{array}{r} 452 \\ \times\ 34 \\ \hline \end{array}$

31. $\begin{array}{r} 728 \\ \times\ 91 \\ \hline \end{array}$
32. $\begin{array}{r} 680 \\ \times\ 76 \\ \hline \end{array}$
33. $\begin{array}{r} 698 \\ \times 400 \\ \hline \end{array}$
34. $\begin{array}{r} 879 \\ \times 600 \\ \hline \end{array}$
35. $\begin{array}{r} 111 \\ \times 111 \\ \hline \end{array}$
36. $\begin{array}{r} 123 \\ \times 321 \\ \hline \end{array}$

37. $\begin{array}{r} 532 \\ \times 200 \\ \hline \end{array}$
38. $\begin{array}{r} 277 \\ \times 900 \\ \hline \end{array}$
39. $\begin{array}{r} 856 \\ \times 232 \\ \hline \end{array}$
40. $\begin{array}{r} 455 \\ \times 248 \\ \hline \end{array}$
41. $\begin{array}{r} 976 \\ \times 628 \\ \hline \end{array}$
42. $\begin{array}{r} 432 \\ \times 555 \\ \hline \end{array}$

43. $\begin{array}{r} 2,468 \\ \times\ 135 \\ \hline \end{array}$
44. $\begin{array}{r} 2,725 \\ \times\ 324 \\ \hline \end{array}$
45. $\begin{array}{r} 24,563 \\ \times\ 735 \\ \hline \end{array}$
46. $\begin{array}{r} 56,728 \\ \times\ 852 \\ \hline \end{array}$
47. $\begin{array}{r} 44,777 \\ \times\ 5,888 \\ \hline \end{array}$
48. $\begin{array}{r} 33,999 \\ \times\ 2,555 \\ \hline \end{array}$

49. $\begin{array}{r} 203,104 \\ \times\ 402 \\ \hline \end{array}$
50. $\begin{array}{r} 601,208 \\ \times\ 304 \\ \hline \end{array}$

Complete the following tables.

51.

FIRST NUMBER a	SECOND NUMBER b	THEIR PRODUCT ab
11	11	
11	22	
22	22	
22	44	

52.

FIRST NUMBER a	SECOND NUMBER b	THEIR PRODUCT ab
11	111	
11	222	
22	111	
22	222	

53.

FIRST NUMBER a	SECOND NUMBER b	THEIR PRODUCT ab
25	15	
25	30	
50	15	
50	30	

54.

FIRST NUMBER a	SECOND NUMBER b	THEIR PRODUCT ab
12	20	
36	20	
12	40	
36	40	

Each diagram below illustrates the distributive property. In each case, use symbols to write the statement that is illustrated by the diagram.

55. ⬜⬜⬜◯◯◯◯ ⬜⬜⬜ ◯◯◯◯
⬜⬜⬜◯◯◯◯ = ⬜⬜⬜ + ◯◯◯◯
⬜⬜⬜◯◯◯◯ ⬜⬜⬜ ◯◯◯◯

56. ⬜⬜⬜⬜⬜◯◯◯◯◯◯◯ ⬜⬜⬜⬜⬜ ◯◯◯◯◯◯◯
⬜⬜⬜⬜⬜◯◯◯◯◯◯◯ ⬜⬜⬜⬜⬜ ◯◯◯◯◯◯◯
⬜⬜⬜⬜⬜◯◯◯◯◯◯◯ = ⬜⬜⬜⬜⬜ + ◯◯◯◯◯◯◯
⬜⬜⬜⬜⬜◯◯◯◯◯◯◯ ⬜⬜⬜⬜⬜ ◯◯◯◯◯◯◯

Applying the Concepts

57. Planning a Trip A family decides to drive their compact car on their vacation. They figure it will require a total of about 230 gallons of gas for the vacation. If each gallon of gas will take them 22 miles, how long is the trip they are planning?

58. Age in Months There are 12 months in every year. How many months have you lived if you are exactly 21 years old?

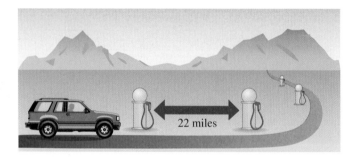

22 miles

59. Farming A rancher has 29 cattle in his herd at the beginning of the year. If 25 new calves are born during the year, and he sells 12 of them, how many animals does he have at the end of the year?

60. Manufacturing A company manufacturing small computers has 256 computers on hand at the beginning of the month. During the month they manufacture 42 more computers and sell 158 computers. How many computers do they have on hand at the end of the month?

61. Rent A student pays $475 rent each month. How much money does she spend on rent in 2 years?

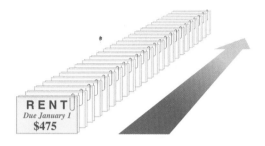

RENT
Due January 1
$475

62. Populations In 1990, there were approximately 31 times as many people living in Texas as were living in Alaska. If there were 551,947 people living in Alaska in 1990, how many people were living in Texas?

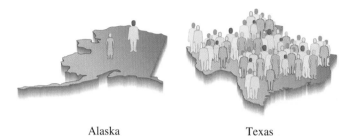

Alaska   Texas

63. Take-Home Pay Sally makes $14 an hour for the first 40 hours she works each week. If she works 27 hours one week and has $128 deducted from her earnings for taxes and retirement, what is her take-home pay?

64. Take-Home Pay Frieda earns $16 an hour for the first 40 hours she works each week. She has $129 deducted from each check for taxes and retirement. If she works 34 hours one week, what is her take-home pay?

Exercise and Calories The table below is an extension of the table we used in Example 14 of this section. It gives the amount of energy expended during 1 hour of various activities for people of different weights. The accompanying figure is a nutrition label from a bag of Doritos tortilla chips. Use the information from the table and the nutrition label to answer Problems 65–70.

TORTILLA CHIPS

Nutrition Facts

Serving Size 1 oz. (28g/About 12 chips)
Servings Per Container About 2

Amount Per Serving	
Calories 140	Calories from fat 60

	% Daily Value*
Total Fat 7g	11%
Saturated Fat 1g	6%
Cholesterol 0mg	0%
Sodium 170mg	7%
Total Carbohydrate 18g	6%
Dietary Fiber 1g	4%
Sugars less than 1g	
Protein 2g	

Vitamin A 0%	●	Vitamin C 0%
Calcium 4%	●	Iron 2%

*Percent Daily Values are based on a 2,000 calorie diet

CALORIES BURNED THROUGH EXERCISE

	CALORIES PER HOUR		
ACTIVITY	120 POUNDS	150 POUNDS	180 POUNDS
Bicycling	299	374	449
Bowling	212	265	318
Handball	544	680	816
Jazzercize	272	340	408
Jogging	544	680	816
Skiing	435	544	653

65. Suppose you weigh 180 pounds. How many calories would you burn if you play handball for 2 hours and then ride your bicycle for 1 hour?

66. How many calories are burned by a 120-lb person who jogs for 1 hour and then goes bike riding for 2 hours?

67. How many calories would you consume if you ate the entire bag of chips?

68. Approximately how many chips are in the bag?

69. If you weigh 180 pounds, will you burn off the calories consumed by eating 3 servings of tortilla chips if you ride your bike 1 hour?

70. If you weigh 120 pounds, will you burn off the calories consumed by eating 3 servings of tortilla chips if you ride your bike for 1 hour?

Estimating

Mentally estimate the answer to each of the following problems by rounding each number to the indicated place and then multiplying.

71. 750 hundred
         × 12 ten

72. 591 hundred
        ×323 hundred

73. 3,472 thousand
         × 511 hundred

74. 399 hundred
        ×298 hundred

75. 2,399 thousand
         × 698 hundred

76. 9,999 thousand
         × 666 hundred

Calculator Problems

Perform each multiplication using a calculator.

77. 3,965(4,282)

78. (1,234)(4,321)

79. 79,428(549)

80. 89,320(476)

81. Jogging Suppose a jogger burns 119 calories for every mile she runs. If that jogger runs 324 miles each month, how many calories does she burn while jogging in 1 year? (Remember, 12 months = 1 year.)

82. Speed of Sound In water, sound travels 4,938 feet every second. How far would it travel if it could travel for 265 seconds?

Extending the Concepts: Number Sequences

A *geometric sequence* is a sequence of numbers in which each number is obtained from the previous number by multiplying by the same number each time. For example, the sequence $3, 6, 12, 24, \ldots$ is a geometric sequence, starting with 3, in which each number comes from multiplying the previous number by 2 each time.

Each sequence below is a geometric sequence. Find the next number in each sequence, and then give a written description of the sequence. (For instance, one way to give a written description of the sequence $3, 6, 12, 24, \ldots$ would be to write "The sequence starts with 3. Then each number comes from the number before it by multiplying by 2 each time.")

83. $5, 10, 20, \ldots$

84. $10, 50, 250, \ldots$

85. $2, 6, 18, \ldots$

86. $12, 24, 48, \ldots$

Recall that in an arithmetic sequence, each number after the first number comes from the previous number by adding the same amount each time. Identify each of the following sequences as either arithmetic, geometric, or neither.

87. $1, 3, 5, 7, \ldots$

88. $1, 3, 9, 27, \ldots$

89. $2, 4, 8, 16, \ldots$

90. $2, 4, 6, 8, \ldots$

91. $1, 3, 6, 10, \ldots$

92. $1, 4, 9, 16, \ldots$

1.8 Division with Whole Numbers

Introduction . . .

Darlene is planning a party and would like to serve 8-ounce glasses of soda. The glasses will be filled from 32-ounce bottles of soda. In order to know how many bottles of soda to buy, she needs to find out how many of the 8-ounce glasses can be filled by one of the 32-ounce bottles. One way to solve this problem is with division: dividing 32 by 8. A diagram of the problem is shown in Figure 1.

32-ounce bottle   8-ounce glasses

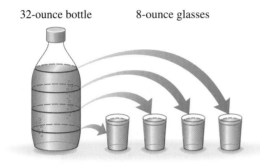

Figure 1

As a division problem: **As a multiplication problem:**

$32 \div 8 = 4$ $4 \cdot 8 = 32$

Notation

As was the case with multiplication, there are many ways to write division statements. All the following statements are equivalent. They all mean 10 divided by 5.

$$10 \div 5, \quad \frac{10}{5}, \quad 10/5, \quad 5\overline{)10}, \quad \tfrac{10}{5}$$

The kind of notation we use to write division problems will depend on the situation. We will use the notation $5\overline{)10}$ mostly with the longer division problems found in this chapter. The notation $\frac{10}{5}$ will be used in the chapter on fractions and in later chapters. The horizontal line used with the notation $\frac{10}{5}$ is called the *fraction bar.*

Vocabulary

The word *quotient* is used to indicate division. If we say "The quotient of 10 and 5 is 2," then we mean

$$10 \div 5 = 2 \quad \text{or} \quad \frac{10}{5} = 2$$

The 10 is called the *dividend,* and the 5 is called the *divisor.* All the expressions, $10 \div 5, \frac{10}{5},$ and 2, are called the *quotient* of 10 and 5.

Some additional examples of division statements are given in Table 1.

Table 1

IN ENGLISH	IN SYMBOLS
The quotient of 15 and 3	$15 \div 3$, or $\dfrac{15}{3}$, or 15/3
The quotient of 3 and 15	$3 \div 15$, or $\dfrac{3}{15}$, or 3/15
The quotient of 8 and n	$8 \div n$, or $\dfrac{8}{n}$, or 8/n
x divided by 2	$x \div 2$, or $\dfrac{x}{2}$, or x/2
The quotient of 21 and 3 is 7	$21 \div 3 = 7$, or $\dfrac{21}{3} = 7$

The Meaning of Division

One way to arrive at an answer to a division problem is by thinking in terms of multiplication. For example, if we want to find the quotient of 32 and 8, we may ask, "What do we multiply by 8 to get 32?"

$$32 \div 8 = ? \quad \text{means} \quad 8 \cdot ? = 32$$

Because we know from our work with multiplication that $8 \cdot 4 = 32$, it must be true that

$$32 \div 8 = 4$$

Table 2 lists some additional examples.

Table 2

DIVISION			MULTIPLICATION
$18 \div 6 = 3$	because	$6 \cdot 3 = 18$
$32 \div 8 = 4$	because	$8 \cdot 4 = 32$
$10 \div 2 = 5$	because	$2 \cdot 5 = 10$
$72 \div 9 = 8$	because	$9 \cdot 8 = 72$

Another way to think of division is in terms of repeated subtraction. When we say "54 divided by 6 is 9," we also mean that if we subtract 9 sixes from 54, the result will be 0.

Division by One-Digit Numbers

Consider the following division problem:

$$465 \div 5$$

We can think of this problem as asking the question, "How many fives can we subtract from 465?" To answer the question we begin subtracting multiples of 5. One way to organize this process is shown below:

$$\begin{array}{r} 90 \leftarrow \textbf{We first guess that there are at least 90 fives in 465} \\ 5{\overline{)465}} \\ \underline{-450} \leftarrow \textbf{90(5) = 450} \\ 15 \leftarrow \textbf{15 is left after we subtract 90 fives from 465} \end{array}$$

What we have done so far is subtract 90 fives from 465 and found that 15 is still left. Because there are 3 fives in 15, we continue the process.

$$\begin{array}{r} 3 \leftarrow \textbf{There are 3 fives in 15} \\ 90 \\ 5\overline{)465} \\ -450 \\ \hline 15 \\ -15 \leftarrow \textbf{3} \cdot \textbf{5 = 15} \\ \hline 0 \leftarrow \textbf{The difference is 0} \end{array}$$

The total number of fives we have subtracted from 465 is

$$90 + 3 = 93 \qquad \text{the number of fives subtracted from 465}$$

We now summarize the results of our work.

$$465 \div 5 = 93 \qquad \begin{array}{l}\text{which we could check}\\ \text{with multiplication} \rightarrow\end{array} \begin{array}{r} \overset{1}{9}3 \\ \times \ 5 \\ \hline 465 \end{array}$$

Notation

The division problem just shown can be shortened by eliminating the subtraction signs, eliminating the zeros in each estimate, and eliminating some of the numbers that are repeated in the problem.

$$\begin{array}{ll} \text{The shorthand} & \begin{array}{r} 3 \\ 90 \\ 5\overline{)465} \\ 450 \\ \hline 15 \\ 15 \\ \hline 0 \end{array} \quad \begin{array}{l}\text{looks like}\\ \text{this.}\end{array} \quad \begin{array}{r} 93 \\ 5\overline{)465} \\ 45{\downarrow} \\ \hline 15 \\ 15 \\ \hline 0 \end{array} \quad \begin{array}{l}\text{The arrow}\\ \text{indicates that}\\ \text{we bring down}\\ \text{the 5 after}\\ \text{we subtract.}\end{array} \end{array}$$

(The shorthand form for this problem)

The problem shown above on the right is the shortcut form of what is called *long division.* Here is an example showing this shortcut form of long division from start to finish.

EXAMPLE 1 Divide: $595 \div 7$

SOLUTION Because $7(8) = 56$, our first estimate of the number of sevens that can be subtracted from 595 is 80:

$$\begin{array}{r} 8 \leftarrow \textbf{The 8 is placed above the tens column} \\ 7\overline{)595} \quad \textbf{so we know our first estimate is 80} \\ 56{\downarrow} \leftarrow \textbf{8(7) = 56} \\ \hline 35 \leftarrow \textbf{59 − 56 = 3; then bring down the 5} \end{array}$$

Since $7(5) = 35$, we have

$$\begin{array}{r} 85 \leftarrow \textbf{There are 5 sevens in 35} \\ 7\overline{)595} \\ 56{\downarrow} \\ \hline 35 \\ 35 \leftarrow \textbf{5(7) = 35} \\ \hline 0 \leftarrow \textbf{35 − 35 = 0} \end{array}$$

Practice Problems

1. Divide: $288 \div 4$

Answer

1. 72

Our result is 595 ÷ 7 = 85, which we can check with multiplication:

$$\begin{array}{r} \overset{3}{85} \\ \times\ 7 \\ \hline 595 \end{array}$$

∎

Division by Two-Digit Numbers

2. Divide: 6,792 ÷ 24

EXAMPLE 2   Divide: 9,380 ÷ 35

SOLUTION   In this case our divisor, 35, is a two-digit number. The process of division is the same. We still want to find the number of thirty-fives we can subtract from 9,380.

```
          2    ← The 2 is placed above the hundreds column
    35)9,380
       7 0↓    ← 2(35) = 70
       2 38    ← 93 − 70 = 23; then bring down the 8
```

We can make a few preliminary calculations to help estimate how many thirty-fives are in 238:

$$5 \times 35 = 175 \qquad 6 \times 35 = 210 \qquad 7 \times 35 = 245$$

Because 210 is the closest to 238 without being larger than 238, we use 6 as our next estimate:

```
         26    ← 6 in the tens column means this estimate is 60
    35)9,380
       7 0
       2 38
       2 10    ← 6(35) = 210
        280    ← 238 − 210 = 28; bring down the 0
```

Because 35(8) = 280, we have

```
        268
    35)9,380
       7 0
       2 38
       2 10
        280
        280    ← 8(35) = 280
          0    ← 280 − 280 = 0
```

We can check our result with multiplication:

$$\begin{array}{r} 268 \\ \times\ 35 \\ \hline 1{,}340 \\ 8{,}040 \\ \hline 9{,}380 \end{array}$$

∎

Answer
2. 283

Division with Remainders

Suppose Darlene were planning to use 6-ounce glasses instead of 8-ounce glasses for her party. To see how many glasses she could fill from the 32-ounce bottle, she would divide 32 by 6. If she did so, she would find that she could fill 5 glasses, but after doing so she would have 2 ounces of soda left in the bottle. A diagram of this problem is shown in Figure 2.

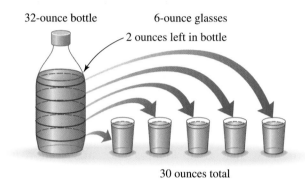

Figure 2

Writing the results in the diagram as a division problem looks like this:

$$
\begin{array}{r}
5 \leftarrow \textbf{Quotient} \\
\text{Divisor} \rightarrow 6\overline{)32} \leftarrow \textbf{Dividend} \\
\underline{30} \\
2 \leftarrow \textbf{Remainder}
\end{array}
$$

EXAMPLE 3 Divide: $1{,}690 \div 67$

SOLUTION   Dividing as we have previously, we get

$$
\begin{array}{r}
25 \\
67\overline{)1{,}690} \\
1\ 34\downarrow \\
\hline
350 \\
335 \\
\hline
15 \leftarrow \textbf{15 is left over}
\end{array}
$$

We have 15 left, and because 15 is less than 67, no more sixty-sevens can be subtracted. In a situation like this we call 15 the *remainder* and write

These indicate that the remainder is 15

$$
\begin{array}{r}
25 \text{ R } \textbf{15} \\
67\overline{)1{,}690} \\
1\ 34 \\
\hline
350 \\
335 \\
\hline
15
\end{array}
\quad \text{or} \quad
\begin{array}{r}
25\frac{15}{67} \\
67\overline{)1{,}690} \\
1\ 34 \\
\hline
350 \\
335 \\
\hline
15
\end{array}
$$

Both forms of notation shown above indicate that 15 is the remainder. The notation R 15 is the notation we will use in this chapter. The notation $\frac{15}{67}$ will be useful in Chapter 3.

3. Divide: $1{,}883 \div 27$

Calculator Note

Here is how we would work the problem shown in Example 3 on a calculator:

Scientific Calculator:
1690 $\boxed{\div}$ 67 $\boxed{=}$

Graphing Calculator:
1690 $\boxed{\div}$ 67 $\boxed{\text{ENT}}$

In both cases the calculator will display 25.223881 (give or take a few digits at the end), which gives the remainder in decimal form. We will discuss decimals later in the book.

Answer
3. 69 R 20, or $69\frac{20}{27}$

To check a problem like this, we multiply the divisor and the quotient as usual, and then add the remainder to this result:

$$
\begin{array}{r}
67 \\
\times 25 \\
\hline
335 \\
1{,}340 \\
\hline
1{,}675
\end{array}
$$
⟵ **Product of divisor and quotient**

$$1{,}675 + 15 = 1{,}690$$

↑ **Remainder**   ↖ **Dividend**

■

4. A family spends $1,872 on a 12-day vacation. How much did they spend each day?

EXAMPLE 4 A family has an annual income of $11,520. How much is their monthly income?

SOLUTION   Because there are 12 months in a year and the yearly (annual) income is $11,520, we want to know what $11,520 divided into 12 equal parts is. Therefore we have

$$
\begin{array}{r}
960 \\
12\overline{)11{,}520} \\
\underline{10\ 8} \\
72 \\
\underline{72} \\
00
\end{array}
$$

Because $11{,}520 \div 12 = 960$, the monthly income for this family is $960.   ■

Note

To estimate the answer to Example 4 quickly, we can replace 11,520 with 12,000 and mentally calculate

$$12{,}000 \div 12$$

which gives an estimate of 1,000. Our actual answer, 960, is close enough to our estimate to convince us that we have not made a major error in our calculation.

Division by Zero

We cannot divide by 0. That is, we cannot use 0 as a divisor in any division problem. Here's why.

Suppose there was an answer to the problem

$$\frac{8}{0} = ?$$

That would mean that

$$0 \cdot ? = 8$$

But we already know that multiplication by 0 always produces 0. There is no number we can use for the ? to make a true statement out of

$$0 \cdot ? = 8$$

Because this was equivalent to the original division problem

$$\frac{8}{0} = ?$$

we have no number to associate with the expression $\frac{8}{0}$. It is undefined.

> **RULE**
>
> Division by 0 is undefined. Any expression with a divisor of 0 is undefined. We cannot divide by 0.

Answer

4. $156

Problem Set 1.8

Write each of the following phrases in symbols.

1. The quotient of 6 and 3

2. The quotient of 3 and 6

3. The quotient of 45 and 9

4. The quotient of 12 and 4

5. The quotient of r and s

6. The quotient of s and r

7. The quotient of 20 and 4 is 5.

8. The quotient of 20 and 5 is 4.

Write a multiplication statement that is equivalent to each of the following division statements.

9. $6 \div 2 = 3$

10. $6 \div 3 = 2$

11. $\dfrac{36}{9} = 4$

12. $\dfrac{36}{4} = 9$

13. $\dfrac{48}{6} = 8$

14. $\dfrac{35}{7} = 5$

15. $28 \div 7 = 4$

16. $81 \div 9 = 9$

Find each of the following quotients. (Divide.)

17. $25 \div 5$

18. $72 \div 8$

19. $40 \div 5$

20. $12 \div 2$

21. $9 \div 0$

22. $7 \div 1$

23. $360 \div 8$

24. $285 \div 5$

25. $\dfrac{138}{6}$

26. $\dfrac{267}{3}$

27. $5\overline{)7,650}$

28. $5\overline{)5,670}$

29. $5\overline{)6,750}$

30. $5\overline{)6,570}$

31. $3\overline{)54,000}$

32. $3\overline{)50,400}$

33. $3\overline{)50,040}$

34. $3\overline{)50,004}$

Estimating

Work Problems 35 and 36 mentally, without using a calculator.

35. The quotient $845 \div 93$ is closest to which of the following numbers.

    a. 10   **b.** 100   **c.** 1,000   **d.** 10,000

36. The quotient $762 \div 43$ is closest to which of the following numbers?

    a. 2   **b.** 20   **c.** 200   **d.** 2,000

37. The quotient $15,208 \div 771$ is closest to which of the following numbers?

    a. 2   **b.** 20   **c.** 200   **d.** 2,000

38. The quotient $24,471 \div 523$ is closest to which of the following numbers?

    a. 5   **b.** 50   **c.** 500   **d.** 5,000

Mentally give a one-digit estimate for each of the following quotients. That is, for each quotient, mentally estimate the answer using one of the digits 1, 2, 3, 4, 5, 6, 7, 8, or 9.

39. $316 \div 289$ **40.** $662 \div 289$ **41.** $728 \div 355$ **42.** $728 \div 177$

43. $921 \div 243$ **44.** $921 \div 442$ **45.** $673 \div 109$ **46.** $673 \div 218$

Divide. You shouldn't have any wrong answers because you can always check your results with multiplication.

47. $1,440 \div 32$ **48.** $1,206 \div 67$ **49.** $\dfrac{2,401}{49}$ **50.** $\dfrac{4,606}{49}$

51. $28\overline{)12,096}$ **52.** $28\overline{)96,012}$ **53.** $63\overline{)90,594}$ **54.** $45\overline{)17,595}$

Complete the following tables.

55.

FIRST NUMBER a	SECOND NUMBER b	THE QUOTIENT OF a AND b $\dfrac{a}{b}$
100	25	
100	26	
100	27	
100	28	

56.

FIRST NUMBER a	SECOND NUMBER b	THE QUOTIENT OF a AND b $\dfrac{a}{b}$
100	25	
101	25	
102	25	
103	25	

The following division problems all have remainders.

57. $6\overline{)370}$ **58.** $8\overline{)390}$ **59.** $3\overline{)271}$ **60.** $3\overline{)172}$

61. $26\overline{)345}$ **62.** $26\overline{)543}$ **63.** $71\overline{)16,620}$ **64.** $71\overline{)33,240}$

65. $23\overline{)10,400}$ **66.** $23\overline{)20,800}$ **67.** $169\overline{)5,950}$ **68.** $391\overline{)34,450}$

Applying the Concepts

The application problems that follow may involve more than merely division. Some may require addition, subtraction, or multiplication, whereas others may use a combination of two or more operations.

69. Monthly Income A family has an annual income of $22,200. How much is their monthly income?

70. Hourly Wages If a man works an 8-hour shift and is paid $48, how much does he make for 1 hour?

71. Price per Pound If 6 pounds of a certain kind of fruit cost 96¢, how much does 1 pound cost?

72. Cost of a Dress A dress shop orders 45 dresses for a total of $675. If they paid the same amount for each dress, how much was each dress?

73. Filling Glasses How many 4-ounce glasses can a person fill from a 32-ounce bottle of soda?

74. Filling Glasses How many 8-ounce glasses can be filled from three 32-ounce bottles of soda?

75. Filling Glasses   How many 5-ounce glasses can be filled from a 32-ounce bottle of milk? How many ounces of milk will be left in the bottle when all the glasses are full?

76. Filling Glasses   How many 3-ounce glasses can be filled from a 28-ounce bottle of milk? How many ounces of milk will be left in the bottle when all the glasses are filled?

77. Filling Glasses   How many 32-ounce bottles of Coke will be needed to fill sixteen 6-ounce glasses?

78. Filling Glasses   How many 28-ounce bottles of 7-Up will be needed to fill fourteen 6-ounce glasses?

79. Cost of Wine   If a person paid $192 for 16 cases of wine, how much did each case cost?

80. Miles per Gallon   A traveling salesman kept track of his mileage for 1 month. He found that he traveled 1,104 miles and used 48 gallons of gas. How many miles did he travel on each gallon of gas?

81. Milligrams of Calcium   Suppose one egg contains 25 milligrams of calcium, a piece of toast contains 40 milligrams of calcium, and a glass of milk contains 215 milligrams of calcium. How many milligrams of calcium are contained in a breakfast that consists of three eggs, two glasses of milk, and four pieces of toast?

82. Milligrams of Iron   Suppose a glass of juice contains 3 milligrams of iron and a piece of toast contains 2 milligrams of iron. If Diane drinks two glasses of juice and has three pieces of toast for breakfast, how much iron is contained in the meal?

Calculator Problems

Find each of the following quotients using a calculator.

83. $305,026 \div 698$

84. $771,537 \div 949$

85. $18,436,466 \div 5,678$

86. $2,492,735 \div 2,345$

87. $695\overline{)603,955}$

88. $876\overline{)875,124}$

89. $4,903\overline{)27,868,652}$

90. $3,090\overline{)2,308,230}$

91. Gallons per Minute   If a 79,768-gallon tank can be filled in 472 minutes, how many gallons enter the tank each minute?

92. Weight per Case   A truckload of 632 crates of motorcycle parts weighs 30,968 pounds. How much does each of the crates weigh, if they each weigh the same amount?

1.9 Exponents and Order of Operations

Exponents are a shorthand way of writing repeated multiplication. In the expression 2^3, 2 is called the *base* and 3 is called the *exponent.* The expression 2^3 is read "2 to the third power" or "2 cubed." The exponent 3 tells us to use the base 2 as a multiplication factor three times.

$$2^3 = 2 \cdot 2 \cdot 2 \quad \textbf{2 is used as a factor three times}$$

We can simplify the expression by multiplication:

$$2^3 = 2 \cdot 2 \cdot 2$$
$$= 4 \cdot 2$$
$$= 8$$

The expression 2^3 is equal to the number 8. We can summarize this discussion with the following definition:

DEFINITION

An **exponent** is a number that indicates how many times the base is to be used as a factor. Exponents indicate repeated multiplication.

For example, in the expression 5^2, 5 is the base and 2 is the exponent. The meaning of the expression is

$$5^2 = 5 \cdot 5 \quad \textbf{5 is used as a factor two times}$$
$$= 25$$

The expression 5^2 is read "5 to the second power" or "5 squared."
Here are some more examples.

EXAMPLE 1 3^2 The base is 3, and the exponent is 2. The expression is read "3 to the second power" or "3 squared." ∎

EXAMPLE 2 3^3 The base is 3, and the exponent is 3. The expression is read "3 to the third power" or "3 cubed." ∎

EXAMPLE 3 2^4 The base is 2, and the exponent is 4. The expression is read "2 to the fourth power." ∎

As you can see from these examples, a base raised to the second power is also said to be *squared,* and a base raised to the third power is also said to be *cubed.* These are the only two exponents (^{2} and ^{3}) that have special names. All other exponents are referred to only as "fourth powers," "fifth powers," "sixth powers," and so on.

The next examples show how we can simplify expressions involving exponents by using repeated multiplication.

EXAMPLE 4 $3^2 = 3 \cdot 3 = 9$ ∎

EXAMPLE 5 $4^2 = 4 \cdot 4 = 16$ ∎

Practice Problems
For each expression, name the base and the exponent, and write the expression in words.
1. 5^2

2. 2^3

3. 1^4

Simplify each of the following by using repeated multiplication.

4. 5^2
5. 9^2

Answers
1.-3. See solutions section.
4. 25 **5.** 81

6. 2^3

7. 1^4

8. 2^5

EXAMPLE 6 $3^3 = 3 \cdot 3 \cdot 3 = 9 \cdot 3 = 27$ ∎

EXAMPLE 7 $3^4 = 3 \cdot 3 \cdot 3 \cdot 3 = 9 \cdot 9 = 81$ ∎

EXAMPLE 8 $2^4 = 2 \cdot 2 \cdot 2 \cdot 2 = 4 \cdot 4 = 16$ ∎

USING TECHNOLOGY

Calculators

Here is how we use a calculator to evaluate exponents, as we did in Example 8:

Scientific Calculator: 2 $\boxed{x^y}$ 4 $\boxed{=}$
Graphing Calculator: 2 $\boxed{\wedge}$ 4 $\boxed{\text{ENT}}$ or 2 $\boxed{x^y}$ 4 $\boxed{\text{ENT}}$
(depending on the calculator)

Finally, we should consider what happens when the numbers 0 and 1 are used as exponents. First of all, any number raised to the first power is itself. That is, if we let the letter a represent any number, then

$$a^1 = a$$

To take care of the cases when 0 is used as an exponent, we must use the following definition:

DEFINITION

Any number other than 0 raised to the 0 power is 1. That is, if a represents any nonzero number, then it is always true that

$$a^0 = 1$$

Simplify each of the following expressions.

9. 7^1

10. 4^1

11. 9^0

12. 1^0

EXAMPLE 9 $5^1 = 5$ ⎫
                                    ⎬ Any number raised to the first power is itself.
EXAMPLE 10 $9^1 = 9$ ⎭

EXAMPLE 11 $4^0 = 1$ ⎫
                                     ⎬ Any nonzero number raised to the 0 power is 1.
EXAMPLE 12 $8^0 = 1$ ⎭

Order of Operations

The symbols we use to specify operations, $+, -, \cdot, \div$, along with the symbols we use for grouping, ( ) and [ ], serve the same purpose in mathematics as punctuation marks in English. They may be called the punctuation marks of mathematics.

Consider the following sentence:

Bob said John is tall.

Answers
6. 8 **7.** 1 **8.** 32 **9.** 7
10. 4 **11.** 1 **12.** 1

It can have two different meanings, depending on how we punctuate it:

    1 "Bob," said John, "is tall."

    2 Bob said, "John is tall."

Without the punctuation marks we don't know which meaning the sentence has. Now, consider the following mathematical expression:

    $4 + 5 \cdot 2$

It can have two different meanings, depending on the order in which we perform the operations. If we add 4 and 5 first, and then multiply 2, we get 18. On the other hand, if we multiply 5 and 2 first, and then add 4, we get 14. There seem to be two different answers. In mathematics we want to avoid situations in which two different results are possible. Therefore we follow the rule for order of operations.

RULE

Order of Operations When evaluating mathematical expressions, we will perform the operations in the following order:

1. If the expression contains grouping symbols, such as parentheses ( ), brackets [ ], or a fraction bar, then we perform the operations inside the grouping symbols, or above and below the fraction bar, first.

2. Then we evaluate, or simplify, any numbers with exponents.

3. Then we do all multiplications and divisions in order, starting at the left and moving right.

4. Finally, we do all additions and subtractions, from left to right.

According to our rule, the expression $4 + 5 \cdot 2$ would have to be evaluated by multiplying 5 and 2 first, and then adding 4. The correct answer—and the only answer—to this problem is 14.

$$4 + 5 \cdot 2 = 4 + 10 \quad \textbf{Multiply first}$$
$$= 14 \qquad \textbf{Then add}$$

Here are some more examples that illustrate how we apply the rule for order of operations to simplify (or evaluate) expressions.

EXAMPLE 13 Simplify: $4 \cdot 8 - 2 \cdot 6$

SOLUTION   We multiply first and then subtract:

$$4 \cdot 8 - 2 \cdot 6 = 32 - 12 \quad \textbf{Multiply first}$$
$$= 20 \qquad \textbf{Then subtract}$$ ■

13. Simplify: $5 \cdot 7 - 3 \cdot 6$

EXAMPLE 14 Simplify: $5 + 2(7 - 1)$

SOLUTION   According to the rule for the order of operations, we must do what is inside the parentheses first:

$$5 + 2(7 - 1) = 5 + 2(6) \quad \textbf{Inside parentheses first}$$
$$= 5 + 12 \qquad \textbf{Then multiply}$$
$$= 17 \qquad \textbf{Then add}$$ ■

14. Simplify: $7 + 3(6 + 4)$

Answers
13. 17   **14.** 37

15. Simplify: $6 \cdot 3^2 + 64 \div 2^4 - 2$

EXAMPLE 15 Simplify: $9 \cdot 2^3 + 36 \div 3^2 - 8$

SOLUTION

$$9 \cdot 2^3 + 36 \div 3^2 - 8 = 9 \cdot 8 + 36 \div 9 - 8 \quad \text{Exponents first}$$

$$= 72 + 4 - 8 \quad \text{Then multiply and divide, left to right}$$

$$\begin{aligned} &= 76 - 8 \\ &= 68 \end{aligned} \right\} \quad \text{Add and subtract, left to right} \quad \blacksquare$$

U S I N G   T E C H N O L O G Y

Calculators

Here is how we use a calculator to work the problem shown in Example 14:

Scientific Calculator: 5 $\boxed{+}$ 2 $\boxed{\times}$ $\boxed{(}$ 7 $\boxed{-}$ 1 $\boxed{)}$ $\boxed{=}$
Graphing Calculator: 5 $\boxed{+}$ 2 $\boxed{(}$ 7 $\boxed{-}$ 1 $\boxed{)}$ $\boxed{\text{ENT}}$

Example 15 on a calculator looks like this:

Scientific Calculator: 9 $\boxed{\times}$ 2 $\boxed{x^y}$ 3 $\boxed{+}$ 36 $\boxed{\div}$ 3 $\boxed{x^y}$ 2 $\boxed{-}$ 8 $\boxed{=}$
Graphing Calculator: 9 $\boxed{\times}$ 2 $\boxed{\wedge}$ 3 $\boxed{+}$ 36 $\boxed{\div}$ 3 $\boxed{\wedge}$ 2 $\boxed{-}$ 8 $\boxed{\text{ENT}}$

---

16. Simplify:
$5 + 3[24 - 5(6 - 2)]$

EXAMPLE 16 Simplify: $3 + 2[10 - 3(5 - 2)]$

SOLUTION   The brackets, [ ], are used in the same way as parentheses. In a case like this we move to the innermost grouping symbols first and begin simplifying:

$$\begin{aligned} 3 + 2[10 - 3(5 - 2)] &= 3 + 2[10 - 3(3)] \\ &= 3 + 2[10 - 9] \\ &= 3 + 2[1] \\ &= 3 + 2 \\ &= 5 \end{aligned}$$

$\blacksquare$

Examples 13–16 all involve expressions that contain more than one operation. We want to be able to translate phrases written in English that involve more than one operation into symbols. Table 1 lists some English expressions and their corresponding mathematical expressions written in symbols.

Table 1	
IN ENGLISH	**MATHEMATICAL EQUIVALENT**
5 times the sum of 3 and 8	$5(3 + 8)$
Twice the difference of 4 and 3	$2(4 - 3)$
6 added to 7 times the sum of 5 and 6	$6 + 7(5 + 6)$
The sum of 4 times 5 and 8 times 9	$4 \cdot 5 + 8 \cdot 9$
3 subtracted from the quotient of 10 and 2	$10 \div 2 - 3$

Descriptive Statistics: Mean, Median, and Mode

Next we turn our attention to averages. If we go online to the Merriam-Webster dictionary at www.m-w.com, we find the following definition for the word *average* when it is used as a noun:

Answers
15. 56   **16.** 17

av·er·age *noun* 1 a : a single value (as a mean, mode, or median) that summarizes or represents the general significance of a set of unequal values ...

In everyday language, the word *average* can refer to the mean, the median, or the mode. The mean is probably the most common average. Here is the definition:

DEFINITION

To find the **mean** for a set of numbers, we add all the numbers and then divide the sum by the number of numbers in the set. The mean is sometimes called the **arithmetic mean.**

EXAMPLE 17 An instructor at Cuesta College as of 1999 earned the following salaries for the first five years of teaching. Find the mean of these salaries.

33,344   35,190   37,049   38,906   40,786

SOLUTION   We add the five numbers and then divide by 5, the number of numbers in the set.

$$\text{Mean} = \frac{33,344 + 35,190 + 37,049 + 38,906 + 40,786}{5} = \frac{185,275}{5} = 37,055$$

The instructor's mean salary for the first five years of work is $37,055 per year. ■

Median The table below appeared in *Parade* magazine in February, 2000. It shows the weekly wages for a number of professions.

WEEKLY WAGES	
All Americans	$549
Butchers	$400
Dietitians	$577
Social workers	$601
Electricians	$645
Clergy	$657
Special ed teachers	$677
Lawyers	$1168

SOURCE: U.S. Bureau of Labor Statistics (all wages are median figures for 1999)

If you look at the type at the bottom of the table, you can see that the numbers are the *median* figures for 1999. The median for a set of numbers is the number such

17. A woman traveled the following distances on a 5-day business trip: 187 miles, 273 miles, 150 miles, 173 miles, and 227 miles. What was the mean distance the woman traveled each day?

Answer
17. 202 miles

that half of the numbers in the set are above it and half are below it. Here is the exact definition.

DEFINITION

To find the median for a set of numbers, we write the numbers in order from smallest to largest. If there is an odd number of numbers, the median is the middle number. If there is an even number of numbers, then the median is the mean of the two numbers in the middle.

---

18. Find the median for the distances in Practice Problem 17.

EXAMPLE 18 Find the median of the numbers given in Example 17.

SOLUTION   The numbers in Example 17, written from smallest to largest, are shown below. Because there are an odd number of numbers in the set, the median is the middle number.

$$33{,}344 \qquad 35{,}190 \qquad 37{,}049 \qquad 38{,}906 \qquad 40{,}786$$
$$\uparrow$$
$$\textbf{median}$$

The instructor's median salary for the first five years she teaches is $37,049. ∎

19. A teacher earns the following amounts for the first 4 years he teaches. Find the median.
$40,763 $42,635 $44,475
$46,320

EXAMPLE 19 A teacher starting at Cuesta College in 1999 with credit for eight years of teaching experience will make the following salaries for the first four years she teaches.

$$51{,}890 \qquad 53{,}745 \qquad 55{,}601 \qquad 57{,}412$$

Find the mean and the median for the four salaries.

SOLUTION   To find the mean, we add the four numbers and then divide by 4:

$$\frac{51{,}890 + 53{,}745 + 55{,}601 + 57{,}412}{4} = \frac{218{,}648}{4} = 54{,}662$$

To find the median, we write the numbers in order from smallest to largest. Then, because there is an even number of numbers, we average the middle two numbers to obtain the median.

$$51{,}890 \qquad \underline{53{,}745 \qquad 55{,}601} \qquad 57{,}412$$
$$\textbf{Median}$$
$$\downarrow$$
$$\frac{53{,}745 + 55{,}601}{2} = 54{,}673$$

The mean is $54,662, and the median is $54,673. ∎

Mode The mode is best used when we are looking for the most common eye color in a group of people, the most popular breed of dog in the United States, and the movie that was seen the most often. When we have a set of numbers in which one number occurs more often than the rest, that number is the mode.

DEFINITION

The mode for a set of numbers is the number that occurs most frequently. If all the numbers in the set occur the same number of times, there is no mode.

Answers
18. 187 miles   **19.** $43,555

EXAMPLE 20 A math class with 18 students had the grades shown below on their first test. Find the mean, the median, and the mode.

$$
\begin{array}{cccccc}
77 & 87 & 100 & 65 & 79 & 87 \\
79 & 85 & 87 & 95 & 56 & 87 \\
56 & 75 & 79 & 93 & 97 & 92
\end{array}
$$

SOLUTION   To find the mean, we add all the numbers and divide by 18:

$$
\text{mean} = \frac{77+87+100+65+79+87+79+85+87+95+56+87+56+75+79+93+97+92}{18}
$$

$$
= \frac{1{,}476}{18} = 82
$$

To find the median, we must put the test scores in order from smallest to largest; then, because there are an even number of test scores, we must find the mean of the middle two scores.

$$
56\ \ 56\ \ 65\ \ 75\ \ 77\ \ 79\ \ 79\ \ 79\ \ 85\ \ 87\ \ 87\ \ 87\ \ 87\ \ 92\ \ 93\ \ 95\ \ 97\ \ 100
$$

$$
\text{Median} = \frac{85 + 87}{2} = 86
$$

The mode is the most frequently occurring score. Because 87 occurs 4 times, and no other scores occur that many times, 87 is the mode.

The mean is 82, the median is 86, and the mode is 87.   ■

More Vocabulary When we used the word *average* in the beginning of this section, we used it as noun. It can also be used as an adjective and a verb. Below is the definition of the word *average* when it is used as a verb.

av·er·age *verb* . . . 2 : to find the arithmetic mean of (a series of unequal quantities) . . .

Merriam-Webster

In everyday language, if you are asked for, or given, the *average* of a set of numbers, the word *average* can represent the mean, the median, or the mode. When used in this way, the word *average* is a noun. However, if you are asked to *average* a set of numbers, then the word *average* is a verb, and you are being asked to find the mean of the numbers.

20. The students in a small math class have the following scores on their final exam. Find the mode.

$$
\begin{array}{cccccc}
56 & 89 & 74 & 68 & 97 & 74 & 68 \\
74 & 88 & 45
\end{array}
$$

Answer
20. 74

Problem Set 1.9

For each of the following expressions, name the base and the exponent.

1. 4^5 **2.** 5^4 **3.** 3^6 **4.** 6^3 **5.** 8^2 **6.** 2^8

7. 9^1 **8.** 1^9 **9.** 4^0 **10.** 0^4

Use the definition of exponents as indicating repeated multiplication to simplify each of the following expressions.

11. 6^2 **12.** 7^2 **13.** 2^3 **14.** 2^4 **15.** 1^4 **16.** 5^1

17. 9^0 **18.** 27^0 **19.** 9^2 **20.** 8^2 **21.** 10^1 **22.** 8^1

23. 12^1 **24.** 16^0 **25.** 45^0 **26.** 3^4

Use the rule for the order of operations to simplify each expression.

27. $3 + 5 \cdot 8$ **28.** $7 + 4 \cdot 9$ **29.** $3 \cdot 6 - 2$ **30.** $5 \cdot 1 + 6$

31. $6 \cdot 2 + 9 \cdot 8$ **32.** $4 \cdot 5 + 9 \cdot 7$ **33.** $8 \cdot 10^2 - 6 \cdot 4^3$ **34.** $5 \cdot 11^2 - 3 \cdot 2^3$

35. $2(3 + 6 \cdot 5)$ **36.** $8(1 + 4 \cdot 2)$ **37.** $19 + 50 \div 5^2$ **38.** $9 + 8 \div 2^2$

39. $9 - 2(4 - 3)$ **40.** $15 - 6(9 - 7)$ **41.** $8 \cdot 2^4 + 25 \div 5 - 3^2$ **42.** $5 \cdot 3^4 + 16 \div 8 - 2^2$

43. $5 + 2[9 - 2(4 - 1)]$ **44.** $6 + 3[8 - 3(1 + 1)]$ **45.** $3 + 4[6 + 8(2 - 0)]$ **46.** $2 + 5[9 + 3(4 - 1)]$

Translate each English expression into an equivalent mathematical expression written in symbols. Then simplify each expression.

47. 8 times the sum of 4 and 2 **48.** 3 times the difference of 6 and 1

49. Twice the sum of 10 and 3

50. 5 times the difference of 12 and 6

51. 4 added to 3 times the sum of 3 and 4

52. 25 added to 4 times the difference of 7 and 5

53. 9 subtracted from the quotient of 20 and 2

54. 7 added to the quotient of 6 and 2

55. The sum of 8 times 5 and 5 times 4

56. The difference of 10 times 5 and 6 times 2

Applying the Concepts

Nutrition Labels Use the three nutrition labels below to work Problems 57–64.

SPAGHETTI

Nutrition Facts

Serving Size 2 oz. (56g/l/8 of pkg) dry
Servings Per Container: 8

Amount Per Serving

Calories 210	Calories from fat 10

	% Daily Value*
Total Fat 1g	2%
Saturated Fat 0g	0%
Poly unsaturated Fat 0.5g	
Monounsaturated Fat 0g	
Cholesterol 0mg	0%
Sodium 0mg	0%
Total Carbohydrate 42g	14%
Dietary Fiber 2g	7%
Sugars 3g	
Protein 7g	

Vitamin A 0%	●	Vitamin C 0%
Calcium 0%	●	Iron 10%

*Percent Daily Values are based on a 2,000 calorie diet

CANNED ITALIAN TOMATOES

Nutrition Facts

Serving Size 1/2 cup (121g)
Servings Per Container: about 3 1/2

Amount Per Serving

Calories 25	Calories from fat 0

	% Daily Value*
Total Fat 0g	0%
Saturated Fat 0g	0%
Cholesterol 0mg	0%
Sodium 300mg	12%
Potassium 145mg	4%
Total Carbohydrate 4g	2%
Dietary Fiber 1g	4%
Sugars 4g	
Protein 1g	

Vitamin A 20%	●	Vitamin C 15%
Calcium 4%	●	Iron 15%

*Percent Daily Values are based on a 2,000 calorie diet. Your daily values may be higher or lower depending on your calorie needs.

SHREDDED ROMANO CHEESE

Nutrition Facts

Serving Size 2 tsp (5g)
Servings Per Container: 34

Amount Per Serving

Calories 20	Calories from fat 10

	% Daily Value*
Total Fat 1.5g	2%
Saturated Fat 1g	5%
Cholesterol 5mg	2%
Sodium 70mg	3%
Total Carbohydrate 0g	0%
Fiber 0g	0%
Sugars 0g	
Protein 2g	

Vitamin A 0%	●	Vitamin C 0%
Calcium 4%	●	Iron 0%

*Percent Daily Values (DV) are based on a 2,000 calorie diet

Find the total number of calories in each of the following meals.

57. Spaghetti   1 serving
Tomatoes   1 serving
Cheese   1 serving

58. Spaghetti   1 serving
Tomatoes   2 servings
Cheese   1 serving

59. Spaghetti   2 servings
Tomatoes   1 serving
Cheese   1 serving

60. Spaghetti   2 servings
Tomatoes   1 serving
Cheese   2 servings

Find the number of calories from fat in each of the following meals.

61. Spaghetti   2 servings
Tomatoes   1 serving
Cheese   1 serving

62. Spaghetti   2 servings
Tomatoes   1 serving
Cheese   2 servings

Use the table on page 42 in Problem Set 1.4 to work Problems 63 and 64.

63. Compare the total number of calories in the meal in Problem 57 with the number of calories in a McDonald's Big Mac.

64. Compare the total number of calories in the meal in Problem 60 with the number of calories in a Burger King hamburger.

Find the mean for each set of numbers.

65. 1, 2, 3, 4, 5

66. 2, 4, 6, 8, 10

67. 1, 3, 9, 11

68. 5, 7, 9, 12, 12

Find the median for each set of numbers.

69. 5, 9, 11, 13, 15

70. 42, 48, 50, 64

71. 10, 20, 50, 90, 100

72. 700, 900, 1,100

Find the mode.

73. 14, 18, 27, 36, 18, 73

74. 11, 27, 18, 11, 72, 11

75. Average Salary Over a 3-year period a woman's annual salaries were $14,000, $16,000, and $18,000. What was her mean annual salary for this 3-year period?

76. Scoring Average During the first five games of the season, a basketball player scores 18, 17, 12, 20, and 13 points. What is the mean?

77. Test Average A student's scores for four exams in a basic math class were 79, 62, 87, and 90. What is the student's median test score?

78. Grade-Point Average A student completed four 3-unit classes in one semester and received one A, two B's, and a C. If an A counts as 4 points, a B as 3 points, and a C as 2 points, what was the student's grade-point average for the semester?

Calculator Problems

Use a calculator to do the following problems.

79. Simplify: $325 + 89 \cdot 63$

80. Simplify: $4,289 \cdot 56 + 2,654 \cdot 95$

81. Simplify: $23,895 + 46(6,342 + 9,987)$

82. Simplify: $16,405 + (26,892 - 24 \cdot 75)$

83. Find the mean of 23,954, 26,421, 95,789, and 106,872.

84. Average Enrollment The number of students enrolled in a community college during a 5-year period was as follows:

YEAR	ENROLLMENT
1987	6,789
1988	6,970
1989	7,242
1990	6,981
1991	6,423

Find the mean enrollment for this 5-year period.

Extending the Concepts: Number Sequences

There is a relationship between the two sequences below. The first sequence is the *sequence of odd numbers.* The second sequence is called the *sequence of squares.*

$1, 3, 5, 7, \ldots$ The sequence of odd numbers
$1, 4, 9, 16, \ldots$ The sequence of squares

85. Add the first two numbers in the sequence of odd numbers.

86. Add the first three numbers in the sequence of odd numbers.

87. Add the first four numbers in the sequence of odd numbers.

88. Add the first five numbers in the sequence of odd numbers.

89. Compare the answers to Problems 85–88 with the numbers in the sequence of squares. In your own words, explain how the sequence of odd numbers can be used to obtain the sequence of squares.

90. Draw a diagram that shows how the sequence of odd numbers can be used to produce the sequence of squares. The diagram can contain numbers, addition symbols, equal signs, and arrows, but it should not contain any words. Your goal is to draw the diagram so that anyone who knows how to add whole numbers can look at it and see the relationship between the numbers in the two sequences.

Below is a translation of the first paragraph of *The Book of Squares* written by a mathematician named Fibonacci in the year 1225. (The first English translation of this book was published in 1987.)

I thought about the origin of all square numbers and discovered that they arise out of the increasing sequence of odd numbers; for the unity is a square and from it is made the first square, namely 1; to this unity is added 3, making the second square, namely 4; if to the sum is added the third odd number, namely 5, the third square is created, namely 9; and thus sums of consecutive odd numbers and a sequence of squares always arise together in order.

Fibonacci enjoyed investigating sequences of numbers. He is credited with discovering the sequence of numbers written below. The sequence is called the *Fibonacci sequence.*

$1, 1, 2, 3, 5, 8, \ldots$

91. Write the next three numbers in the Fibonacci sequence.

92. If you were to describe the Fibonacci sequence in words, you would start this way: "The first two numbers are 1's. After that, each number is found by . . ." Finish the sentence so that someone reading it will know how to find members of the Fibonacci sequence.

1.10 Area and Volume

The *area* of a flat object is a measure of the amount of surface the object has. The rectangle in Figure 1 below has an area of 6 square inches, because that is the number of squares (each of which is 1 inch long and 1 inch wide) it takes to cover the rectangle.

one square inch	one square inch	one square inch
one square inch	one square inch	one square inch

2 inches

3 inches

Figure 1   A rectangle with an area of 6 square inches

It is no coincidence that the area of the rectangle in Figure 1 and the product of the length and the width are the same number. We can calculate the area of the rectangle in Figure 1 by simply multiplying the length and the width together:

$$
\begin{aligned}
\text{Area} &= (\text{length}) \cdot (\text{width}) \\
&= (3 \text{ inches}) \cdot (2 \text{ inches}) \\
&= (3 \cdot 2) \cdot (\text{inches} \cdot \text{inches}) \\
&= 6 \text{ square inches}
\end{aligned}
$$

The unit *square inches* can be abbreviated as *sq. in.* or *in*^{2}.

Facts from Geometry: Area

Figure 2 shows three common geometric figures along with the formulas for their areas. The only figure we have not seen before is the parallelogram.

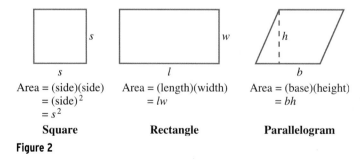

Area = (side)(side)   Area = (length)(width)   Area = (base)(height)
    = (side)^2 = lw = bh
    = s^2

Square   **Rectangle**   **Parallelogram**

Figure 2

Practice Problems

1. Find the total area enclosed by the figure.

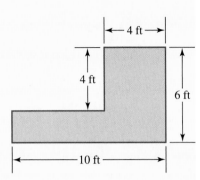

EXAMPLE 1   Find the total area enclosed by Figure 3 below.

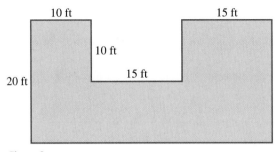

Figure 3

SOLUTION   We begin by drawing in two additional lines (shown as broken lines in Figure 4) so that the original figure is now composed of three individual figures.

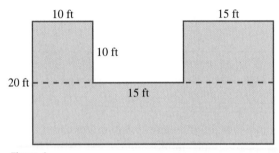

Figure 4

Next we fill in the missing dimensions on the three individual figures (Figure 5). Then we number them for reference.

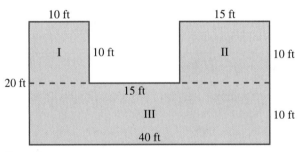

Figure 5

Finally, we calculate the area of the original figure by adding the areas of the three individual figures:

$$
\begin{aligned}
\text{Area} &= \text{Area I} + \text{Area II} + \text{Area III} \\
&= 10 \cdot 10 + 15 \cdot 10 + 40 \cdot 10 \\
&= 100 + 150 + 400 \\
&= 650 \text{ square feet}
\end{aligned}
$$

2. Find the number of square feet in 1 square yard.

EXAMPLE 2   How many square inches are contained in a square foot?

SOLUTION   To solve this problem, we draw a square and label each side with 1 foot = 12 inches, as in Figure 6.

Answers

1. 36 square feet
2. 9 square feet

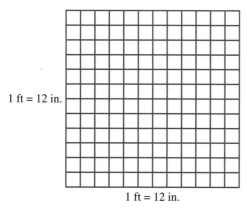

1 ft = 12 in.

1 ft = 12 in.

Figure 6   1 square foot

To find the number of square inches in Figure 6, we can either count them to get 144 square inches, or we can calculate it directly by using the formula for the area of a square:

$$A = (1 \text{ foot}) \cdot (1 \text{ foot})$$
$$= (12 \text{ inches}) \cdot (12 \text{ inches})$$
$$= (12 \cdot 12) \text{ square inches}$$
$$= 144 \text{ square inches} \qquad \blacksquare$$

In addition to the relationship between square inches and square feet as given in Example 2, we would like to know the relationship between other measures of area. The table below gives some relationships between the most common units of area measure. Some of these will be familiar to you, while others may not.

Table 1

UNITS OF AREA IN THE U.S. SYSTEM

THE RELATIONSHIP BETWEEN	IS
square inches (in²) and square feet (ft²)	$144 \text{ in}^2 = 1 \text{ ft}^2$
square yards (yd²) and square feet (ft²)	$9 \text{ ft}^2 = 1 \text{ yd}^2$
acres and square feet	$1 \text{ acre} = 43{,}560 \text{ ft}^2$
acres and square miles (mi²)	$640 \text{ acres} = 1 \text{ mi}^2$

EXAMPLE 3   For the two squares shown in Figure 7, a side in the larger square is 3 times as large as a side in the smaller square. Give a similar comparison of the areas of the two squares.

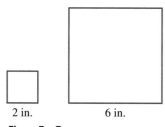

2 in.   6 in.

Figure 7   Two squares

SOLUTION   First we find the area of each square, as shown in Figure 8.

3. One square has a side 4 feet long, while a second square has a side 3 times that, or 12 feet long. How many times larger than the first square is the area of the second square?

Answer
3. 9 times as large

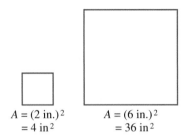

$A = (2 \text{ in.})^2$ $A = (6 \text{ in.})^2$
$\quad = 4 \text{ in}^2$ $\quad = 36 \text{ in}^2$

Figure 8   Two squares with areas calculated

The two areas are 4 square inches and 36 square inches. Because we multiply 4 by 9 to get 36, we say the area of the larger square is 9 times the area of the smaller square.   ■

Facts from Geometry: Volume

Next, we move up one dimension and consider what is called *volume.* Volume is the measure of the space enclosed by a solid. For instance, if each edge of a cube is 3 feet long, as shown in Figure 9, then we can think of the cube as being made up of a number of smaller cubes, each of which is 1 foot long, 1 foot wide, and 1 foot high. Each of these smaller cubes is called a cubic foot. To count the number of them in the larger cube, think of the large cube as having three layers. You can see that the top layer contains 9 cubic feet. Because there are three layers, the total number of cubic feet in the large cube is $9 \cdot 3 = 27$.

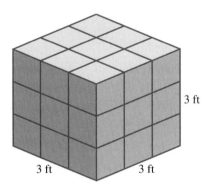

Figure 9   A cube in which each edge is 3 feet long

On the other hand, if we multiply the length, the width, and the height of the cube, we have the same result:

$$\begin{aligned} \text{Volume} &= (3 \text{ feet})(3 \text{ feet})(3 \text{ feet}) \\ &= (3 \cdot 3 \cdot 3)(\text{feet} \cdot \text{feet} \cdot \text{feet}) \\ &= 27 \text{ cubic feet} \end{aligned}$$

For the present we will confine our discussion of volume to volumes of *rectangular solids.* Rectangular solids are the three-dimensional equivalents of rectangles: Opposite sides are parallel, and any two sides that meet, meet at right angles. A rectangular solid is shown in Figure 10, along with the formula used to calculate its volume.

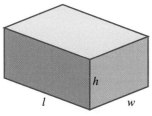

Volume = (length)(width)(height)
$$V = lwh$$

Figure 10   A rectangular solid

EXAMPLE 4   Find the volume of a rectangular solid with length 15 inches, width 3 inches, and height 5 inches.

SOLUTION   To find the volume we apply the formula shown in Figure 10 above:

$$V = l \cdot w \cdot h$$
$$= (15 \text{ in.}) \cdot (3 \text{ in.}) \cdot (5 \text{ in.})$$
$$= 225 \text{ in}^3$$

Note in Example 4 that we used in³ to represent cubic inches. Another abbreviation for cubic inches is cu in.

Table 2

UNITS OF VOLUME IN THE U.S. SYSTEM

THE RELATIONSHIP BETWEEN	IS
cubic inches (in³) and cubic feet (ft³)	1 ft³ = 1,728 in³
cubic feet and cubic yards (yd³)	1 yd³ = 27 ft³
fluid ounces (fl oz) and pints (pt)	1 pt = 16 fl oz
pints and quarts (qt)	1 qt = 2 pt
quarts and gallons (gal)	1 gal = 4 qt

Facts from Geometry: Surface Area

Figure 11 shows a closed box with length l, width w, and height h. The surfaces of the box are labeled as sides, top, bottom, front, and back.

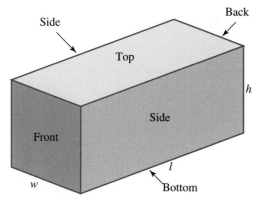

Figure 11   A box with dimensions *l*, *w*, and *h*

4. A home has a dining room that is 12 feet wide and 15 feet long. If the ceiling is 8 feet high, find the volume of the dining room.

Answer
4. 1,440 cubic feet

To find the *surface area* of the box, we add the areas of each of the six surfaces that are labeled in Figure 11.

$$\text{Surface area} = \text{side} + \text{side} + \text{front} + \text{back} + \text{top} + \text{bottom}$$
$$S = l \cdot h + l \cdot h + h \cdot w + h \cdot w + l \cdot w + l \cdot w$$
$$= 2lh + 2hw + 2lw$$

---

5. A family is painting a dining room that is 12 feet wide and 15 feet long.

   a. If the ceiling is 8 feet high, find the surface area of the walls and the ceiling, but not the floor.

   b. If a gallon of paint will cover 400 square feet, how many gallons should they buy to paint the walls and the ceiling?

EXAMPLE 5   Find the surface area of the box shown in Figure 12.

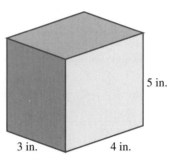

5 in.

3 in.   4 in.

**Figure 12   A box 4 inches long,
3 inches wide, and 5 inches high**

SOLUTION   To find the surface area we find the area of each surface individually, and then we add them together:

$$\text{Surface area} = 2(3 \text{ in.} \cdot 4 \text{ in.}) + 2(3 \text{ in.} \cdot 5 \text{ in.}) + 2(4 \text{ in.} \cdot 5 \text{ in.})$$
$$= 24 \text{ in}^2 + 30 \text{ in}^2 + 40 \text{ in}^2$$
$$= 94 \text{ in}^2$$

The total surface area is 94 square inches. If we calculate the volume enclosed by the box, it is $V = 3$ in. $\cdot$ 4 in. $\cdot$ 5 in. $= 60$ in^3. The surface area measures how much material it takes to make the box, whereas the volume measures how much space the box will hold. ∎

Answers

5. a. 612 square feet

   b. 2 gallons will cover everything, with some paint left over.

Problem Set 1.10

Find the area enclosed by each figure. (Note that some of the units on the figures come from the metric system. The abbreviations are as follows: Meter is abbreviated m, centimeter is abbreviated cm, and millimeter is abbreviated mm. A meter is about 3 inches longer than a yard.)

1.
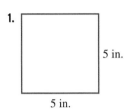
5 in.

5 in.

2.
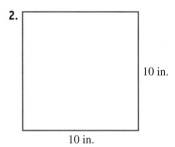
10 in.

10 in.

3.
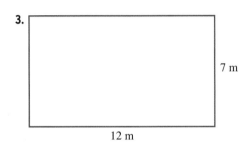
7 m

12 m

4.
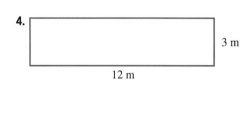
3 m

12 m

5.
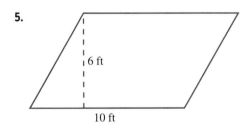
6 ft

10 ft

6.
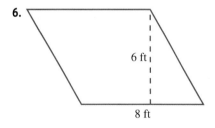
6 ft

8 ft

7.
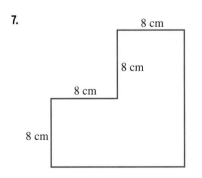
8 cm

8 cm

8 cm

8 cm

8.
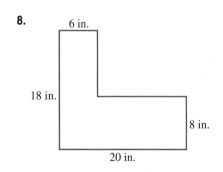
6 in.

18 in.

8 in.

20 in.

9.

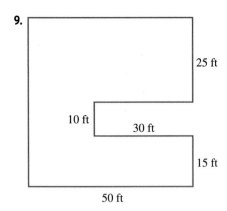

10.

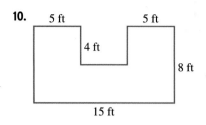

11.

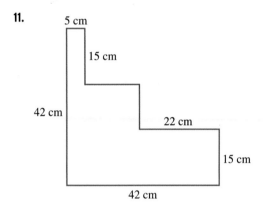

12.

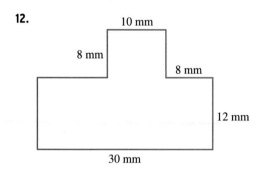

Find the volume of each figure.

13.

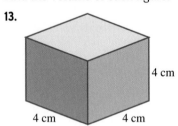

14.

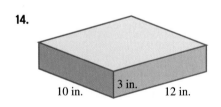

15.

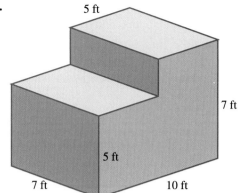

16.

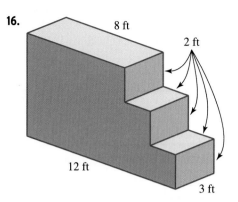

17.

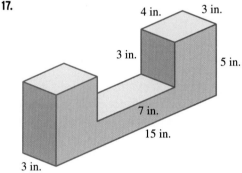

18.

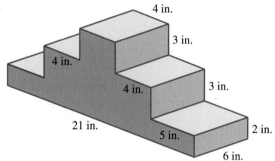

19. Surface Area Find the surface area of the figure in Problem 13, assuming it is closed at the top and bottom.

20. Surface Area Find the surface area of the figure in Problem 14, assuming it is closed at the top and bottom.

21. Surface Area Find the surface area of a rectangular solid with length 5 feet, width 4 feet, and height 6 feet.

22. Surface Area Find the surface area of a rectangular solid with length 6 feet, width 5 feet, and height 7 feet.

23. Area of a Square The area of a square is 49 square feet. What is the length of each side?

24. Area of a Square The area of a square is 144 square feet. How long is each side?

25. Area of a Rectangle A rectangle has an area of 36 square feet. If the width is 4 feet, what is the length?

26. Area of a Rectangle A rectangle has an area of 39 square feet. If the length is 13 feet, what is the width?

Applying the Concepts

27. Area A swimming pool is 20 feet wide and 40 feet long. If it is surrounded by square tiles, each of which is 1 foot by 1 foot, how many tiles are there surrounding the pool?

28. Area A garden is rectangular with a width of 8 feet and a length of 12 feet. If it is surrounded by a walkway 2 feet wide, how many square feet of area does the walkway cover?

29. Surface Area A living room is 10 feet long and 8 feet wide. If the ceiling is 8 feet high, what is the total surface area of the four walls?

30. Surface Area A family room is 12 feet wide and 14 feet long. If the ceiling is 8 feet high, what is the total surface area of the four walls?

31. Comparing Areas The side of a square is 5 feet long. If all four sides are increased by 2 feet, by how much is the area increased?

32. Comparing Areas The length of a side in a square is 20 inches. If all four sides are decreased by 4 inches, by how much is the area decreased?

33. a. Each side of the red square in the corner is 1 centimeter, and all squares are the same size. On the grid below, draw three more squares. Each side of the first one will be 2 centimeters, each side of the second square will be 3 centimeters, and each side of the third square will be 4 centimeters.

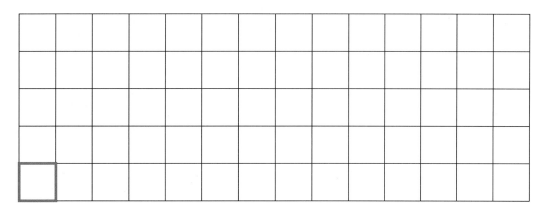

b. Use the squares you have drawn above to complete each of the following tables.

PERIMETERS OF SQUARES	
LENGTH OF EACH SIDE (IN CENTIMETERS)	PERIMETER (IN CENTIMETERS)
1	
2	
3	
4	

AREAS OF SQUARES	
LENGTH OF EACH SIDE (IN CENTIMETERS)	AREA (IN SQUARE CENTIMETERS)
1	
2	
3	
4	

c. Use the figures below to construct bar charts from the information in the tables in part b.

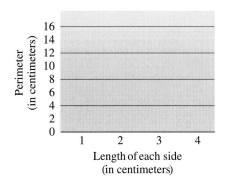

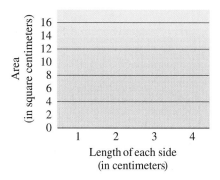

34. a. The lengths of the sides of the squares in the grid below are all 1 centimeter. The red square has a perimeter of 12 centimeters. On the grid below, draw two different rectangles, each with a perimeter of 12 centimeters.

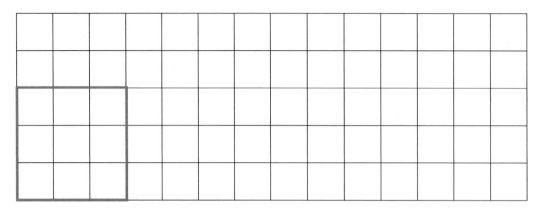

   b. Find the area of each of the three figures in part a.

Calculator Problems

35. Find the area of a square if each of its sides is 111 feet long.

36. Find the area of a square if each of its sides is 1,111 feet long.

37. A cube is 11 feet long, 11 feet wide, and 11 feet high. Find its volume and its surface area.

38. A cube is 111 feet long, 111 feet wide, and 111 feet high. Find its volume and its surface area.

39. A rectangular solid is 11 feet wide, 111 feet long, and 1,111 feet high. Find its volume.

40. A rectangular solid is 11 feet long, 11 feet wide, and 121 feet high. Find its volume.

Chapter **1** Summary

The numbers in brackets indicate the sections in which the topics were discussed.

Place Values for Decimal Numbers [1.1]

The place values for the digits of any base 10 number are as follows:

Billions			Millions			Thousands					
Hundred Billions	Ten Billions	Billions	Hundred Millions	Ten Millions	Millions	Hundred Thousands	Ten Thousands	Thousands	Hundreds	Tens	Ones

1 The number 42,103,045 written in words is "forty-two million, one hundred three thousand, forty-five."

The number 5,745 written in expanded form is

$5,000 + 700 + 40 + 5$

Vocabulary Associated with Addition, Subtraction, Multiplication, and Division [1.2, 1.5, 1.6, 1.8]

The word *sum* indicates addition.

The word *difference* indicates subtraction.

The word *product* indicates multiplication.

The word *quotient* indicates division.

2 The sum of 5 and 2 is $5 + 2$.

The difference of 5 and 2 is $5 - 2$.

The product of 5 and 2 is $5 \cdot 2$.

The quotient of 10 and 2 is $10 \div 2$.

Properties of Addition and Multiplication [1.2, 1.6, 1.7]

If a, b, and c represent any three numbers, then the properties of addition and multiplication used most often are:

Commutative property of addition: $a + b = b + a$
Commutative property of multiplication: $a \cdot b = b \cdot a$
Associative property of addition: $(a + b) + c = a + (b + c)$
Associative property of multiplication: $(a \cdot b) \cdot c = a \cdot (b \cdot c)$
Distributive property: $a(b + c) = a(b) + a(c)$

3 $3 + 2 = 2 + 3$

$3 \cdot 2 = 2 \cdot 3$

$(x + 3) + 5 = x + (3 + 5)$

$(4 \cdot 5) \cdot 6 = 4 \cdot (5 \cdot 6)$

$3(4 + 7) = 3(4) + 3(7)$

Perimeter of a Polygon [1.3]

The *perimeter* of any polygon is the sum of the lengths of the sides, and it is denoted with the letter P.

4 The perimeter of the rectangle below is

$P = 37 + 37 + 24 + 24$

$\quad = 122$ feet

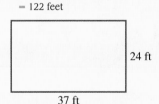

24 ft

37 ft

Steps for Rounding Whole Numbers [1.4]

1. Locate the digit just to the right of the place you are to round to.

2. If that digit is less than 5, replace it and all digits to its right with zeros.

3. If that digit is 5 or more, replace it and all digits to its right with zeros, and add 1 to the digit to its left.

5 5,482 to the nearest ten is 5,480
5,482 to the nearest hundred is 5,500
5,482 to the nearest thousand is 5,000

6 $4 + 6(8 - 2) = 4 + 6(6)$

Inside parentheses first

$= 4 + 36$ **Then multiply**

$= 40$ **Then add**

7 The average of 4, 7, 9 and 12 is

$(4 + 7 + 9 + 12) \div 4 = 32 \div 4$

$= 8$

8 Each expression below is undefined.

$5 \div 0 \quad \frac{7}{0} \quad 4/0$

9 $2^3 = 2 \cdot 2 \cdot 2 = 8$

$5^0 = 1$

Order of Operations [1.9]

To simplify a mathematical expression:

1. We simplify the expression inside the grouping symbols first. Grouping symbols are parentheses ( ), brackets [ ], or a fraction bar.
2. Then we evaluate any numbers with exponents.
3. We then perform all multiplications and divisions in order, starting at the left and moving right.
4. Finally, we do all the additions and subtractions, from left to right.

Average [1.9]

The *average* for a set of numbers is their sum divided by the number of numbers in the set.

Division by 0 (Zero) [1.8]

Division by 0 is undefined. We cannot use 0 as a divisor in any division problem.

Exponents [1.9]

In the expression 2^3, 2 is the *base* and 3 is the *exponent*. An exponent is a shorthand notation for repeated multiplication. The exponent 0 is a special exponent. Any nonzero number to the 0 power is 1.

Formulas for Area [1.10]

Below are three common geometric figures, along with the formulas for their areas.

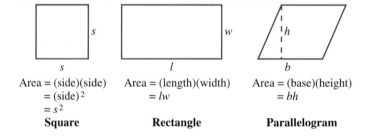

Area = (side)(side) Area = (length)(width) Area = (base)(height)
= (side)^2 = lw = bh
= s^2
Square **Rectangle** **Parallelogram**

Formulas for Volume and Surface Area [1.10]

Below is a rectangular solid, along with the formulas for its volume and surface area.

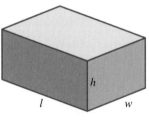

Volume = lwh
Surface area = $2lh + 2hw + 2lw$

COMMON MISTAKE

A common mistake is to write numbers in the wrong order when translating subtraction or division statements from words to symbols.

The difference of 5 and 2 is $5 - 2$,   *not* $2 - 5$.
The quotient of 12 and 4 is $12 \div 4$,   *not* $4 \div 12$.

Chapter **1** Review

The numbers in brackets indicate the sections in which problems of a similar type can be found.

1. As of March 2000, the largest Pacific blue marlin had been caught near Hawaii in 1982. It weighed 1,376 pounds. Write 1,376 in words. [1.1]

2. In 1999 the Colorado Rockies had the highest home attendance in major league baseball. The attendance that year was 3,792,683. Write 3,792,683 in words. [1.1]

For Problems 3 and 4, write each number with digits instead of words. [1.1]

3. Five million, two hundred forty-five thousand, six hundred fifty-two

4. Twelve million, twelve thousand, twelve

5. In 1999 the Montreal Expos has the lowest attendance in major league baseball. The attendance that year was 914,909. Write 914,909 in expanded form. [1.1]

6. In 1998, there were 177,030 female physicians practicing medicine in the United States. Write 177,030 in expanded form. [1.1]

Identify each of the statements in Problems 7–14 as an example of one of the following properties. [1.2, 1.6]

a. Addition property of 0
b. Multiplication property of 0
c. Multiplication property of 1
d. Commutative property of addition

e. Commutative property of multiplication
f. Associative property of addition
g. Associative property of multiplication

7. $5 + 7 = 7 + 5$

8. $(4 + 3) + 2 = 4 + (3 + 2)$

9. $6 \cdot 1 = 6$

10. $8 + 0 = 8$

11. $5 \cdot 0 = 0$

12. $4 \cdot 6 = 6 \cdot 4$

13. $5 \cdot (3 \cdot 2) = (5 \cdot 3) \cdot 2$

14. $(6 + 2) + 3 = (2 + 6) + 3$

Find each of the following sums. (Add.) [1.3]

15.
$$\begin{array}{r} 498 \\ +251 \\ \hline \end{array}$$

16.
$$\begin{array}{r} 784 \\ +598 \\ \hline \end{array}$$

17.
$$\begin{array}{r} 7,384 \\ 251 \\ +637 \\ \hline \end{array}$$

18.
$$\begin{array}{r} 4,901 \\ 648 \\ +3,592 \\ \hline \end{array}$$

Find each of the following differences. (Subtract.) [1.5]

19.   789
       -475

20.   792
       -178

21.   5,908
       $-2,759$

22.   3,527
       $-1,789$

Find each of the following products. (Multiply.) [1.7]

23. 8(73)

24. 7(984)

25. 63(59)

26. 49(876)

Find each of the following quotients. (Divide.) [1.8]

27. 692 ÷ 4

28. 1,020 ÷ 15

29. $36\overline{)15,408}$

30. $286\overline{)21,736}$

Round the number 3,781,092 to the nearest: [1.4]

31. Ten

32. Hundred

33. Hundred thousand

34. Million

Use the rule for the order of operations to simplify each expression as much as possible. [1.9]

35. $4 + 3 \cdot 5^2$

36. $7(9)^2 - 6(4)^3$

37. $3(2 + 8 \cdot 9)$

38. $7 - 2(6 - 4)$

39. A first-year math student had grades of 79, 64, 78, and 95 on the first four tests. What is the student's mean test grade? [1.9]

40. If a person has scores of 205, 222, 174, 236, 185, and 215 for six games of bowling, what is the median score for the six games? [1.9]

Write an expression using symbols that is equivalent to each of the following expressions; then simplify. [1.9]

41. 3 times the sum of 4 and 6

42. 9 times the difference of 5 and 3

43. Twice the difference of 17 and 5

44. The product of 5 and the sum of 8 and 2

Applying the Concepts

45. Income and Expenses A person has a monthly income of $1,783 and monthly expenses of $1,295. What is the difference between the monthly income and the expenses? [1.5]

46. Number of Sheep A rancher bought 395 sheep and then sold 197 of them. How many were left? [1.5]

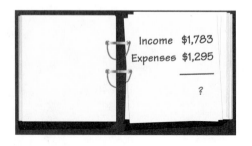
Income $1,783
Expenses $1,295
_____
?

Area and Perimeter The rules for soccer state that the playing field must be from 100 to 120 yards long and 55 to 75 yards wide. The 1999 Women's World Cup was played at the Rose Bowl on a playing field 116 yards long and 72 yards wide. The diagram below shows the smallest possible soccer field, the largest possible soccer field, and the soccer field at the Rose Bowl. [1.3, 1.10]

Soccer Fields

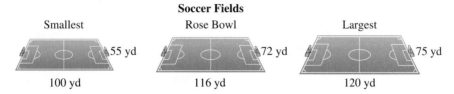

Smallest   Rose Bowl   Largest

55 yd   72 yd   75 yd

100 yd   116 yd   120 yd

47. Find the perimeter of each soccer field.

48. Find the area of each soccer field.

49. Monthly Budget Each month a family budgets $1,150 for rent, $625 for food, and $257 for entertainment. What is the sum of these numbers?

50. Checking Account If a person wrote 23 checks in January, 37 checks in February, 40 checks in March, and 27 checks in April, what is the total number of checks written in the 4-month period?

51. Yearly Income A person has a yearly income of $23,256. What is the person's monthly income? [1.8]

52. Jogging It takes a jogger 126 minutes to run 14 miles. At that rate, how long does it take the jogger to run 1 mile? [1.8]

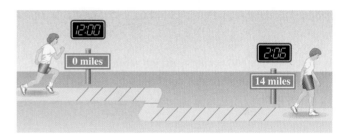

53. Take-Home Pay Jeff makes $6 an hour for the first 40 hours he works in a week and $9 an hour for every hour after that. Each week he has $102 deducted from his check for income taxes and retirement. If he works 45 hours in one week, how much is his take-home pay? [1.7]

54. Take-Home Pay Barbara earns $8 an hour for the first 40 hours she works in a week and $12 an hour for every hour after that. Each week she has $123 deducted from her check for income taxes and retirement. What is her take-home pay for a week in which she works 50 hours? [1.7]

55. Geometry Find the perimeter and the area of the parallelogram below. [1.3, 1.10]

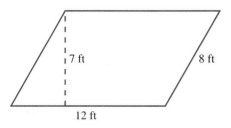

7 ft   8 ft

12 ft

56. Geometry Find the volume and the surface area of the rectangular solid shown below. [1.10]

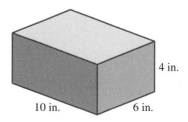

4 in.

10 in.   6 in.

Exercise and Calories The tables below are similar to two of the tables we have worked with in this chapter. Use the information in the tables to work Problems 57–64. [1.7]

NUMBER OF CALORIES IN FAST FOOD

FOOD	CALORIES
McDonald's hamburger	270
Burger King hamburger	260
Jack in the Box hamburger	280
McDonald's Big Mac	510
Burger King Whopper	630
Jack in the Box Colossus burger	940
Roy Rogers roast beef sandwich	260
McDonald's Chicken McNuggets (6)	300
Taco Bell chicken burrito	345
McDonald's french fries (large)	450
Burger King BK Broiler	540
Burger King chicken sandwich	700

NUMBER OF CALORIES BURNED IN 30 MINUTES

	130-POUND PERSON	170-POUND PERSON
INDOOR ACTIVITIES		
Vacuuming	100	130
Mopping floors	105	135
Shopping for food	110	135
Ironing clothes	115	145
OUTDOOR ACTIVITIES		
Chopping wood	150	200
Ice skating	170	235
Cross-country skiing	250	330
Shoveling snow	260	350

57. How many calories do you consume if you eat one large order of McDonald's french fries and 2 Big Macs?

58. How many calories do you consume if you eat one order of Chicken McNuggets and a McDonald's hamburger?

59. How many more calories are in one Colossus burger than in two Taco Bell chicken burritos?

60. What is the difference in calories between a Whopper and a BK Broiler?

61. If you weigh 170 pounds and ice skate for 1 hour, will you burn all the calories consumed by eating one Whopper?

62. If you weigh 130 pounds and go cross-country skiing for 1 hour, will you burn all the calories consumed by eating one large order of McDonald's french fries?

63. Suppose you eat a Big Mac and a large order of fries for lunch. If you weigh 130 pounds, what combination of 30-minute activities could you do to burn all the calories you consumed at lunch?

64. Suppose you weigh 170 pounds and you eat two Taco Bell chicken burritos for lunch. What combination of 30-minute activities could you do to burn all the calories in the burritos?

Chapter **1** Test

1. Write the number 20,347 in words.

2. Write the number two million, forty-five thousand six with digits instead of words.

3. Write the number 123,407 in expanded form.

Identify each of the statements in Problems 4–7 as an example of one of the following properties.

a. Addition property of 0
b. Multiplication property of 0
c. Multiplication property of 1
d. Commutative property of addition

e. Commutative property of multiplication
f. Associative property of addition
g. Associative property of multiplication

4. $(5 + 6) + 3 = 5 + (6 + 3)$ **5.** $7 \cdot 1 = 7$

6. $9 + 0 = 9$

7. $5 \cdot 6 = 6 \cdot 5$

Find each of the following sums. (Add.)

8. $\begin{array}{r} 135 \\ +741 \\ \hline \end{array}$

9. $\begin{array}{r} 5,401 \\ 329 \\ +10,653 \\ \hline \end{array}$

Find each of the following differences. (Subtract.)

10. $\begin{array}{r} 937 \\ -413 \\ \hline \end{array}$

11. $\begin{array}{r} 7,052 \\ -3,967 \\ \hline \end{array}$

Find each of the following products. (Multiply.)

12. 9(186)

13. 62(359)

Find each of the following quotients. (Divide.)

14. $1,105 \div 13$

15. $583\overline{)12,243}$

16. Round the number 516,249 to the nearest ten thousand.

Use the rule for the order of operations to simplify each expression as much as possible.

17. $8(5)^2 - 7(3)^3$

18. $8 - 2(5 - 3)$

19. Home sales Below are listed the prices paid for 10 homes that sold during the month of February in the city of Houston.

$210,000	$139,000	$122,000	$145,000	$120,000
$540,000	$167,000	$125,000	$125,000	$950,000

a. Find the mean housing price for the month.

b. Find the median housing price for the month.

c. Find the mode of the housing prices for the month.

d. Which measure of "average" best describes the average housing price for the month? Explain your answer.

Write an expression using symbols that is equivalent to each of the following expressions; then simplify.

20. Twice the sum of 11 and 7

21. The quotient of 20 and 5 increased by 9

22. Hours of Commuting In 1997 the Texas Transportation Institute conducted a study of the number of hours a year commuters spent in gridlock in the country's 68 largest urban areas. The top five areas from that study are listed in the bar chart below. Use the information in the bar chart to complete the table.

Time Spent in Gridlock

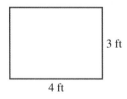

URBAN AREA	AVERAGE HOURS IN GRIDLOCK PER YEAR
Los Angeles	82
Washington	
Seattle-Everett	
Atlanta	68
Boston	

23. Take-Home Pay Karen earns $7 an hour for the first 40 hours she works in a week and $10 an hour for every hour after that. Each week she has $115 deducted from her check for income taxes and retirement. If she works 47 hours in one week, how much is her take-home pay?

24. Geometry Find the perimeter and the area of the rectangle below.

3 ft

4 ft

25. Geometry Find the volume and the surface area of the rectangular solid shown below.

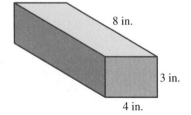

8 in.

3 in.

4 in.

Introduction to Algebra

INTRODUCTION

The table below gives the record low temperature, in degrees Fahrenheit, for each month of the year in the city of Jackson, Wyoming. The diagram that follows is drawn from the information in the table and is called a line graph. Line graphs are one of the items that we will study in this chapter.

MONTH	TEMPERATURE	MONTH	TEMPERATURE
January	−50°F	July	24°F
February	−44°F	August	18°F
March	−32°F	September	14°F
April	−5°F	October	2°F
May	12°F	November	−27°F
June	19°F	December	−49°F

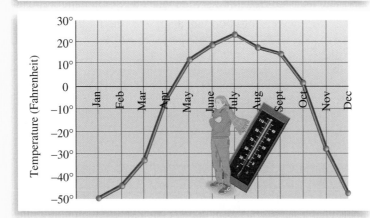

From these graphics, you can see that the temperatures that are below zero are given as negative numbers. Most of this chapter deals with negative numbers. Here is the type of question we will ask in this chapter: Can you find the difference in record low temperatures between September and November from the table or the graph above? The table gives the record low temperature for September as 14°F and the record low for November as −27°F. Because the word *difference* is associated with subtraction, our question is answered with the following problem:

$$\begin{pmatrix} \text{Difference in September} \\ \text{and November temperatures} \end{pmatrix} = \begin{pmatrix} \text{September low} \\ \text{temperature} \end{pmatrix} - \begin{pmatrix} \text{November low} \\ \text{temperature} \end{pmatrix} = 14 - (-27)$$

In order to work problems such as these, we need to develop rules for addition, subtraction, multiplication, and division with negative numbers. And as we have said previously, that is our main goal for this chapter.

STUDY SKILLS

If you have successfully completed Chapter 1, then you have made a good start at developing the study skills necessary to succeed in all math classes. Here is the list of study skills for this chapter. Some are a continuation of the skills from Chapter 1, while others are new to this chapter.

1. Continue to Set and Keep a Schedule. Sometimes I find that students who do well in Chapter 1 become overconfident. They begin to put in less time with their homework. Don't make that mistake. Keep to the same schedule.

2. Increase Effectiveness. You want to become more and more effective with the time you spend on your homework. You want to increase the amount of learning you obtain in the time you have set aside. Increase those activities that you feel are the most beneficial and decrease those that have not given you the results you wanted. As time goes by you will get more out of the time you spend studying mathematics. But you must make an effort to have that happen by constantly evaluating the effectiveness of what you are doing.

3. List Difficult Problems. Begin to make lists of problems that give you the most difficulty—those that you are repeatedly making mistakes with.

4. Look for Difficulties Caused by Your Intuition. You will probably find, as you progress through this chapter, that some of the problems we work will be contrary to your own intuition. For example, many people will answer 6, if they are asked for the difference between 8 and -2. As you will see when we get to Section 2.3, this difference is actually 10. However, even though you may understand the rules we will introduce for subtraction, you may still have a tendency to use your intuition, rather than the rules we develop, to answer questions. The solution to this is simply to notice when your intuition is getting in the way and make it a point to work problems the way you see them done in the book and the way your instructor does them. The more problems you work correctly, the more closely your intuition will align with the mathematics you see here.

Many of my students keep a notebook that contains everything that they need for the course: class notes, homework, quizzes, tests, and other projects. A three-ring binder with tabs is ideal. Your main goal with the notebook is to organize it so that you can easily get to any item you need.

2.1 Positive and Negative Numbers

Introduction . . .

Suppose you have a balance of $20 in your checkbook and then write a check for $30. You are now overdrawn by $10. How will you write this new balance? One way is with a negative number. You could write the balance as −$10, which is a negative number.

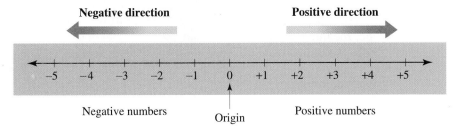

Negative numbers can be used to describe other situations as well—for instance, temperature below zero and distance below sea level.

To see the relationship between negative and positive numbers, we can extend the number line as shown in Figure 1. We first draw a straight line and label a convenient point with 0. This is called the *origin*, and it is usually in the middle of the line. We then label positive numbers to the right (as we have done previously), and negative numbers to the left.

Negative direction **Positive direction**

Negative numbers   Positive numbers
Origin

Figure 1

The numbers increase going from left to right. If we move to the right, we are moving in the positive direction. If we move to the left, we are moving in the negative direction. *Any number to the left of another number is considered to be smaller than the number to its right.*

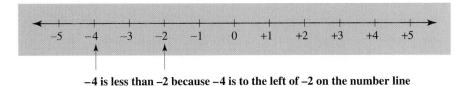

−4 is less than −2 because −4 is to the left of −2 on the number line

Figure 2

We see from the line that every negative number is less than every positive number.

In algebra we can use inequality symbols when comparing numbers.

> **Notation**
>
> If a and b are any two numbers on the number line, then
>
> $a < b$ is read "a is less than b"
> $a > b$ is read "a is greater than b"

As you can see, the inequality symbols always point to the smaller of the two numbers being compared. Here are some examples that illustrate how we use the inequality symbols.

EXAMPLE 1 $3 < 5$ is read "3 is less than 5." Note that it would also be correct to write $5 > 3$. Both statements, "3 is less than 5" and "5 is greater than 3," have the same meaning. The inequality symbols always point to the smaller number. ∎

EXAMPLE 2 $0 > 100$ is a false statement, because 0 is less than 100, not greater than 100. To write a true inequality statement using the numbers 0 and 100, we would have to write either $0 < 100$ or $100 > 0$. ∎

EXAMPLE 3 $-3 < 5$ is a true statement, because -3 is to the left of 5 on the number line, and, therefore, it must be less than 5. Another statement that means the same thing is $5 > -3$. ∎

EXAMPLE 4 $-5 < -2$ is a true statment, because -5 is to the left of -2 on the number line, meaning that -5 is less than -2. Both statements $-5 < -2$ and $-2 > -5$ have the same meaning; they both say that -5 is a smaller number than -2. ∎

It is sometimes convenient to talk about only the numerical part of a number and disregard the sign (+ or −) in front of it. The following definition gives us a way of doing this.

> **DEFINITION**
>
> The **absolute value** of a number is its distance from 0 on the number line. We denote the absolute value of a number with vertical lines. For example, the absolute value of -3 is written $|-3|$.

The absolute value of a number is never negative because it is a distance, and a distance is always measured in positive units (unless it happens to be 0).

Here are some examples of absolute value problems.

EXAMPLE 5 $|5| = 5$ The number 5 is 5 units from 0. ∎

EXAMPLE 6 $|-3| = 3$ The number -3 is 3 units from 0. ∎

EXAMPLE 7 $|-7| = 7$ The number -7 is 7 units from 0. ∎

Practice Problems

Write each statement in words.

1. $2 < 8$

2. $5 > 10$ (Is this a true statement?)

3. $-4 < 4$

4. $-7 < -2$

Give the absolute value of each of the following.

5. $|6|$

6. $|-5|$

7. $|-8|$

Answers

1. 2 is less than 8.
2. 5 is greater than 10. (No.)
3. -4 is less than 4.
4. -7 is less than -2.
5. 6   6. 5   7. 8

DEFINITION

Two numbers that are the same distance from 0 but in opposite directions from 0 are called **opposites.*** The notation for the opposite of a is $-a$.

EXAMPLE 8 Give the opposite of each of the following numbers:

5, 7, 1, −5, −8

SOLUTION The opposite of 5 is −5.
The opposite of 7 is −7.
The opposite of 1 is −1.
The opposite of −5 is −(−5), or 5.
The opposite of −8 is −(−8), or 8. ◼

8. Give the opposite of each of the following numbers: 8, 10, 0, −4.

We see from this example that the opposite of every positive number is a negative number, and, likewise, the opposite of every negative number is a positive number. The last two parts of Example 8 illustrate the following property:

Property

If a represents any positive number, then it is always true that

$$-(-a) = a$$

In other words, this property states that the opposite of a negative number is a positive number.

It should be evident now that the symbols + and − can be used to indicate several different ideas in mathematics. In the past we have used them to indicate addition and subtraction. They can also be used to indicate the direction a number is from 0 on the number line. For instance, the number +3 (read "positive 3") is the number that is 3 units from zero in the positive direction. On the other hand, the number −3 (read "negative 3") is the number that is 3 units from 0 in the negative direction. The symbol − can also be used to indicate the opposite of a number, as in −(−2) = 2. The interpretation of the symbols + and − depends on the situation in which they are used. For example:

3 + 5 The + sign indicates addition.
7 − 2 The − sign indicates subtraction.
−7 The − sign is read "negative" 7.
−(−5) The first − sign is read "the opposite of." The second − sign is read "negative" 5.

This may seem confusing at first, but as you work through the problems in this chapter you will get used to the different interpretations of the symbols + and −.

We should mention here that the set of whole numbers along with their opposites forms the set of *integers*. That is:

Integers = {. . . , −3, −2, −1, 0, 1, 2, 3, . . .}

*In some books opposites are called *additive inverses*.

Answer
8. −8, −10, 0, 4

Displaying Information

In the United States, temperature is measured on the Fahrenheit temperature scale. On this scale, water boils at 212 degrees and freezes at 32 degrees. To denote a temperature of 32 degrees on the Fahrenheit scale, we write

32°F, which is read "32 degrees Fahrenheit"

In the introduction to this chapter, we showed a table of temperatures in which the temperatures below zero were represented by negative numbers. The table from that introduction is repeated here for reference.

Table 1

**RECORD LOW TEMPERATURES
FOR JACKSON, WYOMING**

MONTH	TEMPERATURE
January	−50°F
February	−44°F
March	−32°F
April	−5°F
May	12°F
June	19°F
July	24°F
August	18°F
September	14°F
October	2°F
November	−27°F
December	−49°F

Scatter Diagrams and Line Graphs

Figure 3 shows one way to visualize the information in Table 1. It is known as a *scatter diagram*. The dots represent the record low temperature for each of the 12 months of the year. If we connect the dots in Figure 3 with straight lines, we produce the diagram in Figure 4, which is known as a *line graph*.

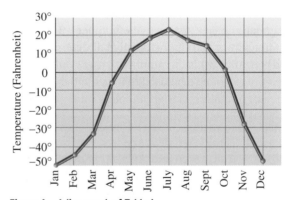

Figure 3   A scatter diagram of Table 1

Figure 4   A line graph of Table 1

Problem Set 2.1

Write each of the following statements in words.

1. $4 < 7$

2. $0 < 10$

3. $5 > -2$

4. $8 > -8$

5. $-10 < -3$

6. $-20 < -5$

7. $0 > -4$

8. $0 > -100$

Write each of the following in symbols.

9. 30 is greater than -30.

10. -30 is less than 30.

11. -10 is less than 0.

12. 0 is greater than -10.

13. -3 is greater than -15.

14. -15 is less than -3.

Give the opposite of each of the following numbers.

15. 3

16. -6

17. -2

18. 15

19. 75

20. -32

21. 0

22. 1

23. -5

24. -100

25. 100

26. 5

Place either $<$ or $>$ between each of the following pairs of numbers so that the resulting statement is true.

27. 3   7

28. 17   0

29. 7 -5

30. 2 -13

31. -6 0

32. -14 0

33. -12 -2

34. -20 -1

35. -10 -100

36. -100 -10

37. -4 4

38. 5 -5

39. -3 -2

40. -3 -4

41. -3 $|6|$

42. $|8|$ -2

43. 15 $|-4|$

44. 20 $|-6|$

45. $|-2|$ $|-7|$

46. $|-3|$ $|-1|$

Find each of the following absolute values.

47. $|2|$ **48.** $|7|$ **49.** $|100|$ **50.** $|10,000|$ **51.** $|-8|$ **52.** $|-9|$

53. $|-231|$ **54.** $|-457|$ **55.** $|-6|$ **56.** $|-17|$ **57.** $|-200|$ **58.** $|-350|$

59. $|8|$ **60.** $|9|$ **61.** $|231|$ **62.** $|457|$

Simplify each of the following.

63. $-(-2)$ **64.** $-(-5)$ **65.** $-(-8)$ **66.** $-(-3)$

67. $-|2|$ **68.** $-|5|$ **69.** $-|8|$ **70.** $-|3|$

71. What number is its own opposite?

72. Is $|a| = a$ is always a true statement?

73. If n is a negative number, is $-n$ positive or negative?

74. If n is a positive number, is $-n$ positive or negative?

Estimating

Work Problems 75–80 mentally, without pencil and paper or a calculator.

75. Is -60 closer to 0 or -100?

76. Is -20 closer to 0 or -30?

77. Is -10 closer to -20 or 20?

78. Is -20 closer to -40 or 10?

79. Is -362 closer to -360 or -370?

80. Is -368 closer to -360 or -370?

Applying the Concepts

81. Temperature and Altitude Yamina is flying from Phoenix to San Francisco on a Boeing 737 jet. When the plane reaches an altitude of 33,000 feet, the temperature outside the plane is 61 degrees below zero Fahrenheit. Represent this temperature with a negative number. If the temperature outside the plane gets warmer by 10 degrees, what will the new temperature be?

82. Temperature Change At 11:00 in the morning in Superior, Wisconsin, Jim notices the temperature is 15 degrees below zero Fahrenheit. Write this temperature as a negative number. At noon it has warmed up by 8 degrees. What is the temperature at noon?

83. Temperature Change At 10:00 in the morning in White Bear Lake, Wisconsin, Zach notices the temperature is 5 degrees below zero Fahrenheit. Write this temperature as a negative number. By noon the temperature has dropped another 10 degrees. What is the temperature at noon?

84. Snorkeling Steve is snorkeling in the ocean near his home in Maui. At one point he is 6 feet below the surface. Represent this situation with a negative number. If he descends another 6 feet, what negative number will represent his new position?

Table 2 lists various wind chill temperatures. The top row gives air temperature, while the first column gives wind speed, in miles per hour. The numbers within the table indicate how cold the weather will feel. For example, if the thermometer reads 30°F and the wind is blowing at 15 miles per hour, the wind chill temperature is 9°F.

Table 2

WIND CHILL TEMPERATURES

WIND SPEED	AIR TEMPERATURE (°F)							
	30°	25°	20°	15°	10°	5°	0°	−5°
10 mph	16°	10°	3°	−3°	−9°	−15°	−22°	−27°
15 mph	9°	2°	−5°	−11°	−18°	−25°	−31°	−38°
20 mph	4°	−3°	−10°	−17°	−24°	−31°	−39°	−46°
25 mph	1°	−7°	−15°	−22°	−29°	−36°	−44°	−51°
30 mph	−2°	−10°	−18°	−25°	−33°	−41°	−49°	−56°

85. Wind Chill Find the wind chill temperature if the thermometer reads 25°F and the wind is blowing at 25 miles per hour.

86. Wind Chill Find the wind chill temperature if the thermometer reads 10°F and the wind is blowing at 25 miles per hour.

87. Wind Chill Which will feel colder: a day with an air temperature of 10°F and a 25-mph wind, or a day with an air temperature of −5°F and a 10-mph wind?

88. Wind Chill Which will feel colder: a day with an air temperature of 15°F and a 20-mph wind, or a day with an air temperature of 5°F and a 10-mph wind?

Table 3 lists the record low temperatures for each month of the year for Lake Placid, New York. Table 4 lists the record high temperatures for the same city.

Table 3

RECORD LOW TEMPERATURES FOR LAKE PLACID, NEW YORK

MONTH	TEMPERATURE
January	−36°F
February	−30°F
March	−14°F
April	−2°F
May	19°F
June	22°F
July	35°F
August	30°F
September	19°F
October	15°F
November	−11°F
December	−26°F

Table 4

RECORD HIGH TEMPERATURES FOR LAKE PLACID, NEW YORK

MONTH	TEMPERATURE
January	54°F
February	59°F
March	69°F
April	82°F
May	90°F
June	93°F
July	97°F
August	93°F
September	90°F
October	87°F
November	67°F
December	60°F

89. Temperature Figure 5 is a bar chart of the information in Table 3. Use the template in Figure 6 to construct a scatter diagram of the same information. Then connect the dots in the scatter diagram to obtain a line graph of that same information. (Notice that we have used the numbers 1 through 12 to represent the months January through December.)

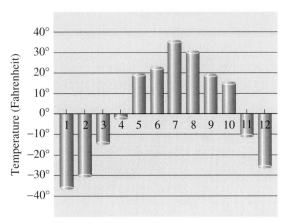

Figure 5   A bar chart of Table 3

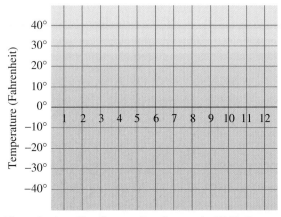

Figure 6   A scatter diagram, then line graph of Table 3

90. Temperature Figure 7 is a bar chart of the information in Table 4. Use the template in Figure 8 to construct a scatter diagram of the same information. Then connect the dots in the scatter diagram to obtain a line graph of that same information. (Again, we have used the numbers 1 through 12 to represent the months January through December.)

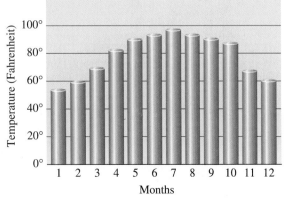

Figure 7   A bar chart of Table 4

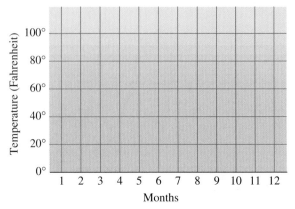

Figure 8   A scatter diagram, then line graph of Table 4

Calculator Problems

Display each of the following numbers on your calculator.

91. -18 **92.** -6 **93.** -144 **94.** -5

Perform each of the following additions on your calculator.

95. $-18 + 18$ **96.** $-6 + 6$ **97.** $-144 + 144$ **98.** $-5 + 5$

Review Problems

From here on, each problem set will contain some review problems. In mathematics it is very important to review the material you have studied. The more you do so, the better you will understand the topics we cover and the longer you will remember them.

The problems below review the basic facts and properties of addition with the whole numbers from Chapter 1.

Complete each statement using the commutative property of addition.

99. $3 + 5 =$ **100.** $9 + x =$

Complete each statement using the associative property of addition.

101. $7 + (2 + 6) =$ **102.** $(x + 3) + 5 =$

Write each of the following in symbols.

103. The sum of x and 4   **104.** The sum of x and 4 is 9.

105. 5 more than y   **106.** x increased by 8

Add.

107. $276 + 32 + 4,005$ **108.** $17 + 3 + 152 + 1,200$

109. $(43,567 + 76,543) + 11,111$ **110.** $43,567 + (76,543 + 11,111)$

Extending the Concepts

111. There are two numbers that are 5 units from 2 on the number line. One of them is 7. What is the other one?

112. There are two numbers that are 5 units from -2 on the number line. One of them is 3. What is the other one?

113. In your own words and in complete sentences, explain what the opposite of a number is.

114. In your own words and in complete sentences, explain what the absolute value of a number is.

115. The expression $-(-3)$ is read "the opposite of negative 3," and it simplifies to just 3. Give a similar written description of the expression $-|-3|$, and then simplify it.

116. Give written descriptions of the expressions $-(-4)$ and $-|-4|$, and then simplify each of them.

2.2 Addition with Negative Numbers

Introduction . . .

Suppose you are in Las Vegas playing blackjack and you lose $3 on the first hand and then you lose $5 on the next hand. If you represent winning with positive numbers and losing with negative numbers, how will you represent the results from your first two hands? Since you lost $3 and $5 for a total of $8, one way to represent the situation is with addition of negative numbers:

$$(-\$3) + (-\$5) = -\$8$$

From this example we see that the sum of two negative numbers is a negative number. To generalize addition of positive and negative numbers, we can use our number line.

We can think of each number on the number line as having two characteristics: (1) a *distance* from 0 (absolute value) and (2) a *direction* from 0 (positive or negative). The distance from 0 is represented by the numerical part of the number (like the 5 in the number −5), and its direction is represented by the + or − sign in front of the number.

We can visualize addition of numbers on the number line by thinking in terms of distance and direction from 0. Let's begin with a simple problem we know the answer to. We interpret the sum 3 + 5 on the number line as follows:

1. The first number is 3, which tells us "start at the origin, and move 3 units in the positive direction."

2. The + sign is read "and then move."

3. The 5 means "5 units in the positive direction."

Note
This method of adding numbers may seem a little complicated at first, but it will allow us to add numbers we couldn't otherwise add.

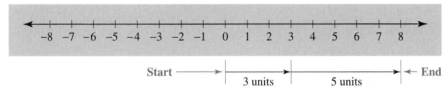

Figure 1

Figure 1 shows these steps. To summarize, 3 + 5 means to start at the origin (0), move 3 units in the *positive* direction, and then move 5 units in the *positive* direction. We end up at 8, which is the sum we are looking for: 3 + 5 = 8.

EXAMPLE 1   Add 3 + (−5) using the number line.

SOLUTION   We start at the origin, move 3 units in the positive direction, and then move 5 units in the negative direction, as shown in Figure 2. The last arrow ends at −2, which must be the sum of 3 and −5. That is:

$$3 + (-5) = -2$$

Practice Problems
1.   Add: 2 + (−5)

Answer
1.   −3

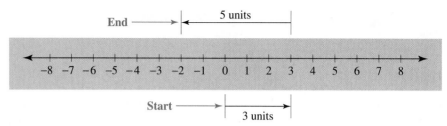

Figure 2

2. Add: $-2 + 5$

EXAMPLE 2 Add $-3 + 5$ using the number line.

SOLUTION We start at the origin, move 3 units in the negative direction, and then move 5 units in the positive direction, as shown in Figure 3. We end up at 2, which is the sum of -3 and 5. That is:

$$-3 + 5 = 2$$

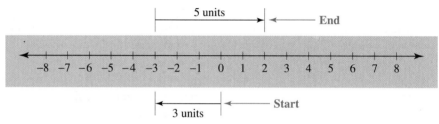

Figure 3

3. Add: $-2 + (-5)$

EXAMPLE 3 Add $-3 + (-5)$ using the number line.

SOLUTION We start at the origin, move 3 units in the negative direction, and then move 5 more units in the negative direction. This is shown on the number line in Figure 4. As you can see, the last arrow ends at -8. We must conclude that the sum of -3 and -5 is -8. That is:

$$-3 + (-5) = -8$$

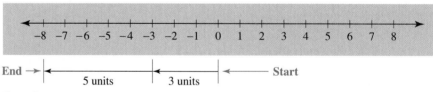

Figure 4

Adding numbers on the number line as we have done in these first three examples gives us a way of visualizing addition of positive and negative numbers. We eventually want to be able to write a rule for addition of positive and negative numbers that doesn't involve the number line. The number line is a way of justifying the rule we will eventually write. Here is a summary of the results we have so far:

$$3 + 5 = 8 \qquad -3 + 5 = 2$$
$$3 + (-5) = -2 \qquad -3 + (-5) = -8$$

Examine these results to see if you notice any pattern in the answers.

Answers
2. 3   **3.** -7

EXAMPLE 4 $4 + 7 = 11$

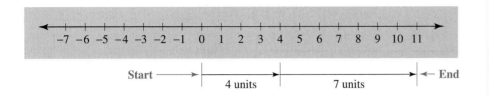

4. Add: $2 + 6$

EXAMPLE 5 $4 + (-7) = -3$

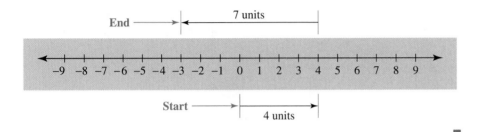

5. Add: $2 + (-6)$

EXAMPLE 6 $-4 + 7 = 3$

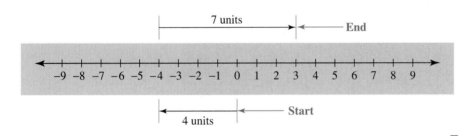

6. Add: $-2 + 6$

EXAMPLE 7 $-4 + (-7) = -11$

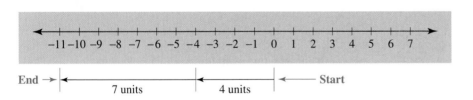

7. Add: $-2 + (-6)$

A summary of the results of these last four examples looks like this:

$$
\begin{aligned}
4 + 7 &= 11 \\
4 + (-7) &= -3 \\
-4 + 7 &= 3 \\
-4 + (-7) &= -11
\end{aligned}
$$

Answers
4. 8 **5.** -4 **6.** 4 **7.** -8

Looking over all the examples in this section, and noticing how the results in the problems are related, we can write the following rule for adding any two numbers:

RULE

1. To add two numbers with the *same* sign: Simply add their absolute values, and use the common sign. If both numbers are positive, the answer is positive. If both numbers are negative, the answer is negative.

2. To add two numbers with *different* signs: Subtract the smaller absolute value from the larger absolute value. The answer will have the sign of the number with the larger absolute value.

The following examples show how the rule is used. You will find that the rule for addition is consistent with all the results obtained using the number line.

EXAMPLE 8 Add all combinations of positive and negative 10 and 15.

SOLUTION
$$10 + 15 = 25$$
$$10 + (-15) = -5$$
$$-10 + 15 = 5$$
$$-10 + (-15) = -25$$

Notice that when we add two numbers with the same sign, the answer also has that sign. When the signs are not the same, the answer has the sign of the number with the larger absolute value. ∎

Once you have become familiar with the rule for adding positive and negative numbers, you can apply it to more complicated sums.

EXAMPLE 9 Simplify: $10 + (-5) + (-3) + 4$

SOLUTION   Adding left to right, we have:

$$10 + (-5) + (-3) + 4 = \mathbf{5} + (-3) + 4 \qquad \mathbf{10 + (-5) = 5}$$
$$= 2 + 4 \qquad \mathbf{5 + (-3) = 2}$$
$$= 6$$ ∎

EXAMPLE 10 Simplify: $[-3 + (-10)] + [8 + (-2)]$

SOLUTION   We begin by adding the numbers inside the brackets.

$$[-3 + (-10)] + [8 + (-2)] = [-13] + [6]$$
$$= -7$$ ∎

Note

This rule covers all possible addition problems involving positive and negative numbers. You *must* memorize it. After you have worked some problems, the rule will seem almost automatic.

8. Add all combinations of positive and negative 12 and 15.

9. Simplify:
   $12 + (-3) + (-7) + 5$

10. Simplify:
   $[-2 + (-12)] + [7 + (-5)]$

U S I N G   T E C H N O L O G Y

Calculator Note

There are a number of different ways in which calculators display negative numbers. Some calculators use a key labeled $\boxed{+/-}$, whereas others use a key labeled $\boxed{(-)}$. You will need to consult with the manual that came with your calculator to see how your calculator does the job.

Here are a couple of ways to find the sum $-10 + (-15)$ on a calculator:

Scientific Calculator: $10 \boxed{+/-} \boxed{+} 15 \boxed{+/-} \boxed{=}$
Graphing Calculator: $\boxed{(-)} 10 \boxed{+} \boxed{(-)} 15 \boxed{ENT}$

Answers
8. See solutions section.
9. 7   **10.** −12

Problem Set 2.2

Draw a number line from −10 to +10 and use it to add the following numbers.

1. 2 + 3 **2.** 2 + (−3) **3.** −2 + 3 **4.** −2 + (−3) **5.** 5 + (−7) **6.** −5 + 7

7. −4 + (−2) **8.** −8 + (−2) **9.** 10 + (−6) **10.** −9 + 3 **11.** 7 + (−3) **12.** −7 + 3

13. −4 + (−5) **14.** −2 + (−7)

Combine the following by using the rule for addition of positive and negative numbers. (Your goal is to be fast and accurate at addition, with the latter being more important.)

15. 7 + 8 **16.** 9 + 12 **17.** 5 + (−8) **18.** 4 + (−11)

19. −6 + (−5) **20.** −7 + (−2) **21.** −10 + 3 **22.** −14 + 7

23. −1 + (−2) **24.** −5 + (−4) **25.** −11 + (−5) **26.** −16 + (−10)

27. 4 + (−12) **28.** 9 + (−1) **29.** −85 + (−42) **30.** −96 + (−31)

31. −121 + 170 **32.** −130 + 158 **33.** −375 + 409 **34.** −765 + 213

Complete the following tables.

35.

FIRST NUMBER	SECOND NUMBER	THEIR SUM
a	b	$a + b$
5	−3	
5	−4	
5	−5	
5	−6	
5	−7	

36.

FIRST NUMBER	SECOND NUMBER	THEIR SUM
a	b	$a + b$
−5	3	
−5	4	
−5	5	
−5	6	
−5	7	

37.

FIRST NUMBER	SECOND NUMBER	THEIR SUM
x	y	$x + y$
−5	−3	
−5	−4	
−5	−5	
−5	−6	
−5	−7	

38.

FIRST NUMBER	SECOND NUMBER	THEIR SUM
x	y	$x + y$
30	−20	
−30	20	
−30	−20	
30	20	
−30	0	

Add the following numbers left to right.

39. $10 + (-18) + 4$

40. $12 + (-10) + 3$

41. $24 + (-6) + (-8)$

42. $35 + (-5) + (-30)$

43. $-201 + (-143) + (-101)$

44. $-27 + (-56) + (-89)$

45. $-321 + 752 + (-324)$

46. $-571 + 437 + (-502)$

47. $-8 + 3 + (-5) + 9$

48. $-9 + 2 + (-10) + 3$

49. $-2 + (-5) + (-6) + (-7)$

50. $-8 + (-3) + (-4) + (-7)$

51. $15 + (-30) + 18 + (-20)$

52. $20 + (-15) + 30 + (-18)$

53. $-78 + (-42) + 57 + 13$

54. $-89 + (-51) + 65 + 17$

Use the rule for order of operations to simplify each of the following.

55. $(-8 + 5) + (-6 + 2)$

56. $(-3 + 1) + (-9 + 4)$

57. $(-10 + 4) + (-3 + 12)$

58. $(-11 + 5) + (-3 + 2)$

59. $20 + (-30 + 50) + 10$

60. $30 + (-40 + 20) + 50$

61. $108 + (-456 + 275)$

62. $106 + (-512 + 318)$

63. $[5 + (-8)] + [3 + (-11)]$

64. $[8 + (-2)] + [5 + (-7)]$

65. $[57 + (-35)] + [19 + (-24)]$

66. $[63 + (-27)] + [18 + (-24)]$

67. Find the sum of -8, -10, and -3.

68. Find the sum of -4, 17, and -6.

69. What number do you add to 8 to get 3?

70. What number do you add to 10 to get 4?

71. What number do you add to −3 to get −7?

72. What number do you add to −5 to get −8?

73. What number do you add to −4 to get 3?

74. What number do you add to −7 to get 2?

75. If the sum of −3 and 5 is increased by 8, what number results?

76. If the sum of −9 and −2 is increased by 10, what number results?

Estimating

Work Problems 77–84 mentally, without pencil and paper or a calculator.

77. The answer to the problem 251 + 249 is closest to which of the following numbers?
  a. 500   **b.** 0   **c.** −500

78. The answer to the problem 251 + (−249) is closest to which of the following numbers?
  a. 500   **b.** 0   **c.** −500

79. The answer to the problem −251 + 249 is closest to which of the following numbers?
  a. 500   **b.** 0   **c.** −500

80. The answer to the problem −251 + (−249) is closest to which of the following numbers?
  a. 500   **b.** 0   **c.** −500

81. The sum of 77 and 22 is closest to which of the following numbers?
  a. −100   **b.** −60   **c.** 60   **d.** 100

82. The sum of −77 and 22 is closest to which of the following numbers?
  a. −100   **b.** −60   **c.** 60   **d.** 100

83. The sum of 77 and −22 is closest to which of the following numbers?
  a. −100   **b.** −60   **c.** 60   **d.** 100

84. The sum of −77 and −22 is closest to which of the following numbers?
  a. −100   **b.** −60   **c.** 60   **d.** 100

Applying the Concepts

85. Checkbook Balance Ethan has a balance of −$40 in his checkbook. If he deposits $100 and then writes a check for $50, what is the new balance in his checkbook?

		RECORD ALL CHARGES OR CREDITS THAT AFFECT YOUR ACCOUNT						BALANCE	
NUMBER	DATE	DESCRIPTION OF TRANSACTION	PAYMENT/DEBIT (−)	√ T	FEE (IF ANY) (−)	DEPOSIT/CREDIT (+)	$	−40	00
	9/20	Deposit from work	$		$	$ 100	00		
1502	9/21	Vons market	50	00					

86. Checkbook Balance Kendra has a balance of −$20 in her checkbook. If she deposits $45 and then writes a check for $15, what is the new balance in her checkbook?

87. Gambling While gambling in Las Vegas, a person wins $74 playing blackjack and then loses $141 on roulette. Use positive and negative numbers to write this situation in symbols. Then give the person's net loss or gain.

88. Gambling While playing blackjack, a person loses $17 on his first hand, then wins $14, and then loses $21. Write this situation using positive and negative numbers and addition; then simplify.

89. Stock Gain/Loss Suppose a certain stock gains 3 points on the stock exchange on Monday and then loses 5 points on Tuesday. Express the situation using positive and negative numbers, and then give the net gain or loss of the stock for this 2-day period.

90. Stock Gain/Loss A stock gains 2 points on Wednesday, then loses 1 on Thursday, and gains 3 on Friday. Use positive and negative numbers and addition to write this situation in symbols, and then simplify.

91. Distance The distance between two numbers on the number line is 10. If one of the numbers is 3, what are the two possibilities for the other number?

92. Distance The distance between two numbers is 8. If one of the numbers is -5, what are the two possibilities for the other number?

Review Problems

The problems below review the basic facts and properties of subtraction with whole numbers from Chapter 1.

Write each of the following statements in symbols.

93. The difference of 10 and x

94. The difference of x and 10

95. 17 subtracted from y

96. y subtracted from 17

Subtract.

97. $763 - 289$

98. $894 - 299$

99. $20 - 9 - 3$

100. $12 - 8 - 2$

101. $20 - (9 - 3)$

102. $12 - (8 - 2)$

103. Subtract 238 from 963, and round your answer to the nearest ten.

104. Subtract 456 from 762, and round your answer to the nearest hundred.

2.3 Subtraction with Negative Numbers

Introduction . . .

At the beginning of this chapter we asked how we would represent the final balance in a checkbook if the original balance was $20 and we wrote a check for $30. We decided that the final balance would be −$10. We can summarize the whole situation with subtraction:

$$\$20 - \$30 = -\$10$$

From this we see that subtracting 30 from 20 gives us −10. Another example that gives the same answer but involves addition is this:

$$20 + (-30) = -10$$

From the two examples above, we find that subtracting 30 gives the same result as adding −30. We use this kind of reasoning to give a definition for subtraction that will allow us to use the rules we developed for addition to do our subtraction problems. Here is that definition:

DEFINITION

Subtraction If *a* and *b* represent any two numbers, then it is always true that

$$a - b = a + (-b)$$

To subtract *b*   Add its opposite, −*b*

In words: Subtracting a number is equivalent to adding its opposite.

Let's see if this definition conflicts with what we already know to be true about subtraction.

EXAMPLE 1   Subtract: 5 − 2

SOLUTION   From previous experience we know that

$$5 - 2 = 3$$

We can get the same answer by using the definition we just gave for subtraction. Instead of subtracting 2, we can add its opposite, −2. Here is how it looks:

$$5 - 2 = 5 + (-2) \quad \textbf{Change subtraction to}$$
$$\textbf{addition of the opposite}$$
$$= 3 \quad \textbf{Apply the rule for addition of}$$
$$\textbf{positive and negative numbers} \quad \blacksquare$$

The result is the same whether we use our previous knowledge of subtraction or the new definition. The new definition is essential when the problems begin to get more complicated.

Note
This definition of subtraction may seem a little strange at first. In Example 1 you will notice that using the definition gives us the same results we are used to getting with subtraction. As we progress further into the section, we will use the definition to subtract numbers we haven't been able to subtract before.

Practice Problems
1. Subtract: 7 − 3

Answer
1. 4

2. Subtract: $-7 - 3$

Note

A real-life analogy to Example 2 would be: "If the temperature were 7° below 0 and then it dropped another 2°, what would the temperature be then?"

3. Subtract: $-8 - 6$

4. Subtract: $10 - (-6)$

5. Subtract: $-10 - (-15)$

Note

Examples 4 and 5 may give results you are not used to getting. But you must realize that the results are correct. That is, $12 - (-6)$ is 18, and $-20 - (-30)$ is 10. If you think these results should be different, then you are not thinking of subtraction correctly.

6. Subtract each of the following.
  a. $8 - 5$

  b. $-8 - 5$

  c. $8 - (-5)$

  d. $-8 - (-5)$

  e. $12 - 10$

  f. $-12 - 10$

  g. $12 - (-10)$

  h. $-12 - (-10)$

Answers
2. -10 **3.** -14 **4.** 16   **5.** 5
6. a. 3   **b.** -13 **c.** 13
    d. -3 **e.** 2   **f.** -22
    g. 22   **h.** -2

EXAMPLE 2   Subtract: $-7 - 2$

SOLUTION   We have never subtracted a positive number from a negative number before. We must apply our definition of subtraction:

$$-7 - 2 = -7 + (-2)$$ **Instead of subtracting 2, we add its opposite, -2**

$$= -9$$ **Apply the rule for addition** ■

EXAMPLE 3   Subtract: $-10 - 5$

SOLUTION   We apply the definition of subtraction (if you don't know the definition of subtraction yet, go back and read it) and add as usual.

$$-10 - 5 = -10 + (-5)$$ **Definition of subtraction**

$$= -15$$ **Addition** ■

EXAMPLE 4   Subtract: $12 - (-6)$

SOLUTION   The first $-$ sign is read "subtract," and the second one is read "negative." The problem in words is "12 subtract negative 6." We can use the definition of subtraction to change this to the addition of positive 6:

$$12 - (-6) = 12 + 6$$ **Subtracting -6 is equivalent to adding $+6$**

$$= 18$$ **Addition** ■

EXAMPLE 5   Subtract: $-20 - (-30)$

SOLUTION   Instead of subtracting -30, we can use the definition of subtraction to write the problem again as the addition of 30:

$$-20 - (-30) = -20 + 30$$ **Definition of subtraction**

$$= 10$$ **Addition** ■

Examples 1–5 illustrate all the possible combinations of subtraction with positive and negative numbers. There are no new rules for subtraction. We apply the definition to change each subtraction problem into an equivalent addition problem. The rule for addition can then be used to obtain the correct answer.

EXAMPLE 6   The following table shows the relationship between subtraction and addition:

SUBTRACTION	ADDITION OF THE OPPOSITE	ANSWER
$7 - 9$	$7 + (-9)$	-2
$-7 - 9$	$-7 + (-9)$	-16
$7 - (-9)$	$7 + 9$	16
$-7 - (-9)$	$-7 + 9$	2
$15 - 10$	$15 + (-10)$	5
$-15 - 10$	$-15 + (-10)$	-25
$15 - (-10)$	$15 + 10$	25
$-15 - (-10)$	$-15 + 10$	-5

■

EXAMPLE 7 Combine: $-3 + 6 - 2$

SOLUTION The first step is to change subtraction to addition of the opposite. After that has been done, we add left to right.

$$-3 + 6 - 2 = -3 + 6 + (-2)$$ **Subtracting 2 is equivalent to adding −2**
$$= 3 + (-2)$$ **Add left to right**
$$= 1$$ ∎

EXAMPLE 8 Combine: $10 - (-4) - 8$

SOLUTION Changing subtraction to addition of the opposite, we have

$$10 - (-4) - 8 = 10 + 4 + (-8)$$
$$= 14 + (-8)$$
$$= 6$$ ∎

EXAMPLE 9 Subtract 3 from -5.

SOLUTION Subtracting 3 is equivalent to adding -3:

$$-5 - 3 = -5 + (-3) = -8$$

Subtracting 3 from -5 gives us -8. ∎

EXAMPLE 10 Subtract -4 from 9.

SOLUTION Subtracting -4 is the same as adding $+4$:

$$9 - (-4) = 9 + 4 = 13$$

Subtracting -4 from 9 gives us 13. ∎

EXAMPLE 11 Find the difference of 8 and -2.

SOLUTION Subtracting -2 from 8, we have

$$8 - (-2) = 8 + 2 = 10$$

The difference of 8 and -2 is 10. ∎

EXAMPLE 12 Find the difference of -7 and -4.

SOLUTION Subtracting -4 from -7 looks like this:

$$-7 - (-4) = -7 + 4 = -3$$

The difference of -7 and -4 is -3. ∎

7. Combine: $-4 + 6 - 7$

8. Combine: $15 - (-5) - 8$

9. Subtract 2 from -8.

10. Subtract -5 from 7.

11. Find the difference of 7 and -3.

12. Find the difference of -8 and -6.

USING TECHNOLOGY

Calculator Note

Here is how we work the subtraction problem shown in Example 12 on a calculator.

Scientific Calculator: 7 $\boxed{+/-}$ $\boxed{-}$ 4 $\boxed{+/-}$ $\boxed{=}$

Graphing Calculator: $\boxed{(-)}$ 7 $\boxed{-}$ $\boxed{(-)}$ 4 $\boxed{ENT}$

Answers
7. -5 **8.** 12 **9.** -10
10. 12 **11.** 10 **12.** -2

13. Suppose the temperature is 42°F at takeoff and then drops to −42°F when the plane reaches its cruising altitude. Find the difference in temperature at takeoff and at 28,000 feet.

EXAMPLE 13 Many of the planes used by the United States during World War II were not pressurized or sealed from outside air. As a result, the temperature inside these planes was the same as the surrounding air temperature outside. Suppose the temperature inside a B-17 Flying Fortress is 50°F at takeoff and then drops to −30°F when the plane reaches its cruising altitude of 28,000 feet. Find the difference in temperature inside this plane at takeoff and at 28,000 feet.

SOLUTION   The temperature at takeoff is 50°F, whereas the temperature at 28,000 feet is −30°F. To find the difference we subtract, with the numbers in the same order as they are given in the problem:

$$50 - (-30) = 50 + 30 = 80$$

The difference in temperature is 80°F.   ■

Subtraction and Taking Away

For some people studying algebra for the first time, subtraction of positive and negative numbers can be a problem. These people may believe that the answer to $-5 - 9$ should be −4 or 4, not −14. If this is happening to you, you are probably thinking of subtraction in terms of taking one number away from another. Thinking of subtraction in this way works well with positive numbers if you always subtract the smaller number from the larger. In algebra, however, we encounter many situations other than this. The definition of subtraction, that $a - b = a + (-b)$, clearly indicates the correct way to use subtraction. That is, when working subtraction problems, you should think "addition of the opposite," not "taking one number away from another." To be successful in algebra, you need to apply these properties and definitions exactly as they are presented here.

Problem Set 2.3

Subtract.

1. $7 - 5$ **2.** $5 - 7$ **3.** $8 - 6$ **4.** $6 - 8$

5. $-3 - 5$ **6.** $-5 - 3$ **7.** $-4 - 1$ **8.** $-1 - 4$

9. $5 - (-2)$ **10.** $2 - (-5)$ **11.** $3 - (-9)$ **12.** $9 - (-3)$

13. $-4 - (-7)$ **14.** $-7 - (-4)$ **15.** $-10 - (-3)$ **16.** $-3 - (-10)$

17. $15 - 18$ **18.** $20 - 32$ **19.** $100 - 113$ **20.** $121 - 21$

21. $-30 - 20$ **22.** $-50 - 60$ **23.** $-79 - 21$ **24.** $-86 - 31$

25. $156 - (-243)$ **26.** $292 - (-841)$ **27.** $-35 - (-14)$ **28.** $-29 - (-4)$

Complete the following tables.

29.

FIRST NUMBER x	SECOND NUMBER y	THE DIFFERENCE OF x AND y $x - y$
8	6	
8	7	
8	8	
8	9	
8	10	

30.

FIRST NUMBER x	SECOND NUMBER y	THE DIFFERENCE OF x AND y $x - y$
10	12	
10	11	
10	10	
10	9	
10	8	

31.

FIRST NUMBER x	SECOND NUMBER y	THE DIFFERENCE OF x AND y $x - y$
8	-6	
8	-7	
8	-8	
8	-9	
8	-10	

32.

FIRST NUMBER x	SECOND NUMBER y	THE DIFFERENCE OF x AND y $x - y$
-10	-12	
-10	-11	
-10	-10	
-10	-9	
-10	-8	

Simplify as much as possible by first changing all subtractions to addition of the opposite and then adding left to right.

33. $-8 + 3 - 4$ **34.** $-10 - 1 + 16$ **35.** $-8 - 4 - 2$ **36.** $-7 - 3 - 6$

37. $3 - (-2) - 6$ **38.** $4 - (-1) - 5$ **39.** $-9 + 4 - (-10)$ **40.** $-3 + 6 - (-2)$

41. Subtract -6 from 5.

42. Subtract 8 from -2.

43. Find the difference of -5 and -1.

44. Find the difference of -7 and -3.

45. Subtract -4 from the sum of -8 and 12.

46. Subtract -7 from the sum of 7 and -12.

47. What number do you subtract from -3 to get -9?

48. What number do you subtract from 5 to get 8?

Estimating

Work Problems 49–58 mentally, without pencil and paper or a calculator.

49. The answer to the problem $52 - 49$ is closest to which of the following numbers?
  a. 100   **b.** 0   **c.** -100

50. The answer to the problem $-52 - 49$ is closest to which of the following numbers?
  a. 100   **b.** 0   **c.** -100

51. The answer to the problem $52 - (-49)$ is closest to which of the following numbers?
  a. 100   **b.** 0   **c.** -100

52. The answer to the problem $-52 - (-49)$ is closest to which of the following numbers?
  a. 100   **b.** 0   **c.** -100

53. Is the difference $-161 - (-62)$ closer to -200 or -100?

54. Is the difference of $-553 - 50$ closer to -600 or -500?

55. The difference between 37 and 61 is closest to which of the following numbers?
  a. -100 **b.** -20 **c.** 20   **d.** 100

56. The difference between 37 and -61 is closest to which of the following numbers?
  a. -100 **b.** -20 **c.** 20   **d.** 100

57. The difference between -37 and 61 is closest to which of the following numbers?
  a. -100 **b.** -20 **c.** 20   **d.** 100

58. The difference between -37 and -61 is closest to which of the following numbers?
  a. -100 **b.** -20 **c.** 20   **d.** 100

Applying the Concepts

59. Temperature On Monday the temperature reached a high of 28° above 0. That night it dropped to 16° below 0. What is the difference between the high and the low temperatures for Monday?

60. Gambling A gambler loses $35 playing poker one night and then loses another $25 the next night. Express this situation with numbers. How much did the gambler lose?

61. Checkbook Balance Felicia has a balance of $68 in her checking account when she writes a check for $37 and then another check for $53. Write a subtraction problem that gives the new balance in her checking account. What is the new balance in her checking account?

62. Checkbook Balance Susan has a balance of $572 in her checking account when she writes a check for $435 to pay the rent. Then she writes another check for $172 for textbooks. Write a subtraction problem that gives the new balance in her checking account. What is the new balance in her checking account?

Repeated below is the table of wind chill temperatures that we used in Problem Set 2.1. Use it for Problems 63–66.

	AIR TEMPERATURE (°F)							
WIND SPEED	**30°**	**25°**	**20°**	**15°**	**10°**	**5°**	**0°**	**−5°**
10 mph	16°	10°	3°	−3°	−9°	−15°	−22°	−27°
15 mph	9°	2°	−5°	−11°	−18°	−25°	−31°	−38°
20 mph	4°	−3°	−10°	−17°	−24°	−31°	−39°	−46°
25 mph	1°	−7°	−15°	−22°	−29°	−36°	−44°	−51°
30 mph	−2°	−10°	−18°	−25°	−33°	−41°	−49°	−56°

63. Wind Chill If the temperature outside is 15°F, what is the difference in wind chill temperature between a 15-mile-per-hour wind and a 25-mile-per-hour wind?

64. Wind Chill If the temperature outside is 0°F, what is the difference in wind chill temperature between a 15-mile-per-hour wind and a 25-mile-per-hour wind?

65. Wind Chill Find the difference in temperature between a day in which the air temperature is 20°F and the wind is blowing at 10 miles per hour and a day in which the air temperature is 10°F and the wind is blowing at 20 miles per hour.

66. Wind Chill Find the difference in temperature between a day in which the air temperature is 0°F and the wind is blowing at 10 miles per hour and a day in which the air temperature is −5°F and the wind is blowing at 20 miles per hour.

Use the tables below to work Problems 67–70.

Table 1

RECORD LOW TEMPERATURES
FOR LAKE PLACID, NEW YORK

MONTH	TEMPERATURE
January	−36°F
February	−30°F
March	−14°F
April	−2°F
May	19°F
June	22°F
July	35°F
August	30°F
September	19°F
October	15°F
November	−11°F
December	−26°F

Table 2

RECORD HIGH TEMPERATURES
FOR LAKE PLACID, NEW YORK

MONTH	TEMPERATURE
January	54°F
February	59°F
March	69°F
April	82°F
May	90°F
June	93°F
July	97°F
August	93°F
September	90°F
October	87°F
November	67°F
December	60°F

67. Temperature Difference Find the difference between the record high temperature and the record low temperature for the month of December.

68. Temperature Difference Find the difference between the record high temperature and the record low temperature for the month of March.

69. Temperature Difference Find the difference between the record low temperatures of March and December.

70. Temperature Difference Find the difference between the record high temperatures of March and December.

Review Problems

The problems below review the basic facts and properties of multiplication with whole numbers from Chapter 1.

Write each of the following statements in symbols.

71. The product of 3 and 5.

72. The product of 5 and 3.

73. The product of 7 and x.

74. The product of 2 and y.

Rewrite the following statements using the commutative property of multiplication.

75. $3(5) =$

76. $7(x) =$

Rewrite the following statements using the associative property of multiplication.

77. $5(7 \cdot 8) =$

78. $4(6 \cdot y)$

Apply the distributive property to each expression and then simplify the result.

79. $2(3 + 4)$

80. $5(6 + 7)$

Multiply.

81. $42(56)$

82. $247(531)$

Extending the Concepts

83. Give an example that shows that subtraction is not a commutative operation.

84. Why is the expression "two negatives make a positive" not correct?

85. Give an example of an everyday situation that is modeled by the subtraction problem $\$10 - \$12 = -\$2$.

86. Give an example of an everyday situation that is modeled by the subtraction problem $-\$10 - \$12 = -\$22$.

In Chapter 1 we defined an arithmetic sequence as a sequence of numbers in which each number, after the first number, is obtained from the previous number by adding the same amount each time.

Find the next two numbers in each arithmetic sequence below.

87. $10, 5, 0, \ldots$

88. $8, 3, -2, \ldots$

89. $-10, -6, -2, \ldots$

90. $-4, -1, 2, \ldots$

2.4 Multiplication with Negative Numbers

Introduction . . .

Suppose you own three shares of a stock and the price per share drops $5. How much money have you lost? The answer is $15. Because it is a loss, we can express it as −$15. The multiplication problem below can be used to describe the relationship among the numbers.

3 shares   each loses $5   for a total of −$15

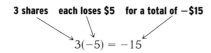

$$3(-5) = -15$$

From this we conclude that it is reasonable to say that the product of a positive number and a negative number is a negative number.

In order to generalize multiplication with negative numbers, recall that we first defined multiplication by whole numbers to be repeated addition. That is:

$$3 \cdot 5 = 5 + 5 + 5$$

   ↑   ↑ ↑ ↑

Multiplication   Repeated addition

This concept is very helpful when it comes to developing the rule for multiplication problems that involve negative numbers. For the first example we look at what happens when we multiply a negative number by a positive number.

EXAMPLE 1 Multiply: $3(-5)$

SOLUTION Writing this product as repeated addition, we have

$$3(-5) = (-5) + (-5) + (-5)$$
$$= -10 + (-5)$$
$$= -15$$

The result, −15, is obtained by adding the three negative 5's. ∎

EXAMPLE 2 Multiply: $-3(5)$

SOLUTION In order to write this multiplication problem in terms of repeated addition, we will have to reverse the order of the two numbers. This is easily done, because multiplication is a commutative operation.

$$-3(5) = 5(-3)$$ **Commutative property**
$$= (-3) + (-3) + (-3) + (-3) + (-3)$$ **Repeated addition**
$$= -15$$ **Addition**

The product of −3 and 5 is −15. ∎

EXAMPLE 3 Multiply: $-3(-5)$

SOLUTION It is impossible to write this product in terms of repeated addition. We will find the answer to $-3(-5)$ by solving a different problem. Look at the following problem:

$$-3[5 + (-5)] = -3[0] = 0$$

The result is 0, because multiplying by 0 always produces 0. Now we can work the same problem another way and in the process find the answer to $-3(-5)$. Applying the distributive property to the same expression, we have

$$-3[5 + (-5)] = -3(5) + (-3)(-5) \quad \textbf{Distributive property}$$
$$= -15 + (?) \qquad\qquad \mathbf{-3(5) = -15}$$

The question mark must be $+15$, because we already know that the answer to the problem is 0, and $+15$ is the only number we can add to -15 to get 0. So, our problem is solved:

$$-3(-5) = +15 \qquad\qquad\qquad\qquad ■$$

Table 1 gives a summary of what we have done so far in this section.

Table 1

ORIGINAL NUMBERS HAVE	FOR EXAMPLE	THE ANSWER IS
Same signs	$3(5) = 15$	Positive
Different signs	$-3(5) = -15$	Negative
Different signs	$3(-5) = -15$	Negative
Same signs	$-3(-5) = 15$	Positive

From the examples we have done so far in this section and their summaries in Table 1, we write the following rule for multiplication of positive and negative numbers:

RULE

To multiply any two numbers, we multiply their absolute values.

1. The answer is *positive* if both the original numbers have the same sign. That is, the product of two numbers with the same sign is positive.

2. The answer is *negative* if the original two numbers have different signs. The product of two numbers with different signs is negative.

This rule should be memorized. By the time you have finished reading this section and working the problems at the end of the section, you should be fast and accurate at multiplication with positive and negative numbers.

Multiply.

4. $3(2)$

5. $-3(-2)$

6. $3(-2)$

7. $-3(2)$

8. $8(-9)$

EXAMPLE 4 $2(4) = 8$ Like signs; positive answer   ■

EXAMPLE 5 $-2(-4) = 8$ Like signs; positive answer   ■

EXAMPLE 6 $2(-4) = -8$ Unlike signs; negative answer   ■

EXAMPLE 7 $-2(4) = -8$ Unlike signs; negative answer   ■

EXAMPLE 8 $7(-6) = -42$ Unlike signs; negative answer   ■

Answers

4. 6   **5.** 6   **6.** -6 **7.** -6
8. -72

EXAMPLE 9 $-5(-8) = 40$ Like signs; positive answer ■

9. $-6(-4)$

EXAMPLE 10 $-3(2)(-5) = -6(-5)$ **Multiply −3 and 2 to get −6**
$$= 30$$ ■

10. $-5(2)(-4)$

EXAMPLE 11 $(-5)^3 = -5(-5)(-5)$ **Expand exponent 3**
$$= 25(-5)$$ **Multiply −5 and −5 to get 25**
$$= -125$$ ■

11. $(-5)^2$

EXAMPLE 12 Simplify: $-6[3 + (-5)]$

SOLUTION   We begin inside the brackets and work our way out:
$$-6[3 + (-5)] = -6[-2]$$
$$= 12$$ ■

12. Simplify: $-2[5 + (-8)]$

EXAMPLE 13 Simplify: $-4 + 5(-6 + 2)$

SOLUTION   Simplifying inside the parentheses first, we have
$$-4 + 5(-6 + 2) = -4 + 5(-4)$$ **Simplify inside parentheses**
$$= -4 + (-20)$$ **Multiply**
$$= -24$$ **Add** ■

13. Simplify: $-3 + 4(-7 + 3)$

EXAMPLE 14 Simplify: $-2(7) + 3(-6)$

SOLUTION   Multiplying left to right before we add gives us
$$-2(7) + 3(-6) = -14 + (-18)$$
$$= -32$$ ■

14. Simplify: $-3(5) + 4(-4)$

EXAMPLE 15 Simplify: $-3(2 - 9) + 4(-7 - 2)$

SOLUTION   We begin by subtracting inside the parentheses:
$$-3(2 - 9) + 4(-7 - 2) = -3(-7) + 4(-9)$$
$$= 21 + (-36)$$
$$= -15$$ ■

15. Simplify: $-2(3 - 5) - 7(-2 - 4)$

EXAMPLE 16 Simplify: $(-3 - 7)(2 - 6)$

SOLUTION   Again, we begin by simplifying inside the parentheses:
$$(-3 - 7)(2 - 6) = (-10)(-4)$$
$$= 40$$ ■

16. Simplify: $(-6 - 1)(4 - 9)$

USING TECHNOLOGY

Calculator Note

Here is how we work the problem shown in Example 16 on a calculator. (The $\boxed{\times}$ key on the first line may, or may not, be necessary. Try your calculator without it and see.)

Scientific Calculator: $\boxed{(}\ 3\ \boxed{+/-}\ \boxed{-}\ 7\ \boxed{)}\ \boxed{\times}\ \boxed{(}\ 2\ \boxed{-}\ 6\ \boxed{)}\ \boxed{=}$

Graphing Calculator: $\boxed{(}\ \boxed{(-)}\ 3\ \boxed{-}\ 7\ \boxed{)}\ \boxed{(}\ 2\ \boxed{-}\ 6\ \boxed{)}\ \boxed{ENT}$

Answers
9. 24   **10.** 40   **11.** 25   **12.** 6
13. −19   **14.** −31   **15.** 46
16. 35

Problem Set 2.4

Find each of the following products. (Multiply.)

1. $7(-8)$

2. $-3(5)$

3. $-6(10)$

4. $4(-8)$

5. $-7(-8)$

6. $-4(-7)$

7. $-9(-9)$

8. $-6(-3)$

9. $4(-6)$

10. $5(-2)$

11. $-6(-5)$

12. $-8(-3)$

13. $-5(10)$

14. $-6(11)$

15. $3(-2)(4)$

16. $5(-1)(3)$

17. $-4(3)(-2)$

18. $-4(5)(-6)$

19. $-1(-2)(-3)$

20. $-2(-3)(-4)$

Use the definition of exponents to expand each of the following expressions. Then multiply according to the rule for multiplication.

21. $(-4)^2$ **22.** $(-5)^2$ **23.** $(-5)^3$ **24.** $(-4)^3$ **25.** $(-2)^4$ **26.** $(-1)^4$

Complete the following tables.

27.

NUMBER	SQUARE
x	x^2
-3	
-2	
-1	
0	
1	
2	
3	

28.

NUMBER	CUBE
x	x^3
-3	
-2	
-1	
0	
1	
2	
3	

29.

FIRST NUMBER	SECOND NUMBER	THEIR PRODUCT
x	y	xy
6	2	
6	1	
6	0	
6	-1	
6	-2	

30.

FIRST NUMBER	SECOND NUMBER	THEIR PRODUCT
x	y	xy
7	4	
7	2	
7	0	
7	-2	
7	-4	

31.

FIRST NUMBER	SECOND NUMBER	THEIR PRODUCT
a	b	ab
−5	3	
−5	2	
−5	1	
−5	0	
−5	−1	
−5	−2	
−5	−3	

32.

FIRST NUMBER	SECOND NUMBER	THEIR PRODUCT
a	b	ab
−9	6	
−9	4	
−9	2	
−9	0	
−9	−2	
−9	−4	
−9	−6	

Use the rule for order of operations along with the rules for addition, subtraction, and multiplication to simplify each of the following expressions.

33. $4(-3 + 2)$

34. $7(-6 + 3)$

35. $-10(-2 - 3)$

36. $-5(-6 - 2)$

37. $-3 + 2(5 - 3)$

38. $-7 + 3(6 - 2)$

39. $-7 + 2[-5 - 9]$

40. $-8 + 3[-4 - 1]$

41. $2(-5) + 3(-4)$

42. $6(-1) + 2(-7)$

43. $3(-2)4 + 3(-2)$

44. $2(-1)(-3) + 4(-6)$

45. $(8 - 3)(2 - 7)$

46. $(9 - 3)(2 - 6)$

47. $(2 - 5)(3 - 6)$

48. $(3 - 7)(2 - 8)$

49. $3(5 - 8) + 4(6 - 7)$

50. $-2(8 - 10) + 3(4 - 9)$

51. $-3(4 - 7) - 2(-3 - 2)$

52. $-5(-2 - 8) - 4(6 - 10)$

53. $3(-2)(6 - 7)$

54. $4(-3)(2 - 5)$

55. Find the product of −3, −2, and −1.

56. Find the product of −7, −1, and 0.

57. What number do you multiply by −3 to get 12?

58. What number do you multiply by −7 to get −21?

59. Subtract −3 from the product of −5 and 4.

60. Subtract 5 from the product of −8 and 1.

Applying the Concepts

61. Day Trading Larry is buying and selling stock from his home computer. He owns 100 shares of Oracle Corporation and 50 shares of Gadzoox Networks Inc. On February 28, 2000, those stocks had the gain and loss shown in the table below. What was Larry's net gain or loss for the day on those two stocks?

STOCK	NUMBER OF SHARES	GAIN/LOSS
Oracle	100	−2
Gadzoox	50	+8

62. Stock Gain/Loss Amy owns stock that she keeps in her retirement account. She owns 200 shares of Apple Computer and 100 shares of Gap Inc. For the month of February 2000, those stocks had the gain and loss shown in the table below. What was Amy's net gain or loss for the month of February on those two stocks?

STOCK	NUMBER OF SHARES	GAIN/LOSS
Apple	200	+14
Gap	100	−5

63. Temperature Change A hot-air balloon is rising to its cruising altitude. Suppose the air temperature around the balloon drops 4 degrees each time the balloon rises 1,000 feet. What is the net change in air temperature around the balloon as it rises from 2,000 feet to 6,000 feet?

64. Temperature Change A small airplane is rising to its cruising altitude. Suppose the air temperature around the plane drops 4 degrees each time the plane increases its altitude by 1,000 feet. What is the net change in air temperature around the plane as it rises from 5,000 feet to 12,000 feet?

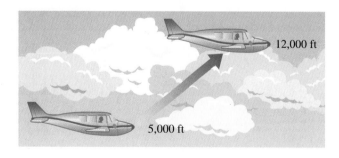

Baseball Major league baseball has various player awards at the end of the year. One of the awards for relief pitchers for 1998 was the Rolaids Relief Man of the Year. Rolaids points are awarded as follows:

Each win (W) earns 2 points   Each loss (L) earns −2 points
Each save (S) earns 3 points   Each blown save (BS) earns −2 points

The Rolaids points for someone with 8 wins, 7 losses, 6 saves, and 5 blown saves would be

$$8(2) + 7(-2) + 6(3) + 5(-2) = 10$$

Use information above to complete the tables in Problems 65 and 66.

65.

NATIONAL LEAGUE

PITCHER, TEAM	W	L	S	BS	PTS
Trevor Hoffman, S.D.	3	1	39	1	
Robb Nen, S.F.	7	3	31	3	
Rod Beck, Chi.	1	2	36	6	
Jeff Shaw, L.A.	2	5	36	6	
Ugueth Urbina, Mon.	4	2	26	4	
Billy Wagner, Hou.	3	3	23	2	
Kerry Ligtenberg, Atl.	3	2	20	2	
Mark Leiter, Phi.	6	3	22	8	
Gregg Olson, Ari.	1	3	21	3	
Bob Wickman, Mil.	5	6	20	4	

66.

AMERICAN LEAGUE

PITCHER, TEAM	W	L	S	BS	PTS
Tom Gordon, Bos.	6	3	34	1	
Mariano Rivera, N.Y.	2	0	32	3	
John Wetteland, Texas	3	1	32	4	
Troy Percival, Ana.	2	5	33	4	
Mike Jackson, Cle.	1	1	29	5	
Randy Myers, Tor.	3	4	28	5	
Rick Aguilera, Min.	3	7	30	8	
Jeff Montgomery, K.C.	2	4	26	4	
Billy Taylor, Oak.	4	9	24	4	
Armando Benitez, Bal.	5	3	17	2	

Golf One way to give scores in golf is in relation to par, the number of strokes considered necessary to complete a hole or course at the expert level. Scoring this way, if you complete a hole in one stroke less than par, your score is −1, which is called a *birdie*. If you shoot 2 under par, your score is −2, which is called an *eagle*. Shooting 1 over par is a score of +1, which is a *bogie*. A *double bogie* is 2 over par, and results in a score of +2.

67. Sergio Garcia's Scorecard The table below shows the scores Sergio Garcia had on the first round of the PGA Masters tournament in April 2000. Fill in the last column by multiplying each value by the number of times it occurs. Then add the numbers in the last column to find the total. If par for the course was 72, what was Sergio Garcia's score?

	VALUE	NUMBER	PRODUCT
Eagle	−2	0	
Birdie	−1	7	
Par	0	7	
Bogie	+1	3	
Double Bogie	+2	1	
		Total:	

68. **Karrie Webb's Scorecard** The table below shows the scores Karrie Webb had on the final round of the LPGA Standard Register Ping Tournament in March 2000. Fill in the last column by multiplying each value by the number of times it occurs. Then add the numbers in the last column to find the total. If par for the course was 72, what was Karrie Webb's score?

	VALUE	NUMBER	PRODUCT
Eagle	−2	1	
Birdie	−1	5	
Par	0	8	
Bogie	+1	3	
Double Bogie	+2	1	
		Total:	

Estimating

Work Problems 69–76 mentally, without pencil and paper or a calculator.

69. The product $-31(-522)$ is closest to which of the following numbers?
    a. 15,000   **b.** −500   **c.** −1,500   **d.** −15,000

70. The product $31(-522)$ is closest to which of the following numbers?
    a. 15,000   **b.** −500   **c.** −1,500   **d.** −15,000

71. The product $-47(470)$ is closest to which of the following numbers?
    a. 25,000   **b.** 420   **c.** −2,500   **d.** −25,000

72. The product $-47(-470)$ is closest to which of the following numbers?
    a. 25,000   **b.** 420   **c.** −2,500   **d.** −25,000

73. The product $-222(-987)$ is closest to which of the following numbers?
    a. 200,000   **b.** 800   **c.** −800   **d.** −1,200

74. The sum $-222 + (-987)$ is closest to which of the following numbers?
    a. 200,000   **b.** 800   **c.** −800   **d.** −1,200

75. The difference $-222 - (-987)$ is closest to which of the following numbers?
    a. 200,000   **b.** 800   **c.** −800   **d.** −1,200

76. The difference $-222 - 987$ is closest to which of the following numbers?
    a. 200,000   **b.** 800   **c.** −800   **d.** −1,200

Review Problems

The problems below review the basic facts of division with whole numbers from Chapter 1.

Write each of the following statements in symbols.

77. The quotient of 12 and 6

78. The quotient of x and 5

Rewrite each of the following multiplication problems as an equivalent division problem.

79. $2(3) = 6$

80. $5 \cdot 4 = 20$

Rewrite each of the following division problems as an equivalent multiplication problem.

81. $10 \div 5 = 2$

82. $\dfrac{63}{9} = 7$

Divide.

83. $4{,}984 \div 56$

84. $4{,}994 \div 56$

Extending the Concepts

85. Of the following sentences, which are always true?
   a. The sum of two negative numbers is a negative number.
   b. Two negatives make a positive.
   c. The product of two negative numbers is always a positive number.

86. In your own words, explain why $2(-7)$ is -14. (Think in terms of repeated addition.)

87. Give an example of an everyday situation that is modeled by the multiplication problem $3(-\$4) = -\$12$.

88. In your own words, explain why $-2(-7)$ is 14.

In Chapter 1 we defined a geometric sequence to be a sequence of numbers in which each number, after the first number, is obtained from the previous number by multiplying by the same amount each time.

Find the next two terms in each of the following geometric sequences.

89. $2, -6, 18, \ldots$

90. $1, -4, 16, \ldots$

91. $-2, 6, -18, \ldots$

92. $-1, 4, -16, \ldots$

Simplify each of the following according to the rule for order of operations.

93. $5(-2)^2 - 3(-2)^3$

94. $8(-1)^3 - 6(-3)^2$

95. $7 - 3(4 - 8)$

96. $6 - 2(9 - 11)$

97. $5 - 2[3 - 4(6 - 8)]$

98. $7 - 4[6 - 3(2 - 9)]$

2.5 Division with Negative Numbers

Introduction . . .

If you and three friends bought equal shares of an investment for a total of $10,000, and then sold them later for only $9,000, how much did each person lose? Since the total loss can be represented by $-\$1,000$ and there are four people with equal shares, each person's loss can be found by division:

$$(-\$1,000) \div 4 = -\$250$$

From this example it seems reasonable to assume that a negative number divided by a positive number will give a negative answer.

To cover all the possible situations we can encounter with division of negative numbers, we use the relationship between multiplication and division. If we let n be the answer to the problem $12 \div (-2)$, then we know that

$$12 \div (-2) = n \quad \text{and} \quad -2(n) = 12$$

From our work with multiplication, we know that n must be -6 in the multiplication problem above, because -6 is the only number we can multiply -2 by to get 12. Because of the relationship between the two problems above, it must be true that 12 divided by -2 is -6.

The following pairs of problems show more quotients of positive and negative numbers. In each case the multiplication problem on the right justifies the answer to the division problem on the left.

$$
\begin{array}{llr}
6 \div 3 = 2 & \text{because} & 3(2) = 6 \\
6 \div (-3) = -2 & \text{because} & -3(-2) = 6 \\
-6 \div 3 = -2 & \text{because} & 3(-2) = -6 \\
-6 \div (-3) = 2 & \text{because} & -3(2) = -6
\end{array}
$$

The results given above can be used to write the rule for division with negative numbers.

RULE

To divide two numbers, we divide their absolute values.

1. The answer is *positive* if both the original numbers have the same sign. That is, the quotient of two numbers with the same signs is positive.

2. The answer is *negative* if the original two numbers have different signs. That is, the quotient of two numbers with different signs is negative.

EXAMPLE 1 $-12 \div 4 = -3$ Unlike signs; negative answer ■

EXAMPLE 2 $12 \div (-4) = -3$ Unlike signs; negative answer ■

EXAMPLE 3 $-12 \div (-4) = 3$ Like signs; positive answer ■

Practice Problems

Divide.

1. $-8 \div 2$

2. $8 \div (-2)$

3. $-8 \div (-2)$

Answers

1. -4 2. -4 3. 4

4. $\dfrac{20}{-5}$

EXAMPLE 4 $\dfrac{15}{-5} = -3$ Unlike signs; negative answer ■

5. $\dfrac{-30}{-5}$

EXAMPLE 5 $\dfrac{-20}{-4} = 5$ Like signs; positive answer ■

From the examples we have done so far, we can make the following generalization about quotients that contain negative signs:

If a and b are numbers and b is not equal to 0, then

$$\frac{-a}{b} = \frac{a}{-b} = -\frac{a}{b} \quad \text{and} \quad \frac{-a}{-b} = \frac{a}{b}$$

The last examples in this section involve more than one operation. We use the rules developed previously in this chapter and the rule for order of operations to simplify each.

6. Simplify: $\dfrac{8(-5)}{-4}$

EXAMPLE 6 Simplify: $\dfrac{6(-3)}{-2}$

SOLUTION We begin by multiplying 6 and -3:

$$\frac{6(-3)}{-2} = \frac{-18}{-2} \quad \textbf{Multiplication; } 6(-3) = -18$$
$$= 9 \quad \textbf{Like signs; positive answer} \quad ■$$

7. Simplify: $\dfrac{-20 + 6(-2)}{7 - 11}$

EXAMPLE 7 Simplify: $\dfrac{-15 + 5(-4)}{12 - 17}$

SOLUTION Simplifying above and below the fraction bar, we have

$$\frac{-15 + 5(-4)}{12 - 17} = \frac{-15 + (-20)}{-5}$$
$$= \frac{-35}{-5}$$
$$= 7 \quad ■$$

8. Simplify: $-3(4^2) + 10 \div (-5)$

EXAMPLE 8 Simplify: $-4(10^2) + 20 \div (-4)$

SOLUTION Applying the rule for order of operations, we have

$$-4(10^2) + 20 \div (-4) = -4(100) + 20 \div (-4) \quad \textbf{Exponents first}$$
$$= -400 + (-5) \quad \textbf{Multiply and divide}$$
$$= -405 \quad \textbf{Add} \quad ■$$

9. Simplify: $-80 \div 2 \div 10$

EXAMPLE 9 Simplify: $-80 \div 10 \div 2$

SOLUTION In a situation like this, the rule for order of operations states that we are to divide left to right.

$$-80 \div 10 \div 2 = -8 \div 2 \quad \textbf{Divide } -80 \textbf{ by 10}$$
$$= -4 \quad ■$$

Answers
4. -4 **5.** 6   **6.** 10   **7.** 8
8. -50 **9.** -4

Problem Set 2.5

Find each of the following quotients. (Divide.)

1. $-15 \div 5$

2. $15 \div (-3)$

3. $20 \div (-4)$

4. $-20 \div 4$

5. $-30 \div (-10)$

6. $-50 \div (-25)$

7. $\dfrac{-14}{-7}$

8. $\dfrac{-18}{-6}$

9. $\dfrac{12}{-3}$

10. $\dfrac{12}{-4}$

11. $\dfrac{-22}{11}$

12. $\dfrac{-35}{7}$

13. $\dfrac{0}{-3}$

14. $\dfrac{0}{-5}$

15. $-125 \div (-25)$

16. $-144 \div (-9)$

Complete the following tables.

17.

FIRST NUMBER	SECOND NUMBER	THE QUOTIENT OF a AND b
a	b	$\dfrac{a}{b}$
100	-5	
100	-10	
100	-25	
100	-50	

18.

FIRST NUMBER	SECOND NUMBER	THE QUOTIENT OF a AND b
a	b	$\dfrac{a}{b}$
24	-4	
24	-3	
24	-2	
24	-1	

19.

FIRST NUMBER	SECOND NUMBER	THE QUOTIENT OF a AND b
a	b	$\dfrac{a}{b}$
-100	-5	
-100	5	
100	-5	
100	5	

20.

FIRST NUMBER	SECOND NUMBER	THE QUOTIENT OF a AND b
a	b	$\dfrac{a}{b}$
-24	-2	
-24	-4	
-24	-6	
-24	-8	

Use any of the rules developed in this chapter and the rule for order of operations to simplify each of the following expressions as much as possible.

21. $\dfrac{4(-7)}{-28}$

22. $\dfrac{6(-3)}{-18}$

23. $\dfrac{-3(-10)}{-5}$

24. $\dfrac{-4(-12)}{-6}$

25. $\dfrac{2(-3)}{6-3}$

26. $\dfrac{2(-3)}{3-6}$

27. $\dfrac{4-8}{8-4}$

28. $\dfrac{9-5}{5-9}$

29. $\dfrac{2(-3)+10}{-4}$

30. $\dfrac{7(-2)-6}{-10}$

31. $\dfrac{2+3(-6)}{4-12}$

32. $\dfrac{3+9(-1)}{5-7}$

33. $\dfrac{6(-7) + 3(-2)}{20 - 4}$

34. $\dfrac{9(-8) + 5(-1)}{12 - 1}$

35. $\dfrac{3(-7)(-4)}{6(-2)}$

36. $\dfrac{-2(4)(-8)}{(-2)(-2)}$

37. $(-5)^2 + 20 \div 4$

38. $6^2 + 36 \div 9$

39. $100 \div (-5)^2$

40. $400 \div (-4)^2$

41. $-100 \div 10 \div 2$

42. $-500 \div 50 \div 10$

43. $-100 \div (10 \div 2)$

44. $-500 \div (50 \div 10)$

45. $(-100 \div 10) \div 2$

46. $(-500 \div 50) \div 10$

47. Find the quotient of -25 and 5.

48. Find the quotient of -38 and -19.

49. What number do you divide by -5 to get -7?

50. What number do you divide by 6 to get -7?

51. Subtract -3 from the quotient of 27 and 9.

52. Subtract -7 from the quotient of -72 and -9.

Estimating

Work Problems 53–60 mentally, without pencil and paper or a calculator.

53. Is $397 \div (-401)$ closer to 1 or -1?

54. Is $-751 \div (-749)$ closer to 1 or -1?

55. The quotient $-121 \div 27$ is closest to which of the following numbers?
   a. -150 **b.** -100 **c.** -4 **d.** 6

56. The quotient $1,000 \div (-337)$ is closest to which of the following numbers?
   a. 663 **b.** -3 **c.** -30 **d.** -663

57. Which number is closest to the sum $-151 + (-49)$?
   a. -200 **b.** -100 **c.** 3 **d.** $7,500$

58. Which number is closest to the difference $-151 - (-49)$?
   a. -200 **b.** -100 **c.** 3 **d.** $7,500$

59. Which number is closest to the product of $-151(-49)$?
   a. -200 **b.** -100 **c.** 3 **d.** $7,500$

60. Which number is closest to the quotient $-151 \div (-49)$?
   a. -200 **b.** -100 **c.** 3 **d.** $7,500$

Applying the Concepts

61. Mean Find the mean of the numbers −5, 0, 5.

62. Median Find the median of the numbers −5, 0, 5.

63. Averages For the numbers −5, −4, 0, 2, 2, 2, 3, find
   a. the mean
   b. the median
   c. the mode

64. Averages For the numbers −5, −2, −2, 0, 2, 7, find
   a. the mean
   b. the median
   c. the mode

65. Temperature Line Graph The table below gives the low temperature for each day of one week in White Bear Lake, Minnesota. Use the diagram in the figure to draw a line graph of the information in the table.

LOW TEMPERATURES IN WHITE BEAR LAKE, MINNESOTA	
DAY	TEMPERATURE
Monday	10°F
Tuesday	8°F
Wednesday	−5°F
Thursday	−3°F
Friday	−8°F
Saturday	5°F
Sunday	7°F

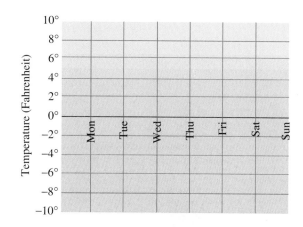

66. Temperature Line Graph The table below gives the low temperature for each day of one week in Fairbanks, Alaska. Use the diagram in the figure to draw a line graph of the information in the table.

LOW TEMPERATURES IN FAIRBANKS, ALASKA	
DAY	TEMPERATURE
Monday	−26°F
Tuesday	−5°F
Wednesday	9°F
Thursday	12°F
Friday	3°F
Saturday	−15°F
Sunday	−20°F

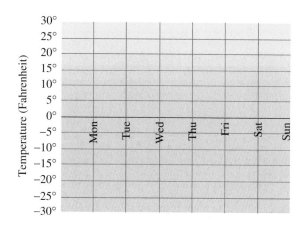

67. Average Temperatures Use the information in the table in Problem 65 to find
   a. the mean low temperature for the week
   b. the median low temperature for the week

68. Average Temperatures Use the information in the table in Problem 66 to find
   a. the mean low temperature for the week
   b. the median low temperature for the week

Review Problems

The problems below review some of the properties of addition and multiplication we covered in Chapter 1.

Rewrite each expression using the commutative property of addition or multiplication.

69. $3 + x$

70. $4y$

Rewrite each expression using the associative property of addition or multiplication.

71. $5 + (7 + a)$

72. $(x + 4) + 6$

73. $3(4y)$

74. $(3y)8$

Apply the distributive property to each expression.

75. $5(3 + 7)$

76. $8(4 + 2)$

Simplify.

77. $a + 0 =$

78. $0 + a =$

79. $1 \cdot x =$

80. $1 \cdot y =$

Extending the Concepts

81. Give an example that shows that division is not a commutative operation.

82. Give an example that shows that division is not an associative operation.

83. Give an example of an everyday situation that is modeled by the division problem $-12 \div 2 = -6$.

84. In your own words, explain why the expression $\frac{5}{0}$ is not defined.

Find the next term in each sequence below.

85. $32, -16, 8, \ldots$

86. $243, -81, 27, \ldots$

87. $-32, 16, -8, \ldots$

88. $-243, 81, -27, \ldots$

Simplify each of the following expressions.

89. $\dfrac{-5 - 3}{-5 + 3}$

90. $\dfrac{5 - 3 - 2}{-5 - 3 + 2}$

91. $\dfrac{6 - 3(2 - 11)}{6 - 3(2 + 11)}$

92. $\dfrac{8 + 4(3 - 5)}{8 - 4(3 + 5)}$

93. $\dfrac{6 - (3 - 4) - 3}{1 - 2 - 3}$

94. $\dfrac{7 - (3 - 6) - 4}{-1 - 2 - 3}$

2.6 Simplifying Algebraic Expressions

In this section we want to combine some of the work we have done in this chapter and in Chapter 1 to simplify expressions containing variables—that is, algebraic expressions.

To begin let's review how we use the associative properties for addition and multiplication to simplify expressions.

Consider the expression $4(5x)$. We can apply the associative property of multiplication to this expression to change the grouping so that the 4 and the 5 are grouped together, instead of the 5 and the x. Here's how it looks:

$$4(5x) = (4 \cdot 5)x \quad \textbf{Associative property}$$
$$= 20x \quad \textbf{Multiplication: } 4 \cdot 5 = 20$$

We have simplified the expression to $20x$, which in most cases in algebra will be easier to work with than the original expression.

Here are some more examples.

EXAMPLE 1 $\quad 7(3a) = (7 \cdot 3)a \quad$ **Associative property**
$$= 21a \quad \textbf{7 times 3 is 21}$$

EXAMPLE 2 $\quad -2(5x) = (-2 \cdot 5)x \quad$ **Associative property**
$$= -10x \quad \textbf{The product of } -2 \text{ and 5 is } -10$$

EXAMPLE 3 $\quad 3(-4y) = [3(-4)]y \quad$ **Associative property**
$$= -12y \quad \textbf{3 times } -4 \text{ is } -12$$

We can use the associative property of addition to simplify expressions also.

EXAMPLE 4 $\quad 3 + (8 + x) = (3 + 8) + x \quad$ **Associative property**
$$= 11 + x \quad \textbf{The sum of 3 and 8 is 11}$$

EXAMPLE 5 $\quad (2x + 5) + 10 = 2x + (5 + 10) \quad$ **Associative property**
$$= 2x + 15 \quad \textbf{Addition}$$

In Chapter 1 we introduced the distributive property. In symbols it looks like this:

$$a(b + c) = ab + ac$$

Because subtraction is defined as addition of the opposite, the distributive property holds for subtraction as well as addition. That is,

$$a(b - c) = ab - ac$$

We say that multiplication distributes over addition and subtraction. Here are some examples that review how the distributive property is applied to expressions that contain variables.

EXAMPLE 6 $\quad 4(x + 5) = 4(x) + 4(5) \quad$ **Distributive property**
$$= 4x + 20 \quad \textbf{Multiplication}$$

Practice Problems

Multiply.
1. $5(7a)$

2. $-3(9x)$

3. $5(-8y)$

Simplify.
4. $6 + (9 + x)$

5. $(3x + 7) + 4$

Apply the distributive property.
6. $6(x + 4)$

Answers
1. $35a$ 2. $-27x$ 3. $-40y$
4. $15 + x$ 5. $3x + 11$
6. $6x + 24$

7. $7(a - 5)$

8. $6(4x + 5)$

9. $3(8a - 4)$

10. $8(3x + 4y)$

EXAMPLE 7 $2(a - 3) = 2(a) - 2(3)$ **Distributive property**
$$= 2a - 6 \qquad \textbf{Multiplication}$$

In Examples 1–3 we simplified expressions such as $4(5x)$ by using the associative property. Here are some examples that use a combination of the associative property and the distributive property.

EXAMPLE 8 $4(5x + 3) = 4(5x) + 4(3)$ **Distributive property**
$$= (4 \cdot 5)x + 4(3) \qquad \textbf{Associative property}$$
$$= 20x + 12 \qquad \textbf{Multiplication}$$

EXAMPLE 9 $7(3a - 6) = 7(3a) - 7(6)$ **Distributive property**
$$= 21a - 42 \qquad \textbf{Associative property and multiplication}$$

EXAMPLE 10 $5(2x + 3y) = 5(2x) + 5(3y)$ **Distributive property**
$$= 10x + 15y \qquad \textbf{Associative property and multiplication}$$

We can also use the distributive property to simplify expressions like $4x + 3x$. Because multiplication is a commutative operation, we can also rewrite the distributive property like this:

$$b \cdot a + c \cdot a = (b + c)a$$

Applying the distributive property in this form to the expression $4x + 3x$, we have

$$4x + 3x = (4 + 3)x \qquad \textbf{Distributive property}$$
$$= 7x \qquad \textbf{Addition}$$

Similar Terms

Expressions like $4x$ and $3x$ are called *similar terms* because the variable parts are the same. Some other examples of similar terms are $5y$ and $-6y$ and the terms $7a$, $-13a$, $\frac{3}{4}a$. To simplify an algebraic expression (an expression that involves both numbers and variables), we combine similar terms by applying the distributive property. Table 1 shows several pairs of similar terms and how they can be combined using the distributive property.

Table 1

ORIGINAL EXPRESSION		APPLY DISTRIBUTIVE PROPERTY		SIMPLIFIED EXPRESSION
$4x + 3x$	=	$(4 + 3)x$	=	$7x$
$7a + a$	=	$(7 + 1)a$	=	$8a$
$-5x + 7x$	=	$(-5 + 7)x$	=	$2x$
$8y - y$	=	$(8 - 1)y$	=	$7y$
$-4a - 2a$	=	$(-4 - 2)a$	=	$-6a$
$3x - 7x$	=	$(3 - 7)x$	=	$-4x$

As you can see from the table, the distributive property can be applied to any combination of positive and negative terms so long as they are similar terms.

Answers

7. $7a - 35$ **8.** $24x + 30$
9. $24a - 12$ **10.** $24x + 32y$

Exponents and Variables

Next, we want to simplify products like $2x(3x)$. Because the problem involves only multiplication, the commutative property tells us we can change the order of the numbers and variables. Likewise, the associative property tells us we can change the grouping of the numbers and variables. We use the two properties together to group the numbers separately from the variables.

$$2x(3x) = (2 \cdot 3)(x \cdot x) \qquad \textbf{Commutative and associative properties}$$
$$= 6x^2 \qquad \textbf{2 · 3 = 6 and } x \cdot x = x^2$$

Note that the expression $6x^2$ means $6 \cdot x \cdot x$, not $6x \cdot 6x$. The exponent 2 in the expression $6x^2$ is associated only with the number directly to its left, unless parentheses are used. That is:

$$6x^2 = 6 \cdot x \cdot x \qquad \text{and} \qquad (6x)^2 = (6x)(6x)$$

EXAMPLE 11 Expand and simplify: $(5a)^3$

SOLUTION We begin by writing the expression as $(5a)(5a)(5a)$:

$$(5a)^3 = (5a)(5a)(5a) \qquad \textbf{Definition of exponents}$$
$$= (5 \cdot 5 \cdot 5)(a \cdot a \cdot a) \qquad \textbf{Commutative and associative properties}$$
$$= 125a^3 \qquad \textbf{5 · 5 · 5 = 125; } a \cdot a \cdot a = a^3 \qquad ■$$

EXAMPLE 12 Expand and simplify: $(8xy)^2$

SOLUTION Proceeding as we did above, we have

$$(8xy)^2 = (8xy)(8xy) \qquad \textbf{Definition of exponents}$$
$$= (8 \cdot 8)(x \cdot x)(y \cdot y) \qquad \textbf{Commutative and associative properties}$$
$$= 64x^2y^2 \qquad \textbf{8 · 8 = 64; } x \cdot x = x^2; y \cdot y = y^2 \qquad ■$$

EXAMPLE 13 Simplify: $(7x)^2(8xy)^2$

SOLUTION We begin by applying the definition of exponents:

$$(7x)^2(8xy)^2 = (7x)(7x)(8xy)(8xy)$$
$$= (7 \cdot 7 \cdot 8 \cdot 8)(x \cdot x \cdot x \cdot x)(y \cdot y) \qquad \textbf{Commutative and associative properties}$$
$$= 3{,}136x^4y^2 \qquad ■$$

EXAMPLE 14 Simplify: $(2x)^3(4x)^2$

SOLUTION Proceeding as we did above, we have

$$(2x)^3(4x)^2 = (2x)(2x)(2x)(4x)(4x)$$
$$= (2 \cdot 2 \cdot 2 \cdot 4 \cdot 4)(x \cdot x \cdot x \cdot x \cdot x)$$
$$= 128x^5 \qquad ■$$

11. Expand and simplify: $(2a)^3$

12. Expand and simplify: $(7xy)^2$

13. Simplify: $(3x)^2(7xy)^2$

14. Simplify: $(5x)^3(2x)^2$

Answers
11. $8a^3$ **12.** $49x^2y^2$ **13.** $441x^4y^2$
14. $500x^5$

Algebraic Expressions Representing Area and Perimeter

Below are a square with a side of length s and a rectangle with a length of l and a width of w. The table that follows the figures gives the formulas for the area and perimeter of each.

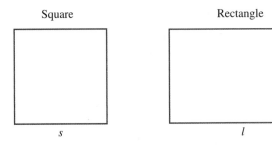

Square

Rectangle

w

s

l

	SQUARE	RECTANGLE
Area A	s^2	lw
Perimeter P	$4s$	$2l + 2w$

15. Find the area and perimeter of a square if its side is 12 feet long.

EXAMPLE 15 Find the area and perimeter of a square with a side 6 inches long.

SOLUTION   Substituting 6 for s in the formulas for area and perimeter of a square, we have

$$\text{Area} = A = s^2 = 6^2 = 36 \text{ square inches}$$
$$\text{Perimeter} = P = 4s = 4(6) = 24 \text{ inches}$$ ∎

16. A football field is 100 yards long and approximately 53 yards wide. Find the area and perimeter.

EXAMPLE 16 A soccer field is 100 yards long and 75 yards wide. Find the area and perimeter.

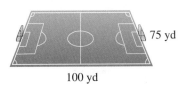

75 yd

100 yd

SOLUTION   Substituting 100 for l and 75 for w in the formulas for area and perimeter of a rectangle, we have

$$\text{Area} = A = lw = 100(75) = 7,500 \text{ square yards}$$
$$\text{Perimeter} = P = 2l + 2w = 2(100) + 2(75) = 200 + 150 = 350 \text{ yards}$$ ∎

Problem Set 2.6

Apply the associative property to each expression, and then simplify the result.

1. $5(4a)$

2. $8(9a)$

3. $6(8a)$

4. $3(2a)$

5. $-6(3x)$

6. $-2(7x)$

7. $-3(9x)$

8. $-4(6x)$

9. $5(-2y)$

10. $3(-8y)$

11. $6(-10y)$

12. $5(-5y)$

13. $2 + (3 + x)$

14. $9 + (6 + x)$

15. $5 + (8 + x)$

16. $3 + (9 + x)$

17. $4 + (6 + y)$

18. $2 + (8 + y)$

19. $7 + (1 + y)$

20. $4 + (1 + y)$

21. $(5x + 2) + 4$

22. $(8x + 3) + 10$

23. $(6y + 4) + 3$

24. $(3y + 7) + 8$

25. $(12a + 2) + 19$

26. $(6a + 3) + 14$

27. $(7x + 8) + 20$

28. $(14x + 3) + 15$

Apply the distributive property to each expression, and then simplify.

29. $7(x + 5)$

30. $8(x + 3)$

31. $6(a - 7)$

32. $4(a - 9)$

33. $2(x - y)$

34. $5(x - a)$

35. $4(5 + x)$

36. $8(3 + x)$

37. $3(2x + 5)$

38. $8(5x + 4)$

39. $6(3a + 1)$

40. $4(8a + 3)$

41. $2(6x - 3y)$

42. $7(5x - y)$

43. $5(7 - 4y)$

44. $8(6 - 3y)$

Use the distributive property to combine similar terms. (See Table 1.)

45. $3x + 5x$

46. $7x + 8x$

47. $3a + a$

48. $8a + a$

49. $-2x + 6x$

50. $-3x + 9x$

51. $6y - y$

52. $3y - y$

53. $-8a - 2a$

54. $-7a - 5a$

55. $4x - 9x$

56. $5x - 11x$

Use the definition of exponents to expand each of the following expressions. Apply the commutative and associative properties, and simplify the results in each case.

57. $(6x)^2$

58. $(9x)^2$

59. $(4x)^2$

60. $(10x)^2$

61. $(3a)^3$

62. $(6a)^3$

63. $(2ab)^3$

64. $(5ab)^3$

65. $(9xy)^2$

66. $(5xy)^2$

67. $(5xyz)^2$

68. $(7xyz)^2$

69. $(4x)^2(9xy)^2$

70. $(10x)^2(5xy)^2$

71. $(2x)^2(3x)^2(4x)^2$

72. $(5x)^2(2x)^2(10x)^2$

73. $(2x)^3(5x)^2$

74. $(3x)^3(4x)^2$

75. $(2a)^3(3a)^2(10a)^2$

76. $(3a)^2(2a)^2(10a)^2$

Applying the Concepts

Area and Perimeter Find the area and perimeter of each square if the length of each side is as given below.

77. $s = 6$ feet

78. $s = 14$ yards

79. $s = 9$ inches

80. $s = 15$ meters

Area and Perimeter Find the area and perimeter for a rectangle if the length and width are as given below.

81. $l = 20$ inches, $w = 10$ inches

82. $l = 40$ yards, $w = 20$ yards

83. $l = 25$ feet, $w = 12$ feet

84. $l = 210$ meters, $w = 120$ meters

Temperature Scales In the metric system, the scale we use to measure temperature is the Celsius scale. On this scale water boils at 100 degrees and freezes at 0 degrees. When we write 100 degrees measured on the Celsius scale, we use the notation 100°C, which is read "100 degrees Celsius." If we know the temperature in degrees Fahrenheit, we can convert to degrees Celsius by using the formula

$$\frac{5(F - 32)}{9}$$

where F is the temperature in degrees Fahrenheit. Use this formula to find the temperature in degrees Celsius for each of the following Fahrenheit temperatures.

85. 68°F

86. 59°F

87. 41°F

88. 23°F

89. 14°F

90. 32°F

Review Problems

The problems below review addition, subtraction, multiplication, and division of positive and negative numbers, as covered in this chapter.

Perform the indicated operations.

91. $8 + (-4)$

92. $-8 + 4$

93. $-8 + (-4)$

94. $-8 - 4$

95. $8 - (-4)$

96. $-8 - (-4)$

97. $8(-4)$

98. $-8(4)$

99. $-8(-4)$

100. $8 \div (-4)$

101. $-8 \div 4$

102. $-8 \div (-4)$

2.7 Multiplication Properties of Exponents

In this section we will extend our work with exponents to develop some general properties of exponents that we can use to simplify expressions. Each of the properties we will develop has its foundation in the definition for exponents.

EXAMPLE 1   Multiply: $x^2 \cdot x^4$

SOLUTION   We write x^2 as $x \cdot x$ and x^4 as $x \cdot x \cdot x \cdot x$ and then use the definition of exponents to write the answer with just one exponent.

$$
\begin{aligned}
x^2 \cdot x^4 &= (x \cdot x)(x \cdot x \cdot x \cdot x) && \textbf{Expand exponents} \\
&= x \cdot x \cdot x \cdot x \cdot x \cdot x \\
&= x^6 && \textbf{Write with a single exponent} \qquad \blacksquare
\end{aligned}
$$

Notice that the two exponents in the original problem given in Example 1 add up to the exponent in the answer: $2 + 4 = 6$. We use this result as justification for writing our first property of exponents.

Property 1 for Exponents

If r and s are any two whole numbers and a is an integer, then

$$a^r \cdot a^s = a^{r+s}$$

In words: To multiply two expressions with the same base, add exponents and use the common base.

EXAMPLE 2   Multiply: $5x^3 \cdot 3x^2$

SOLUTION   We apply the commutative and associative properties so that the numbers 5 and 3 are together, and the x's are also. Then we use Property 1 for exponents to add exponents.

$$
\begin{aligned}
5x^3 \cdot 3x^2 &= (5 \cdot 3)(x^3 \cdot x^2) && \textbf{Commutative and associative properties} \\
&= (5 \cdot 3)(x^{3+2}) && \textbf{Property 1 for exponents} \\
&= 15x^5 && \textbf{Multiply 3 and 5; add 3 and 2} \qquad \blacksquare
\end{aligned}
$$

EXAMPLE 3   Multiply: $4x^3 \cdot 3x^4 \cdot 5x^2$

SOLUTION   Here we have the product of three expressions with exponents. Using the same steps shown in Example 2, we have

$$
\begin{aligned}
4x^3 \cdot 3x^4 \cdot 5x^2 &= (4 \cdot 3 \cdot 5)(x^3 \cdot x^4 \cdot x^2) \\
&= (4 \cdot 3 \cdot 5)(x^{3+4+2}) \\
&= 60x^9 \qquad \blacksquare
\end{aligned}
$$

Practice Problems

1. Multiply: $x^3 \cdot x^5$

2. Multiply: $4x^2 \cdot 6x^5$

3. Multiply: $2x^5 \cdot 3x^4 \cdot 4x^3$

Answers

1. x^8 **2.** $24x^7$ **3.** $24x^{12}$

4. Multiply: $(10xy^2)(9x^3y^5)$

EXAMPLE 4   Multiply: $(2x^2y^3)(7x^4y)$

SOLUTION   This time we have two different variables to work with. We use the commutative and associative properties to group the numbers together, the x's together, and the y's together.

$$(2x^2y^3)(7x^4y) = (2 \cdot 7)(x^2 \cdot x^4)(y^3 \cdot y)$$
$$= (2 \cdot 7)(x^{2+4})(y^{3+1})$$
$$= 14x^6y^4 \qquad \blacksquare$$

Another common expression in algebra is that of a power raised to another power—for example, $(x^2)^3$, $(10^4)^5$, and $(y^6)^7$. To see how these expressions can be written with a single exponent, we return to the definition of exponents.

$$(x^2)^3 = x^2 \cdot x^2 \cdot x^2 \qquad \textbf{Definition of exponents}$$
$$= x^{2+2+2} \qquad \textbf{Property 1 for exponents}$$
$$= x^6 \qquad \textbf{Add exponents}$$

As you can see, to obtain the exponent in the answer, we multiply the exponents in the original problem. The result leads us to our second property of exponents.

Property 2 for Exponents

If r and s are any two whole numbers and a is an integer, then

$$(a^r)^s = a^{r \cdot s}$$

In words: A power raised to another power is the base raised to the product of the powers.

---

5. Simplify the expression: $(2^2)^3$

EXAMPLE 5   Simplify the expression $(2^3)^2$.

SOLUTION   We apply Property 2 by multiplying the exponents; then we simplify if we can.

$$(2^3)^2 = 2^{3 \cdot 2} \qquad \textbf{Property 2 for exponents}$$
$$= 2^6 \qquad \textbf{Multiply exponents}$$
$$= 64 \qquad \blacksquare$$

6. Simplify: $(x^3)^4 \cdot (x^5)^2$

EXAMPLE 6   Simplify: $(x^5)^6 \cdot (x^4)^7$

SOLUTION   In this case we apply Property 2 for exponents first to write each expression with a single exponent. Then we apply Property 1 to simplify further.

$$(x^5)^6 \cdot (x^4)^7 = x^{5 \cdot 6} \cdot x^{4 \cdot 7} \qquad \textbf{Property 2 for exponents}$$
$$= x^{30} \cdot x^{28} \qquad \textbf{Multiply exponents}$$
$$= x^{30+28} \qquad \textbf{Property 1 for exponents}$$
$$= x^{58} \qquad \textbf{Add exponents} \qquad \blacksquare$$

Answers
4. $90x^4y^7$ **5.** 64   **6.** x^{22}

The third, and final, property of exponents for this section covers the situation that occurs when we have the product of more than one number or variable, all raised to a power. Expressions such as $(4x)^2$, $(2ab)^3$, and $(5x^2y^3)^5$ all fall into this category.

$$
\begin{aligned}
(4x)^2 &= (4x)(4x) && \textbf{Definition of exponents} \\
&= (4 \cdot 4)(x \cdot x) && \textbf{Commutative and associative properties} \\
&= 4^2 \cdot x^2 && \textbf{Definition of exponents}
\end{aligned}
$$

As you can see, both the numbers, 4 and x, contained within the parentheses end up raised to the second power.

Property 3 for Exponents

If r is a whole number and a and b are integers, then

$$(a \cdot b)^r = a^r \cdot b^r$$

EXAMPLE 7 Simplify: $(3xy)^4$

SOLUTION Applying our new property, we have

$$
\begin{aligned}
(3xy)^4 &= 3^4 \cdot x^4 \cdot y^4 && \textbf{Property 3 for exponents} \\
&= 81x^4y^4 && \textbf{Simplify} \quad \blacksquare
\end{aligned}
$$

EXAMPLE 8 Simplify: $(2x^3y^2)^5$

SOLUTION In this case we will have to apply Property 3 first and then Property 2, in order to simplify the expression.

$$
\begin{aligned}
(2x^3y^2)^5 &= 2^5(x^3)^5(y^2)^5 && \textbf{Property 3 for exponents} \\
&= 32x^{15}y^{10} && \textbf{Property 2 for exponents} \quad \blacksquare
\end{aligned}
$$

EXAMPLE 9 Simplify: $(4x^3y)^2(5x^2y^4)^3$

SOLUTION This simplification will require all three properties of exponents, as well as the commutative and associative properties of multiplication. Here are all the steps:

$$
\begin{aligned}
(4x^3y)^2(5x^2y^4)^3 &= 4^2(x^3)^2y^2 \cdot 5^3(x^2)^3(y^4)^3 && \textbf{Property 3 for exponents} \\
&= 16x^6y^2 \cdot 125x^6y^{12} && \textbf{Property 2 for exponents} \\
&= (16 \cdot 125)(x^6x^6)(y^2y^{12}) && \textbf{Commutative and associative} \\
&&& \textbf{properties} \\
&= 2{,}000x^{12}y^{14} && \textbf{Property 1 for exponents} \quad \blacksquare
\end{aligned}
$$

7. Simplify: $(5xy)^3$

8. Simplify: $(4x^5y^2)^3$

9. Simplify: $(2x^3y^4)^3(3xy^5)^2$

Answers
7. $125x^3y^3$ **8.** $64x^{15}y^6$
9. $72x^{11}y^{22}$

Problem Set 2.7

Multiply.

1. $x^3 \cdot x^5$

2. $x^4 \cdot x^7$

3. $a^3 \cdot a^5$

4. $a^7 \cdot a^5$

5. $y^{10} \cdot y^5$

6. $y^{12} \cdot y^8$

7. $x^4 \cdot x$

8. $x^6 \cdot x$

9. $x^2 \cdot x^3 \cdot x^4$

10. $x^3 \cdot x^5 \cdot x^7$

11. $a^6 \cdot a^4 \cdot a^2$

12. $a^5 \cdot a^4 \cdot a^3$

13. $y^3 \cdot y^2 \cdot y$

14. $y^5 \cdot y^3 \cdot y$

15. $2x^4 \cdot 7x^5$

16. $6x^2 \cdot 9x^4$

17. $3y^5 \cdot 8y^3$

18. $2y^4 \cdot 7y^4$

19. $7x \cdot 3x$

20. $6x \cdot 2x$

21. $4x^3 \cdot 6x^5 \cdot 8x^7$

22. $3x^4 \cdot 5x^6 \cdot 7x^8$

23. $2a \cdot 3a \cdot 4a$

24. $10a \cdot 9a \cdot 8a$

25. $5y^4 \cdot 5y^4 \cdot 5y^4$

26. $4y^2 \cdot 4y^2 \cdot 4y^2$

27. $(3x^2y^7)(2x^3y)$

28. $(4x^5y^2)(3x^2y^2)$

29. $(7a^3b^2)(8ab^3)$

30. $(9a^4b^4)(10ab^3)$

31. $(2ab^3)(3a^2b^2)(4a^3b)$

32. $(8a^4b^4)(5a^3b^5)(a^2b^6)$

33. $(7a^3b^5)(4a^4b^4)(a^5b^3)$

34. $(6a^6b^3)(3a^3b^6)(ab)$

35. $(x^2)^5$

36. $(x^5)^2$

37. $(a^3)^7$

38. $(a^7)^3$

39. $(5y^4)^3$

40. $(4y^2)^3$

41. $(9x^4)^2$

42. $(10x^5)^2$

43. $(2ab)^3$

44. $(3ab)^2$

45. $(5x^2y^3)^4$

46. $(3x^3y^2)^4$

47. $(x^2)^3 \cdot (x^4)^2$

48. $(x^3)^5 \cdot (x^2)^6$

49. $(a^3)^7 \cdot (a^4)^2$

50. $(a^4)^5 \cdot (a^3)^6$

51. $(3x^3y^5)^2(2x^2y^4)^3$

52. $(5x^4y^3)^2(2x^3y^2)^3$

53. $(5a^2b^4)^2(7a^8b)$

54. $(9a^5b^6)^2(8a^5b)$

55. $(a^2b^4)^3(ab^5)^4(a^4b)^2$

56. $(a^5b^3)^2(ab^3)^4(a^3b)^5$

Review Problems

The problems that follow review material we covered in Section 1.3.

Find the perimeter of each figure.

57.

Square

13 ft

58.

Rectangle

8 ft

12 ft

59.

Parallelogram   18 ft

24 ft

60.

10 ft

8 ft   Trapezoid   8 ft

12 ft

61.

3 in.   4 in.

Triangle

2 in.

62.

8 in.   8 in.

Triangle

8 in.

2.8 Adding and Subtracting Polynomials

Previously we wrote numbers in expanded form to show the place value of each of the digits. For example, the number 456 in expanded form looks like this:

$$456 = 4 \cdot 100 + 5 \cdot 10 + 6 \cdot 1$$

If we replace 100 with 10^2, we have

$$456 = 4 \cdot 10^2 + 5 \cdot 10 + 6 \cdot 1$$

If we replace the 10's with x's, we get what is called a *polynomial*. It looks like this:

$$4x^2 + 5x + 6$$

Polynomials are to algebra what whole numbers written in expanded form are to arithmetic. As in other expressions in algebra, we can use any variable we choose. Here are some other examples of polynomials:

$$5a + 3 \qquad y^2 - 2y + 4 \qquad x^3 - 2x^2 + 5x - 1$$

Adding Polynomials

When we add two whole numbers, we add in columns. That is, if we add 234 and 345, we write one number under the other and add the numbers in the ones column, then the numbers in the tens column, and finally the numbers in the hundreds column. Here is how it looks:

$$
\begin{array}{ll}
234 & 2 \cdot 10^2 + 3 \cdot 10 + 4 \\
\underline{345} & \underline{3 \cdot 10^2 + 4 \cdot 10 + 5} \\
579 & 5 \cdot 10^2 + 7 \cdot 10 + 9
\end{array}
$$

We add polynomials in the same manner. If we want to add $2x^2 + 3x + 4$ and $3x^2 + 4x + 5$, we write one polynomial under the other, and then add in columns:

$$
\begin{array}{l}
2x^2 + 3x + 4 \\
\underline{3x^2 + 4x + 5} \\
5x^2 + 7x + 9
\end{array}
$$

The sum of the two polynomials is the polynomial $5x^2 + 7x + 9$. We add only the digits. Notice that the variable parts (the x's) stay the same, just as the powers of 10 did when we added 234 and 345. The reason we add the numbers, while the variable parts of each term stay the same, can be explained with the commutative, associative, and distributive properties. Here is the same problem again, but this time we show the properties in use:

$$
\begin{aligned}
(2x^2 + 3x &+ 4) + (3x^2 + 4x + 5) \\
&= (2x^2 + 3x^2) + (3x + 4x) + (4 + 5) \qquad &&\textbf{Commutative and} \\
& &&\textbf{associative properties} \\
&= (2 + 3)x^2 + (3 + 4)x + (4 + 5) \qquad &&\textbf{Distributive property} \\
&= 5x^2 + 7x + 9 \qquad &&\textbf{Addition}
\end{aligned}
$$

Practice Problems

1. Add $5x^2 - 3x + 2$ and $2x^2 + 10x - 9$.

EXAMPLE 1   Add $3x^2 - 2x + 1$ and $5x^2 + 3x - 4$.

SOLUTION   We will work the problem two ways. First, we write one polynomial under the other, and add in columns:

$$\begin{array}{r} 3x^2 - 2x + 1 \\ \underline{5x^2 + 3x - 4} \\ 8x^2 + 1x - 3 \end{array}$$

The sum of the two polynomials is $8x^2 + x - 3$.

Next we add horizontally, showing the commutative and associative properties in the first step, then the distributive property in the second step:

$$(3x^2 - 2x + 1) + (5x^2 + 3x - 4)$$
$$= (3x^2 + 5x^2) + (-2x + 3x) + (1 - 4)$$
$$= (3 + 5)x^2 + (-2 + 3)x + (1 - 4)$$
$$= 8x^2 + 1x + (-3)$$
$$= 8x^2 + x - 3 \qquad \blacksquare$$

2. Add $3y^2 + 9y - 5$ and $6y^2 - 4$.

EXAMPLE 2   Add $6y^2 + 4y - 3$ and $2y^2 - 4$.

SOLUTION   We write one polynomial under the other, so that the terms with y^2 line up, and the terms without any y's line up:

$$\begin{array}{r} 6y^2 + 4y - 3 \\ \underline{2y^2 \qquad - 4} \\ 8y^2 + 4y - 7 \end{array}$$

The same problem, written horizontally, looks like this:

$$(6y^2 + 4y - 3) + (2y^2 - 4) = (6y^2 + 2y^2) + 4y + (-3 - 4)$$
$$= (6 + 2)y^2 + 4y + (-3 - 4)$$
$$= 8y^2 + 4y + (-7)$$
$$= 8y^2 + 4y - 7 \qquad \blacksquare$$

3. Add $8x^3 + 4x^2 + 3x + 2$ and $4x^2 + 5x + 6$.

EXAMPLE 3   Add $2x^3 + 5x^2 + 3x + 4$ and $3x^2 + 2x + 1$.

SOLUTION   Showing only the vertical method, we line up terms with the same variable part and add:

$$\begin{array}{r} 2x^3 + 5x^2 + 3x + 4 \\ \underline{3x^2 + 2x + 1} \\ 2x^3 + 8x^2 + 5x + 5 \end{array} \qquad \blacksquare$$

Subtracting Polynomials

If there is a negative sign directly preceding the parentheses surrounding a polynomial, we may remove the parentheses and preceding negative sign by applying the distributive property. For example, to remove the parentheses from the expression

$$-(3x + 4)$$

we think of the negative sign as representing -1. Doing so allows us to apply the distributive property:

$$-(3x + 4) = -1(3x + 4)$$
$$= -1(3x) + (-1)(4) \qquad \textbf{Distributive property}$$
$$= -3x + (-4) \qquad \textbf{Multiply}$$
$$= -3x - 4 \qquad \textbf{Simplify}$$

Answers

1. $7x^2 + 7x - 7$
2. $9y^2 + 9y - 9$
3. $8x^3 + 8x^2 + 8x + 8$

As a result, the sign of each term inside the parentheses changes. Without showing all the steps involved in removing the parentheses, here are some more examples:

$$-(2x - 8) = -2x + 8$$
$$-(x^2 + 2x + 3) = -x^2 - 2x - 3$$
$$-(-4x^2 + 5x - 7) = 4x^2 - 5x + 7$$
$$-(3y^3 - 6y^2 + 7y - 3) = -3y^3 + 6y^2 - 7y + 3$$

In each case we remove the parentheses and preceding negative sign by changing the sign of each term that is found within the parentheses.

EXAMPLE 4 Subtract: $(6x^2 - 3x + 5) - (3x^2 + 2x - 3)$

SOLUTION Because subtraction is addition of the opposite, we simply change the sign of each term of the second polynomial and then add. With subtraction, there is less chance of making mistakes if we add horizontally, rather than in columns.

$$(6x^2 - 3x + 5) - (3x^2 + 2x - 3) \qquad \text{\bf Subtraction}$$
$$= 6x^2 - 3x + 5 - 3x^2 - 2x + 3 \qquad \text{\bf Addition of the opposite}$$
$$= (6x^2 - 3x^2) + (-3x - 2x) + (5 + 3)$$
$$= (6 - 3)x^2 + (-3 - 2)x + (5 + 3) \qquad \text{\bf Distributive property}$$
$$= 3x^2 - 5x + 8 \qquad \blacksquare$$

EXAMPLE 5 Subtract: $(2y^3 - 3y^2 - 4y - 2) - (3y^3 - 6y^2 + 7y - 3)$

SOLUTION Again, to subtract, we add the opposite of the polynomial that follows the subtraction sign. That is, we change the sign of each term in the second polynomial; then we combine similar terms.

$$(2y^3 - 3y^2 - 4y - 2) - (3y^3 - 6y^2 + 7y - 3) \qquad \text{\bf Subtraction}$$
$$= 2y^3 - 3y^2 - 4y - 2 - 3y^3 + 6y^2 - 7y + 3$$
$$= -y^3 + 3y^2 - 11y + 1 \qquad \blacksquare$$

EXAMPLE 6 Subtract $4x^2 - 3x + 1$ from $-3x^2 + 5x - 2$.

SOLUTION We must supply our own subtraction sign and write the two polynomials in the correct order.

$$(-3x^2 + 5x - 2) - (4x^2 - 3x + 1)$$
$$= -3x^2 + 5x - 2 - 4x^2 + 3x - 1$$
$$= -7x^2 + 8x - 3 \qquad \blacksquare$$

Finding the Value of a Polynomial

The last topic we want to consider in this section is finding the value of a polynomial for a given value of the variable.

EXAMPLE 7 Find the value of $4x^2 - 7x + 2$ when $x = -2$.

SOLUTION When $x = -2$, the polynomial $4x^2 - 7x + 2$ becomes

$$4(-2)^2 - 7(-2) + 2 = 4(4) + 14 + 2$$
$$= 16 + 14 + 2$$
$$= 32 \qquad \blacksquare$$

4. Subtract:
$(5x^2 - 2x + 7) - (4x^2 + 8x - 4)$

5. Subtract: $(3y^3 - 2y^2 + 7y - 6) - (8y^3 - 6y^2 + 4y - 8)$

Note
Examples 5 and 6 show the minimum number of steps needed to subtract two polynomials.

6. Subtract $6x^2 - 2x + 5$ from $-2x^2 + 5x - 1$.

7. Find the value of $5x^2 - 3x + 8$ when $x = -3$.

Answers
4. $x^2 - 10x + 11$
5. $-5y^3 + 4y^2 + 3y + 2$
6. $-8x^2 + 7x - 6$ **7.** 62

More about Sequences

As the next example indicates, when we substitute the counting numbers, in order, into algebraic expressions, we form some of the sequences of numbers that we studied in the "Extending the Concepts" sections of Chapter 1. To review, recall that the sequence of counting numbers (also called the sequence of positive integers) is

$$\text{Counting numbers} = 1, 2, 3, \ldots$$

8. Substitute 1, 2, 3, and 4 for n in the expression $2n + 1$.

EXAMPLE 8 Substitute $1, 2, 3$, and 4 for n in the expression $2n - 1$.

SOLUTION Substituting as indicated, we have

When $n = 1$, $2n - 1 = 2 \cdot 1 - 1 = 1$.
When $n = 2$, $2n - 1 = 2 \cdot 2 - 1 = 3$.
When $n = 3$, $2n - 1 = 2 \cdot 3 - 1 = 5$.
When $n = 4$, $2n - 1 = 2 \cdot 4 - 1 = 7$.

As you can see, substituting the first four counting numbers into $2n - 1$ produces the first four numbers in the sequence of odd numbers. ∎

The next example is similar to Example 8 but uses tables to display the information.

9. Fill in the tables

a.

n	1	2	3	4
$3n$				

b.

n	1	2	3	4
n^3				

EXAMPLE 9 Fill in the tables below to find the sequences formed by substituting the first four counting numbers into the expressions $2n$ and n^2.

(a)

n	1	2	3	4
$2n$				

(b)

n	1	2	3	4
n^2				

SOLUTION Proceeding as we did in the previous example, we substitute the numbers 1, 2, 3, and 4 into the given expressions.

(a) When $n = 1$, $2n = 2 \cdot 1 = 2$.
When $n = 2$, $2n = 2 \cdot 2 = 4$.
When $n = 3$, $2n = 2 \cdot 3 = 6$.
When $n = 4$, $2n = 2 \cdot 4 = 8$.

As you can see, the expression $2n$ produces the sequence of even numbers when n is replaced by the counting numbers. Placing these results into our first table gives us

n	1	2	3	4
$2n$	2	4	6	8

(b) The expression n^2 produces the sequence of squares when n is replaced by 1, 2, 3, and 4. In table form we have

n	1	2	3	4
n^2	1	4	9	16

∎

Answers

8. 3, 5, 7, 9

9.

a.

n	1	2	3	4
$3n$	3	6	9	12

b.

n	1	2	3	4
n^3	1	8	27	64

Problem Set **2.8**

Add.

1. $(3x^2 + 2x + 5) + (3x^2 + 4x + 3)$

2. $(x^2 + 3x + 7) + (x^2 + 4x + 5)$

3. $(3a^2 - 4a + 2) + (2a^2 - 5a + 6)$

4. $(5a^2 - 2a + 1) + (3a^2 - 4a + 2)$

5. $(6x^3 - 5x^2 + 3x) + (9x^2 - 6x + 2)$

6. $(5x^3 + 2x^2 + 4x) + (2x^2 + 5x + 1)$

7. $(7t^2 - 4t - 5) + (6t^2 + 4t + 3)$

8. $(8t^2 - 3t - 2) + (7t^2 + 3t + 9)$

9. $(20x^3 - 12x^2 + x - 5) + (13x^3 + 2x^2 - 5x + 1)$

10. $(30x^3 + 3x^2 - 10x + 3) + (12x^3 - 3x^2 + 10x - 3)$

Subtract.

11. $(4x^2 + 3x + 2) - (2x^2 + 5x + 1)$

12. $(8x^2 + 4x + 1) - (6x^2 + 2x + 3)$

13. $(9y^3 - 8y^2 - 4) - (5y^3 - 3y + 8)$

14. $(12y^2 - 2y + 9) - (10y^3 + 3y^2 - 9)$

15. $(4x - 5) - (3x + 2) - (5x - 3)$

16. $(8x + 4) - (5x - 1) - (7x - 5)$

17. Subtract $10x^2 + 20x - 50$ from $11x^2 - 10x + 14$.

18. Subtract $2x^2 - 3x + 5$ from $4x^2 - 6x + 12$.

19. Subtract $3y^2 + 7y - 15$ from $10y^2 + 10y + 10$.

20. Subtract $15y^2 - 7y + 3$ from $15y^2 - 7y - 3$.

21. Find the value of the polynomial $x^2 - 10x + 25$ when x is 2.

22. Find the value of $(x - 5)^2$ when x is 2.

23. Find the value of $a^2 + 6a + 9$ when a is -3.

24. Find the value of $(a + 3)^2$ when a is -3.

25. Find the value of $4y^2 + 12y + 9$ when $y = -1$.

26. Find the value of $(2y + 3)^2$ when $y = -1$.

Substitute 1, 2, 3, and 4 for n in each of the following expressions.

27. $3n - 1$ **28.** $3n + 1$ **29.** $(n + 1)^2$ **30.** $n^2 + 1$

31. $n^2 + 9$ **32.** $(n + 3)^2$

Fill in each table.

33.

n	1	2	3	4
$4n$				

34.

n	1	2	3	4
n^4				

35.

n	1	2	3	4
$n - 3$				

36.

n	1	2	3	4
$2n - 5$				

Review Problems

The problems that follow review material we covered in Section 1.10.

Find the area of each figure.

37.

13 ft

38.

7 ft

Squares

39.

10 in.

21 in.

40.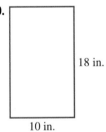

18 in.

10 in.

Rectangles

41.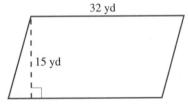

32 yd

15 yd

42.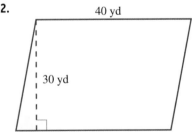

40 yd

30 yd

Parallelograms

2.9 Multiplying Polynomials: An Introduction

Recall that the distributive property allows us to multiply across parentheses when a sum or difference is enclosed within the parentheses. That is:

$$a(b + c) = a \cdot b + a \cdot c$$

We can use the distributive property to multiply polynomials.

EXAMPLE 1 Multiply: $x^2(x^3 + x^4)$

SOLUTION Applying the distributive property, we have

$$\boldsymbol{x^2}(x^3 + x^4) = \boldsymbol{x^2} \cdot x^3 + \boldsymbol{x^2} \cdot x^4 \quad \textbf{Distributive property}$$
$$= x^5 + x^6 \qquad \blacksquare$$

The distributive property works for multiplication from the right as well as the left. That is, we can also write the distributive property this way:

$$(b + c)a = b \cdot a + c \cdot a$$

EXAMPLE 2 Multiply: $(x^3 + x^4)x^2$

SOLUTION Because multiplication is a commutative operation, we should expect to obtain the same answer as in Example 1 above.

$$(x^3 + x^4)\boldsymbol{x^2} = x^3 \cdot \boldsymbol{x^2} + x^4 \cdot \boldsymbol{x^2}$$
$$= x^5 + x^6 \qquad \blacksquare$$

EXAMPLE 3 Multiply: $4x^3(6x^2 - 8)$

SOLUTION The distributive property allows us to multiply $4x^3$ by both $6x^2$ and 8:

$$\boldsymbol{4x^3}(6x^2 - 8) = \boldsymbol{4x^3} \cdot 6x^2 - \boldsymbol{4x^3} \cdot 8$$
$$= (4 \cdot 6)(x^3 \cdot x^2) - (4 \cdot 8)x^3$$
$$= 24x^5 - 32x^3 \qquad \blacksquare$$

EXAMPLE 4 Multiply: $2a^4b^2(3a^3 + 4b^5)$

SOLUTION Applying the distributive property as we did in the previous two examples, we have

$$\boldsymbol{2a^4b^2}(3a^3 + 4b^5) = \boldsymbol{2a^4b^2} \cdot 3a^3 + \boldsymbol{2a^4b^2} \cdot 4b^5$$
$$= (2 \cdot 3)(a^4 \cdot a^3)b^2 + (2 \cdot 4)(a^4)(b^2 \cdot b^5)$$
$$= 6a^7b^2 + 8a^4b^7 \qquad \blacksquare$$

Multiplying Binomials

Polynomials with exactly two terms are called *binomials*. We multiply binomials by applying the distributive property.

EXAMPLE 5 Multiply: $(x + 3)(x + 5)$

SOLUTION We can think of the first binomial, $x + 3$, as a single number. (Remember, for any value of x, $x + 3$ will be just a number.) We apply the distributive property by multiplying $x + 3$ times both x and 5.

Practice Problems
1. Multiply: $x^3(x^5 + x^7)$

2. Multiply: $(x^5 + x^7)x^3$

3. Multiply: $5x^2(6x^3 - 4)$

4. Multiply: $5a^3b^5(2a^2 + 7b^2)$

5. Multiply: $(x + 2)(x + 6)$

Answers
1. $x^8 + x^{10}$ 2. $x^8 + x^{10}$
3. $30x^5 - 20x^2$
4. $10a^5b^5 + 35a^3b^7$
5. $x^2 + 8x + 12$

$$(x + 3)(x + 5) = (x + 3) \cdot x + (x + 3) \cdot 5$$

Next, we apply the distributive property again to multiply x times both x and 3 and 5 times both x and 3.

$$= x \cdot x + 3 \cdot x + x \cdot 5 + 3 \cdot 5$$
$$= x^2 + 3x + 5x + 15$$

The last thing to do is to combine the similar terms $3x$ and $5x$ to get $8x$. (Remember, this is also an application of the distributive property.)

$$= x^2 + 8x + 15 \qquad \blacksquare$$

6. Multiply: $(x - 2)(x + 6)$

EXAMPLE 6   Multiply: $(x - 3)(x + 5)$

SOLUTION   The only difference between the binomials in this example and those in Example 5 is the subtraction sign in $x - 3$. The steps in multiplying are exactly the same.

$$(x - 3)(x + 5) = (x - 3) \cdot x + (x - 3) \cdot 5 \qquad \textbf{Multiply } \boldsymbol{x - 3} \textbf{ times both } \boldsymbol{x} \textbf{ and 5}$$
$$= x \cdot x - 3 \cdot x + x \cdot 5 - 3 \cdot 5 \qquad \textbf{Distributive property two more times}$$
$$= x^2 - 3x + 5x - 15 \qquad \textbf{Simplify each term}$$
$$= x^2 + 2x - 15 \qquad \boldsymbol{-3x + 5x = 2x} \quad \blacksquare$$

7. Multiply: $(3x - 2)(5x + 4)$

EXAMPLE 7   Multiply: $(2x - 3)(4x + 7)$

SOLUTION   Using the same steps shown in Examples 5 and 6, we have

$$(2x - 3)(4x + 7) = (2x - 3) \cdot 4x + (2x - 3) \cdot 7$$
$$= 2x \cdot 4x - 3 \cdot 4x + 2x \cdot 7 - 3 \cdot 7$$
$$= 8x^2 - 12x + 14x - 21$$
$$= 8x^2 + 2x - 21 \qquad \blacksquare$$

Our next two examples show how we raise binomials to the second power.

8. Expand and multiply: $(x + 3)^2$

EXAMPLE 8   Expand and multiply: $(x + 5)^2$

SOLUTION   We use the definition of exponents to write $(x + 5)^2$ as $(x + 5)(x + 5)$. Then we multiply as we did in the previous examples.

$$(x + 5)^2 = (x + 5)(x + 5) \qquad \textbf{Definition of exponents}$$
$$= (x + 5) \cdot x + (x + 5) \cdot 5 \qquad \textbf{Distributive property}$$
$$= x \cdot x + 5 \cdot x + x \cdot 5 + 5 \cdot 5 \qquad \textbf{Distributive property}$$
$$= x^2 + 5x + 5x + 25 \qquad \textbf{Simplify each term}$$
$$= x^2 + 10x + 25 \qquad \boldsymbol{5x + 5x = 10x} \quad \blacksquare$$

9. Expand and multiply: $(3x - 5)^2$

EXAMPLE 9   Expand and multiply: $(3x - 7)^2$

SOLUTION   We know that $(3x - 7)^2 = (3x - 7)(3x - 7)$. It will be easier to apply the distributive property to this last expression if we think of the second $3x - 7$ as $3x + (-7)$. In doing so we will also be less likely to make a mistake in our signs. (Try the problem without changing subtraction to addition of the opposite, and see how your answer compares to the answer in this example.)

Answers
6. $x^2 + 4x - 12$ **7.** $15x^2 + 2x - 8$
8. $x^2 + 6x + 9$
9. $9x^2 - 30x + 25$

$$(3x - 7)(3x - 7) = (3x - 7)[3x + (-7)]$$
$$= (3x - 7) \cdot 3x + (3x - 7)(-7)$$
$$= 3x \cdot 3x - 7 \cdot 3x + 3x(-7) - 7(-7)$$
$$= 9x^2 - 21x - 21x + 49$$
$$= 9x^2 - 42x + 49 \qquad \blacksquare$$

Multiplying Polynomials Geometrically

Suppose we have a rectangle with length $x + 3$ and width $x + 2$. Remember, the letter x is used to represent a number, so $x + 3$ and $x + 2$ are just numbers. Here is a diagram:

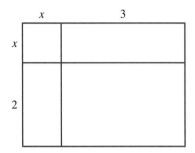

The area of the whole rectangle is the length times the width, or

$$\text{Total area} = (x + 3)(x + 2)$$

But we can also find the total area by first finding the area of each smaller rectangle and then adding these smaller areas together. The area of each rectangle is its length times its width, as shown in the following diagram:

Because the total area $(x + 3)(x + 2)$ must be the same as the sum of the smaller areas, we have:

$$(x + 3)(x + 2) = x^2 + 2x + 3x + 6$$
$$= x^2 + 5x + 6 \qquad \textbf{Add 2\textit{x} and 3\textit{x} to get 5\textit{x}}$$

The polynomial $x^2 + 5x + 6$ is the product of the two polynomials $x + 3$ and $x + 2$. Here are some more examples.

10. Find the product of $x + 4$ and $x + 2$ by using the following diagram:

$$
\begin{array}{c|c|c}
 & x & 4 \\
\hline
x & & \\
\hline
2 & & \\
\end{array}
$$

EXAMPLE 10 Find the product of $x + 5$ and $x + 2$ by using the following diagram:

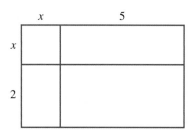

SOLUTION The total area is given by $(x + 5)(x + 2)$. We can fill in the smaller areas by multiplying length times width in each case:

$$
\begin{array}{c|c|c}
 & x & 5 \\
\hline
x & x^2 & 5x \\
\hline
2 & 2x & 10 \\
\end{array}
$$

The product of $(x + 5)$ and $(x + 2)$ is

$$(x + 5)(x + 2) = x^2 + 2x + 5x + 10$$
$$= x^2 + 7x + 10 \qquad \blacksquare$$

11. Find the product of $3x + 7$ and $2x + 5$ by using the following diagram:

$$
\begin{array}{c|c|c}
 & 3x & 7 \\
\hline
2x & & \\
\hline
5 & & \\
\end{array}
$$

EXAMPLE 11 Find the product of $2x + 5$ and $3x + 2$ by using the following diagram:

$$
\begin{array}{c|c|c}
 & 2x & 5 \\
\hline
3x & & \\
\hline
2 & & \\
\end{array}
$$

SOLUTION We fill in each of the smaller rectangles by multiplying length times width in each case:

$$
\begin{array}{c|c|c}
 & 2x & 5 \\
\hline
3x & 6x^2 & 15x \\
\hline
2 & 4x & 10 \\
\end{array}
$$

Using the information from the diagram, we have:

$$(2x + 5)(3x + 2) = 6x^2 + 4x + 15x + 10$$
$$= 6x^2 + 19x + 10 \qquad \blacksquare$$

Answer
10. $x^2 + 6x + 8$
11. $6x^2 + 29x + 35$

Problem Set 2.9

Multiply.

1. $x^2(x^4 + x^3)$

2. $x^3(x^2 + x^4)$

3. $(x^4 + x^3)x^2$

4. $(x^4 + x^2)x^3$

5. $2x^4(3x^2 - 7)$

6. $5x^3(2x^2 - 8)$

7. $4x^2y^3(3x^4 + 2y^3)$

8. $8x^2y^2(5x^3 + 4y^3)$

9. $(3x^2 - 7)2x^4$

10. $(2x^2 - 8)5x^3$

11. $3xy(4xy - 9)$

12. $2xy(3xy - 10)$

13. $(4xy - 9)3xy$

14. $(3xy - 10)2xy$

15. $2x^4y^5(3xy^2 + 2x^2y + 5)$

16. $5x^2y^2(2x^3y + 4xy^3 + 7)$

17. $(x + 2)(x + 3)$

18. $(x + 3)(x + 4)$

19. $(x - 2)(x + 3)$

20. $(x - 3)(x + 4)$

21. $(2x + 5)(3x + 2)$

22. $(3x + 5)(2x + 3)$

23. $(3x - 6)(x + 4)$

24. $(5x - 2)(x + 6)$

25. $(4x - 3)(4x + 3)$

26. $(7x - 3)(7x + 3)$

27. $(3x - 1)(4x + 1)$

28. $(6x - 1)(3x + 1)$

29. $(2x - 5)(3x - 4)$

30. $(7x - 1)(4x - 5)$

Use the diagram in each problem to help multiply the polynomials.

31. $(x + 4)(x + 2)$

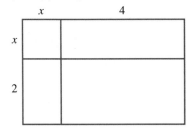

32. $(x + 1)(x + 3)$

33. $(2x + 3)(3x + 2)$

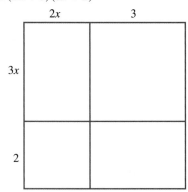

34. $(5x + 4)(6x + 1)$

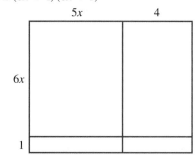

35. $(7x + 2)(3x + 4)$

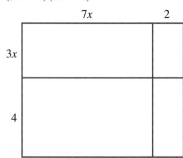

36. $(3x + 5)(2x + 5)$

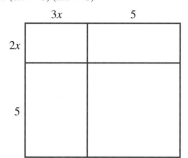

Expand and multiply.

37. $(x + 3)^2$

38. $(x + 4)^2$

39. $(x + 6)^2$

40. $(x + 7)^2$

41. $(3x + 2)^2$

42. $(4x + 5)^2$

43. $(2x + 3)^2$

44. $(5x + 4)^2$

45. $(7x + 6)^2$

46. $(6x + 7)^2$

47. $(4x + 1)^2$

48. $(3x + 1)^2$

49. $(x - 5)^2$

50. $(x - 4)^2$

51. $(2x - 4)^2$

52. $(5x - 2)^2$

Fill in the following tables.

53.

x	$(x + 3)^2$	$x^2 + 9$	$x^2 + 6x + 9$
1			
2			
3			
4			

54.

x	$(x - 5)^2$	$x^2 + 25$	$x^2 - 10x + 25$
1			
2			
3			
4			

Simplify each expression by filling in the diagrams and then using the results to find areas.

55. $(x + 3)^2$

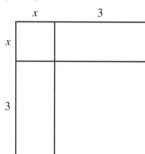

56. $(x + 4)^2$

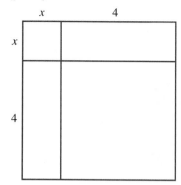

57. $(x + 5)^2$

58. $(x + 1)^2$

59. $(2x + 3)^2$

60. $(3x + 2)^2$

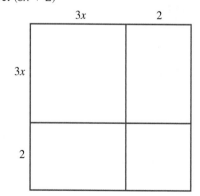

Review Problems

The problems that follow review material we covered in Section 1.10.

Find the volume and the surface area of each figure.

61.

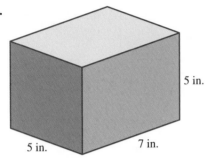

5 in.

5 in.   7 in.

62.

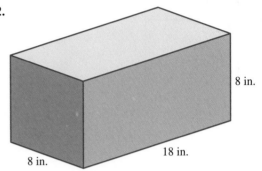

8 in.

8 in.   18 in.

63.

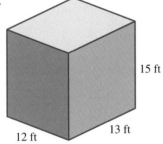

15 ft

12 ft   13 ft

64.

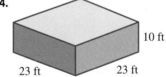

10 ft

23 ft   23 ft

Chapter 2 Summary

Absolute Value [2.1]

The absolute value of a number is its distance from 0 on the number line. It is the numerical part of a number. The absolute value of a number is never negative.

Opposites [2.1]

Two numbers are called opposites if they are the same distance from 0 on the number line but in opposite directions from 0. The opposite of a positive number is a negative number, and the opposite of a negative number is a positive number.

Addition of Positive and Negative Numbers [2.2]

1. To add two numbers with *the same sign:* Simply add absolute values and use the common sign. If both numbers are positive, the answer is positive. If both numbers are negative, the answer is negative.

2. To add two numbers with *different signs:* Subtract the smaller absolute value from the larger absolute value. The answer has the same sign as the number with the larger absolute value.

Subtraction [2.3]

Subtracting a number is equivalent to adding its opposite. If a and b represent numbers, then subtraction is defined in terms of addition as follows:

$$a - b = a + (-b)$$

$\uparrow$ $\uparrow$

Subtraction   Addition of the opposite

Multiplication with Positive and Negative Numbers [2.4]

To multiply two numbers, multiply their absolute values.

1. The answer is *positive* if both numbers have the same sign.

2. The answer is *negative* if the numbers have different signs.

Division [2.5]

The rule for assigning the correct sign to the answer in a division problem is the same as the rule for multiplication. That is, like signs give a positive answer, and unlike signs give a negative answer.

Property 1 for Exponents [2.7]

If r and s are any two whole numbers and a is an integer, then

$$a^r \cdot a^s = a^{r+s}$$

In words: To multiply two expressions with the same base, add exponents and use the common base.

Property 2 for Exponents [2.7]

If r and s are any two whole numbers and a is an integer, then

$$(a^r)^s = a^{r \cdot s}$$

In words: A power raised to another power is the base raised to the product of the powers.

Examples

1. $|3| = 3$ and $|-3| = 3$

2. $-(5) = -5$ and $-(-5) = 5$

3. $\quad 3 + 5 = 8$
$-3 + (-5) = -8$

$\quad 5 + (-3) = 2$
$-5 + 3 = -2$

4. $\quad 3 - 5 = 3 + (-5) = -2$
$-3 - 5 = -3 + (-5) = -8$
$\quad 3 - (-5) = 3 + 5 = 8$
$-3 - (-5) = -3 + 5 = 2$

5. $\quad 3(5) = 15$
$\quad 3(-5) = -15$
$-3(5) = -15$
$-3(-5) = 15$

6. $\dfrac{12}{4} = 3$

$\dfrac{-12}{4} = -3$

$\dfrac{12}{-4} = -3$

$\dfrac{-12}{-4} = 3$

7. $x^3 \cdot x^4 \cdot x^2 = x^{3+4+2}$
$\quad\quad\quad\quad = x^9$

8. $(2^3)^2 = 2^{3 \cdot 2} = 2^6$

9. $(3xy)^4 = 3^4 \cdot x^4 \cdot y^4$
    $\qquad = 81x^4y^4$

10. $2x^2 + 3x + 4$
    $\underline{3x^2 + 4x + 5}$
    $5x^2 + 7x + 9$

11. $(6x^2 - 3x + 5) - (3x^2 + 2x - 3)$
    $= 6x^2 - 3x + 5 - 3x^2 - 2x + 3$
    $= (6x^2 - 3x^2) + (-3x - 2x) + (5 + 3)$
    $= (6 - 3)x^2 + (-3 - 2)x + (5 + 3)$
    $= 3x^2 - 5x + 8$

12. $x^2(x^3 + x^4) = x^2 \cdot x^3 + x^2 \cdot x^4$
    $\qquad\qquad = x^5 + x^6$

    $(2x - 3)(4x + 7)$
    $= (2x - 3) \cdot 4x + (2x - 3) \cdot 7$
    $= 2x \cdot 4x - 3 \cdot 4x + 2x \cdot 7 - 3 \cdot 7$
    $= 8x^2 - 12x + 14x - 21$
    $= 8x^2 + 2x - 21$

Property 3 for Exponents [2.7]

If r is a whole number and a and b are integers, then

$$(a \cdot b)^r = a^r \cdot b^r$$

Adding Polynomials [2.8]

We add polynomials by writing one polynomial under the other and then adding in columns.

Subtracting Polynomials [2.8]

Because subtraction is addition of the opposite, we simply change the sign of each term of the second polynomial and then add. With subtraction, there is less chance of making mistakes if we add horizontally, rather than in columns.

Multiplying Polynomials [2.9]

We use the distributive property to multiply polynomials.

Chapter **2** Review

Give the opposite of each number. [2.1]

1. 17

2. -32

Place an inequality symbol (< or >) between each pair of numbers so that the resulting statement is true. [2.1]

3. 6 -6

4. -8 -3

5. $|-3|$ 2

6. $|-4|$ $|6|$

Simplify each expression. [2.1]

7. $-(-4)$

8. $|-6|$

Perform the indicated operations. [2.2, 2.3, 2.4, 2.5]

9. $5 + (-7)$

10. $-3 + 8$

11. $-345 + (-626)$

12. $7 - 9 - 4 - 6$

13. $-7 - 5 - 2 - 3$

14. $4 - (-3)$

15. $5(-4)$

16. $-4(-3)$

17. $\dfrac{48}{-16}$

18. $\dfrac{-20}{5}$

19. $\dfrac{-14}{-7}$

Simplify the following expressions as much as possible. [2.2, 2.3, 2.4, 2.5]

20. $(-6)^2$

21. $(-2)^3$

22. $7 + 4(6 - 9)$

23. $(-3)(-4) + 2(-5)$

24. $(7 - 3)(7 - 9)$

25. $3(-6) + 8(2 - 5)$

26. $\dfrac{8 - 4}{-8 + 4}$

27. $\dfrac{-4 + 2(-5)}{6 - 4}$

28. $\dfrac{8(-2) + 5(-4)}{12 - 3}$

29. $\dfrac{-2(5) + 4(-3)}{10 - 8}$

30. Give the sum of -19 and -23. [2.2]

31. Give the sum of -78 and -51. [2.2]

32. Find the difference of -6 and 5. [2.3]

33. Subtract -8 from -10. [2.3]

34. What is the product of -9 and 3? [2.4]

35. What is -3 times the sum of -9 and -4? [2.4]

36. Divide the product of 8 and -4 by -16. [2.5]

37. Give the quotient of -38 and 2. [2.5]

Indicate whether each statement is *True* or *False*. [2.2, 2.3, 2.4, 2.5]

38. $\dfrac{-10}{-5} = -2$

39. $10 - (-5) = 15$

40. $2(-3) = -3 + (-3)$

41. $-6 - (-2) = -8$

42. $3 - 5 = 5 - 3$

43. Reaction Distance The table below shows how many feet your car will travel from the time you decide you want to stop to the time it takes you to hit the brake pedal. Use the template to construct a line graph of the information in the table. [2.1]

REACTION DISTANCES	
SPEED (MI/HR)	DISTANCE (FT)
0	0
10	11
20	22
30	33
40	44
50	55
60	66
70	77
80	88

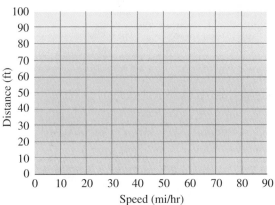

44. Gambling A gambler wins $58 Saturday night and then loses $86 on Sunday. Use positive and negative numbers to describe this situation. Then give the gambler's net loss or gain as a positive or negative number. [2.2]

45. Name two numbers that are 7 units from −8 on the number line. [2.1]

46. Temperature On Wednesday, the temperature reaches a high of 17° above 0 and a low of 7° below 0. What is the difference between the high and low temperatures for Wednesday? [2.3]

47. If the difference between two numbers is −3, and one of the numbers is 5, what is the other number? [2.3]

Use the associative properties to simplify each expression. [2.6]

48. $(3x + 4) + 8$

49. $8(3x)$

50. $-3(7a)$

51. $6(-5y)$

Apply the distributive property and then simplify if possible. [2.6]

52. $4(x + 3)$

53. $2(x - 5)$

54. $7(3y - 8)$

55. $3(2a + 5b)$

Expand and simplify. [2.6]

56. $(4x)(5x)$

57. $(5x)^3$

58. $(3x)^3(4x)^2$

Use the properties of exponents to simplify each expression. [2.7]

59. $x^3 \cdot x^5$

60. $(x^3)^6$

61. $(3xy)^4$

62. $4x^2 \cdot 7x^7$

63. $(5x^6y^2)^3$

64. $(2x^4y^5)^3 \cdot (4x^5y^8)^2$

65. Add $4x^2 + 4x - 5$ and $8x^2 - 7x + 2$. [2.8]

66. Add $6x^3 - 2x^2 + 5x - 2$ and $9x^3 + 4x - 3$. [2.8]

67. Subtract $5x + 3$ from $8x - 9$. [2.8]

68. Subtract $6x^2 - 3x - 2$ from $10x^2 + 3x - 6$. [2.8]

Multiply. [2.9]

69. $x^2(x^3 + x^5)$

70. $4a^4(2a^3 - 5)$

71. $(x + 2)(x + 7)$

72. $(3x + 5)(2x - 7)$

Expand and multiply. [2.9]

73. $(x + 5)^2$

74. $(4x - 3)^2$

Chapter **2** Test

Give the opposite of each number.

1. 14   **2.** -5

Place an inequality symbol ($<$ or $>$) between each pair of numbers so that the resulting statement is true.

3. $-1 \quad -4$ **4.** $|-4| \quad |2|$

Simplify each expression.

5. $-(-7)$ **6.** $-|-2|$

Perform the indicated operations.

7. $8 + (-17)$ **8.** $-4 - 2$ **9.** $-9 + (-12)$ **10.** $-65 - (-29)$

11. $(-6)(-7)$ **12.** $-3(-18)$ **13.** $\dfrac{-80}{16}$ **14.** $\dfrac{-35}{-7}$

Simplify the following expressions as much as possible.

15. $(-3)^2$ **16.** $(-2)^3$ **17.** $(-7)(3) + (-2)(-5)$ **18.** $(8 - 5)(6 - 11)$

19. $\dfrac{-5 + 3(-3)}{5 - 7}$ **20.** $\dfrac{-3(2) + 5(-2)}{7 - 3}$

21. Give the sum of -15 and -46.   **22.** Subtract -5 from -12.

23. What is the product of -8 and -3?   **24.** Give the quotient of 45 and -9.

25. Garbage Production   The table and bar chart below give the annual production of garbage in the United States for some specific years.

YEAR	GARBAGE (MILLIONS OF TONS)
1960	88
1970	121
1980	152
1990	205
1997	217

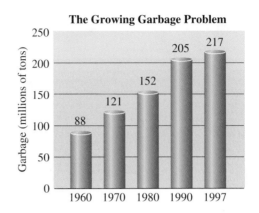

Use the information from the table and bar chart to construct a line graph using the template below.

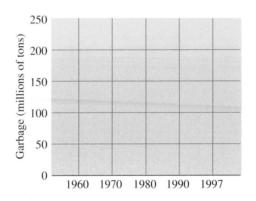

26. Gambling   A gambler loses $100 Saturday night and wins $65 on Sunday. Give the gambler's net loss or gain as a positive or negative number.

27. Temperature   On Friday, the temperature reaches a high of 21° above 0 and a low of 4° below 0. What is the difference between the high and low temperatures for Friday?

28. Use the properties of exponents to simplify $6x^3 \cdot 5x^4$.

29. Multiply $2a^2(4a^3 - 6)$.

30. Multiply $(2x + 4)(3x - 5)$.

31. Expand and multiply $(x + 3)^2$.

32. Add $2x^2 + 5x - 7$ and $3x^2 - 8x + 2$.

33. Subtract $4x + 2$ from $7x - 6$.

Chapters **1-2** Cumulative Review

Add.

1. 6,801
      539
   + 374

2. $(2x^2 + 5x - 8) + (5x^2 - 6x + 4)$

Subtract.

3. 5,038
      −2,769

4. $(5x - 9) - (6x + 7)$

Multiply.

5. 52(867)

6. $(-3)(-14)$

7. $6a^2(7a^3 - 9)$

8. $(x + 4)(x + 6)$

9. $(2x - 5)(3x + 8)$

10. $4x^3 \cdot 8x^5$

11. $(8b^3)^2$

12. $(x^5)^2 \cdot (x^3)^4$

Divide.

13. $1023 \div 15$

14. $473\overline{)15,609}$

15. $\dfrac{-24}{-6}$

16. $\dfrac{-75}{15}$

Simplify.

17. $9 - 3(8 - 5)$

18. $7(4)^2 - 5(2)^3$

19. $-|-11|$

20. $-(-9)$

21. $(-3)^3$

22. $(-7)^2$

23. $(6 - 3)(8 - 13)$

24. $\dfrac{-7 + 3(-5)}{9 - 20}$

25. $\dfrac{-8(6)}{-4}$

Expand and multiply.

26. $(x - 7)^2$

27. $(3x + 2)^2$

Place an inequality symbol (< or >) between the following pairs of numbers so that the resulting statement is true.

28. −4   −6

29. |3| |−5|

30. Identify the property or properties used in the following:
   $3y + (5 + 2y) = (3y + 2y) + 5$

31. Translate into symbols; then simplify: twice the difference of 19 and 7

32. Baseball The table below shows the top four records for the highest attended opening-day games in major league baseball. Complete the table by rounding each number to the nearest hundred.

OPENING DAY ATTENDANCE RECORDS			
GAME	DATE	ATTENDANCE	TO THE NEAREST HUNDRED
Montreal at Colorado	4/9/93	80,227	
San Francisco at Los Angeles	4/7/58	78,672	
Detroit at Cleveland	4/7/73	74,420	
St. Louis at Cleveland	4/20/48	73,163	

33. Stopping Distances The bar chart below shows how many feet it takes to stop a car traveling at different rates of speed, once the brakes are applied. Use this information in the bar chart to fill in the table.

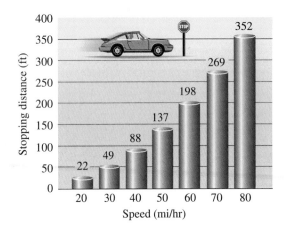

SPEED (MI/HR)	DISTANCE (FT)
20	22
30	
40	
50	
	198
	269
80	

34. Stock Market A stock gains 3 points on Tuesday, then loses 4 points on Wednesday, and gains 2 points on Thursday. What is the net gain or loss of the stock for this 3-day period?

35. Hourly Pay Jean tutors in the math lab and earns $56 in one week. If she works 8 hours that week, what is her hourly pay?

36. Average If a basketball team has scores of 64, 76, 98, 55, and 102 in their first five games, what is the average number of points scored for the five games?

37. Temperature Change On Monday, Jack notices that the temperature reaches a high of 18° above 0 and a low of 5° below 0. What is the difference between the high and and low temperatures for Monday?

38. Geometry Find the perimeter and area of a square with side 8 inches.

39. Movie Tickets A movie theater has a total of 250 seats. If they have a sell-out crowd for a matinee, and each ticket costs $7, how much money will ticket sales bring in that afternoon?

40. Number Problem The product of 6 and 8 is how much larger than the sum of 6 and 8?

Fractions and Mixed Numbers

INTRODUCTION

The table below shows the amount of gain or loss each day of the week of March 6, 2000, for the stock price of eCollege.com, an Internet company specializing in distance learning for college students. The bar chart was constructed from the information in the table.

DATE	GAIN/LOSS
Monday, March 6, 2000	$\frac{3}{4}$
Tuesday, March 7, 2000	$-\frac{9}{16}$
Wednesday, March 8, 2000	$\frac{3}{32}$
Thursday, March 9, 2000	$\frac{7}{32}$
Friday, March 10, 2000	$\frac{1}{16}$

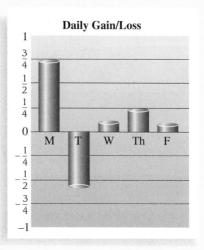

To find the mean gain or loss for the week, we would add the gain/loss values for each of the 5 days and then divide the sum by 5. In symbols, the mean is

$$\text{mean} = \frac{\frac{3}{4} + \left(-\frac{9}{16}\right) + \frac{3}{32} + \frac{7}{32} + \frac{1}{16}}{5}$$

This expression is called a complex fraction, and it is one of the topics we will study in this chapter.

We begin this chapter with the definition of a fraction and the properties that we can apply to fractions. Then we discuss how to reduce fractions to lowest terms. The remainder of the chapter is spent learning how to add, subtract, multiply, and divide fractions and mixed numbers. In addition, the chapter has been written so that, in the process of learning mathematical skills with fractions and mixed numbers, you will also sharpen your "common sense" with fractions.

STUDY SKILLS

The study skills for this chapter are concerned with getting ready to take an exam.

1. Get Ready to Take an Exam. Try to arrange your daily study habits so that you have very little studying to do the night before your next exam. The next two steps will help you achieve this goal.

2. Review with the Exam in Mind. Review material that will be covered on the next exam every day. Your review should consist of working problems. Preferably, the problems you work should be problems from your list of difficult problems.

3. Continue to List Difficult Problems. This study skill was started in the previous chapter. You should continue to list and rework the problems that give you the most difficulty. It is this list that you will use to study for the next exam. Your goal is to go into the next exam knowing that you can successfully work any problem from your list of difficult problems.

4. Pay Attention to Instructions. Taking a test is not like doing homework. On a test, the problems will be varied. When you do your homework, you usually work a number of similar problems. I have some students who do very well on their homework but become confused when they see the same problems on a test. The reason for their confusion is that they have not paid attention to the instructions on their homework. If a test problem asks for the *average* of some numbers, then you must know what the word *average* means. Likewise, if a test problem asks you to find a *sum* and then to *round* your answer to the nearest hundred, then you must know that the word *sum* indicates addition, and, after you have added, you must round your answer as indicated.

If you train yourself to pay attention to the instructions that accompany a problem as you work through the assigned problems, you will not find yourself confused about what to do with a problem when you see it on a test.

3.1   The Meaning and Properties of Fractions

Introduction . . .

The information in the table below was taken from the Web site for Cal Poly State University in California. The pie chart was created from the table. Both the table and pie chart use fractions to specify how the students at Cal Poly are distributed among the different schools within the university.

CAL POLY STATE UNIVERSITY ENROLLMENT FOR FALL 1999	
SCHOOL	**FRACTION of STUDENTS**
Agriculture	$\frac{11}{50}$
Architecture and Environmental Design	$\frac{1}{10}$
Business	$\frac{3}{20}$
Engineering	$\frac{1}{4}$
Liberal Arts	$\frac{4}{25}$
Science and Mathematics	$\frac{3}{25}$

Cal Poly enrollment for Fall 1999

Liberal Arts $\frac{4}{25}$ Science and Mathematics $\frac{3}{25}$

Agriculture $\frac{11}{50}$

Engineering $\frac{1}{4}$

Business $\frac{3}{20}$

Architecture and Environmental Design $\frac{1}{10}$

From the table, we see that $\frac{1}{4}$ (one-fourth) of the students are enrolled in the School of Engineering. This means that one out of every four students at Cal Poly is studying Engineering. The fraction $\frac{1}{4}$ tells us we have 1 part of 4 equal parts. That is, the students at Cal Poly could be divided into 4 equal groups, so that one of the groups contained all the engineering students and only engineering students.

Figure 1 at the right shows a rectangle that has been divided into equal parts, four different ways. The shaded area for each rectangle is $\frac{1}{2}$ the total area.

Now that we have an intuitive idea of the meaning of fractions, here are the more formal definitions and vocabulary associated with fractions.

DEFINITION

A **fraction** is any number that can be put in the form $\frac{a}{b}$ (also sometimes written a/b), where a and b are integers and b cannot be 0.

Some examples of fractions are:

$$\frac{1}{2} \qquad\qquad \frac{3}{4} \qquad\qquad \frac{7}{8} \qquad\qquad \frac{9}{5}$$

One-half   Three-fourths   Seven-eighths   Nine-fifths

DEFINITION

For the fraction $\frac{a}{b}$, a and b are called the **terms** of the fraction. More specifically, a is called the **numerator**, and b is called the **denominator**.

a $\frac{1}{2}$ is shaded

b $\frac{2}{4}$ are shaded

c $\frac{3}{6}$ are shaded

d $\frac{4}{8}$ are shaded

Figure 1   Four Ways to Visualize $\frac{1}{2}$

EXAMPLE 1   The terms of the fraction $\frac{3}{4}$ are 3 and 4. The 3 is called the numerator, and the 4 is called the denominator.   ■

EXAMPLE 2   The numerator of the fraction $\frac{a}{5}$ is a. The denominator is 5. Both a and 5 are called terms.   ■

EXAMPLE 3   The number 7 may also be put in fraction form, because it can be written as $\frac{7}{1}$. In this case, 7 is the numerator and 1 is the denominator.   ■

> **DEFINITION**
>
> A **proper fraction** is a fraction in which the numerator is less than the denominator. If the numerator is greater than or equal to the denominator, the fraction is called an **improper fraction.**

EXAMPLE 4   The fractions $\frac{3}{4}$, $\frac{1}{8}$, and $\frac{9}{10}$ are all proper fractions, because in each case the numerator is less than the denominator.   ■

EXAMPLE 5   The numbers $\frac{9}{5}$, $\frac{10}{10}$, and 6 are all improper fractions, because in each case the numerator is greater than or equal to the denominator. (Remember that 6 can be written as $\frac{6}{1}$, in which case 6 is the numerator and 1 is the denominator.)   ■

Fractions on the Number Line

We can give meaning to the fraction $\frac{2}{3}$ by using a number line. If we take that part of the number line from 0 to 1 and divide it into *three equal parts,* we say that we have divided it into *thirds* (see Figure 2). Each of the three segments is $\frac{1}{3}$ (one third) of the whole segment from 0 to 1.

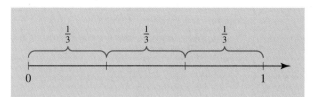

Figure 2

Two of these smaller segments together are $\frac{2}{3}$ (two thirds) of the whole segment. And three of them would be $\frac{3}{3}$ (three thirds), or the whole segment, as indicated in Figure 3.

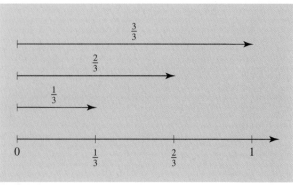

Figure 3

Let's do the same thing again with six and twelve equal divisions of the segment from 0 to 1 (see Figure 4). The same point that we labeled with $\frac{1}{3}$ in Figure 3 is now labeled with $\frac{2}{6}$ and with $\frac{4}{12}$. It must be true then that

$$\frac{4}{12} = \frac{2}{6} = \frac{1}{3}$$

Although these three fractions look different, each names the same point on the number line, as shown in Figure 4. All three fractions have the same *value*, because they all represent the same number.

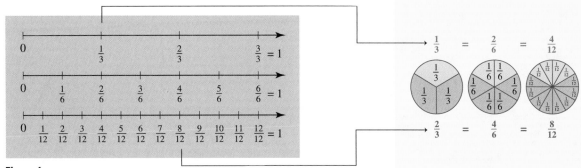

Figure 4

DEFINITION

Fractions that represent the same number are said to be **equivalent.** Equivalent fractions may look different, but they must have the same value.

It is apparent that every fraction has many different representations, each of which is equivalent to the original fraction. The next two properties give us a way of changing the terms of a fraction without changing its value.

Property 1 for Fractions

If *a*, *b*, and *c* are integers and *b* and *c* are not 0, then it is always true that

$$\frac{a}{b} = \frac{a \cdot c}{b \cdot c}$$

In words: If the numerator and the denominator of a fraction are multiplied by the same nonzero number, the resulting fraction is equivalent to the original fraction.

6. Write $\frac{2}{3}$ as an equivalent fraction with denominator 12.

EXAMPLE 6 Write $\frac{3}{4}$ as an equivalent fraction with denominator 20.

SOLUTION   The denominator of the original fraction is 4. The fraction we are trying to find must have a denominator of 20. We know that if we multiply 4 by 5, we get 20. Property 1 indicates that we are free to multiply the denominator by 5 so long as we do the same to the numerator.

$$\frac{3}{4} = \frac{3 \cdot \mathbf{5}}{4 \cdot \mathbf{5}} = \frac{15}{20}$$

The fraction $\frac{15}{20}$ is equivalent to the fraction $\frac{3}{4}$. ■

7. Write $\frac{2}{3}$ as an equivalent fraction with denominator $12x$.

EXAMPLE 7 Write $\frac{3}{4}$ as an equivalent fraction with denominator $12x$.

SOLUTION   If we multiply 4 by $3x$, we will have $12x$:

$$\frac{3}{4} = \frac{3 \cdot \mathbf{3x}}{4 \cdot \mathbf{3x}} = \frac{9x}{12x}$$

■

Note

When we encounter fractions that have variables in their denominator, like $\frac{9x}{12x}$, we will assume that those variables cannot be 0, because division by 0 is undefined.

Property 2 for Fractions

If a, b, and c are integers and b and c are not 0, then it is always true that

$$\frac{a}{b} = \frac{a \div c}{b \div c}$$

In words: If the numerator and the denominator of a fraction are divided by the same nonzero number, the resulting fraction is equivalent to the original fraction.

---

8. Write $\frac{15}{20}$ as an equivalent fraction with denominator 4.

EXAMPLE 8 Write $\frac{10}{12}$ as an equivalent fraction with denominator 6.

SOLUTION   If we divide the original denominator 12 by 2, we obtain 6. Property 2 indicates that if we divide both the numerator and the denominator by 2, the resulting fraction will be equal to the original fraction:

$$\frac{10}{12} = \frac{10 \div \mathbf{2}}{12 \div \mathbf{2}} = \frac{5}{6}$$

■

The Number 1 and Fractions

There are two situations involving fractions and the number 1 that occur frequently in mathematics. The first is when the denominator of a fraction is 1. In this case, if we let a represent any number, then

$$\frac{a}{1} = a \qquad \text{for any number } a$$

The second situation occurs when the numerator and the denominator of a fraction are the same nonzero number:

$$\frac{a}{a} = 1 \qquad \text{for any nonzero number } a$$

9. Simplify.

a. $\frac{18}{1}$ b. $\frac{18}{18}$

c. $\frac{36}{18}$ d. $\frac{72}{18}$

EXAMPLE 9 Simplify each expression.

a. $\frac{24}{1}$ b. $\frac{24}{24}$ c. $\frac{48}{24}$ d. $\frac{72}{24}$

SOLUTION   In each case we divide the numerator by the denominator:

a. $\frac{24}{1} = 24$ b. $\frac{24}{24} = 1$ c. $\frac{48}{24} = 2$ d. $\frac{72}{24} = 3$ ■

Answers

6. $\frac{8}{12}$ **7.** $\frac{8x}{12x}$ **8.** $\frac{3}{4}$

9. a. 18   **b.** 1   **c.** 2   **d.** 4

Problem Set **3.1**

Name the numerator of each fraction.

1. $\dfrac{1}{3}$ **2.** $\dfrac{1}{4}$ **3.** $\dfrac{2}{3}$ **4.** $\dfrac{2}{4}$

5. $\dfrac{x}{8}$ **6.** $\dfrac{y}{10}$ **7.** $\dfrac{a}{b}$ **8.** $\dfrac{x}{y}$

Name the denominator of each fraction.

9. $\dfrac{2}{5}$ **10.** $\dfrac{3}{5}$ **11.** 6 **12.** 2

13. $\dfrac{a}{12}$ **14.** $\dfrac{b}{14}$

Complete the following tables.

15.

NUMERATOR a	DENOMINATOR b	FRACTION $\dfrac{a}{b}$
3	5	
1		$\dfrac{1}{7}$
	y	$\dfrac{x}{y}$
$x + 1$	x	

16.

NUMERATOR a	DENOMINATOR b	FRACTION $\dfrac{a}{b}$
2	9	
	3	$\dfrac{4}{3}$
1		$\dfrac{1}{x}$
x		$\dfrac{x}{x + 1}$

17. For the set of numbers $\left\{ \dfrac{3}{4}, \dfrac{6}{5}, \dfrac{12}{3}, \dfrac{1}{2}, \dfrac{9}{10}, \dfrac{20}{10} \right\}$, list all the proper fractions.

18. For the set of numbers $\left\{ \dfrac{1}{8}, \dfrac{7}{9}, \dfrac{6}{3}, \dfrac{18}{6}, \dfrac{3}{5}, \dfrac{9}{8} \right\}$, list all the improper fractions.

Indicate whether each of the following is *True* or *False*.

19. Every whole number greater than 1 can also be expressed as an improper fraction.

20. Some improper fractions are also proper fractions.

21. Adding the same number to the numerator and the denominator of a fraction will not change its value.

22. The fractions $\dfrac{3}{4}$ and $\dfrac{9}{16}$ are equivalent.

23. The pie chart below shows the fraction of workers who responded to a survey about sending non-work-related e-mail from the office. Use the pie chart to fill in the table.

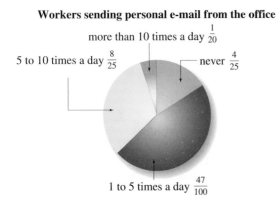

Workers sending personal e-mail from the office

more than 10 times a day $\frac{1}{20}$

5 to 10 times a day $\frac{8}{25}$

never $\frac{4}{25}$

1 to 5 times a day $\frac{47}{100}$

HOW OFTEN WORKERS SEND NON-WORK-RELATED E-MAIL FROM THE OFFICE	FRACTION OF RESPONDENTS SAYING YES
never	
1 to 5 times a day	
5 to 10 times a day	
more than 10 times a day	

24. The pie chart below shows the fraction of workers who responded to a survey about viewing non-work-related sites during working hours. Use the pie chart to fill in the table.

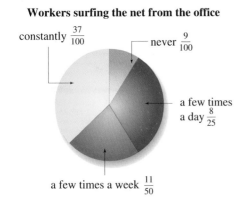

Workers surfing the net from the office

constantly $\frac{37}{100}$

never $\frac{9}{100}$

a few times a day $\frac{8}{25}$

a few times a week $\frac{11}{50}$

HOW OFTEN WORKERS VIEW NON-WORK-RELATED SITES FROM THE OFFICE?	FRACTION OF RESPONDENTS SAYING YES
never	
a few times a week	
a few times a day	
constantly	

Divide the numerator and the denominator of each of the following fractions by 2.

25. $\frac{6}{8}$

26. $\frac{10}{12}$

27. $\frac{86}{94}$

28. $\frac{106}{142}$

Divide the numerator and the denominator of each of the following fractions by 3.

29. $\frac{12}{9}$

30. $\frac{33}{27}$

31. $\frac{39}{51}$

32. $\frac{57}{69}$

Write each of the following fractions as an equivalent fraction with denominator 6.

33. $\frac{2}{3}$

34. $\frac{1}{2}$

35. $\frac{55}{66}$

36. $\frac{65}{78}$

Write each of the following fractions as an equivalent fraction with denominator 12.

37. $\frac{2}{3}$ **38.** $\frac{5}{6}$ **39.** $\frac{56}{84}$ **40.** $\frac{143}{156}$

41. Adults Overweight The table below shows the fraction of adults age 25 and over who say they are 1/5 or more above their recommended weight. Complete the table by writing each fraction as an equivalent fraction with denominator 100.

YEAR	FRACTION	EQUIVALENT FRACTION
1985	$\frac{3}{20}$	$\frac{}{100}$
1990	$\frac{4}{25}$	$\frac{}{100}$
1995	$\frac{11}{50}$	$\frac{}{100}$
2000	$\frac{8}{25}$	$\frac{}{100}$

42. Displaying Information Use the template below to construct a line graph of the information in the table in Problem 41.

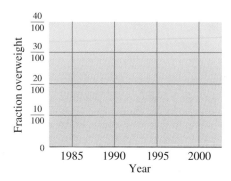

Write each fraction as an equivalent fraction with denominator 12x.

43. $\frac{1}{6}$ **44.** $\frac{3}{4}$

Write each fraction as an equivalent fraction with denominator 24a.

45. $\frac{3}{8}$ **46.** $\frac{5}{12}$ **47.** $\frac{5}{6}$ **48.** $\frac{2}{3}$

49. One-fourth of the first circle below is shaded. Use the other three circles to show three other ways to shade one-fourth of the circle.

50. The objects below are hexagons, six-sided figures. One-third of the first hexagon is shaded. Shade the other three hexagons to show three other ways to represent one-third.

Simplify by dividing the numerator by the denominator.

51. $\frac{3}{1}$ **52.** $\frac{3}{3}$ **53.** $\frac{6}{3}$ **54.** $\frac{12}{3}$

55. $\frac{37}{1}$ **56.** $\frac{37}{37}$

57. Number of Children If there are 3 girls in a family with 5 children, then we say that $\frac{3}{5}$ of the children are girls. If there are 4 girls in a family with 5 children, what fraction of the children are girls?

58. Medical School If 3 out of every 7 people who apply to medical school actually get accepted, what fraction of the people who apply get accepted?

59. Spending Money If a man spends $2 out of every $15 he earns on clothes, what fraction of his income is spent on clothes?

60. Television Owners In 1996, 98 out of every 100 households in the United States owned a television set. What fraction of the households owned television sets in 1996?

61. Number of Students Of the 43 people who started a math class meeting at 10:00 each morning, only 29 finished the class. What fraction of the people finished the class?

62. Number of Students In a class of 51 students, 23 are freshmen and 28 are juniors. What fraction of the students are freshmen?

63. Expenses If your monthly income is $1,791 and your house payment is $1,121, what fraction of your monthly income must go to pay your house payment?

64. Expenses If you spend $623 on food each month and your monthly income is $2,599, what fraction of your monthly income do you spend on food?

65. Running In a recent marathon, 238 runners finished in under 3 hours. If 527 runners started the race, what fraction of the starters finished in under 3 hours?

66. Running In a recent 10-kilometer run, 757 runners started the race and 599 finished it. What fraction of the runners finished the race?

67. For each square below, what fraction of the area is given by the shaded region?

a.   b.   c. d.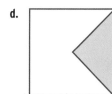

68. For each square below, what fraction of the area is given by the shaded region?

a.   b.   c.   d.

The number line at the right extends from 0 to 2, with the segment from 0 to 1 and the segment from 1 to 2 each divided into 8 equal parts. Locate each of the following numbers on this number line.

69. $\frac{1}{4}$ **70.** $\frac{1}{8}$ **71.** $\frac{1}{16}$ **72.** $\frac{5}{8}$ **73.** $\frac{3}{4}$ **74.** $\frac{15}{16}$

75. $\dfrac{3}{2}$ **76.** $\dfrac{5}{4}$ **77.** $\dfrac{31}{16}$ **78.** $\dfrac{15}{8}$

Estimating

79. Which of the following fractions is closest to the number 0?

    a. $\dfrac{1}{2}$ **b.** $\dfrac{1}{3}$ **c.** $\dfrac{1}{4}$ **d.** $\dfrac{1}{5}$

80. Which of the following fractions is closest to the number 1?

    a. $\dfrac{1}{2}$ **b.** $\dfrac{1}{3}$ **c.** $\dfrac{1}{4}$ **d.** $\dfrac{1}{5}$

81. Which of the following fractions is closest to the number 0?

    a. $\dfrac{1}{8}$ **b.** $\dfrac{3}{8}$ **c.** $\dfrac{5}{8}$ **d.** $\dfrac{7}{8}$

82. Which of the following fractions is closest to the number 1?

    a. $\dfrac{1}{8}$ **b.** $\dfrac{3}{8}$ **c.** $\dfrac{5}{8}$ **d.** $\dfrac{7}{8}$

Calculator Problems

Use your calculator to answer the following problems.

83. Is the fraction $\dfrac{596}{1,788}$ equivalent to the fraction $\dfrac{1}{3}$?

84. Is the fraction $\dfrac{1}{7}$ equivalent to the fraction $\dfrac{396}{2,772}$?

85. Write $\dfrac{3}{9}$ as an equivalent fraction with denominator 2,205.

86. Write $\dfrac{7}{11}$ as an equivalent fraction with numerator 5,761.

For Problems 87–90, fill in the missing term so that each pair of fractions is equal.

87. $\dfrac{1}{4} = \dfrac{?}{384}$ **88.** $\dfrac{1}{9} = \dfrac{?}{756}$ **89.** $\dfrac{5}{6} = \dfrac{1,580}{?}$ **90.** $\dfrac{11}{12} = \dfrac{5,951}{?}$

Review Problems

The problems below review material covered in Section 1.9.

Simplify.

91. $3 + 4 \cdot 5$ **92.** $20 - 8 \cdot 2$ **93.** $5 \cdot 2^4 - 3 \cdot 4^2$ **94.** $7 \cdot 8^2 + 2 \cdot 5^2$

95. $4 \cdot 3 + 2(5 - 3)$ **96.** $6 \cdot 8 + 3(4 - 1)$ **97.** $18 + 12 \div 4 - 3$ **98.** $20 + 16 \div 2 - 5$

Extending the Concepts

99. If three fractions all have the same denominator, then the largest fraction has the largest numerator and the smallest fraction has the smallest numerator. Correct?

100. If three fractions all have the same numerator, then the largest fraction has the largest denominator and the smallest fraction has the smallest denominator. Correct?

3.2 Prime Numbers, Factors, and Reducing to Lowest Terms

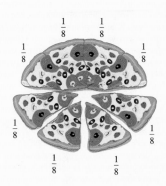

Introduction . . .

Suppose you and a friend decide to split a medium-sized pizza for lunch. When the pizza is delivered you find that it has been cut into eight equal pieces. If you eat four pieces, you have eaten $\frac{4}{8}$ of the pizza, but you also know that you have eaten $\frac{1}{2}$ of the pizza. The fraction $\frac{4}{8}$ is equivalent to the fraction $\frac{1}{2}$; that is, they both have the same value. The mathematical process we use to rewrite $\frac{4}{8}$ as $\frac{1}{2}$ is called *reducing to lowest terms.* Before we look at that process, we need to define some new terms. Here is our first one.

> **DEFINITION**
>
> A **prime number** is any whole number greater than 1 that has exactly two divisors—itself and 1.
> (A number is a divisor of another number if it divides it without a remainder.)

$$\text{Prime numbers} = \{2, 3, 5, 7, 11, 13, 17, 19, 23, 29, 31, 37, \dots\}$$

The list goes on indefinitely. Each number in the list has exactly two distinct divisors—itself and 1.

> **DEFINITION**
>
> Any whole number greater than 1 that is not a prime number is called a **composite number.**
> A composite number always has at least one divisor other than itself and 1.

EXAMPLE 1 Identify each of the numbers below as either a prime number or a composite number. For those that are composite, give two divisors other than the number itself or 1.

    a. 43   **b.** 12

SOLUTION   **a.** 43 is a prime number, because the only numbers that divide it without a remainder are 43 and 1.

          b. 12 is a composite number, because it can be written as $12 = 4 \cdot 3$, which means that 4 and 3 are divisors of 12. (These are not the only divisors of 12; other divisors are 1, 2, 6, and 12.) ∎

Every composite number can be written as the product of prime factors. Let's look at the composite number 108. We know we can write 108 as $2 \cdot 54$. The number 2 is a prime number, but 54 is not prime. Because 54 can be written as $2 \cdot 27$, we have

$$108 = 2 \cdot 54$$
$$= 2 \cdot 2 \cdot 27$$

Practice Problems

1. Which of the numbers below are prime numbers, and which are composite? For those that are composite, give two divisors other than the number itself and 1.
37, 39, 51, 59

Note

You may have already noticed that the word *divisor* as we are using it here means the same as the word *factor.* A divisor and a factor of a number are the same thing. A number can't be a divisor of another number without also being a factor of it.

Answer

1. See solutions section.

Note
This process works by writing the original composite number as the product of any two of its factors and then writing any factor that is not prime as the product of any two of its factors. The process is continued until all factors are prime numbers.

2. Factor 90 into a product of prime factors.

Note
There are some shortcuts to finding the divisors of a number. For instance, if a number ends in 0 or 5, then it is divisible by 5. If a number ends in an even number (0, 2, 4, 6, or 8), then it is divisible by 2. A number is divisible by 3 if the sum of its digits is divisible by 3. For example, 921 is divisible by 3 because the sum of its digits is $9 + 2 + 1 = 12$, which is divisible by 3.

3. Which of the following fractions are in lowest terms?
$$\frac{1}{6}, \frac{2}{8}, \frac{15}{25}, \frac{9}{13}$$

4. Reduce $\frac{12}{18}$ to lowest terms by dividing the numerator and the denominator by 6.

Answers

2. $2 \cdot 3^2 \cdot 5$ **3.** $\frac{1}{6}, \frac{9}{13}$ **4.** $\frac{2}{3}$

Now the number 27 can be written as $3 \cdot 9$ or $3 \cdot 3 \cdot 3$ (because $9 = 3 \cdot 3$), so

$$108 = 2 \cdot 54$$
$$108 = 2 \cdot 2 \cdot 27$$
$$108 = 2 \cdot 2 \cdot 3 \cdot 9$$
$$108 = 2 \cdot 2 \cdot 3 \cdot 3 \cdot 3$$

This last line is the number 108 written as the product of prime factors. We can use exponents to simplify the last line:

$$108 = 2^2 \cdot 3^3$$

EXAMPLE 2 Factor 60 into a product of prime factors.

SOLUTION   We begin by writing 60 as $6 \cdot 10$ and continue factoring until all factors are prime numbers:

$$60 = 6 \cdot 10$$
$$= 2 \cdot 3 \cdot 2 \cdot 5$$
$$= 2^2 \cdot 3 \cdot 5$$

Notice that if we had started by writing 60 as $3 \cdot 20$, we would have achieved the same result:

$$60 = 3 \cdot 20$$
$$= 3 \cdot 2 \cdot 10$$
$$= 3 \cdot 2 \cdot 2 \cdot 5$$
$$= 2^2 \cdot 3 \cdot 5$$

We can use the method of factoring numbers into prime factors to help reduce fractions to lowest terms. Here is the definition for lowest terms.

DEFINITION

A fraction is said to be in **lowest terms** if the numerator and the denominator have no factors in common other than the number 1.

EXAMPLE 3 The fractions $\frac{1}{2}, \frac{1}{3}, \frac{2}{3}, \frac{1}{4}, \frac{3}{4}, \frac{1}{5}, \frac{2}{5}, \frac{3}{5}$, and $\frac{4}{5}$ are all in lowest terms, because in each case the numerator and the denominator have no factors other than 1 in common. That is, in each fraction, no number other than 1 divides both the numerator and the denominator exactly (without a remainder).

EXAMPLE 4 The fraction $\frac{6}{8}$ is not written in lowest terms, because the numerator and the denominator are both divisible by 2. To write $\frac{6}{8}$ in lowest terms, we apply Property 2 from Section 3.1 and divide both the numerator and the denominator by 2:

$$\frac{6}{8} = \frac{6 \div \mathbf{2}}{8 \div \mathbf{2}} = \frac{3}{4}$$

The fraction $\frac{3}{4}$ is in lowest terms, because 3 and 4 have no factors in common except the number 1.

Reducing a fraction to lowest terms is simply a matter of dividing the numerator and the denominator by all the factors they have in common. We know from Property 2 of Section 3.1 that this will produce an equivalent fraction.

EXAMPLE 5 Reduce the fraction $\frac{12}{15}$ to lowest terms by first factoring the numerator and the denominator into prime factors and then dividing both the numerator and the denominator by the factor they have in common.

SOLUTION   The numerator and the denominator factor as follows:

$$12 = 2 \cdot 2 \cdot 3 \quad \text{and} \quad 15 = 3 \cdot 5$$

The factor they have in common is 3. Property 2 tells us that we can divide both terms of a fraction by 3 to produce an equivalent fraction. So

$$\frac{12}{15} = \frac{2 \cdot 2 \cdot 3}{3 \cdot 5} \qquad \textbf{Factor the numerator and the} \\ \textbf{denominator completely}$$

$$= \frac{2 \cdot 2 \cdot 3 \div \mathbf{3}}{3 \cdot 5 \div \mathbf{3}} \qquad \textbf{Divide by 3}$$

$$= \frac{2 \cdot 2}{5} = \frac{4}{5}$$

The fraction $\frac{4}{5}$ is equivalent to $\frac{12}{15}$ and is in lowest terms, because the numerator and the denominator have no factors other than 1 in common. ∎

We can shorten the work involved in reducing fractions to lowest terms by using a slash to indicate division. For example, we can write the above problem this way:

$$\frac{12}{15} = \frac{2 \cdot 2 \cdot \cancel{3}}{\cancel{3} \cdot 5} = \frac{4}{5}$$

So long as we understand that the slashes through the 3's indicate that we have divided both the numerator and the denominator by 3, we can use this notation.

EXAMPLE 6 Laura is having a party. She puts 4 six-packs of diet soda in a cooler for her guests. At the end of the party she finds that only 4 sodas have been consumed. What fraction of the sodas are left? Write your answer in lowest terms.

SOLUTION   She had 4 six-packs of soda, which is 4(6) = 24 sodas. Only 4 were consumed at the party, so 20 are left. The fraction of sodas left is

$$\frac{20}{24}$$

Factoring 20 and 24 completely and then dividing out both the factors they have in common gives us

$$\frac{20}{24} = \frac{\cancel{2} \cdot \cancel{2} \cdot 5}{\cancel{2} \cdot \cancel{2} \cdot 2 \cdot 3} = \frac{5}{6}$$ ∎

Note The slashes in Example 6 indicate that we have divided both the numerator and the denominator by 2 · 2, which is equal to 4. With some fractions it is apparent at the start what number divides the numerator and the denominator. For instance, you may have recognized that both 20 and 24 in Example 6 are divisible by 4. We can divide both terms by 4 without factoring first, just as we did in Section 3.1. Property 2 guarantees that dividing both terms of a fraction by 4 will produce an equivalent fraction:

$$\frac{20}{24} = \frac{20 \div \mathbf{4}}{24 \div \mathbf{4}} = \frac{5}{6}$$

5. Reduce the fraction $\frac{15}{20}$ to lowest terms by first factoring the numerator and the denominator into prime factors and then dividing out the factors they have in common.

6. Reduce $\frac{30}{35}$ to lowest terms.

Answers

5. $\frac{3}{4}$ **6.** $\frac{6}{7}$

7. Reduce $\frac{8}{72}$ to lowest terms.

EXAMPLE 7   Reduce $\frac{6}{42}$ to lowest terms.

SOLUTION   We begin by factoring both terms. We then divide through by any factors common to both terms:

$$\frac{6}{42} = \frac{\cancel{2} \cdot \cancel{3}}{\cancel{2} \cdot \cancel{3} \cdot 7} = \frac{1}{7}$$

We must be careful in a problem like this to remember that the slashes indicate division. They are used to indicate that we have divided both the numerator and the denominator by $2 \cdot 3 = 6$. The result of dividing the numerator 6 by $2 \cdot 3$ is 1. It is a very common mistake to call the numerator 0 instead of 1 or to leave the numerator out of the answer. ■

In Examples 8 and 9, reduce each fraction to lowest terms.

Reduce each fraction to lowest terms.

8. $\frac{5}{50}$

EXAMPLE 8 $\frac{4}{40} = \frac{\cancel{2} \cdot \cancel{2} \cdot 1}{\cancel{2} \cdot \cancel{2} \cdot 2 \cdot 5}$

$$= \frac{1}{10}$$ ■

9. $\frac{120}{25}$

EXAMPLE 9 $\frac{105}{30} = \frac{3 \cdot \cancel{5} \cdot 7}{2 \cdot 3 \cdot \cancel{5}}$

$$= \frac{7}{2}$$ ■

If a fraction contains variables (letters) in its numerator and/or denominator, we treat the variables in the same way we treat the other factors in the numerator or denominator. If a variable factor is common to the numerator and the denominator, we can reduce to lowest terms by dividing the numerator and the denominator by that variable. (Remember also that any variable that appears in a denominator is assumed to be nonzero.)

10. $\frac{54x}{90xy}$

EXAMPLE 10   Reduce $\frac{36xy}{120x}$ to lowest terms.

SOLUTION   First we factor 36 and 120 completely. Then we divide out all the factors that are common to the numerator and the denominator, including any variable factors.

$$\frac{36xy}{120x} = \frac{\cancel{2} \cdot \cancel{2} \cdot \cancel{3} \cdot 3 \cdot \cancel{x} \cdot y}{\cancel{2} \cdot \cancel{2} \cdot 2 \cdot \cancel{3} \cdot 5 \cdot \cancel{x}}$$
$$= \frac{3y}{10}$$ ■

11. $\frac{306a^2}{228a}$

EXAMPLE 11   Reduce $\frac{204a^2}{342a}$ to lowest terms.

SOLUTION   We factor both the numerator and the denominator completely and then divide out any factors common to both.

$$\frac{204a^2}{342a} = \frac{\cancel{2} \cdot 2 \cdot \cancel{3} \cdot 17 \cdot \cancel{a} \cdot a}{\cancel{2} \cdot \cancel{3} \cdot 3 \cdot 19 \cdot \cancel{a}} = \frac{34a}{57}$$ ■

Answers

7. $\frac{1}{9}$ **8.** $\frac{1}{10}$ **9.** $\frac{24}{5}$

10. $\frac{3}{5y}$ **11.** $\frac{51a}{38}$

Problem Set 3.2

Identify each of the numbers below as either prime numbers or composite numbers. For those that are composite, give at least one divisor (factor) other than the number itself or the number 1.

1. 11

2. 23

3. 105

4. 41

5. 81

6. 50

7. 13

8. 219

Factor each of the following into a product of prime factors.

9. 12

10. 8

11. 81

12. 210

13. 215

14. 75

15. 15

16. 42

Reduce each fraction to lowest terms.

17. $\dfrac{5}{10}$

18. $\dfrac{3}{6}$

19. $\dfrac{4}{6}$

20. $\dfrac{4}{10}$

21. $\dfrac{8x}{10x}$

22. $\dfrac{6x}{10x}$

23. $\dfrac{36}{20}$

24. $\dfrac{32}{12}$

25. $\dfrac{42}{66}$

26. $\dfrac{36}{60}$

27. $\dfrac{24xy}{40y}$

28. $\dfrac{50xy}{75x}$

29. $\dfrac{14}{98}$

30. $\dfrac{12}{84}$

31. $\dfrac{70}{90}$

32. $\dfrac{80}{90}$

33. $\dfrac{42x^2}{30x}$

34. $\dfrac{60x^2}{36x}$

35. $\dfrac{18a^2b}{90ab}$

36. $\dfrac{150ab^2}{210ab}$

37. $\dfrac{110}{70}$

38. $\dfrac{45}{75}$

39. $\dfrac{180xyz}{108xy}$

40. $\dfrac{105xyz}{30yz}$

41. $\dfrac{96x^2y}{108xy^2}$

42. $\dfrac{66x^2y}{84xy^2}$

43. $\dfrac{126}{165}$

44. $\dfrac{210}{462}$

45. $\dfrac{102a^2bc^3}{114ab^2c^2}$

46. $\dfrac{255a^3b^2c}{285ab^2c^3}$

47. $\dfrac{294}{693}$

48. $\dfrac{273}{385}$

49. Reduce each fraction to lowest terms.

    a. $\dfrac{6}{51}$ **b.** $\dfrac{6}{52}$ **c.** $\dfrac{6}{54}$ **d.** $\dfrac{6}{56}$ **e.** $\dfrac{6}{57}$

50. Reduce each fraction to lowest terms.

    a. $\dfrac{6}{42}$ **b.** $\dfrac{6}{44}$ **c.** $\dfrac{6}{45}$ **d.** $\dfrac{6}{46}$ **e.** $\dfrac{6}{48}$

51. Reduce each fraction to lowest terms.

    a. $\dfrac{2}{90}$ **b.** $\dfrac{3}{90}$ **c.** $\dfrac{5}{90}$ **d.** $\dfrac{6}{90}$ **e.** $\dfrac{9}{90}$

52. Reduce each fraction to lowest terms.

    a. $\dfrac{3}{105}$ **b.** $\dfrac{5}{105}$ **c.** $\dfrac{7}{105}$ **d.** $\dfrac{15}{105}$ **e.** $\dfrac{21}{105}$

53. The answer to each problem below is wrong. Give the correct answer.

    a. $\dfrac{5}{15} = \dfrac{\cancel{5}}{3 \cdot \cancel{5}} = \dfrac{0}{3}$ **b.** $\dfrac{5}{6} = \dfrac{3+2}{4+2} = \dfrac{3}{4}$ **c.** $\dfrac{6}{30} = \dfrac{2 \cdot 3}{2 \cdot 3 \cdot 5} = 5$

54. The answer to each problem below is wrong. Give the correct answer.

    a. $\dfrac{10}{20} = \dfrac{7+\cancel{3}}{17+\cancel{3}} = \dfrac{7}{17}$ **b.** $\dfrac{9}{36} = \dfrac{\cancel{3} \cdot \cancel{3}}{2 \cdot 2 \cdot \cancel{3} \cdot \cancel{3}} = \dfrac{0}{4}$ **c.** $\dfrac{4}{12} = \dfrac{2 \cdot 2}{2 \cdot 2 \cdot 3} = 3$

55. Which of the fractions $\frac{6}{8}, \frac{15}{20}, \frac{9}{16}$, and $\frac{21}{28}$ does not reduce to $\frac{3}{4}$?

56. Which of the fractions $\frac{4}{9}, \frac{10}{15}, \frac{8}{12}$, and $\frac{6}{12}$ does not reduce to $\frac{2}{3}$?

Applying the Concepts

57. Income A family's monthly income is $2,400, and they spend $600 each month on food. Write the amount they spend on food as a fraction of their monthly income in lowest terms.

58. Hours and Minutes There are 60 minutes in 1 hour. What fraction of an hour is 20 minutes? Write your answer in lowest terms.

59. Hours and Days There are 24 hours in 1 day. What fraction of a day is 4 hours? Write your answer in lowest terms.

60. Students The table below shows the number of male and female students enrolled at a small community college. What fraction of the students is male? Write your answer in lowest terms.

COMMUNITY COLLEGE ENROLLMENT	
Men	1,400
Women	1,800

61. Final Exam Suppose 33 people took the final exam in a math class. If 11 people got an A on the final exam, what fraction of the students did not get an A on the exam? Write your answer in lowest terms.

62. Income Tax A person making $21,000 a year pays $3,000 in income tax. What fraction of the person's income is paid as income tax? Write your answer in lowest terms.

The number line at the right extends from 0 to 2, with the segment from 0 to 1 and the segment from 1 to 2 each divided into 8 equal parts. Locate each of the following numbers on this number line.

63. $\frac{1}{2}$, $\frac{2}{4}$, $\frac{4}{8}$, and $\frac{8}{16}$ **64.** $\frac{3}{2}$, $\frac{6}{4}$, $\frac{12}{8}$, and $\frac{24}{16}$ **65.** $\frac{5}{4}$, $\frac{10}{8}$, and $\frac{20}{16}$ **66.** $\frac{1}{4}$, $\frac{2}{8}$, and $\frac{4}{16}$

Estimating

67. Which of the following fractions is closest to the number $\frac{1}{2}$?

   a. $\frac{5}{4}$ **b.** $\frac{5}{8}$ **c.** $\frac{5}{16}$

68. Which of the following fractions is closest to the number $\frac{1}{2}$?

   a. $\frac{3}{8}$ **b.** $\frac{5}{8}$ **c.** $\frac{7}{8}$

69. Which of the following fractions is closest to the number $\frac{1}{2}$?

   a. $\frac{3}{4}$ **b.** $\frac{5}{8}$ **c.** $\frac{9}{16}$

70. Which of the following fractions is closest to the number 1?

   a. $\frac{3}{4}$ **b.** $\frac{7}{8}$ **c.** $\frac{15}{16}$

The nutrition labels below are from two different granola bars. Use them to work Problems 71–76.

Granola Bar 1

Nutrition Facts
Serving Size 2 bars (47g)
Servings Per Container: 6

Amount Per Serving

Calories 210	Calories from fat 70

	% Daily Value*
Total Fat 8g	12%
Saturated Fat 1g	5%
Cholesterol 0mg	0%
Sodium 150mg	6%
Total Carbohydrate 32g	11%
Fiber 2g	10%
Sugars 12g	
Protein 4g	

*Percent Daily Values are based on a 2,000 calorie diet. Your daily values may be higher or lower depending on your calorie needs.

Granola Bar 2

Nutrition Facts
Serving Size 1 bar (21g)
Servings Per Container: 8

Amount Per Serving

Calories 80	Calories from fat 15

	% Daily Value*
Total Fat 1.5g	2%
Saturated Fat 0g	0%
Cholesterol 0mg	0%
Sodium 60mg	3%
Total Carbohydrate 16g	5%
Fiber 1g	4%
Sugars 5g	
Protein 2g	

*Percent Daily Values are based on a 2,000 calorie diet. Your daily values may be higher or lower depending on your calorie needs.

71. What fraction of the calories in Bar 1 comes from fat?

72. What fraction of the calories in Bar 2 comes from fat?

73. For Bar 1, what fraction of the total fat is from saturated fat?

74. For Bar 2, what fraction of the total fat is from saturated fat?

75. What fraction of the total carbohydrates in Bar 1 is from sugar?

76. What fraction of the total carbohydrates in Bar 2 is from sugar?

Calculator Problems

Use your calculator to help fill in the blanks for the following problems.

77. $\dfrac{5}{7} = \dfrac{?}{574}$

78. $\dfrac{2}{3} = \dfrac{?}{2,223}$

79. $\dfrac{8}{9} = \dfrac{728}{?}$

80. $\dfrac{7}{8} = \dfrac{34,972}{?}$

81. $\dfrac{3,787}{4,869} = \dfrac{?}{9}$

82. $\dfrac{684}{836} = \dfrac{?}{11}$

83. $\dfrac{262}{1,965} = \dfrac{?}{15}$

84. $\dfrac{616}{1,672} = \dfrac{?}{19}$

Review Problems

The problems below review material we covered in Section 1.7.

Use the distributive property to rewrite each of the following expressions. Once the distributive property has been used, simplify as much as possible.

85. $6(4 + 8)$

86. $8(2 + 3)$

87. $9(3 + 2)$

88. $6(5 + 1)$

89. $5(x + 4)$

90. $4(x + 5)$

91. $3(4x + 5)$

92. $5(4x + 3)$

Extending the Concepts

93. In your own words, explain the process of reducing a fraction to lowest terms. If you want to include an example in your explanation, use the fraction $\frac{36}{42}$.

94. In your own words, compare the fractions $\frac{2}{3}$ and $\frac{3}{2}$. Include in your comparison the words *proper*, *improper*, *numerator*, and *denominator*. Use the number line to show which of the numbers is larger.

95. How many numbers between 1 and 20 have the prime number 3 as one of their factors? List them.

96. How many numbers between 20 and 40 have the prime number 3 as one of their factors? List them.

97. Can the square of a prime number be a prime number also? Explain your answer.

98. Is it possible for the sum of two composite numbers to be a prime number? Explain your answer.

3.3 Multiplication with Fractions and the Area of a Triangle

Introduction . . .

A recipe calls for $\frac{3}{4}$ cup of flour. If you are making only $\frac{1}{2}$ the recipe, how much flour do you use? This question can be answered by multiplying $\frac{1}{2}$ and $\frac{3}{4}$. Here is the problem written in symbols:

$$\frac{1}{2} \cdot \frac{3}{4} = \frac{3}{8}$$

As you can see from this example, to multiply two fractions, we multiply the numerators and then multiply the denominators. We begin this section with the rule for multiplication of fractions.

> **RULE**
>
> The product of two fractions is the fraction whose numerator is the product of the two numerators and whose denominator is the product of the two denominators. We can write this rule in symbols as follows:
>
> If a, b, c, and d represent any numbers and b and d are not zero, then
>
> $$\frac{a}{b} \cdot \frac{c}{d} = \frac{a \cdot c}{b \cdot d}$$

EXAMPLE 1 Multiply: $\frac{3}{5} \cdot \frac{2}{7}$

SOLUTION Using our rule for multiplication, we multiply the numerators and multiply the denominators:

$$\frac{3}{5} \cdot \frac{2}{7} = \frac{3 \cdot 2}{5 \cdot 7} = \frac{6}{35}$$

The product of $\frac{3}{5}$ and $\frac{2}{7}$ is the fraction $\frac{6}{35}$. The numerator 6 is the product of 3 and 2, and the denominator 35 is the product of 5 and 7. ∎

EXAMPLE 2 Multiply: $-\frac{3}{8} \cdot 5$

SOLUTION The number 5 can be written as $\frac{5}{1}$. That is, 5 can be considered a fraction with numerator 5 and denominator 1. Writing 5 this way enables us to apply the rule for multiplying fractions. Remember also that the product of two numbers with different signs is negative:

$$-\frac{3}{8} \cdot 5 = -\frac{3}{8} \cdot \frac{5}{1}$$
$$= -\frac{3 \cdot 5}{8 \cdot 1} \quad \textbf{Unlike signs give a negative answer}$$
$$= -\frac{15}{8}$$
∎

Practice Problems

1. Multiply: $\frac{2}{3} \cdot \frac{5}{9}$

2. Multiply: $-\frac{2}{5} \cdot 7$

Answers

1. $\frac{10}{27}$ 2. $-\frac{14}{5}$

3. Multiply: $\dfrac{1}{3}\left(\dfrac{4}{5} \cdot \dfrac{1}{3}\right)$

EXAMPLE 3 Multiply: $\dfrac{1}{2}\left(\dfrac{3}{4} \cdot \dfrac{1}{5}\right)$

SOLUTION   We find the product inside the parentheses first and then multiply the result by $\frac{1}{2}$:

$$\frac{1}{2}\left(\frac{3}{4} \cdot \frac{1}{5}\right) = \frac{1}{2}\left(\frac{3 \cdot 1}{4 \cdot 5}\right)$$
$$= \frac{1}{2}\left(\frac{3}{20}\right)$$
$$= \frac{1 \cdot 3}{2 \cdot 20} = \frac{3}{40}$$
∎

The properties of multiplication that we developed in Chapter 1 for whole numbers apply to fractions as well. That is, if a, b, and c are fractions, then

$$a \cdot b = b \cdot a$$ **Multiplication with fractions is commutative**

$$a \cdot (b \cdot c) = (a \cdot b) \cdot c$$ **Multiplication with fractions is associative**

To demonstrate the associative property for fractions, let's do Example 3 again, but this time we will apply the associative property first:

$$\frac{1}{2}\left(\frac{3}{4} \cdot \frac{1}{5}\right) = \left(\frac{1}{2} \cdot \frac{3}{4}\right) \cdot \frac{1}{5}$$ **Associative property**
$$= \left(\frac{1 \cdot 3}{2 \cdot 4}\right) \cdot \frac{1}{5}$$
$$= \left(\frac{3}{8}\right) \cdot \frac{1}{5}$$
$$= \frac{3 \cdot 1}{8 \cdot 5} = \frac{3}{40}$$

The result is identical to that of Example 3.

Here is another example that involves the associative property. Problems like this will be useful in the next chapter when we solve equations.

4. Multiply: $\dfrac{1}{4}(4y)$

EXAMPLE 4 Multiply: $\dfrac{1}{3}(3x)$

SOLUTION   Remember that $3x$ means 3 times x. The 3 and the x are not combined. If we apply the associative property so that we can group the $\frac{1}{3}$ and 3 together, we will be able to multiply them:

$$\frac{1}{3}(3x) = \left(\frac{1}{3} \cdot 3\right)x$$ **Associative property**
$$= 1x$$ **The product of $\dfrac{1}{3}$ and 3 is 1**
$$= x$$ **The product of 1 and x is x**
∎

Answers

3. $\dfrac{4}{45}$ **4.** y

The answers to all the examples so far in this section have been in lowest terms. Let's see what happens when we multiply two fractions to get a product that is not in lowest terms.

EXAMPLE 5 Multiply: $\dfrac{15}{8} \cdot \dfrac{4}{9}$

SOLUTION   Multiplying the numerators and multiplying the denominators, we have

$$\frac{15}{8} \cdot \frac{4}{9} = \frac{15 \cdot 4}{8 \cdot 9}$$
$$= \frac{60}{72}$$

The product is $\frac{60}{72}$, which can be reduced to lowest terms by factoring 60 and 72 and then dividing out any factors they have in common:

$$\frac{60}{72} = \frac{2 \cdot 2 \cdot 3 \cdot 5}{2 \cdot 2 \cdot 2 \cdot 3 \cdot 3}$$
$$= \frac{5}{6}$$

We can actually save ourselves some time by factoring before we multiply. Here's how it is done:

$$\frac{15}{8} \cdot \frac{4}{9} = \frac{15 \cdot 4}{8 \cdot 9}$$
$$= \frac{(3 \cdot 5) \cdot (2 \cdot 2)}{(2 \cdot 2 \cdot 2) \cdot (3 \cdot 3)}$$
$$= \frac{3 \cdot 5 \cdot 2 \cdot 2}{2 \cdot 2 \cdot 2 \cdot 3 \cdot 3}$$
$$= \frac{5}{6}$$ ∎

The result is the same in both cases. Reducing to lowest terms before we actually multiply takes less time. Here are some additional examples.

EXAMPLE 6 $-\dfrac{9}{2} \cdot \left(-\dfrac{8}{18}\right) = \dfrac{9 \cdot 8}{2 \cdot 18}$ **Like signs give a positive product**

$$= \frac{(3 \cdot 3) \cdot (2 \cdot 2 \cdot 2)}{2 \cdot (2 \cdot 3 \cdot 3)}$$
$$= \frac{3 \cdot 3 \cdot 2 \cdot 2 \cdot 2}{2 \cdot 2 \cdot 3 \cdot 3}$$
$$= \frac{2}{1}$$
$$= 2$$ ∎

5. Multiply: $\dfrac{12}{25} \cdot \dfrac{5}{6}$

6. Find the product: $\dfrac{8}{3} \cdot \dfrac{9}{24}$

Note
Although $\frac{2}{1}$ is in lowest terms, it is still simpler to write the answer as just 2. We will always do this when the denominator is the number 1.

Find each product.

7. $\dfrac{yz^2}{x} \cdot \dfrac{x^3}{yz}$

Note

Remember, we are assuming that any variables that appear in a denominator are not 0.

8. $\dfrac{3}{4} \cdot \dfrac{8}{3} \cdot \dfrac{1}{6}$

Apply the definition of exponents, and then multiply.

9. $\left(\dfrac{2}{3}\right)^2$

10. $\left(\dfrac{3}{4}\right)^2 \cdot \dfrac{1}{2}$

EXAMPLE 7
$$\dfrac{x^2y}{z} \cdot \dfrac{z^3}{xy} = \dfrac{x^2y \cdot z^3}{z \cdot xy}$$
$$= \dfrac{(x \cdot x \cdot y)(z \cdot z \cdot z)}{z \cdot x \cdot y}$$
$$= \dfrac{x \cdot x \cdot \cancel{y} \cdot \cancel{z} \cdot z \cdot z}{\cancel{z} \cdot x \cdot \cancel{y}}$$
$$= \dfrac{x \cdot z \cdot z}{1}$$
$$= xz^2 \qquad\qquad\blacksquare$$

EXAMPLE 8
$$\dfrac{2}{3} \cdot \dfrac{6}{5} \cdot \dfrac{5}{8} = \dfrac{2 \cdot 6 \cdot 5}{3 \cdot 5 \cdot 8}$$
$$= \dfrac{2 \cdot (2 \cdot 3) \cdot 5}{3 \cdot 5 \cdot (2 \cdot 2 \cdot 2)}$$
$$= \dfrac{\cancel{2} \cdot 2 \cdot \cancel{3} \cdot \cancel{5}}{\cancel{3} \cdot \cancel{5} \cdot \cancel{2} \cdot 2 \cdot 2}$$
$$= \dfrac{1}{2} \qquad\qquad\blacksquare$$

In Chapter 1 we did some work with exponents. We can extend our work with exponents to include fractions, as the following examples indicate.

EXAMPLE 9
$$\left(-\dfrac{3}{4}\right)^2 = -\dfrac{3}{4}\left(-\dfrac{3}{4}\right)$$
$$= \dfrac{3 \cdot 3}{4 \cdot 4} \qquad \textbf{Like signs give a positive product}$$
$$= \dfrac{9}{16} \qquad\qquad\blacksquare$$

EXAMPLE 10
$$\left(\dfrac{5}{6}\right)^2 \cdot \dfrac{1}{2} = \dfrac{5}{6} \cdot \dfrac{5}{6} \cdot \dfrac{1}{2}$$
$$= \dfrac{5 \cdot 5 \cdot 1}{6 \cdot 6 \cdot 2}$$
$$= \dfrac{25}{72} \qquad\qquad\blacksquare$$

The word *of* used in connection with fractions indicates multiplication. If we want to find $\frac{1}{2}$ of $\frac{2}{3}$, then what we do is multiply $\frac{1}{2}$ and $\frac{2}{3}$.

Answers

7. x^2z **8.** $\dfrac{1}{3}$ **9.** $\dfrac{4}{9}$ **10.** $\dfrac{9}{32}$

EXAMPLE 11 Find $\frac{1}{2}$ of $\frac{2}{3}$.

SOLUTION Knowing the word *of* as used here indicates multiplication, we have

$$\frac{1}{2} \text{ of } \frac{2}{3} = \frac{1}{2} \cdot \frac{2}{3}$$

$$= \frac{1 \cdot 2}{2 \cdot 3} = \frac{1}{3}$$

This seems to make sense. Logically, $\frac{1}{2}$ of $\frac{2}{3}$ should be $\frac{1}{3}$, as Figure 1 shows.

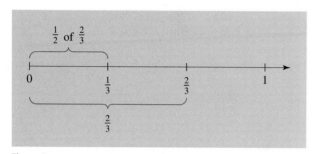

Figure 1

EXAMPLE 12 What is $\frac{3}{4}$ of -12?

SOLUTION Again, *of* means multiply.

$$\frac{3}{4} \text{ of } -12 = \frac{3}{4}(-12)$$

$$= \frac{3}{4}\left(-\frac{12}{1}\right)$$

$$= -\frac{3 \cdot 12}{4 \cdot 1} \qquad \textbf{Unlike signs give a negative product}$$

$$= -\frac{3 \cdot 2 \cdot 2 \cdot 3}{2 \cdot 2 \cdot 1}$$

$$= -\frac{9}{1} = -9$$

Facts from Geometry: The Area of a Triangle

The formula for the area of a triangle is one application of multiplication with fractions. Figure 2 shows a triangle with base b and height h. Below the triangle is the formula for its area. As you can see, it is a product containing the fraction $\frac{1}{2}$.

$$\text{Area} = \frac{1}{2} \text{(base)(height)}$$

$$A = \frac{1}{2} bh$$

Figure 2 The area of a triangle

11. Find $\frac{2}{3}$ of $\frac{1}{2}$.

12. What is $\frac{2}{3}$ of -12?

Note on Shortcuts
As you become familiar with multiplying fractions, you may notice shortcuts that reduce the number of steps in the problems. It's okay to use these shortcuts if you understand why they work and are consistently getting correct answers. If you are using shortcuts and not consistently getting correct answers, then go back to showing all the work until you completely understand the process.

Answers

11. $\frac{1}{3}$ **12.** -8

13. Find the area of the triangle below.

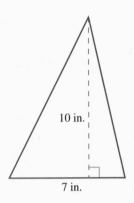

14. Find the total area enclosed by the figure.

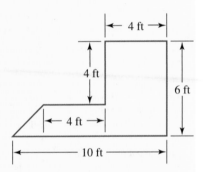

EXAMPLE 13 Find the area of the triangle in Figure 3.

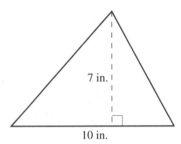

Figure 3   A triangle with base 10 inches and height 7 inches

SOLUTION   Applying the formula for the area of a triangle, we have

$$A = \frac{1}{2}bh = \frac{1}{2} \cdot 10 \cdot 7 = 5 \cdot 7 = 35 \text{ in}^2$$

∎

EXAMPLE 14 Find the area of the figure in Figure 4.

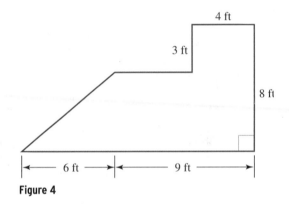

Figure 4

SOLUTION   We divide the figure into three parts and then find the area of each part (see Figure 5). The area of the whole figure is the sum of the areas of its parts.

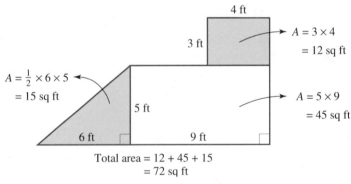

Total area = 12 + 45 + 15
= 72 sq ft

Figure 5   ∎

Problem Set (3.3)

Find each of the following products. (Multiply.)

1. $\dfrac{2}{3} \cdot \dfrac{4}{5}$

2. $\dfrac{5}{6} \cdot \dfrac{7}{4}$

3. $\dfrac{1}{2} \cdot \dfrac{7}{4}$

4. $\dfrac{3}{5} \cdot \dfrac{4}{7}$

5. $-\dfrac{5}{3} \cdot \dfrac{3}{5}$

6. $-\dfrac{4}{7} \cdot \dfrac{7}{4}$

7. $\dfrac{3}{4} \cdot 9$

8. $\dfrac{2}{3} \cdot 5$

9. $-\dfrac{6}{7}\left(-\dfrac{7}{6}\right)$

10. $-\dfrac{2}{9}\left(-\dfrac{9}{2}\right)$

11. $\dfrac{1}{2} \cdot \dfrac{1}{3} \cdot \dfrac{1}{4}$

12. $\dfrac{2}{3} \cdot \dfrac{4}{5} \cdot \dfrac{1}{3}$

13. $\dfrac{2}{5} \cdot \dfrac{3}{5} \cdot \dfrac{4}{5}$

14. $\dfrac{1}{4} \cdot \dfrac{3}{4} \cdot \dfrac{3}{4}$

15. $\dfrac{x}{y} \cdot \dfrac{y}{z} \cdot \dfrac{z}{x}$

16. $\dfrac{y}{x} \cdot \dfrac{x}{z} \cdot \dfrac{z}{y}$

Complete the following tables.

17.

FIRST NUMBER	SECOND NUMBER	THEIR PRODUCT
x	y	xy
$\dfrac{1}{2}$	$\dfrac{2}{3}$	
$\dfrac{2}{3}$	$\dfrac{3}{4}$	
$\dfrac{3}{4}$	$\dfrac{4}{5}$	
$\dfrac{5}{a}$	$\dfrac{a}{6}$	

18.

FIRST NUMBER	SECOND NUMBER	THEIR PRODUCT
x	y	xy
12	$\dfrac{1}{2}$	
12	$\dfrac{1}{3}$	
12	$\dfrac{1}{4}$	
12	$\dfrac{1}{6}$	

19.

FIRST NUMBER	SECOND NUMBER	THEIR PRODUCT
x	y	xy
$\dfrac{1}{2}$	30	
$\dfrac{1}{5}$	30	
$\dfrac{1}{6}$	30	
$\dfrac{1}{15}$	30	

20.

FIRST NUMBER	SECOND NUMBER	THEIR PRODUCT
x	y	xy
$\dfrac{1}{3}$	$\dfrac{3}{5}$	
$\dfrac{3}{5}$	$\dfrac{5}{7}$	
$\dfrac{5}{7}$	$\dfrac{7}{9}$	
$\dfrac{7}{b}$	$\dfrac{b}{11}$	

Multiply each of the following. Be sure all answers are written in lowest terms.

21. $\dfrac{196}{27} \cdot \dfrac{9}{392}$

22. $\dfrac{135}{16} \cdot \dfrac{2}{45}$

23. $\dfrac{3}{4} \cdot 12$

24. $\dfrac{3}{4} \cdot 20$

25. $-\dfrac{1}{3}(-3)$

26. $-\dfrac{1}{5}(-5)$

27. $\dfrac{2}{5} \cdot 20$

28. $\dfrac{3}{5} \cdot 15$

29. $\dfrac{72}{35} \cdot \dfrac{55}{108} \cdot \dfrac{7}{110}$

30. $\dfrac{32}{27} \cdot \dfrac{72}{49} \cdot \dfrac{1}{40}$

31. $\dfrac{a^2 b}{c} \cdot \dfrac{c^3}{ab^2}$

32. $\dfrac{ab^2}{c} \cdot \dfrac{c^2}{a^2 b}$

Use the associative property to rewrite each of the following expressions, and then simplify as much as possible.

33. $\dfrac{3}{4}\left(\dfrac{4}{3}\cdot 9\right)$ **34.** $\dfrac{7}{6}\left(\dfrac{1}{7}\cdot 12\right)$ **35.** $\dfrac{3}{5}\left(\dfrac{1}{3}\cdot 20\right)$ **36.** $\dfrac{2}{3}\left(\dfrac{1}{2}\cdot 27\right)$ **37.** $\dfrac{1}{2}(2x)$ **38.** $\dfrac{1}{4}(4x)$

39. $\dfrac{1}{5}(5y)$ **40.** $\dfrac{1}{6}(6y)$ **41.** $\dfrac{1}{8}(8a)$ **42.** $\dfrac{1}{3}(3a)$

Expand and simplify each of the following.

43. $\left(-\dfrac{1}{2}\right)^2$ **44.** $\left(-\dfrac{1}{3}\right)^2$ **45.** $\left(-\dfrac{2}{3}\right)^3$ **46.** $\left(-\dfrac{3}{5}\right)^3$

47. $\left(\dfrac{3}{4}\right)^2\cdot\dfrac{8}{9}$ **48.** $\left(\dfrac{5}{6}\right)^2\cdot\dfrac{12}{15}$ **49.** $\left(\dfrac{1}{2}\right)^2\left(\dfrac{3}{5}\right)^2$ **50.** $\left(\dfrac{3}{8}\right)^2\left(\dfrac{4}{3}\right)^2$

51. $\left(\dfrac{1}{2}\right)^2\cdot 8+\left(\dfrac{1}{3}\right)^2\cdot 9$ **52.** $\left(\dfrac{2}{3}\right)^2\cdot 9+\left(\dfrac{1}{2}\right)^2\cdot 4$

53. Find $\frac{3}{8}$ of 64.   **54.** Find $\frac{2}{3}$ of 18.

55. What is $\frac{1}{3}$ of the sum of 8 and 4?   **56.** What is $\frac{3}{5}$ of the sum of 8 and 7?

57. Find $\frac{1}{2}$ of $\frac{3}{4}$ of 24.   **58.** Find $\frac{3}{5}$ of $\frac{1}{3}$ of 15.

Find the mistakes in Problems 59–60. Correct the right-hand side of each one.

59. $\dfrac{1}{2}\cdot\dfrac{3}{5}=\dfrac{4}{10}$ **60.** $\dfrac{2}{7}\cdot\dfrac{3}{5}=\dfrac{5}{35}$

61. a. Complete the following table.

NUMBER	SQUARE
x	x^2
1	
2	
3	
4	
5	
6	
7	

b. Using the results of part a, fill in the blank in the following statement:

For numbers larger than 1, the square of the number is _____ than the number.

62. a. Complete the following table.

NUMBER	SQUARE
x	x^2
$\frac{1}{2}$	
$\frac{1}{3}$	
$\frac{1}{4}$	
$\frac{1}{5}$	
$\frac{1}{6}$	
$\frac{1}{7}$	
$\frac{1}{8}$	

b. Using the results of part a, fill in the blank in the following statement:

For numbers between 0 and 1, the square of the number is _____ than the number.

63. Find the area of the triangle with base 19 inches and height 14 inches.

64. Find the area of the triangle with base 13 inches and height 8 inches.

65. The base of a triangle is $\frac{4}{3}$ feet and the height is $\frac{2}{3}$ feet. Find the area.

66. The base of a triangle is $\frac{8}{7}$ feet and the height is $\frac{14}{5}$ feet. Find the area.

Find the area of each figure below.

67.

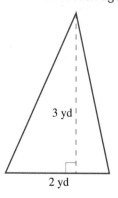

3 yd

2 yd

68.

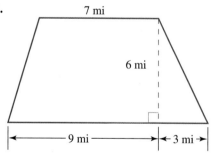

7 mi

6 mi

9 mi

3 mi

69.

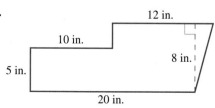

12 in.

10 in.

8 in.

5 in.

20 in.

70.

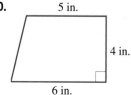

5 in.

4 in.

6 in.

Applying the Concepts

The pie chart below was introduced at the beginning of Section 3.1. Use the information in the pie chart to answer questions 71 and 72.

Cal Poly enrollment for Fall 1999

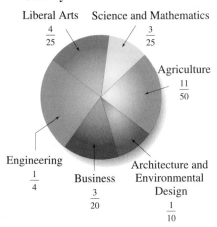

71. Reading a Pie Chart If there are approximately 15,800 students attending Cal Poly, approximately how many of them are studying agriculture?

72. Reading a Pie Chart If there are exactly 15,828 students attending Cal Poly, exactly how many of them are studying engineering?

73. Cooking A recipe calls for $\frac{3}{4}$ cup of flour. If you were going to make only $\frac{1}{2}$ the recipe, how much flour would you use?

74. Cooking A recipe calls for $\frac{2}{3}$ cup of sugar. If the recipe is to be doubled, how much sugar should be used?

75. Perimeter and Area

a. The lengths of the sides of each of the squares in the grid below are all $\frac{1}{4}$ inch, as is each side of the dark square in the corner. Draw three more squares on the grid. Each side in the first one will be $\frac{1}{2}$ inch, each side of the second square will be $\frac{3}{4}$ inch, and each side of the third square will be 1 inch.

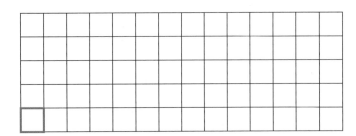

b. Use the squares you have drawn above to complete the following tables. (You need not add fractions to find the perimeters; just multiply the length of each side by 4.)

Table 1

PERIMETERS OF SQUARES

LENGTH OF EACH SIDE (IN INCHES)	PERIMETER (IN INCHES)
$\frac{1}{4}$	
$\frac{2}{4} = \frac{1}{2}$	
$\frac{3}{4}$	
$\frac{4}{4} = 1$	

Table 2

AREAS OF SQUARES

LENGTH OF EACH SIDE (IN INCHES)	AREA (IN SQUARE INCHES)
$\frac{1}{4}$	
$\frac{1}{2}$	
$\frac{3}{4}$	
1	

c. Use Figures 6 and 7 to construct line graphs from the information in Tables 1 and 2.

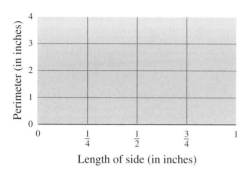

Figure 6   A line graph of Table 1

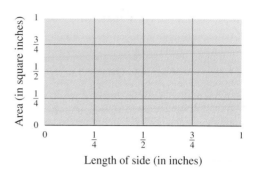

Figure 7   A line graph of Table 2

76. Perimeter and Area

a. The base and the height of the triangle shown below are $\frac{1}{4}$ inch each. Draw three more triangles on the grid. The base and the height in the first one will be $\frac{1}{2}$ inch each, the base and the height for the second one will be $\frac{3}{4}$ inch each, and the base and the height of the third one will be 1 inch each.

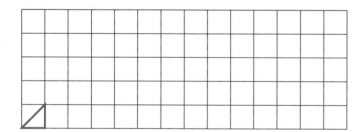

b. Use the triangles you have drawn above to complete the following table.

Table 3

AREAS OF TRIANGLES

LENGTH OF BASE AND HEIGHT (IN INCHES)	AREA (IN SQUARE INCHES)
$\frac{1}{4}$	
$\frac{1}{2}$	
$\frac{3}{4}$	
1	

c. Use Figure 8 to construct a line graph from the information in Table 3.

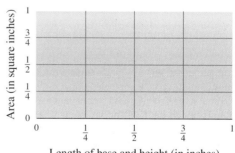

Figure 8   A line graph of Table 3

Estimating

For each problem below, mentally estimate which of the numbers 0, 1, 2, or 3 is closest to the answer. Make your estimate without using pencil and paper or a calculator.

77. $\dfrac{11}{5} \cdot \dfrac{19}{20}$

78. $\dfrac{3}{5} \cdot \dfrac{1}{20}$

79. $\dfrac{16}{5} \cdot \dfrac{23}{24}$

80. $\dfrac{9}{8} \cdot \dfrac{31}{32}$

81. $\dfrac{1}{8} \cdot \dfrac{15}{32}$

82. $\dfrac{7}{8} \cdot \dfrac{17}{16}$

Calculator Problems

Use a calculator to find the following products. Do *not* reduce your answers to lowest terms.

83. $\dfrac{31}{32} \cdot \dfrac{41}{42} \cdot \dfrac{43}{44}$

84. $\dfrac{16}{17} \cdot \dfrac{18}{19} \cdot \dfrac{20}{21}$

85. $\dfrac{93}{97} \cdot 18 \cdot \dfrac{46}{53}$

86. $\dfrac{32}{79} \cdot 25 \cdot \dfrac{87}{91}$

Review Problems

In the next section we will do division with fractions. As you already know, division and multiplication are closely related. These review problems are intended to let you see more of the relationship between multiplication and division.

Perform the indicated operations.

87. $6 \div 2$

88. $6 \cdot \dfrac{1}{2}$

89. $8 \div 4$

90. $8 \cdot \dfrac{1}{4}$

91. $15 \div 3$

92. $15 \cdot \dfrac{1}{3}$

93. $18 \div 6$

94. $18 \cdot \dfrac{1}{6}$

Extending the Concepts

Recall that a geometric sequence is a sequence in which each term comes from the previous term by multiplying by the same number each time. For example, the sequence $1, \frac{1}{2}, \frac{1}{4}, \frac{1}{8}, \ldots$ is a geometric sequence in which each term is found by multiplying the previous term by $\frac{1}{2}$. By observing this fact, we know that the next term in the sequence will be $\frac{1}{8} \cdot \frac{1}{2} = \frac{1}{16}$.

Find the next number in each of the geometric sequences below.

95. $1, \dfrac{1}{3}, \dfrac{1}{9}, \ldots$

96. $1, \dfrac{1}{4}, \dfrac{1}{16}, \ldots$

97. $\dfrac{3}{2}, 1, \dfrac{2}{3}, \dfrac{4}{9}, \ldots$

98. $\dfrac{2}{3}, 1, \dfrac{3}{2}, \dfrac{9}{4}, \ldots$

3.4 Division with Fractions

Introduction . . .

A few years ago our 4-H club was making blankets to keep their lambs clean at the county fair. Each blanket required $\frac{3}{4}$ yard of material. We had 9 yards of material left over from the year before. To see how many blankets we could make, we divided 9 by $\frac{3}{4}$. The result was 12, meaning that we could make 12 lamb blankets for the fair.

Before we define division with fractions, we must first introduce the idea of *reciprocals*. Look at the following multiplication problems:

$$\frac{3}{4} \cdot \frac{4}{3} = \frac{12}{12} = 1 \qquad \frac{7}{8} \cdot \frac{8}{7} = \frac{56}{56} = 1$$

In each case the product is 1. Whenever the product of two numbers is 1, we say the two numbers are *reciprocals*.

> **DEFINITION**
>
> Two numbers whose product is 1 are said to be **reciprocals.** In symbols, the reciprocal of $\frac{a}{b}$ is $\frac{b}{a}$, because
>
> $$\frac{a}{b} \cdot \frac{b}{a} = \frac{a \cdot b}{b \cdot a} = \frac{a \cdot b}{a \cdot b} = 1 \qquad (a \neq 0, b \neq 0)$$

Every number has a reciprocal except 0. The reason 0 does not have a reciprocal is because the product of *any* number with 0 is 0. It can never be 1. Reciprocals of whole numbers are fractions with 1 as the numerator. For example, the reciprocal of 5 is $\frac{1}{5}$, because

$$5 \cdot \frac{1}{5} = \frac{5}{1} \cdot \frac{1}{5} = \frac{5}{5} = 1$$

Table 1 lists some numbers and their reciprocals.

Table 1

NUMBER	RECIPROCAL	REASON
$\frac{3}{4}$	$\frac{4}{3}$	Because $\frac{3}{4} \cdot \frac{4}{3} = \frac{12}{12} = 1$
$-\frac{9}{5}$	$-\frac{5}{9}$	Because $-\frac{9}{5}\left(-\frac{5}{9}\right) = \frac{45}{45} = 1$
$\frac{1}{3}$	3	Because $\frac{1}{3} \cdot 3 = \frac{1}{3} \cdot \frac{3}{1} = \frac{3}{3} = 1$
-7	$-\frac{1}{7}$	Because $-7\left(-\frac{1}{7}\right) = -\frac{7}{1}\left(-\frac{1}{7}\right) = \frac{7}{7} = 1$

Division with fractions is accomplished by using reciprocals. More specifically, we can define division by a fraction to be the same as multiplication by its reciprocal. Here is the precise definition:

> **DEFINITION**
>
> If a, b, c, and d are integers and b, c, and d are all not equal to 0, then
>
> $$\frac{a}{b} \div \frac{c}{d} = \frac{a}{b} \cdot \frac{d}{c}$$

Note
Defining division to be the same as multiplication by the reciprocal does make sense. If we divide 6 by 2, we get 3. On the other hand, if we multiply 6 by $\frac{1}{2}$ (the reciprocal of 2), we also get 3. Whether we divide by 2 or multiply by $\frac{1}{2}$, we get the same result.

FRACTIONS AND MIXED NUMBERS

This definition states that dividing by the fraction $\frac{c}{d}$ is exactly the same as multiplying by its reciprocal $\frac{d}{c}$. Because we developed the rule for multiplying fractions in Section 3.3, we do not need a new rule for division. We simply *replace the divisor by its reciprocal* and multiply. Here are some examples to illustrate the procedure.

Practice Problems

1. Divide: $\frac{1}{3} \div \frac{1}{6}$

EXAMPLE 1   Divide: $\frac{1}{2} \div \frac{1}{4}$

SOLUTION   The divisor is $\frac{1}{4}$, and its reciprocal is $\frac{4}{1}$. Applying the definition of division for fractions, we have

$$\frac{1}{2} \div \frac{1}{4} = \frac{1}{2} \cdot \frac{4}{1}$$
$$= \frac{1 \cdot 4}{2 \cdot 1}$$
$$= \frac{1 \cdot 2 \cdot 2}{2 \cdot 1}$$
$$= \frac{2}{1}$$
$$= 2$$

The quotient of $\frac{1}{2}$ and $\frac{1}{4}$ is 2. Or, $\frac{1}{4}$ "goes into" $\frac{1}{2}$ two times. Logically, our definition for division of fractions seems to be giving us answers that are consistent with what we know about fractions from previous experience. Because 2 times $\frac{1}{4}$ is $\frac{2}{4}$ or $\frac{1}{2}$, it seems logical that $\frac{1}{2}$ divided by $\frac{1}{4}$ should be 2. ■

2. Divide: $\frac{5}{9} \div \frac{10}{3}$

EXAMPLE 2   Divide: $\frac{3}{8} \div \frac{9}{4}$

SOLUTION   Dividing by $\frac{9}{4}$ is the same as multiplying by its reciprocal, which is $\frac{4}{9}$:

$$\frac{3}{8} \div \frac{9}{4} = \frac{3}{8} \cdot \frac{4}{9}$$
$$= \frac{3 \cdot 2 \cdot 2}{2 \cdot 2 \cdot 2 \cdot 3 \cdot 3}$$
$$= \frac{1}{6}$$

The quotient of $\frac{3}{8}$ and $\frac{9}{4}$ is $\frac{1}{6}$.

3. Divide: $-\frac{3}{4} \div 3$

EXAMPLE 3   Divide: $-\frac{2}{3} \div 2$

SOLUTION   The reciprocal of 2 is $\frac{1}{2}$. Applying the definition for division of fractions, we have

$$-\frac{2}{3} \div 2 = -\frac{2}{3} \cdot \frac{1}{2}$$
$$= -\frac{2 \cdot 1}{3 \cdot 2}$$
$$= -\frac{1}{3}$$

■

Answers

1. 2   2. $\frac{1}{6}$ 3. $-\frac{1}{4}$

EXAMPLE 4 Divide: $2 \div \left(-\dfrac{1}{3}\right)$

SOLUTION We replace $-\frac{1}{3}$ by its reciprocal, which is -3, and multiply:

$$2 \div \left(-\frac{1}{3}\right) = 2(-3)$$
$$= -6 \qquad \blacksquare$$

Here are some further examples of division with fractions. Notice in each case that the first step is the only new part of the process.

EXAMPLE 5 $\dfrac{4}{27} \div \dfrac{16}{9} = \dfrac{4}{27} \cdot \dfrac{9}{16}$

$$= \frac{\cancel{4} \cdot \cancel{9}}{3 \cdot \cancel{9} \cdot \cancel{4} \cdot 4}$$
$$= \frac{1}{12} \qquad \blacksquare$$

In this example we did not factor the numerator and the denominator completely in order to reduce to lowest terms because, as you have probably already noticed, it is not necessary to do so. We need to factor only enough to show what numbers are common to the numerator and the denominator. If we factored completely in the second step, it would look like this:

$$= \frac{2 \cdot 2 \cdot \cancel{3} \cdot \cancel{3}}{\cancel{3} \cdot \cancel{3} \cdot 3 \cdot \cancel{2} \cdot \cancel{2} \cdot 2 \cdot 2}$$
$$= \frac{1}{12}$$

The result is the same in both cases. From now on we will factor numerators and denominators only enough to show the factors we are dividing out.

EXAMPLE 6 $\dfrac{16}{35} \div 8 = \dfrac{16}{35} \cdot \dfrac{1}{8}$

$$= \frac{2 \cdot \cancel{8} \cdot 1}{35 \cdot \cancel{8}}$$
$$= \frac{2}{35} \qquad \blacksquare$$

EXAMPLE 7 $-27 \div \left(-\dfrac{3}{2}\right) = -27\left(-\dfrac{2}{3}\right)$

$$= \frac{\cancel{3} \cdot 9 \cdot 2}{\cancel{3}} \qquad \textbf{Like signs give a positive answer}$$
$$= 18 \qquad \blacksquare$$

EXAMPLE 8 $\dfrac{x^2}{y} \div \dfrac{x}{y^3} = \dfrac{x^2}{y} \cdot \dfrac{y^3}{x}$

$$= \frac{\cancel{x} \cdot x \cdot \cancel{y} \cdot y \cdot y}{\cancel{y} \cdot \cancel{x}}$$
$$= \frac{x \cdot y \cdot y}{1}$$
$$= xy^2 \qquad \blacksquare$$

---

4. Divide: $-4 \div \left(-\dfrac{1}{5}\right)$

Find each quotient.

5. $\dfrac{5}{32} \div \dfrac{10}{42}$

6. $\dfrac{12}{25} \div 6$

7. $-12 \div \left(-\dfrac{4}{3}\right)$

8. $\dfrac{x^3}{y} \div \dfrac{x^2}{y^2}$

Answers

4. 20   **5.** $\dfrac{21}{32}$ **6.** $\dfrac{2}{25}$ **7.** 9

8. xy

The next two examples combine what we have learned about division of fractions with the rule for order of operations.

9. The quotient of $\frac{5}{4}$ and $\frac{1}{8}$ is increased by 8. What number results?

EXAMPLE 9   The quotient of $\frac{8}{3}$ and $\frac{1}{6}$ is increased by 5. What number results?

SOLUTION   Translating to symbols, we have

$$\frac{8}{3} \div \frac{1}{6} + 5 = \frac{8}{3} \cdot \frac{6}{1} + 5$$
$$= 16 + 5$$
$$= 21$$ ∎

10. Simplify:

$$18 \div \left(\frac{3}{5}\right)^2 + 48 \div \left(\frac{2}{5}\right)^2$$

EXAMPLE 10 Simplify: $32 \div \left(\frac{4}{3}\right)^2 + 75 \div \left(\frac{5}{2}\right)^2$

SOLUTION   According to the rule for order of operations, we must first evaluate the numbers with exponents, then we divide, and finally we add.

$$32 \div \left(\frac{4}{3}\right)^2 + 75 \div \left(\frac{5}{2}\right)^2 = 32 \div \frac{16}{9} + 75 \div \frac{25}{4}$$
$$= 32 \cdot \frac{9}{16} + 75 \cdot \frac{4}{25}$$
$$= 18 + 12$$
$$= 30$$ ∎

11. How many blankets can the 4-H Club make with 12 yards of material?

EXAMPLE 11   A 4-H Club is making blankets to keep their lambs clean at the county fair. If each blanket requires $\frac{3}{4}$ yard of material, how many blankets can they make from 9 yards of material?

SOLUTION   To answer this question we must divide 9 by $\frac{3}{4}$.

$$9 \div \frac{3}{4} = 9 \cdot \frac{4}{3}$$
$$= 3 \cdot 4$$
$$= 12$$

They can make 12 blankets from the 9 yards of material. ∎

Answers
9. 18   **10.** 350   **11.** 16

Problem Set ⬤3.4

Find the quotient in each case by replacing the divisor by its reciprocal and multiplying.

1. $\dfrac{3}{4} \div \dfrac{1}{5}$

2. $\dfrac{1}{3} \div \dfrac{1}{2}$

3. $-\dfrac{2}{3} \div \dfrac{1}{2}$

4. $-\dfrac{5}{8} \div \dfrac{1}{4}$

5. $6 \div \left(-\dfrac{2}{3}\right)$

6. $8 \div \left(-\dfrac{3}{4}\right)$

7. $20 \div \dfrac{1}{10}$

8. $16 \div \dfrac{1}{8}$

9. $\dfrac{3}{4} \div (-2)$

10. $\dfrac{3}{5} \div (-2)$

11. $\dfrac{7}{8} \div \dfrac{7}{8}$

12. $\dfrac{4}{3} \div \dfrac{4}{3}$

13. $\dfrac{7}{8} \div \dfrac{8}{7}$

14. $\dfrac{4}{3} \div \dfrac{3}{4}$

15. $\dfrac{9}{16} \div \left(-\dfrac{3}{4}\right)$

16. $-\dfrac{25}{36} \div \left(-\dfrac{5}{6}\right)$

17. $\dfrac{25}{46} \div \dfrac{40}{69}$

18. $\dfrac{25}{24} \div \dfrac{15}{36}$

19. $\dfrac{13}{28} \div \dfrac{39}{14}$

20. $\dfrac{28}{125} \div \dfrac{5}{2}$

21. $\dfrac{27}{196} \div \dfrac{9}{392}$

22. $\dfrac{16}{135} \div \dfrac{2}{45}$

23. $\dfrac{x^2}{y^3} \div \dfrac{x}{y^2}$

24. $\dfrac{x}{y^2} \div \dfrac{x^3}{y}$

25. $\dfrac{ab^2}{c} \div \dfrac{b}{c}$

26. $\dfrac{a^2b}{c^2} \div \dfrac{a}{c^3}$

27. $\dfrac{10a^2}{3b} \div \dfrac{5a}{6b}$

28. $\dfrac{12a^2}{5b} \div \dfrac{6a}{5b^2}$

Simplify each expression as much as possible.

29. $10 \div \left(\dfrac{1}{2}\right)^2$

30. $12 \div \left(\dfrac{1}{4}\right)^2$

31. $\dfrac{18}{35} \div \left(\dfrac{6}{7}\right)^2$

32. $\dfrac{48}{55} \div \left(\dfrac{8}{11}\right)^2$

33. $\dfrac{4}{5} \div \dfrac{1}{10} + 5$

34. $\dfrac{3}{8} \div \dfrac{1}{16} + 4$

35. $10 + \dfrac{11}{12} \div \dfrac{11}{24}$

36. $15 + \dfrac{13}{14} \div \dfrac{13}{42}$

37. $24 \div \left(\dfrac{2}{5}\right)^2 + 25 \div \left(\dfrac{5}{6}\right)^2$

38. $18 \div \left(\dfrac{3}{4}\right)^2 + 49 \div \left(\dfrac{7}{9}\right)^2$

39. $100 \div \left(\dfrac{5}{7}\right)^2 + 200 \div \left(\dfrac{2}{3}\right)^2$

40. $64 \div \left(\dfrac{8}{11}\right)^2 + 81 \div \left(\dfrac{9}{11}\right)^2$

41. What is the quotient of $\frac{3}{8}$ and $\frac{5}{8}$?

42. Find the quotient of $\frac{4}{5}$ and $\frac{16}{25}$.

43. If the quotient of 18 and $\frac{3}{5}$ is increased by 10, what number results?

44. If the quotient of 50 and $\frac{5}{3}$ is increased by 8, what number results?

45. Show that multiplying 3 by 5 is the same as dividing 3 by $\frac{1}{5}$.

46. Show that multiplying 8 by $\frac{1}{2}$ is the same as dividing 8 by 2.

Applying the Concepts

Although many of the application problems that follow involve division with fractions, some do not. Be sure to read the problems carefully.

47. Sewing If $\frac{6}{7}$ yard of material is needed to make a blanket, how many blankets can be made from 12 yards of material?

48. Manufacturing A clothing manufacturer is making scarves that require $\frac{3}{8}$ yard of material each. How many can be made from 27 yards of material?

49. Capacity Suppose a bag of candy holds exactly $\frac{1}{4}$ pound of candy. How many of these bags can be filled from 12 pounds of candy?

50. Capacity A certain size bottle holds exactly $\frac{4}{5}$ pint of liquid. How many of these bottles can be filled from a 20-pint container?

51. Cooking A man is making cookies from a recipe that calls for $\frac{3}{4}$ teaspoon of oil. If the only measuring spoon he can find is a $\frac{1}{8}$ teaspoon, how many of these will he have to fill with oil in order to have a total of $\frac{3}{4}$ teaspoon of oil?

52. Cooking A cake recipe calls for $\frac{1}{2}$ cup of sugar. If the only measuring cup available is a $\frac{1}{8}$ cup, how many of these will have to be filled with sugar to make a total of $\frac{1}{2}$ cup of sugar?

53. Student Population If 14 of every 32 students attending Cuesta College are female, what fraction of the students is female? (Simplify your answer.)

54. Population If 27 of every 48 residents of a small town are male, what fraction of the population is male? (Simplify your answer.)

55. Student Population If 14 of every 32 students attending Cuesta College are female, and the total number of students at the school is 4,064, how many of the students are female?

56. Population If 27 of every 48 residents of a small town are male, and the total population of the town is 17,808, how many of the residents are male?

57. Cartons of Milk If a small carton of milk holds exactly $\frac{1}{2}$ pint, how many of the $\frac{1}{2}$-pint cartons can be filled from a 14-pint container?

58. Pieces of Pipe How many pieces of pipe that are $\frac{2}{3}$ foot long must be laid together to make a pipe 16 feet long?

Estimating

For each problem below, mentally estimate which of the numbers 0, 1, or 2 is closest to the answer. Make your estimate without using pencil and paper or a calculator.

59. $\dfrac{11}{5} \div \dfrac{19}{20}$

60. $\dfrac{3}{5} \div 20$

61. $\dfrac{15}{8} \div \dfrac{23}{24}$

62. $\dfrac{9}{8} \div \dfrac{25}{24}$

63. $\dfrac{1}{2} \div 40$

64. $1 \div \dfrac{7}{16}$

Calculator Problems

Use a calculator to do the following division problems. Do *not* reduce to lowest terms.

65. $\dfrac{49}{97} \div \dfrac{37}{71}$

66. $\dfrac{37}{51} \div \dfrac{58}{43}$

67. $\dfrac{17}{83} \div \dfrac{15}{29}$

68. $\dfrac{101}{347} \div \dfrac{58}{67}$

69. $\dfrac{491}{307} \div \dfrac{63}{43}$

70. $\dfrac{763}{247} \div \dfrac{29}{91}$

Review Problems

The problems below review material we covered in Section 3.1. Reviewing these problems will help you understand the next section.

Write each fraction as an equivalent fraction with denominator 6.

71. $\dfrac{1}{2}$

72. $\dfrac{1}{3}$

73. $\dfrac{3}{2}$

74. $\dfrac{2}{3}$

Write each fraction as an equivalent fraction with denominator 12.

75. $\dfrac{1}{3}$

76. $\dfrac{1}{2}$

77. $\dfrac{2}{3}$

78. $\dfrac{3}{4}$

Write each fraction as an equivalent fraction with denominator 12x.

79. $\dfrac{5}{6}$

80. $\dfrac{7}{6}$

81. $\dfrac{2}{3}$

82. $\dfrac{3}{4}$

3.5 Addition and Subtraction with Fractions

Adding and subtracting fractions is actually just another application of the distributive property. The distributive property looks like this:

$$a(b + c) = a(b) + a(c)$$

where a, b, and c may be whole numbers or fractions. We will want to apply this property to expressions like

$$\frac{2}{7} + \frac{3}{7}$$

But before we do, we must make one additional observation about fractions.

The fraction $\frac{2}{7}$ can be written as $2 \cdot \frac{1}{7}$, because

$$2 \cdot \frac{1}{7} = \frac{2}{1} \cdot \frac{1}{7} = \frac{2}{7}$$

Likewise, the fraction $\frac{3}{7}$ can be written as $3 \cdot \frac{1}{7}$, because

$$3 \cdot \frac{1}{7} = \frac{3}{1} \cdot \frac{1}{7} = \frac{3}{7}$$

In general, we can say that the fraction $\frac{a}{b}$ can always be written as $a \cdot \frac{1}{b}$, because

$$a \cdot \frac{1}{b} = \frac{a}{1} \cdot \frac{1}{b} = \frac{a}{b}$$

To add the fractions $\frac{2}{7}$ and $\frac{3}{7}$, we simply rewrite each of them as we have done above and apply the distributive property. Here is how it works:

$$\frac{2}{7} + \frac{3}{7} = 2 \cdot \frac{1}{7} + 3 \cdot \frac{1}{7} \qquad \textbf{Rewrite each fraction}$$

$$= (2 + 3) \cdot \frac{1}{7} \qquad \textbf{Apply the distributive property}$$

$$= 5 \cdot \frac{1}{7} \qquad \textbf{Add 2 and 3 to get 5}$$

$$= \frac{5}{7} \qquad \textbf{Rewrite } 5 \cdot \frac{1}{7} \textbf{ as } \frac{5}{7}$$

We can visualize the process shown above by using circles that are divided into 7 equal parts:

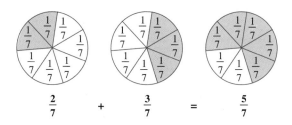

$$\frac{2}{7} \qquad + \qquad \frac{3}{7} \qquad = \qquad \frac{5}{7}$$

The fraction $\frac{5}{7}$ is the sum of $\frac{2}{7}$ and $\frac{3}{7}$. The steps and diagrams above show why we add numerators *but do not add denominators*. Using this example as justification, we can write a rule for adding two fractions that have the same denominator.

Note

Most people who have done any work with adding fractions know that you add fractions that have the same denominator by adding their numerators, but not their denominators. However, most people don't know why this works. The reason why we add numerators but not denominators is because of the distributive property. And that is what the discussion at the left is all about. If you really want to understand addition of fractions, pay close attention to this discussion.

> **RULE**
>
> To add two fractions that have the same denominator, we add their numerators to get the numerator of the answer. The denominator in the answer is the same denominator as in the original fractions.

What we have here is the sum of the numerators placed over the *common denominator*. In symbols we have the following:

If a, b, and c are integers, and c is not equal to 0, then

$$\frac{a}{c} + \frac{b}{c} = \frac{a+b}{c}$$

This rule holds for subtraction as well. That is,

$$\frac{a}{c} - \frac{b}{c} = \frac{a-b}{c}$$

In Examples 1–4, find the sum or difference. (Add or subtract as indicated.) Reduce all answers to lowest terms. (Assume all variables represent nonzero numbers.)

Practice Problems

Find the sum or difference. Reduce all answers to lowest terms.

1. $\dfrac{3}{10} + \dfrac{1}{10}$

EXAMPLE 1 $\quad \dfrac{3}{8} + \dfrac{1}{8} = \dfrac{3+1}{8} \qquad$ **Add numerators; keep the same denominator**

$\qquad\qquad\qquad = \dfrac{4}{8} \qquad$ **The sum of 3 and 1 is 4**

$\qquad\qquad\qquad = \dfrac{1}{2} \qquad$ **Reduce to lowest terms** ∎

2. $\dfrac{a-5}{12} + \dfrac{3}{12}$

EXAMPLE 2 $\quad \dfrac{a+5}{8} - \dfrac{3}{8} = \dfrac{a+5-3}{8} \qquad$ **Combine numerators; keep the same denominator**

$\qquad\qquad\qquad\quad = \dfrac{a+2}{8} \qquad$ **The difference of 5 and 3 is 2** ∎

3. $\dfrac{8}{x} - \dfrac{5}{x}$

EXAMPLE 3 $\quad \dfrac{9}{x} - \dfrac{3}{x} = \dfrac{9-3}{x} \qquad$ **Subtract numerators; keep the same denominator**

$\qquad\qquad\qquad = \dfrac{6}{x} \qquad$ **The difference of 9 and 3 is 6** ∎

4. $\dfrac{5}{9} - \dfrac{8}{9} + \dfrac{5}{9}$

EXAMPLE 4 $\quad \dfrac{3}{7} + \dfrac{2}{7} - \dfrac{9}{7} = \dfrac{3+2-9}{7}$

$\qquad\qquad\qquad\qquad = \dfrac{-4}{7}$

$\qquad\qquad\qquad\qquad = -\dfrac{4}{7} \qquad$ **Unlike signs give a negative answer** ∎

As Examples 1–4 indicate, addition and subtraction are simple, straightforward processes when all the fractions have the same denominator. We will now turn our attention to the process of adding fractions that have different denominators. In order to get started, we need the following definition:

Answers

1. $\dfrac{2}{5}$ **2.** $\dfrac{a-2}{12}$ **3.** $\dfrac{3}{x}$ **4.** $\dfrac{2}{9}$

> **DEFINITION**
>
> The **least common denominator** (LCD) for a set of denominators is the smallest number that is exactly divisible by each denominator. (Note that, in some books, the least common denominator is also called the **least common multiple**.)

In other words, all the denominators of the fractions involved in a problem must divide into the least common denominator exactly. That is, they divide it without leaving a remainder.

EXAMPLE 5 Find the LCD for the fractions $\frac{5}{12}$ and $\frac{7}{18}$.

SOLUTION The least common denominator for the denominators 12 and 18 must be the smallest number divisible by both 12 and 18. We can factor 12 and 18 completely and then build the LCD from these factors. Factoring 12 and 18 completely gives us

$$12 = 2 \cdot 2 \cdot 3 \qquad 18 = 2 \cdot 3 \cdot 3$$

Now, if 12 is going to divide the LCD exactly, then the LCD must have factors of $2 \cdot 2 \cdot 3$. If 18 is to divide it exactly, it must have factors of $2 \cdot 3 \cdot 3$. We don't need to repeat the factors that 12 and 18 have in common:

12 divides the LCD

$$\left. \begin{array}{l} 12 = 2 \cdot 2 \cdot 3 \\ 18 = 2 \cdot 3 \cdot 3 \end{array} \right\} \quad \text{LCD} = 2 \cdot 2 \cdot 3 \cdot 3 = 36$$

18 divides the LCD

The LCD for 12 and 18 is 36. It is the smallest number that is divisible by both 12 and 18; 12 divides it exactly three times, and 18 divides it exactly two times. ∎

We can visualize the results in Example 5 with the diagram below. It shows that 36 is the smallest number that both 12 and 18 divide evenly. As you can see, 12 divides 36 exactly 3 times, and 18 divides 36 exactly 2 times.

12	12	12

18	18

36

EXAMPLE 6 Add: $\frac{5}{12} + \frac{7}{18}$

SOLUTION We can add fractions only when they have the same denominators. In Example 5, we found the LCD for $\frac{5}{12}$ and $\frac{7}{18}$ to be 36. We change $\frac{5}{12}$ and $\frac{7}{18}$ to equivalent fractions that have 36 for a denominator by applying Property 1 for fractions:

$$\frac{5}{12} = \frac{5 \cdot \mathbf{3}}{12 \cdot \mathbf{3}} = \frac{15}{36}$$

$$\frac{7}{18} = \frac{7 \cdot \mathbf{2}}{18 \cdot \mathbf{2}} = \frac{14}{36}$$

5. Find the LCD for the fractions $\frac{5}{18}$ and $\frac{3}{14}$.

Note
The ability to find least common denominators is very important in mathematics. The discussion here is a detailed explanation of how to do it.

6. Add: $\frac{5}{18} + \frac{3}{14}$

Answers

5. 126   **6.** $\frac{31}{63}$

The fraction $\frac{15}{36}$ is equivalent to $\frac{5}{12}$, because it was obtained by multiplying both the numerator and the denominator by 3. Likewise, $\frac{14}{36}$ is equivalent to $\frac{7}{18}$, because it was obtained by multiplying the numerator and the denominator by 2. All we have left to do is to add numerators.

$$\frac{15}{36} + \frac{14}{36} = \frac{29}{36}$$

The sum of $\frac{5}{12}$ and $\frac{7}{18}$ is the fraction $\frac{29}{36}$. Let's write the complete problem again step by step.

$$\frac{5}{12} + \frac{7}{18} = \frac{5 \cdot \mathbf{3}}{12 \cdot \mathbf{3}} + \frac{7 + \mathbf{2}}{18 \cdot \mathbf{2}} \qquad \text{Rewrite each fraction as an equivalent}$$
$$\text{fraction with denominator 36}$$
$$= \frac{15}{36} + \frac{14}{36}$$
$$= \frac{29}{36} \qquad \text{Add numerators; keep the}$$
$$\text{common denominator} \qquad ■$$

7. Find the LCD for $\frac{2}{9}$ and $\frac{4}{15}$.

EXAMPLE 7 Find the LCD for $\frac{3}{4}$ and $\frac{1}{6}$.

SOLUTION   We factor 4 and 6 into products of prime factors and build the LCD from these factors.

$$\left.\begin{array}{l} 4 = 2 \cdot 2 \\ 6 = 2 \cdot 3 \end{array}\right\} \qquad \text{LCD} = 2 \cdot 2 \cdot 3 = 12$$

The LCD is 12. Both denominators divide it exactly; 4 divides 12 exactly 3 times, and 6 divides 12 exactly 2 times.   ■

8. Add: $\frac{2}{9} + \frac{4}{15}$

EXAMPLE 8 Add: $\frac{3}{4} + \frac{1}{6}$

SOLUTION   In Example 7, we found that the LCD for these two fractions is 12. We begin by changing $\frac{3}{4}$ and $\frac{1}{6}$ to equivalent fractions with denominator 12:

$$\frac{3}{4} = \frac{3 \cdot \mathbf{3}}{4 \cdot \mathbf{3}} = \frac{9}{12}$$
$$\frac{1}{6} = \frac{1 \cdot \mathbf{2}}{6 \cdot \mathbf{2}} = \frac{2}{12}$$

The fraction $\frac{9}{12}$ is equal to the fraction $\frac{3}{4}$, because it was obtained by multiplying the numerator and the denominator of $\frac{3}{4}$ by 3. Likewise, $\frac{2}{12}$ is equivalent to $\frac{1}{6}$, because it was obtained by multiplying the numerator and the denominator of $\frac{1}{6}$ by 2. To complete the problem we add numerators:

$$\frac{9}{12} + \frac{2}{12} = \frac{11}{12}$$

The sum of $\frac{3}{4}$ and $\frac{1}{6}$ is $\frac{11}{12}$. Here is how the complete problem looks:

$$\frac{3}{4} + \frac{1}{6} = \frac{3 \cdot \mathbf{3}}{4 \cdot \mathbf{3}} + \frac{1 \cdot \mathbf{2}}{6 \cdot \mathbf{2}} \qquad \text{Rewrite each fraction as an equivalent}$$
$$\text{fraction with denominator 12}$$
$$= \frac{9}{12} + \frac{2}{12}$$
$$= \frac{11}{12} \qquad \text{Add numerators; keep}$$
$$\text{the same denominator} \qquad ■$$

Note

We can visualize the work in Example 8 using circles and shading:

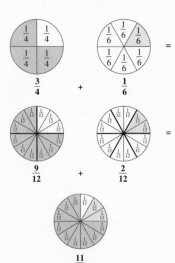

Answers

7. 45   **8.** $\frac{22}{45}$

EXAMPLE 9 Subtract: $\dfrac{7}{15} - \dfrac{3}{10}$

SOLUTION Let's factor 15 and 10 completely and use these factors to build the LCD:

15 divides the LCD

$$\left. \begin{array}{l} 15 = 3 \cdot 5 \\ 10 = 2 \cdot 5 \end{array} \right\} \quad \text{LCD} = 2 \cdot 3 \cdot 5 = 30$$

10 divides the LCD

Changing to equivalent fractions and subtracting, we have

$$\frac{7}{15} - \frac{3}{10} = \frac{7 \cdot \mathbf{2}}{15 \cdot \mathbf{2}} - \frac{3 \cdot \mathbf{3}}{10 \cdot \mathbf{3}} \qquad \text{Rewrite as equivalent fractions with the LCD for the denominator}$$

$$= \frac{14}{30} - \frac{9}{30}$$

$$= \frac{5}{30} \qquad \text{Subtract numerators; keep the LCD}$$

$$= \frac{1}{6} \qquad \text{Reduce to lowest terms} \qquad \blacksquare$$

As a summary of what we have done so far, and as a guide to working other problems, we now list the steps involved in adding and subtracting fractions with different denominators.

To Add or Subtract Any Two Fractions

Step 1 Factor each denominator completely, and use the factors to build the LCD. (Remember, the LCD is the smallest number divisible by each of the denominators in the problem.)

Step 2 Rewrite each fraction as an equivalent fraction that has the LCD for its denominator. This is done by multiplying both the numerator and the denominator of the fraction in question by the appropriate whole number.

Step 3 Add or subtract the numerators of the fractions produced in Step 2. This is the numerator of the sum or difference. The denominator of the sum or difference is the LCD.

Step 4 Reduce the fraction produced in Step 3 to lowest terms if it is not already in lowest terms.

The idea behind adding or subtracting fractions is really very simple. We can only add or subtract fractions that have the same denominators. If the fractions we are trying to add or subtract do not have the same denominators, we rewrite each of them as an equivalent fraction with the LCD for a denominator.

Here are some additional examples of sums and differences of fractions.

EXAMPLE 10 Subtract: $\dfrac{x}{5} - \dfrac{1}{6}$

SOLUTION The LCD for 5 and 6 is their product, 30. We begin by rewriting each fraction with this common denominator:

$$\frac{x}{5} - \frac{1}{6} = \frac{x \cdot \mathbf{6}}{5 \cdot \mathbf{6}} - \frac{1 \cdot \mathbf{5}}{6 \cdot \mathbf{5}}$$

$$= \frac{6x}{30} - \frac{5}{30}$$

$$= \frac{6x - 5}{30} \qquad \blacksquare$$

9. Subtract: $\dfrac{8}{25} - \dfrac{3}{20}$

10. Subtract: $\dfrac{x}{4} - \dfrac{1}{5}$

Answers

9. $\dfrac{17}{100}$ **10.** $\dfrac{5x - 4}{20}$

11. Add: $\dfrac{1}{9} + \dfrac{1}{4} + \dfrac{1}{6}$

EXAMPLE 11 Add: $\dfrac{1}{6} + \dfrac{1}{8} + \dfrac{1}{4}$

SOLUTION   We begin by factoring the denominators completely and building the LCD from the factors that result:

$$\left.\begin{array}{l} 6 = 2 \cdot 3 \\ 8 = 2 \cdot 2 \cdot 2 \\ 4 = 2 \cdot 2 \end{array}\right\}$$

8 divides the LCD

$$\text{LCD} = 2 \cdot 2 \cdot 2 \cdot 3 = 24$$

4 divides the LCD   **6 divides the LCD**

We then change to equivalent fractions and add as usual:

$$\frac{1}{6} + \frac{1}{8} + \frac{1}{4} = \frac{1 \cdot \mathbf{4}}{6 \cdot \mathbf{4}} + \frac{1 \cdot \mathbf{3}}{8 \cdot \mathbf{3}} + \frac{1 \cdot \mathbf{6}}{4 \cdot \mathbf{6}}$$

$$= \frac{4}{24} + \frac{3}{24} + \frac{6}{24}$$

$$= \frac{13}{24} \qquad \blacksquare$$

12. Subtract: $2 - \dfrac{3}{4}$

EXAMPLE 12 Subtract: $3 - \dfrac{5}{6}$

SOLUTION   The denominators are 1 (because $3 = \frac{3}{1}$) and 6. The smallest number divisible by both 1 and 6 is 6.

$$3 - \frac{5}{6} = \frac{3}{1} - \frac{5}{6}$$

$$= \frac{3 \cdot \mathbf{6}}{1 \cdot \mathbf{6}} - \frac{5}{6}$$

$$= \frac{18}{6} - \frac{5}{6}$$

$$= \frac{13}{6} \qquad \blacksquare$$

13. Add: $\dfrac{5}{x} + \dfrac{2}{3}$

Note

In Example 13, it is understood that x cannot be 0. Do you know why?

EXAMPLE 13 Add: $\dfrac{4}{x} + \dfrac{2}{3}$

SOLUTION   The LCD for x and 3 is $3x$. We multiply the numerator and the denominator of the first fraction by 3 and the numerator and the denominator of the second fraction by x to get two fractions with the same denominator. We then add the numerators:

$$\frac{4}{x} + \frac{2}{3} = \frac{4 \cdot \mathbf{3}}{x \cdot \mathbf{3}} + \frac{2 \cdot \mathbf{x}}{3 \cdot \mathbf{x}} \qquad \textbf{Change to equivalent fractions}$$

$$= \frac{12}{3x} + \frac{2x}{3x}$$

$$= \frac{12 + 2x}{3x} \qquad \textbf{Add the numerators} \qquad \blacksquare$$

Answers

11. $\dfrac{19}{36}$ **12.** $\dfrac{5}{4}$ **13.** $\dfrac{15 + 2x}{3x}$

Problem Set 3.5

Find the following sums and differences, and reduce to lowest terms. (Add or subtract as indicated.)

1. $\dfrac{3}{6} + \dfrac{1}{6}$

2. $\dfrac{2}{5} + \dfrac{3}{5}$

3. $\dfrac{3}{8} - \dfrac{5}{8}$

4. $\dfrac{1}{7} - \dfrac{6}{7}$

5. $-\dfrac{1}{4} + \dfrac{3}{4}$

6. $-\dfrac{4}{9} + \dfrac{7}{9}$

7. $\dfrac{x}{3} - \dfrac{1}{3}$

8. $\dfrac{x}{8} - \dfrac{1}{8}$

9. $\dfrac{1}{4} + \dfrac{2}{4} + \dfrac{3}{4}$

10. $\dfrac{2}{5} + \dfrac{3}{5} + \dfrac{4}{5}$

11. $\dfrac{x+7}{2} - \dfrac{1}{2}$

12. $\dfrac{x+5}{4} - \dfrac{3}{4}$

13. $\dfrac{1}{10} - \dfrac{3}{10} - \dfrac{4}{10}$

14. $\dfrac{3}{20} - \dfrac{1}{20} - \dfrac{4}{20}$

15. $\dfrac{1}{a} + \dfrac{4}{a} + \dfrac{5}{a}$

16. $\dfrac{5}{a} + \dfrac{4}{a} + \dfrac{3}{a}$

Complete the following tables.

17.

FIRST NUMBER a	SECOND NUMBER b	THE SUM OF a AND b $a+b$
$\dfrac{1}{2}$	$\dfrac{1}{3}$	
$\dfrac{1}{3}$	$\dfrac{1}{4}$	
$\dfrac{1}{4}$	$\dfrac{1}{5}$	
$\dfrac{1}{5}$	$\dfrac{1}{6}$	

18.

FIRST NUMBER a	SECOND NUMBER b	THE SUM OF a AND b $a+b$
1	$\dfrac{1}{2}$	
1	$\dfrac{1}{3}$	
1	$\dfrac{1}{4}$	
1	$\dfrac{1}{5}$	

19.

FIRST NUMBER a	SECOND NUMBER b	THE SUM OF a AND b $a+b$
$\dfrac{1}{12}$	$\dfrac{1}{2}$	
$\dfrac{1}{12}$	$\dfrac{1}{3}$	
$\dfrac{1}{12}$	$\dfrac{1}{4}$	
$\dfrac{1}{12}$	$\dfrac{1}{6}$	

20.

FIRST NUMBER a	SECOND NUMBER b	THE SUM OF a AND b $a+b$
$\dfrac{1}{8}$	$\dfrac{1}{2}$	
$\dfrac{1}{8}$	$\dfrac{1}{4}$	
$\dfrac{1}{8}$	$\dfrac{1}{16}$	
$\dfrac{1}{8}$	$\dfrac{1}{24}$	

Find the LCD for each of the following; then use the methods developed in this section to add or subtract as indicated.

21. $\dfrac{4}{9} + \dfrac{1}{3}$

22. $\dfrac{1}{2} + \dfrac{1}{4}$

23. $2 + \dfrac{1}{3}$

24. $3 + \dfrac{1}{2}$

25. $-\dfrac{3}{4} + 1$

26. $-\dfrac{3}{4} + 2$

27. $\dfrac{1}{2} + \dfrac{2}{3}$

28. $\dfrac{2}{3} + \dfrac{1}{4}$

29. $\dfrac{x}{4} + \dfrac{1}{5}$

30. $\dfrac{x}{3} + \dfrac{1}{5}$

31. $\dfrac{2}{x} + \dfrac{3}{5}$

32. $\dfrac{3}{x} - \dfrac{2}{5}$

33. $\dfrac{5}{12} - \left(-\dfrac{3}{8}\right)$

34. $\dfrac{9}{16} - \left(-\dfrac{7}{12}\right)$

35. $-\dfrac{1}{20} + \dfrac{8}{30}$

36. $-\dfrac{1}{30} + \dfrac{9}{40}$

37. $\dfrac{a}{10} + \dfrac{1}{100}$

38. $\dfrac{a}{100} + \dfrac{7}{10}$

39. $\dfrac{3}{7} + \dfrac{4}{x}$

40. $\dfrac{2}{9} + \dfrac{5}{x}$

41. $\dfrac{17}{30} + \dfrac{11}{42}$

42. $\dfrac{19}{42} + \dfrac{13}{70}$

43. $\dfrac{25}{84} + \dfrac{41}{90}$

44. $\dfrac{23}{70} + \dfrac{29}{84}$

45. $\dfrac{13}{126} - \dfrac{13}{180}$

46. $\dfrac{17}{84} - \dfrac{17}{90}$

47. $\dfrac{3}{4} + \dfrac{1}{8} + \dfrac{5}{6}$

48. $\dfrac{3}{8} + \dfrac{2}{5} + \dfrac{1}{4}$

49. $\dfrac{4}{y} + \dfrac{2}{3} + \dfrac{1}{2}$

50. $\dfrac{3}{y} + \dfrac{3}{4} + \dfrac{1}{5}$

51. $\dfrac{1}{2} + \dfrac{1}{3} + \dfrac{1}{4} + \dfrac{1}{6}$

52. $\dfrac{1}{8} + \dfrac{1}{4} + \dfrac{1}{5} + \dfrac{1}{10}$

53. Find the sum of $\dfrac{3}{7}$, 2, and $\dfrac{1}{9}$.

54. Find the sum of 6, $\dfrac{6}{11}$, and 11.

55. Give the difference of $\dfrac{7}{8}$ and $\dfrac{1}{4}$.

56. Give the difference of $\dfrac{9}{10}$ and $\dfrac{1}{100}$.

Applying the Concepts

Some of the application problems below involve multiplication and division, while the others involve addition and subtraction.

57. Capacity One carton of milk contains $\frac{1}{2}$ pint while another contains 4 pints. How much milk is contained in both cartons?

58. Baking A recipe calls for $\frac{2}{3}$ cup of flour and $\frac{3}{4}$ cup of sugar. What is the total amount of flour and sugar called for in the recipe?

59. Budget A family decides that they can spend $\frac{5}{8}$ of their monthly income on house payments. If their monthly income is $2,120, how much can they spend for house payments?

60. Savings A family saves $\frac{3}{16}$ of their income each month. If their monthly income is $1,264, how much do they save each month?

Reading a Pie Chart The pie chart below shows how the students at one of the universities in California are distributed among the different schools at the university. Use the information in the pie chart to answer questions 61 and 62.

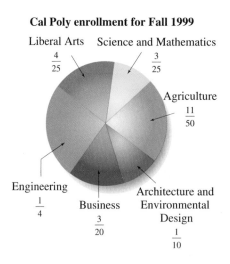

Cal Poly enrollment for Fall 1999

Liberal Arts $\frac{4}{25}$ Science and Mathematics $\frac{3}{25}$

Agriculture $\frac{11}{50}$

Engineering $\frac{1}{4}$ Business $\frac{3}{20}$ Architecture and Environmental Design $\frac{1}{10}$

61. If the students in the Schools of Engineering and Business are combined, what fraction results?

62. What fraction of the university's students are enrolled in the Schools of Agriculture, Engineering, and Business combined?

63. Final Exam Grades The table below gives the fraction of students in a class of 40 that received grades of A, B, or C on the final exam. Fill in all the missing parts of the table.

GRADE	NUMBER OF STUDENTS	FRACTION OF STUDENTS
A		$\frac{1}{8}$
B		$\frac{1}{5}$
C		$\frac{1}{2}$
below C		
Total	40	1

64. Flu During a flu epidemic a company with 200 employees has $\frac{1}{10}$ of their employees call in sick on Monday and another $\frac{3}{10}$ call in sick on Tuesday. What is the total number of employees calling in sick during this 2-day period?

65. Subdivision A 6-acre piece of land is subdivided into $\frac{3}{5}$-acre lots. How many lots are there?

66. Cutting Wood A 12-foot piece of wood is cut into shelves. If each is $\frac{3}{4}$ foot in length, how many shelves are there?

Find the perimeter of each figure.

67.

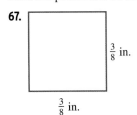

$\frac{3}{8}$ in.

$\frac{3}{8}$ in.

68.

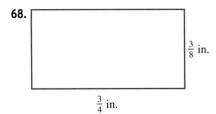

$\frac{3}{8}$ in.

$\frac{3}{4}$ in.

69.

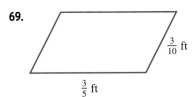

$\frac{3}{10}$ ft

$\frac{3}{5}$ ft

70.

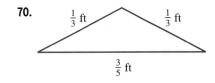

$\frac{1}{3}$ ft $\frac{1}{3}$ ft

$\frac{3}{5}$ ft

Calculator Problems

Use your calculator to help work the following problems.

71. If the LCD for $\frac{7}{71}$ and $\frac{17}{497}$ is 497, find the sum of the two fractions.

72. The LCD for $\frac{12}{17}$ and $\frac{53}{425}$ is 425. Find the sum of the two fractions.

73. The LCD for $\frac{10}{19}$ and $\frac{20}{23}$ is 437. Find the sum of the two fractions.

74. The LCD for $\frac{13}{29}$ and $\frac{18}{43}$ is 1,247. Add the two fractions.

Review Problems

The problems below review multiplication and division with fractions (Sections 3.3 and 3.4).

Multiply or divide as indicated.

75. $\frac{3}{4} \div \frac{5}{6}$

76. $12 \div \frac{1}{2}$

77. $-12 \cdot \frac{2}{3}$

78. $-12 \cdot \frac{3}{4}$

79. $4 \cdot \frac{3}{4}$

80. $4 \cdot \frac{1}{2}$

81. $-\frac{7}{6} \div \left(-\frac{7}{12}\right)$

82. $-\frac{9}{10} \div \left(-\frac{7}{10}\right)$

83. $\frac{x^2}{y} \div \frac{x}{y^3}$

84. $\frac{x^3}{y} \div \frac{x^2}{y}$

85. $\frac{2}{3} \cdot \frac{3}{4} \cdot \frac{4}{5} \cdot \frac{5}{6} \cdot \frac{6}{7}$

86. $\frac{11}{12} \cdot \frac{10}{11} \cdot \frac{9}{10} \cdot \frac{8}{9} \cdot \frac{7}{8}$

87. $\frac{35}{110} \cdot \frac{80}{63} \div \frac{16}{27}$

88. $\frac{20}{72} \cdot \frac{42}{18} \div \frac{20}{16}$

Extending the Concepts

Recall that an arithmetic sequence is a sequence in which each term comes from the previous term by adding the same number each time. For example, the sequence $1, \frac{3}{2}, 2, \frac{5}{2}, \ldots$ is an arithmetic sequence that starts with the number 1. Then each term after that is found by adding $\frac{1}{2}$ to the previous term. By observing this fact, we know that the next term in the sequence will be $\frac{5}{2} + \frac{1}{2} = \frac{6}{2} = 3$.

Find the next number in each arithmetic sequence below.

89. $1, \frac{4}{3}, \frac{5}{3}, 2, \ldots$

90. $1, \frac{5}{4}, \frac{3}{2}, \frac{7}{4}, \ldots$

91. $\frac{3}{2}, 2, \frac{5}{2}, \ldots$

92. $\frac{2}{3}, 1, \frac{4}{3}, \ldots$

3.6 Mixed-Number Notation

Introduction . . .

If you are interested in the stock market, you know that stock prices are given in eighths. For example, on the day I am writing this introduction, one share of Intel Corporation is selling at $73\frac{5}{8}$, or seventy-three and five-eighths dollars. The number $73\frac{5}{8}$ is called a *mixed number*. It is the sum of a whole number and a proper fraction. With mixed-number notation, we leave out the addition sign.

Notation

A number such as $5\frac{3}{4}$ is called a *mixed number* and is equal to $5 + \frac{3}{4}$. It is simply the sum of the whole number 5 and the proper fraction $\frac{3}{4}$, written without a + sign. Here are some further examples:

$$2\frac{1}{8} = 2 + \frac{1}{8}, \qquad 6\frac{5}{9} = 6 + \frac{5}{9}, \qquad 11\frac{2}{3} = 11 + \frac{2}{3}$$

The notation used in writing mixed numbers (writing the whole number and the proper fraction next to each other) must always be interpreted as addition. It is a mistake to read $5\frac{3}{4}$ as meaning 5 times $\frac{3}{4}$. If we want to indicate multiplication, we must use parentheses or a multiplication symbol. That is:

$$5\frac{3}{4} \text{ is not the same as } 5\left(\frac{3}{4}\right)$$

This implies addition   **These imply multiplication**

$$5\frac{3}{4} \text{ is not the same as } 5 \cdot \frac{3}{4}$$

Changing Mixed Numbers to Improper Fractions

To change a mixed number to an improper fraction, we write the mixed number with the + sign showing and then add the two numbers, as we did earlier.

EXAMPLE 1 Change $2\frac{3}{4}$ to an improper fraction.

SOLUTION
$$2\frac{3}{4} = 2 + \frac{3}{4}$$ **Write the mixed number as a sum**

$$= \frac{2}{1} + \frac{3}{4}$$ **Show that the denominator of 2 is 1**

$$= \frac{4 \cdot 2}{4 \cdot 1} + \frac{3}{4}$$ **Multiply the numerator and the denominator of $\frac{2}{1}$ by 4 so both fractions will have the same denominator**

$$= \frac{8}{4} + \frac{3}{4}$$

$$= \frac{11}{4}$$ **Add the numerators; keep the common denominator**

The mixed number $2\frac{3}{4}$ is equal to the improper fraction $\frac{11}{4}$. Figure 1 further illustrates the equivalence of $2\frac{3}{4}$ and $\frac{11}{4}$.

Practice Problems

1. Change $5\frac{2}{3}$ to an improper fraction.

Answer

1. $\frac{17}{3}$

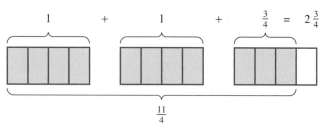

Figure 1

2. Change $3\frac{1}{6}$ to an improper fraction.

EXAMPLE 2   Change $2\frac{1}{8}$ to an improper fraction.

SOLUTION

$$2\frac{1}{8} = 2 + \frac{1}{8}$$ **Write as addition**

$$= \frac{2}{1} + \frac{1}{8}$$ **Write the whole number 2 as a fraction**

$$= \frac{8 \cdot 2}{8 \cdot 1} + \frac{1}{8}$$ **Change $\frac{2}{1}$ to a fraction with denominator 8**

$$= \frac{16}{8} + \frac{1}{8}$$

$$= \frac{17}{8}$$ **Add the numerators**

If we look closely at Examples 1 and 2, we can see a shortcut that will let us change a mixed number to an improper fraction without so many steps. Here is the shortcut:

Shortcut: To change a mixed number to an improper fraction, simply multiply the denominator of the fraction part of the mixed number by the whole number, and add the result to the numerator of the fraction. The result is the numerator of the improper fraction we are looking for. The denominator is the same as the original denominator.

3. Use the shortcut to change $5\frac{2}{3}$ to an improper fraction.

EXAMPLE 3   Use the shortcut to change $5\frac{3}{4}$ to an improper fraction.

SOLUTION

1. First, we multiply 4×5 to get 20.

2. Next, we add 20 to 3 to get 23.

3. The improper fraction equal to $5\frac{3}{4}$ is $\frac{23}{4}$.

Here is a diagram showing what we have done:
Step 1 Multiply $4 \times 5 = 20$.

$$5\frac{3}{4}$$

Step 2 Add $20 + 3 = 23$.
Mathematically, our shortcut is written like this:

$$5\frac{3}{4} = \frac{(4 \cdot 5) + 3}{4} = \frac{20 + 3}{4} = \frac{23}{4}$$ **The result will always have the same denominator as the original mixed number**

The shortcut shown in Example 3 works because the whole-number part of a mixed number can always be written with a denominator of 1. Therefore, the LCD for a whole number and fraction will always be the denominator of the fraction. That is why we multiply the whole number by the denominator of the fraction:

$$5\frac{3}{4} = 5 + \frac{3}{4} = \frac{5}{1} + \frac{3}{4} = \frac{4 \cdot 5}{4 \cdot 1} + \frac{3}{4} = \frac{4 \cdot 5 + 3}{4} = \frac{23}{4}$$

EXAMPLE 4 Change $6\frac{5}{9}$ to an improper fraction.

SOLUTION Using the first method, we have

$$6\frac{5}{9} = 6 + \frac{5}{9} = \frac{6}{1} + \frac{5}{9} = \frac{\mathbf{9} \cdot 6}{\mathbf{9} \cdot 1} + \frac{5}{9} = \frac{54}{9} + \frac{5}{9} = \frac{59}{9}$$

Using the shortcut method, we have

$$6\frac{5}{9} = \frac{(9 \cdot 6) + 5}{9} = \frac{54 + 5}{9} = \frac{59}{9}$$ ∎

Changing Improper Fractions to Mixed Numbers

To change an improper fraction to a mixed number, we divide the numerator by the denominator. The result is used to write the mixed number.

EXAMPLE 5 Change $\frac{11}{4}$ to a mixed number.

SOLUTION Dividing 11 by 4 gives us

$$\begin{array}{r} 2 \\ 4\overline{)11} \\ \underline{8} \\ 3 \end{array}$$

We see that 4 goes into 11 two times with 3 for a remainder. We write this result as

$$\frac{11}{4} = 2 + \frac{3}{4} = 2\frac{3}{4}$$

The improper fraction $\frac{11}{4}$ is equivalent to the mixed number $2\frac{3}{4}$. ∎

An easy way to visualize the results in Example 5 is to imagine having 11 quarters. Your 11 quarters are equivalent to 11/4 dollars. In dollars, your quarters are worth 2 dollars plus 3 quarters, or $2\frac{3}{4}$ dollars.

EXAMPLE 6 $\frac{10}{3}$: $\begin{array}{r} 3 \\ 3\overline{)10} \\ \underline{9} \\ 1 \end{array}$ so $\frac{10}{3} = \mathbf{3} + \frac{\mathbf{1}}{3} = 3\frac{1}{3}$ ∎

EXAMPLE 7 $\frac{208}{24}$: $\begin{array}{r} 8 \\ 24\overline{)208} \\ \underline{192} \\ 16 \end{array}$ so $\frac{208}{24} = \mathbf{8} + \frac{\mathbf{16}}{24} = 8 + \frac{2}{3} = 8\frac{2}{3}$

Reduce to lowest terms ∎

Notice that in Example 7 we had to reduce the fraction part of the mixed number to lowest terms to get the final result. We could have done this to begin with, and the result would be the same:

---

4. Change $6\frac{4}{9}$ to an improper fraction.

Calculator Note
The sequence of keys to press on a calculator to obtain the numerator in Example 4 looks like this:
9 ⊠ 6 ⊞ 5 ⊟

5. Change $\frac{11}{3}$ to a mixed number.

Note
This division process shows us how many ones are in $\frac{11}{4}$ and, when the ones are taken out, how many fourths are left.

Change each improper fraction to a mixed number.

6. $\frac{14}{5}$

7. $\frac{207}{26}$

Answers

4. $\frac{58}{9}$ **5.** $3\frac{2}{3}$ **6.** $2\frac{4}{5}$

7. $7\frac{25}{26}$

$$\frac{208}{24} = \frac{26}{3}: \quad \begin{array}{r} 8 \\ 3\overline{)26} \\ \underline{24} \\ 2 \end{array} \quad \text{so} \quad \frac{26}{3} = 8\frac{2}{3}$$

Reduce

In the first part of this section, we changed mixed numbers to improper fractions. For example we changed $2\frac{3}{4}$ to an improper fraction by adding 2 and $\frac{3}{4}$ (see Example 1). An extension of this concept to algebra would be to add 2 and $\frac{3}{x}$.

8. Add: $3 + \dfrac{2}{x}$

EXAMPLE 8 Add: $2 + \dfrac{3}{x}$

SOLUTION   We can write 2 as $\dfrac{2}{1}$:

$$2 + \frac{3}{x} = \frac{2}{1} + \frac{3}{x}$$

Now, the LCD for 1 and x is $1x$ or just x. To change to equivalent fractions, we multiply the numerator and the denominator of $\frac{2}{1}$ by x.

$$\frac{2}{1} + \frac{3}{x} = \frac{2 \cdot \boldsymbol{x}}{1 \cdot \boldsymbol{x}} + \frac{3}{x} \quad \textbf{The LCD is } \boldsymbol{x}$$

$$= \frac{2x}{x} + \frac{3}{x}$$

$$= \frac{2x + 3}{x} \quad \textbf{Add numerators}$$

9. Subtract: $x - \dfrac{2}{3}$

EXAMPLE 9 Subtract: $x - \dfrac{3}{4}$

SOLUTION   This time we write x as $\dfrac{x}{1}$. The LCD for 1 and for 4 is 4.

$$x - \frac{3}{4} = \frac{x}{1} - \frac{3}{4} \quad \textbf{Write } \boldsymbol{x} \textbf{ as } \dfrac{\boldsymbol{x}}{\boldsymbol{1}}$$

$$= \frac{\boldsymbol{4} \cdot x}{\boldsymbol{4} \cdot 1} - \frac{3}{4} \quad \textbf{The LCD is 4}$$

$$= \frac{4x}{4} - \frac{3}{4}$$

$$= \frac{4x - 3}{4} \quad \textbf{Subtract numerators}$$

As a final note in this section, we should mention that negative mixed numbers are thought of this way:

$$-3\frac{2}{5} = -3 - \frac{2}{5}$$

Both the whole-number part and the fraction part of the mixed number are negative when the mixed number is negative. It would be a mistake to think of mixed numbers this way:

$$-3\frac{2}{5} = -3 + \frac{2}{5} \quad \textbf{Mistake}$$

Because negative mixed numbers can be difficult to work with in algebra, you are more likely to see improper fractions than mixed numbers if you go on to take an algebra class.

Answers

8. $\dfrac{3x + 2}{x}$ **9.** $\dfrac{3x - 2}{3}$

Problem Set 3.6

Change each mixed number to an improper fraction.

1. $4\frac{2}{3}$ **2.** $3\frac{5}{8}$ **3.** $5\frac{1}{4}$ **4.** $7\frac{1}{2}$ **5.** $1\frac{5}{8}$ **6.** $1\frac{6}{7}$

7. $15\frac{2}{3}$ **8.** $17\frac{3}{4}$ **9.** $4\frac{20}{21}$ **10.** $5\frac{18}{19}$ **11.** $12\frac{31}{33}$ **12.** $14\frac{29}{31}$

Change each improper fraction to a mixed number.

13. $\frac{9}{8}$ **14.** $\frac{10}{9}$ **15.** $\frac{19}{4}$ **16.** $\frac{23}{5}$ **17.** $\frac{29}{6}$ **18.** $\frac{7}{2}$

19. $\frac{13}{4}$ **20.** $\frac{41}{15}$ **21.** $\frac{109}{27}$ **22.** $\frac{319}{23}$ **23.** $\frac{428}{15}$ **24.** $\frac{769}{27}$

Add or subtract as indicated.

25. $8 + \frac{3}{x}$ **26.** $7 + \frac{9}{x}$ **27.** $2 - \frac{5}{y}$ **28.** $3 - \frac{4}{y}$ **29.** $x + \frac{5}{6}$ **30.** $x + \frac{2}{7}$

31. $a - \frac{1}{3}$ **32.** $a - \frac{1}{4}$

Applying the Concepts

33. Stocks A certain stock on the New York Stock Exchange is up $2\frac{1}{8}$ points. Change this to an improper fraction.

34. Stocks A stock is listed on the New York Stock Exchange at $24\frac{3}{8}$. Write this number as an improper fraction.

35. Stocks Suppose a stock is selling on a stock exchange for $5\frac{1}{4}$ dollars per share. If the price increases $\frac{3}{4}$ dollar per share, what is the new price of the stock?

36. Stocks Suppose a stock is selling on a stock exchange for $5\frac{1}{4}$ dollars per share. If the price increases 2 dollars per share, what is the new price of the stock?

37. Height If a man is 71 inches tall, then in feet his height is $5\frac{11}{12}$ feet. Change $5\frac{11}{12}$ to an improper fraction.

38. Height If a woman is 63 inches tall, then her height in feet is $\frac{63}{12}$. Write $\frac{63}{12}$ as a mixed number.

39. Sleeping Infants The table below shows the average number of hours an infant sleeps each day. Use the template to construct a line graph from the information in the table.

AGE (MONTHS)	DAILY SLEEP (HOURS)
1	$18\frac{1}{2}$
2	17
3	16
4	15
5	$14\frac{1}{2}$
6	14

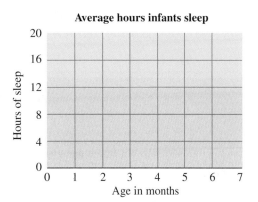

40. Shoe Sales The table below shows the number of shoes sold, in various sizes, during a one-day sale.

SHOE SALE	
Size	Number of Shoes
8	5
$8\frac{1}{2}$	7
9	8
$9\frac{1}{2}$	10
10	15
$10\frac{1}{2}$	12
11	6

SALE
ONE DAY ONLY!

Available in sizes 8-11,
including half sizes

a. Find the mode of the shoe sizes for all shoes sold.

b. Find the median shoe size for all shoes sold. (Do this by looking at the numbers in the second column to decide which shoe size is in the middle of all the sizes of the shoes sold.)

41. Gasoline Prices The price of unleaded gasoline is $135\frac{9}{10}$¢ per gallon. Write this number as an improper fraction.

42. Gasoline Prices Suppose the price of gasoline is $127\frac{9}{10}$¢ if purchased with a credit card, but 5¢ less if purchased with cash. What is the cash price of the gasoline?

Calculator Problems

Use a calculator to work the following problems. Change each mixed number to an improper fraction.

43. $35\dfrac{19}{23}$ **44.** $16\dfrac{29}{31}$ **45.** $147\dfrac{13}{17}$ **46.** $171\dfrac{9}{11}$

47. $5\dfrac{131}{139}$ **48.** $7\dfrac{209}{307}$ **49.** $19\dfrac{49}{51}$ **50.** $25\dfrac{71}{79}$

Review Problems

The problems below review multiplication and division with fractions (Sections 3.3 and 3.4). These are the types of problems you need to know how to do in order to understand the next section.

Find the products. (Multiply.)

51. $\dfrac{3}{8} \cdot \dfrac{3}{5}$ **52.** $\dfrac{5}{4} \cdot \dfrac{8}{15}$ **53.** $-\dfrac{2}{3}\left(-\dfrac{9}{16}\right)$ **54.** $-\dfrac{7}{10}\left(-\dfrac{5}{21}\right)$

Find the quotients. (Divide.)

55. $\dfrac{4}{5} \div \left(-\dfrac{7}{8}\right)$ **56.** $\dfrac{3}{4} \div \left(-\dfrac{1}{2}\right)$ **57.** $\dfrac{9}{10} \div \dfrac{3}{5}$ **58.** $\dfrac{8}{9} \div \dfrac{2}{3}$

Perform the indicated operations.

59. $\dfrac{2}{3} \cdot \dfrac{3}{4} \div \dfrac{5}{8}$ **60.** $4 \cdot \dfrac{7}{6}$ **61.** $\dfrac{a^2c}{b} \cdot \dfrac{b^2}{ac^2}$ **62.** $\dfrac{a^3c}{b^2} \cdot \dfrac{b^3}{a^2c}$

3.7 Multiplication and Division with Mixed Numbers

Introduction . . .

The figure below shows one of the nutrition labels we worked with in Chapter 1. It is from a can of Italian tomatoes. Notice toward the top of the label, the number of servings in the can is $3\frac{1}{2}$. The number $3\frac{1}{2}$ is called a *mixed number*. If we want to know how many calories are in the whole can of tomatoes, we must be able to multiply $3\frac{1}{2}$ by 25 (the number of calories per serving). Multiplication with mixed numbers is one of the topics we will cover in this section.

CANNED ITALIAN TOMATOES

Nutrition Facts
Serving Size 1/2 cup (121g)
Servings Per Container: about 3 1/2

Amount Per Serving

Calories 25	Calories from fat 0

	% Daily Value*
Total Fat 0g	0%
Saturated Fat 0g	0%
Cholesterol 0mg	0%
Sodium 300mg	12%
Potassium 145mg	4%
Total Carbohydrate 4g	2%
Dietary Fiber 1g	4%
Sugars 4g	
Protein 1g	

Vitamin A 20%	•	Vitamin C 15%
Calcium 4%	•	Iron 15%

*Percent Daily Values are based on a 2,000 calorie diet. Your daily values may be higher or lower depending on your calorie needs.

The procedures for multiplying and dividing mixed numbers are the same as those we used in Sections 3.3 and 3.4 to multiply and divide fractions. The only additional work involved is in changing the mixed numbers to improper fractions before we actually multiply or divide.

EXAMPLE 1   Multiply: $2\frac{3}{4} \cdot 3\frac{1}{5}$

SOLUTION   We begin by changing each mixed number to an improper fraction:

$$2\frac{3}{4} = \frac{11}{4} \quad \text{and} \quad 3\frac{1}{5} = \frac{16}{5}$$

Using the resulting improper fractions, we multiply as usual. (That is, we multiply numerators and multiply denominators.)

$$\frac{11}{4} \cdot \frac{16}{5} = \frac{11 \cdot 16}{4 \cdot 5}$$

$$= \frac{11 \cdot \cancel{4} \cdot 4}{\cancel{4} \cdot 5}$$

$$= \frac{44}{5} \quad \text{or} \quad 8\frac{4}{5}$$

∎

Practice Problems

1. Multiply: $2\frac{3}{4} \cdot 4\frac{1}{3}$

Answer

1. $11\frac{11}{12}$

2. Multiply: $2 \cdot 3\frac{5}{8}$

Note

As you can see, once you have changed each mixed number to an improper fraction, you multiply the resulting fractions the same way you did in Section 3.3.

3. Divide: $1\frac{3}{5} \div 3\frac{2}{5}$

4. Divide: $4\frac{5}{8} \div 2$

EXAMPLE 2 Multiply: $3 \cdot 4\frac{5}{8}$

SOLUTION Writing each number as an improper fraction, we have

$$3 = \frac{3}{1} \quad \text{and} \quad 4\frac{5}{8} = \frac{37}{8}$$

The complete problem looks like this:

$$3 \cdot 4\frac{5}{8} = \frac{3}{1} \cdot \frac{37}{8} \qquad \textbf{Change to improper fractions}$$

$$= \frac{111}{8} \qquad \textbf{Multiply numerators and multiply denominators}$$

$$= 13\frac{7}{8} \qquad \textbf{Write the answer as a mixed number} \qquad \blacksquare$$

Dividing mixed numbers also requires that we change all mixed numbers to improper fractions before we actually do the division.

EXAMPLE 3 Divide: $1\frac{3}{5} \div 2\frac{4}{5}$

SOLUTION We begin by rewriting each mixed number as an improper fraction:

$$1\frac{3}{5} = \frac{8}{5} \quad \text{and} \quad 2\frac{4}{5} = \frac{14}{5}$$

We then divide by the same method we used in Section 3.4. Remember? We multiply by the reciprocal of the divisor. Here is the complete problem:

$$1\frac{3}{5} \div 2\frac{4}{5} = \frac{8}{5} \div \frac{14}{5} \qquad \textbf{Change to improper fractions}$$

$$= \frac{8}{5} \cdot \frac{5}{14} \qquad \textbf{To divide by } \tfrac{14}{5}\textbf{, multiply by } \tfrac{5}{14}$$

$$= \frac{8 \cdot 5}{5 \cdot 14} \qquad \textbf{Multiply numerators and multiply denominators}$$

$$= \frac{4 \cdot 2 \cdot \cancel{5}}{\cancel{5} \cdot 2 \cdot 7} \qquad \textbf{Divide out factors common to the numerator and denominator}$$

$$= \frac{4}{7} \qquad \textbf{Answer in lowest terms} \qquad \blacksquare$$

EXAMPLE 4 Divide: $5\frac{9}{10} \div 2$

SOLUTION We change to improper fractions and proceed as usual:

$$5\frac{9}{10} \div 2 = \frac{59}{10} \div \frac{2}{1} \qquad \textbf{Write each number as an improper fraction}$$

$$= \frac{59}{10} \cdot \frac{1}{2} \qquad \textbf{Write division as multiplication by the reciprocal}$$

$$= \frac{59}{20} \qquad \textbf{Multiply numerators and multiply denominators}$$

$$= 2\frac{19}{20} \qquad \textbf{Change to a mixed number} \qquad \blacksquare$$

Answers

2. $7\frac{1}{4}$ **3.** $\frac{8}{17}$ **4.** $2\frac{5}{16}$

Problem Set 3.7

Write your answers as proper fractions or mixed numbers, not as improper fractions.

Find the following products. (Multiply.)

1. $3\frac{2}{5} \cdot 1\frac{1}{2}$

2. $2\frac{1}{3} \cdot 6\frac{3}{4}$

3. $5\frac{1}{8} \cdot 2\frac{2}{3}$

4. $1\frac{5}{6} \cdot 1\frac{4}{5}$

5. $2\frac{1}{10} \cdot 3\frac{3}{10}$

6. $4\frac{7}{10} \cdot 3\frac{1}{10}$

7. $1\frac{1}{4} \cdot 4\frac{2}{3}$

8. $3\frac{1}{2} \cdot 2\frac{1}{6}$

9. $2 \cdot 4\frac{7}{8}$

10. $10 \cdot 1\frac{1}{4}$

11. $\frac{3}{5} \cdot 5\frac{1}{3}$

12. $\frac{2}{3} \cdot 4\frac{9}{10}$

13. $2\frac{1}{2} \cdot 3\frac{1}{3} \cdot 1\frac{1}{2}$

14. $3\frac{1}{5} \cdot 5\frac{1}{6} \cdot 1\frac{1}{8}$

15. $\frac{3}{4} \cdot 7 \cdot 1\frac{4}{5}$

16. $\frac{7}{8} \cdot 6 \cdot 1\frac{5}{6}$

Find the following quotients. (Divide.)

17. $3\frac{1}{5} \div 4\frac{1}{2}$

18. $1\frac{4}{5} \div 2\frac{5}{6}$

19. $6\frac{1}{4} \div 3\frac{3}{4}$

20. $8\frac{2}{3} \div 4\frac{1}{3}$

21. $10 \div 2\frac{1}{2}$

22. $12 \div 3\frac{1}{6}$

23. $8\frac{3}{5} \div 2$

24. $12\frac{6}{7} \div 3$

25. $\left(\frac{3}{4} \div 2\frac{1}{2}\right) \div 3$

26. $\frac{7}{8} \div \left(1\frac{1}{4} \div 4\right)$

27. $\left(8 \div 1\frac{1}{4}\right) \div 2$

28. $8 \div \left(1\frac{1}{4} \div 2\right)$

29. $2\frac{1}{2} \cdot \left(3\frac{2}{5} \div 4\right)$

30. $4\frac{3}{5} \cdot \left(2\frac{1}{4} \div 5\right)$

31. Find the product of $2\frac{1}{2}$ and 3.

32. Find the product of $\frac{1}{5}$ and $3\frac{2}{3}$.

33. What is the quotient of $2\frac{3}{4}$ and $3\frac{1}{4}$?

34. What is the quotient of $1\frac{1}{5}$ and $2\frac{2}{5}$?

Applying the Concepts

35. Cooking A certain recipe calls for $2\frac{3}{4}$ cups of sugar. If the recipe is to be doubled, how much sugar should be used?

36. Cooking If a recipe calls for $3\frac{1}{2}$ cups of flour, how much flour will be needed if the recipe is tripled?

37. Cooking If a recipe calls for $2\frac{1}{2}$ cups of sugar, how much sugar is needed to make $\frac{1}{3}$ of the recipe?

38. Cooking A recipe calls for $3\frac{1}{4}$ cups of flour. If Diane is using only half the recipe, how much flour should she use?

39. Number Problem Find $\frac{3}{4}$ of $1\frac{7}{9}$. (Remember that *of* means multiply.)

40. Number Problem Find $\frac{5}{6}$ of $2\frac{4}{15}$.

41. Cost of Gasoline If a gallon of gas costs $135\frac{9}{10}$¢, how much does 8 gallons cost?

42. Cost of Gasoline If a gallon of gas costs $159\frac{9}{10}$¢, how much does $\frac{1}{2}$ gallon cost?

43. Distance Traveled If a car can travel $32\frac{3}{4}$ miles on a gallon of gas, how far will it travel on 5 gallons of gas?

44. Distance Traveled If a new car can travel $20\frac{3}{10}$ miles on 1 gallon of gas, how far can it travel on $\frac{1}{2}$ gallon of gas?

45. Sewing If it takes $1\frac{1}{2}$ yards of material to make a pillow cover, how much material will it take to make 3 pillow covers?

46. Sewing If the material for the pillow covers in Problem 45 costs $2 a yard, how much will it cost for the material for the 3 pillow covers?

Find the area of each figure.

47.

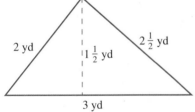

48.

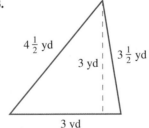

49.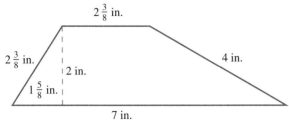

50.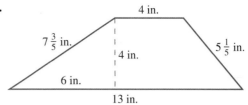

The figure below shows nutrition labels for two different cans of Italian tomatoes.

CANNED TOMATOES 1

Nutrition Facts

Serving Size 1/2 cup (121g)
Servings Per Container: about 3 1/2

Amount Per Serving

Calories 45   Calories from fat 0

	% Daily Value*
Total Fat 0g	**0%**
Saturated Fat 0g	**0%**
Cholesterol 0mg	**0%**
Sodium 560mg	**23%**
Total Carbohydrate 11g	**4%**
Dietary Fiber 2g	**8%**
Sugars 9g	
Protein 1g	

Vitamin A 10%   •   Vitamin C 25%

Calcium 2%   •   Iron 2%

*Percent Daily Values are based on a 2,000 calorie diet.

CANNED TOMATOES 2

Nutrition Facts

Serving Size 1/2 cup (121g)
Servings Per Container: about 3 1/2

Amount Per Serving

Calories 25   Calories from fat 0

	% Daily Value*
Total Fat 0g	**0%**
Saturated Fat 0g	**0%**
Cholesterol 0mg	**0%**
Sodium 300mg	**12%**
Potassium 145mg	**4%**
Total Carbohydrate 4g	**2%**
Dietary Fiber 1g	**4%**
Sugars 4g	
Protein 1g	

Vitamin A 20%   •   Vitamin C 15%

Calcium 4%   •   Iron 15%

*Percent Daily Values are based on a 2,000 calorie diet. Your daily values may be higher or lower depending on your calorie needs.

51. Compare the total number of calories in the two cans of tomatoes.

52. Compare the total amount of sugar in the two cans of tomatoes.

53. Compare the total amount of sodium in the two cans of tomatoes.

54. Compare the total amount of protein in the two cans of tomatoes.

Calculator Problems

Use a calculator to work the following problems. Multiply or divide as indicated. Write your answers as improper fractions. You don't have to reduce to lowest terms.

55. $17 \cdot 23\frac{7}{9}$

56. $18\frac{7}{8} \cdot 19\frac{8}{9}$

57. $\frac{17}{23} \cdot 7\frac{15}{19}$

58. $\frac{19}{29} \cdot 9\frac{7}{9}$

59. $4\frac{31}{37} \div \frac{17}{23}$

60. $5\frac{37}{41} \div 2\frac{41}{43}$

Review Problems

The problems below review addition with fractions (Section 3.5). You need to understand addition with fractions to work the problems in the next section.

Add or subtract the following fractions, as indicated.

61. $\frac{3}{4} + \frac{5}{4}$

62. $\frac{1}{3} + \frac{4}{3}$

63. $\frac{15}{56} - \frac{5}{42}$

64. $\frac{15}{66} - \frac{5}{88}$

65. $\frac{5}{18} + \frac{1}{12} + \frac{3}{4}$

66. $\frac{3}{15} + \frac{1}{10} + \frac{7}{20}$

Extending the Concepts

To find the square of a mixed number, we first change the mixed number to an improper fraction, and then we square the result. For example:

$$\left(2\frac{1}{2}\right)^2 = \left(\frac{5}{2}\right)^2 = \frac{25}{4}$$

If we are asked to write our answers as a mixed number, we write it as $6\frac{1}{4}$.

Find each of the following squares, and write your answers as mixed numbers.

67. $\left(1\frac{1}{2}\right)^2$

68. $\left(3\frac{1}{2}\right)^2$

69. $\left(1\frac{3}{4}\right)^2$

70. $\left(2\frac{3}{4}\right)^2$

71. The area of a square is $2\frac{1}{4}$ ft^2. Write the length of one side as a mixed number.

72. The area of a square is $6\frac{1}{4}$ ft^2. Write the length of one side as a mixed number.

73. The volume of a cube is $3\frac{3}{8}$ ft^3. Write the length of one side as a mixed number.

74. The volume of a cube is $2\frac{10}{27}$ ft^3. Write the length of one side as a mixed number.

3.8 Addition and Subtraction with Mixed Numbers

Introduction . . .

In March 1995, rumors that Michael Jordan would return to basketball sent stock prices for the companies whose products he endorses higher. The price of one share of General Mills, the maker of Wheaties, which Michael Jordan endorses, went from 60\frac{1}{2}$ to 63\frac{3}{8}$. To find the increase in the price of this stock, we must be able to subtract mixed numbers.

The notation we use for mixed numbers is especially useful for addition and subtraction. When adding and subtracting mixed numbers, we will assume you recall how to go about finding a least common denominator (LCD). (If you don't remember, then review Section 3.5.)

EXAMPLE 1   Add: $3\frac{2}{3} + 4\frac{1}{5}$

SOLUTION   We begin by writing each mixed number showing the + sign. We then apply the commutative and associative properties to rearrange the order and grouping:

$$3\frac{2}{3} + 4\frac{1}{5} = 3 + \frac{2}{3} + 4 + \frac{1}{5} \qquad \text{Expand each number to show the + sign}$$

$$= 3 + 4 + \frac{2}{3} + \frac{1}{5} \qquad \text{Commutative property}$$

$$= (3 + 4) + \left(\frac{2}{3} + \frac{1}{5}\right) \qquad \text{Associative property}$$

$$= 7 + \left(\frac{5 \cdot 2}{5 \cdot 3} + \frac{3 \cdot 1}{3 \cdot 5}\right) \qquad \text{Add } 3 + 4 = 7; \text{ then multiply to get the LCD}$$

$$= 7 + \left(\frac{10}{15} + \frac{3}{15}\right) \qquad \text{Write each fraction with the LCD}$$

$$= 7 + \frac{13}{15} \qquad \text{Add the numerators}$$

$$= 7\frac{13}{15} \qquad \text{Write the answer in mixed-number notation}$$

As you can see, we obtain our result by adding the whole-number parts $(3 + 4 = 7)$ and the fraction parts $\left(\frac{2}{3} + \frac{1}{5} = \frac{13}{15}\right)$ of each mixed number. Knowing this, we can save ourselves some writing by doing the same problem in columns:

$$3\frac{2}{3} = 3\frac{2 \cdot 5}{3 \cdot 5} = 3\frac{10}{15} \qquad \text{Add whole numbers}$$

$$+4\frac{1}{5} = 4\frac{1 \cdot 3}{5 \cdot 3} = 4\frac{3}{15} \qquad \text{Then add fractions}$$

$$\overline{\hspace{3.5cm}7\frac{13}{15}}$$

Write each fraction with LCD 15

The second method shown above requires less writing and lends itself to mixed-number notation. We will use this method for the rest of this section.

Practice Problems

1. Add: $3\frac{2}{3} + 2\frac{1}{4}$

Note

You should try both methods given in Example 1 on Practice Problem 1.

Answer

1. $5\frac{11}{12}$

2. Add: $5\dfrac{3}{4} + 6\dfrac{4}{5}$

EXAMPLE 2   Add: $5\dfrac{3}{4} + 9\dfrac{5}{6}$

SOLUTION   The LCD for 4 and 6 is 12. Writing the mixed numbers in a column and then adding looks like this:

$$5\dfrac{3}{4} = 5\dfrac{3\cdot 3}{4\cdot 3} = \ 5\dfrac{9}{12}$$

$$+9\dfrac{5}{6} = 9\dfrac{5\cdot 2}{6\cdot 2} = \ 9\dfrac{10}{12}$$

$$14\dfrac{19}{12}$$

Note

Once you see how to change from a whole number and an improper fraction to a whole number and a proper fraction, you will be able to do this step without showing any work.

The fraction part of the answer is an improper fraction. We rewrite it as a whole number and a proper fraction:

$$14\dfrac{19}{12} = 14 + \dfrac{19}{12} \qquad \textbf{Write the mixed number with a + sign}$$

$$= 14 + 1\dfrac{7}{12} \qquad \textbf{Write } \tfrac{19}{12} \textbf{ as a mixed number}$$

$$= 15\dfrac{7}{12} \qquad \textbf{Add 14 and 1}$$ ■

3. Add: $6\dfrac{3}{4} + 2\dfrac{7}{8}$

EXAMPLE 3   Add: $5\dfrac{2}{3} + 6\dfrac{8}{9}$

SOLUTION $5\dfrac{2}{3} = 5\dfrac{2\cdot 3}{3\cdot 3} = \ 5\dfrac{6}{9}$

$$+6\dfrac{8}{9} = 6\dfrac{8}{9} \ \ = \ 6\dfrac{8}{9}$$

$$11\dfrac{14}{9} = 12\dfrac{5}{9}$$

The last step involves writing $\dfrac{14}{9}$ as $1\dfrac{5}{9}$ and then adding 11 and 1 to get 12.   ■

4. Add: $2\dfrac{1}{3} + 1\dfrac{1}{4} + 3\dfrac{11}{12}$

EXAMPLE 4   Add: $3\dfrac{1}{4} + 2\dfrac{3}{5} + 1\dfrac{9}{10}$

SOLUTION   The LCD is 20. We rewrite each fraction as an equivalent fraction with denominator 20 and add:

$$3\dfrac{1}{4} = 3\dfrac{1\cdot 5}{4\cdot 5} \ = 3\dfrac{5}{20}$$

$$2\dfrac{3}{5} = 2\dfrac{3\cdot 4}{5\cdot 4} \ = 2\dfrac{12}{20} \qquad \textbf{Reduce to lowest terms}$$

$$+1\dfrac{9}{10} = 1\dfrac{9\cdot 2}{10\cdot 2} = 1\dfrac{18}{20}$$

$$6\dfrac{35}{20} = 7\dfrac{15}{20} = 7\dfrac{3}{4}$$

$$\dfrac{35}{20} = 1\dfrac{15}{20}$$

Change to a mixed number   ■

Answers

2. $12\dfrac{11}{20}$ **3.** $9\dfrac{5}{8}$ **4.** $7\dfrac{1}{2}$

We should note here that we could have worked each of the first four examples in this section by first changing each mixed number to an improper fraction and then adding as we did in Section 3.5. To illustrate, if we were to work Example 4 this way, it would look like this:

$$3\frac{1}{4} + 2\frac{3}{5} + 1\frac{9}{10} = \frac{13}{4} + \frac{13}{5} + \frac{19}{10} \qquad \text{Change to improper fractions}$$

$$= \frac{13 \cdot 5}{4 \cdot 5} + \frac{13 \cdot 4}{5 \cdot 4} + \frac{19 \cdot 2}{10 \cdot 2} \qquad \text{LCD is 20}$$

$$= \frac{65}{20} + \frac{52}{20} + \frac{38}{20} \qquad \text{Equivalent fractions}$$

$$= \frac{155}{20} \qquad \text{Add numerators}$$

$$= 7\frac{15}{20} = 7\frac{3}{4} \qquad \text{Change to a mixed number, and reduce}$$

As you can see, the result is the same as the result we obtained in Example 4.

There are advantages to both methods. The method just shown works well when the whole-number parts of the mixed numbers are small. The vertical method shown in Examples 1–4 works well when the whole-number parts of the mixed numbers are large.

Subtraction with mixed numbers is very similar to addition with mixed numbers.

EXAMPLE 5 Subtract: $3\frac{9}{10} - 1\frac{3}{10}$

5. Subtract: $4\frac{7}{8} - 1\frac{5}{8}$

SOLUTION   Because the denominators are the same, we simply subtract the whole numbers and subtract the fractions:

$$
\begin{array}{r}
3\frac{9}{10} \\
-1\frac{3}{10} \\
\hline
2\frac{6}{10} = 2\frac{3}{5}
\end{array}
$$

Reduce to lowest terms

An easy way to visualize the results in Example 5 is to imagine 3 dollar bills and 9 dimes in your pocket. If you spend 1 dollar and 3 dimes, you will have 2 dollars and 6 dimes left.

6. Subtract: $12\dfrac{7}{10} - 7\dfrac{2}{5}$

EXAMPLE 6   Subtract: $12\dfrac{7}{10} - 8\dfrac{3}{5}$

SOLUTION   The common denominator is 10. We must rewrite $\frac{3}{5}$ as an equivalent fraction with denominator 10:

$$
\begin{array}{rcccr}
12\dfrac{7}{10} = & 12\dfrac{7}{10} & = & 12\dfrac{7}{10} \\[2mm]
- \ 8\dfrac{3}{5} = & - \ 8\dfrac{3\cdot 2}{5\cdot 2} & = & - \ 8\dfrac{6}{10} \\[1mm]
\hline
& & & 4\dfrac{1}{10}
\end{array}
$$

When the fraction we are subtracting is larger than the fraction we are subtracting it from, it is necessary to borrow during subtraction of mixed numbers. This is the same idea we used when borrowing was necessary in subtraction of whole numbers.

7. Subtract: $10 - 5\dfrac{4}{7}$

Note

Convince yourself that 10 is the same as $9\frac{7}{7}$. The reason we choose to write the 1 we borrowed as $\frac{7}{7}$ is that the fraction we eventually subtracted from $\frac{7}{7}$ was $\frac{2}{7}$. Both fractions must have the same denominator, 7, so that we can subtract.

EXAMPLE 7   Subtract: $10 - 5\dfrac{2}{7}$

SOLUTION   In order to have a fraction from which to subtract $\frac{2}{7}$, we borrow 1 from 10 and rewrite the 1 we borrow as $\frac{7}{7}$. The process looks like this:

$$
\begin{array}{rcl}
10 & = & 9\dfrac{7}{7} \quad \leftarrow \textbf{We rewrite 10 as 9 + 1, which is } 9 + \frac{7}{7} = 9\frac{7}{7}. \\[2mm]
- \ 5\dfrac{2}{7} & = & -5\dfrac{2}{7} \quad \textbf{Then we can subtract as usual.} \\[1mm]
\hline
& & 4\dfrac{5}{7}
\end{array}
$$

8. Subtract: $6\dfrac{1}{3} - 2\dfrac{2}{3}$

EXAMPLE 8   Subtract: $8\dfrac{1}{4} - 3\dfrac{3}{4}$

SOLUTION   Because $\frac{3}{4}$ is larger than $\frac{1}{4}$, we again need to borrow 1 from the whole number. The 1 that we borrow from the 8 is rewritten as $\frac{4}{4}$, because 4 is the denominator of both fractions:

$$
\begin{array}{rcl}
8\dfrac{1}{4} & = & 7\dfrac{5}{4} \quad \longleftarrow \ \textbf{Borrow 1 in the form } \frac{4}{4}; \\[1mm]
& & \qquad\quad \textbf{then } \frac{4}{4} + \frac{1}{4} = \frac{5}{4} \\[2mm]
-3\dfrac{3}{4} & = & -3\dfrac{3}{4} \\[1mm]
\hline
& & 4\dfrac{2}{4} = 4\dfrac{1}{2} \quad \textbf{Reduce to lowest terms}
\end{array}
$$

9. Subtract: $6\dfrac{3}{4} - 2\dfrac{5}{6}$

EXAMPLE 9   Subtract: $4\dfrac{3}{4} - 1\dfrac{5}{6}$

SOLUTION   This is about as complicated as it gets with subtraction of mixed numbers. We begin by rewriting each fraction with the common denominator 12:

$$
\begin{array}{rcccr}
4\dfrac{3}{4} & = & 4\dfrac{3\cdot 3}{4\cdot 3} & = & 4\dfrac{9}{12} \\[2mm]
-1\dfrac{5}{6} & = & -1\dfrac{5\cdot 2}{6\cdot 2} & = & -1\dfrac{10}{12} \\[1mm]
\hline
\end{array}
$$

Answers

6. $5\dfrac{3}{10}$ **7.** $4\dfrac{3}{7}$ **8.** $3\dfrac{2}{3}$

9. $3\dfrac{11}{12}$

Because $\frac{10}{12}$ is larger than $\frac{9}{12}$, we must borrow 1 from 4 in the form $\frac{12}{12}$ before we subtract:

$$
\begin{array}{l}
4\dfrac{9}{12} = 3\dfrac{21}{12} \\[6pt]
-1\dfrac{10}{12} = -1\dfrac{10}{12} \\[6pt]
\hline
2\dfrac{11}{12}
\end{array}
$$

$\leftarrow$ **4 = 3 + 1 = 3 + $\frac{12}{12}$, so 4$\frac{9}{12}$ = $\left(3 + \frac{12}{12}\right) + \frac{9}{12}$**

$= 3 + \left(\frac{12}{12} + \frac{9}{12}\right)$

$= 3 + \frac{21}{12}$

$= 3\frac{21}{12}$ ∎

EXAMPLE 10 Subtract: $15\dfrac{1}{10} - 11\dfrac{2}{3}$

10. Subtract: $17\dfrac{1}{8} - 12\dfrac{4}{5}$

SOLUTION   The LCD is 30.

$$
\begin{array}{l}
15\dfrac{1}{10} = 15\dfrac{1 \cdot 3}{10 \cdot 3} = 15\dfrac{3}{30} \\[6pt]
-11\dfrac{2}{3} = -11\dfrac{2 \cdot 10}{3 \cdot 10} = -11\dfrac{20}{30}
\end{array}
$$

Because $\frac{20}{30}$ is larger than $\frac{3}{30}$, we must borrow 1 from 15:

$$
\begin{array}{l}
15\dfrac{3}{30} = 14\dfrac{33}{30} \\[6pt]
-11\dfrac{20}{30} = -11\dfrac{20}{30} \\[6pt]
\hline
3\dfrac{13}{30}
\end{array}
$$

Borrow 1 from 15 in the form $\frac{30}{30}$

Then add $\frac{30}{30}$ to $\frac{3}{30}$ to get $\frac{33}{30}$

∎

EXAMPLE 11 The table below shows the selling price of one share of HBJ stock during each trading day of one week in 1990.

11. Using the table in Example 11, how much money would you lose if you bought 10,000 shares of the stock on Tuesday and sold them all on Wednesday?

M	T	W	T	F
$2\dfrac{3}{8}$	$2\dfrac{1}{8}$	$1\dfrac{15}{16}$	2	$2\dfrac{3}{4}$

If you bought 1,000 shares on Monday and sold them all on Friday, how much money would you make?

SOLUTION   The simplest way to work this problem is to find the difference in the price of the stock on Friday and Monday, and then multiply that difference by 1,000:

$$1,000\left(2\dfrac{3}{4} - 2\dfrac{3}{8}\right) = 1,000\left(\dfrac{3}{8}\right) = \dfrac{3,000}{8} = \$375$$

Note that, when you buy and sell stock like this, you must also pay a commission to a stockbroker when you buy the stock and when you sell it. So, your actual profit would be $375, less what you pay in commission. Also, you will have to pay income tax on your profit, which will reduce it even further. ∎

Answers

10. $4\dfrac{13}{40}$

11. You would lose $1,875.

Problem Set 3.8

Add and subtract the following mixed numbers as indicated.

1. $2\frac{1}{5} + 3\frac{3}{5}$

2. $8\frac{2}{9} + 1\frac{5}{9}$

3. $4\frac{3}{10} + 8\frac{1}{10}$

4. $5\frac{2}{7} + 3\frac{3}{7}$

5. $6\frac{8}{9} - 3\frac{4}{9}$

6. $12\frac{5}{12} - 7\frac{1}{12}$

7. $9\frac{1}{6} + 2\frac{5}{6}$

8. $9\frac{1}{4} + 5\frac{3}{4}$

9. $3\frac{5}{8} - 2\frac{1}{4}$

10. $7\frac{9}{10} - 6\frac{3}{5}$

11. $11\frac{1}{3} + 2\frac{5}{6}$

12. $1\frac{5}{8} + 2\frac{1}{2}$

13. $7\frac{5}{12} - 3\frac{1}{3}$

14. $7\frac{3}{4} - 3\frac{5}{12}$

15. $6\frac{1}{3} - 4\frac{1}{4}$

16. $5\frac{4}{5} - 3\frac{1}{3}$

17. $10\frac{5}{6} + 15\frac{3}{4}$

18. $11\frac{7}{8} + 9\frac{1}{6}$

19. $5\frac{2}{3}$
$+6\frac{1}{3}$
___

20. $8\frac{5}{6}$
$+9\frac{5}{6}$
___

21. $10\frac{13}{16}$
$-\ 8\frac{5}{16}$
___

22. $17\frac{7}{12}$
$-\ 9\frac{5}{12}$
___

23. $6\frac{1}{2}$
$+2\frac{5}{14}$
___

24. $9\frac{11}{12}$
$+4\frac{1}{6}$
___

25. $1\frac{5}{8}$
$+1\frac{3}{4}$
___

26. $7\frac{6}{7}$
$+2\frac{3}{14}$
___

27. $4\frac{2}{3}$
$+5\frac{3}{5}$
___

28. $9\frac{4}{9}$
$+1\frac{1}{6}$
___

29. $5\frac{4}{10}$
$-3\frac{1}{3}$
___

30. $12\frac{7}{8}$
$-\ 3\frac{5}{6}$
___

Find the following sums. (Add.)

31. $1\frac{1}{4} + 2\frac{3}{4} + 5$

32. $6 + 5\frac{3}{5} + 8\frac{2}{5}$

33. $7\frac{1}{10} + 8\frac{3}{10} + 2\frac{7}{10}$

34. $5\frac{2}{7} + 8\frac{1}{7} + 3\frac{5}{7}$

35. $\frac{3}{4} + 8\frac{1}{4} + 5$

36. $\frac{5}{8} + 1\frac{1}{8} + 7$

37. $3\frac{1}{2} + 8\frac{1}{3} + 5\frac{1}{6}$

38. $4\frac{1}{5} + 7\frac{1}{3} + 8\frac{1}{15}$

39. $8\frac{2}{3}$

$9\frac{1}{8}$

$+6\frac{1}{4}$

40. $7\frac{3}{5}$

$8\frac{2}{3}$

$+1\frac{1}{5}$

41. $6\frac{1}{7}$

$9\frac{3}{14}$

$+12\frac{1}{2}$

42. $1\frac{5}{6}$

$2\frac{3}{4}$

$+5\frac{1}{2}$

43. $10\frac{1}{20}$

$11\frac{4}{5}$

$+15\frac{3}{10}$

44. $18\frac{7}{12}$

$19\frac{3}{16}$

$+10\frac{2}{3}$

The following problems all involve the concept of borrowing. Subtract in each case.

45. $8 - 1\frac{3}{4}$

46. $5 - 3\frac{1}{3}$

47. $15 - 5\frac{3}{10}$

48. $24 - 10\frac{5}{12}$

49. $8\frac{1}{4} - 2\frac{3}{4}$

50. $12\frac{3}{10} - 5\frac{7}{10}$

51. $9\frac{1}{3} - 8\frac{2}{3}$

52. $7\frac{1}{6} - 6\frac{5}{6}$

53. $4\frac{1}{4} - 2\frac{1}{3}$

54. $6\frac{1}{5} - 1\frac{2}{3}$

55. $9\frac{2}{3} - 5\frac{3}{4}$

56. $12\frac{5}{6} - 8\frac{7}{8}$

57. $16\frac{3}{4} - 10\frac{4}{5}$

58. $18\frac{5}{12} - 9\frac{3}{4}$

59. $10\frac{3}{10} - 4\frac{4}{5}$

60. $9\frac{4}{7} - 7\frac{2}{3}$

61. $13\frac{1}{6} - 12\frac{5}{8}$

62. $21\frac{2}{5} - 20\frac{5}{6}$

63. Find the difference between $6\frac{1}{5}$ and $2\frac{7}{10}$.

64. Give the difference between $5\frac{1}{3}$ and $1\frac{5}{6}$.

65. Find the sum of $3\frac{1}{8}$ and $2\frac{3}{5}$.

66. Find the sum of $1\frac{5}{6}$ and $3\frac{4}{9}$.

Applying the Concepts

67. Building Two pieces of molding $5\frac{7}{8}$ inches and $6\frac{3}{8}$ inches long are placed end to end. What is the total length of the two pieces of molding together?

68. Jogging A jogger runs $2\frac{1}{2}$ miles on Monday, $3\frac{1}{4}$ miles on Tuesday, and $2\frac{2}{5}$ miles on Wednesday. What is the jogger's total mileage for this 3-day period?

69. Length A $2\frac{1}{2}$-inch nail is how much longer than a $1\frac{3}{4}$-inch nail?

70. Length A $3\frac{1}{4}$-inch nail is how much longer than a $\frac{3}{4}$-inch nail?

71. Length of Jeans A pair of jeans is $32\frac{1}{2}$ inches long. How long are the jeans after they have been washed if they shrink $1\frac{1}{3}$ inches?

72. Manufacturing A clothing manufacturer has two rolls of cloth. One roll is $35\frac{1}{2}$ yards, and the other is $62\frac{5}{8}$ yards. What is the total number of yards in the two rolls?

73. Stock Prices A stock selling for $7\frac{1}{8}$¢ a share on Monday sells for $1\frac{1}{4}$¢ less a share on Tuesday. What does it sell for on Tuesday?

74. Stock Prices On Monday, a stock sells for $10\frac{3}{8}$¢ a share. On Tuesday, the same stock sells for $9\frac{1}{4}$¢. Find the difference.

Area and Perimeter The diagrams below show the dimensions of playing fields for the National Football League (NFL), the Canadian Football League, and arena football.

Football Fields

NFL — 100 yd, $53\frac{1}{3}$ yd

Canadian — 110 yd, 65 yd

Arena — 50 yd, $28\frac{1}{3}$ yd

75. Find the perimeter of each football field.

76. Find the area of each football field.

Stock Prices As we mentioned in the introduction to this section, in March 1995, rumors that Michael Jordan would return to basketball sent stock prices for the companies whose products he endorses higher. The table at the right gives some of the details of those increases. Use the table to work Problems 77–80.

STOCK PRICES FOR COMPANIES WITH MICHAEL JORDAN ENDORSEMENTS

COMPANY	PRODUCT ENDORSED	STOCK PRICE (DOLLARS)	
		3/8/95	3/13/95
Nike	Air Jordans	$74\frac{7}{8}$	$77\frac{3}{8}$
Quaker Oats	Gatorade	$32\frac{1}{4}$	$32\frac{5}{8}$
General Mills	Wheaties	$60\frac{1}{2}$	$63\frac{3}{8}$
McDonald's		$32\frac{7}{8}$	$34\frac{3}{8}$

77. a. Find the difference in the price of Nike stock between March 13 and March 8.
  b. If you owned 100 shares of Nike stock, how much more is the 100 shares worth on March 13 than on March 8?

78. a. Find the difference in price of General Mills stock between March 13 and March 8.
  b. If you owned 1,000 shares of General Mills stock on March 8, how much more would it be worth on March 13?

79. If you owned 200 shares of McDonald's stock on March 8, how much more would it be worth on March 13?

80. If you owned 100 shares of McDonald's stock on March 8, how much more would it be worth on March 13?

Find the perimeter of each figure.

81.

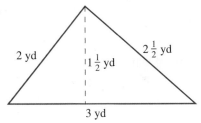

82.

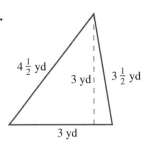

83.

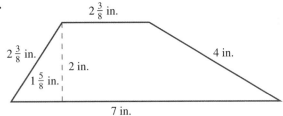

84.

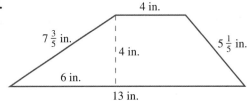

Calculator Problems

Use a calculator to help find the following sums and differences.

85. $796\dfrac{18}{41} + 357\dfrac{49}{53}$ (LCD = 2,173)

86. $248\dfrac{16}{23} + 159\dfrac{41}{47}$ (LCD = 1,081)

87. $499\dfrac{29}{53} - 219\dfrac{99}{109}$ (LCD = 5,777)

88. $926\dfrac{3}{19} - 621\dfrac{39}{41}$ (LCD = 779)

Review Problems

To work the problems in the next section, you need to know the rule for order of operations. The problems below are intended to review the rule for order of operations from Section 1.9.

Use the rule for order of operations to combine the following.

89. $3 + 2 \cdot 7$ **90.** $8 \cdot 3 - 2$ **91.** $4 \cdot 5 - 3 \cdot 2$ **92.** $9 \cdot 7 + 6 \cdot 5$

93. $3 \cdot 2^3 + 5 \cdot 4^2$ **94.** $6 \cdot 5^2 + 2 \cdot 3^3$ **95.** $3[2 + 5(6)]$ **96.** $4[2(3) + 3(5)]$

97. $(7 - 3)(8 + 2)$ **98.** $(9 - 5)(9 + 5)$

3.9 Combinations of Operations and Complex Fractions

The problems in this section all involve combinations of the four basic operations we've already discussed—addition, subtraction, multiplication, and division—involving mixed numbers and fractions.

EXAMPLE 1 Simplify the expression: $5 + \left(2\frac{1}{2}\right)\left(3\frac{2}{3}\right)$

SOLUTION   The rule for order of operations indicates that we should multiply $2\frac{1}{2}$ times $3\frac{2}{3}$ and then add 5 to the result:

$$5 + \left(2\frac{1}{2}\right)\left(3\frac{2}{3}\right) = 5 + \left(\frac{5}{2}\right)\left(\frac{11}{3}\right)$$ **Change the mixed numbers to improper fractions**

$$= 5 + \frac{55}{6}$$ **Multiply the improper fractions**

$$= \frac{30}{6} + \frac{55}{6}$$ **Write 5 as $\frac{30}{6}$ so both numbers have the same denominator**

$$= \frac{85}{6}$$ **Add fractions by adding their numerators**

$$= 14\frac{1}{6}$$ **Write the answer as a mixed number** ■

EXAMPLE 2 Simplify: $\left(\frac{3}{4} + \frac{5}{8}\right)\left(2\frac{3}{8} + 1\frac{1}{4}\right)$

SOLUTION   We begin by combining the numbers inside the parentheses:

$$\frac{3}{4} + \frac{5}{8} = \frac{3 \cdot 2}{4 \cdot 2} + \frac{5}{8}$$ **Write each fraction as an equivalent fraction with LCD = 8**

$$= \frac{6}{8} + \frac{5}{8}$$

$$= \frac{11}{8}$$

$$2\frac{3}{8} = \quad 2\frac{3}{8} \quad = \quad 2\frac{3}{8}$$
$$+1\frac{1}{4} = +1\frac{1 \cdot 2}{4 \cdot 2} = +1\frac{2}{8}$$
$$\overline{\qquad\qquad\qquad\qquad 3\frac{5}{8}}$$

Now that we have combined the expressions inside the parentheses, we can complete the problem by multiplying the results:

$$\left(\frac{3}{4} + \frac{5}{8}\right)\left(2\frac{3}{8} + 1\frac{1}{4}\right) = \left(\frac{11}{8}\right)\left(3\frac{5}{8}\right)$$

$$= \frac{11}{8} \cdot \frac{29}{8}$$ **Change $3\frac{5}{8}$ to an improper fraction**

$$= \frac{319}{64}$$ **Multiply fractions**

$$= 4\frac{63}{64}$$ **Write the answer as a mixed number** ■

Practice Problems

1. Simplify the expression:

$$4 + \left(1\frac{1}{2}\right)\left(2\frac{3}{4}\right)$$

2. Simplify:

$$\left(\frac{2}{3} + \frac{1}{6}\right)\left(2\frac{5}{6} + 1\frac{1}{3}\right)$$

Answers

1. $8\frac{1}{8}$ 2. $3\frac{17}{36}$

3. Simplify:

$$\frac{3}{7}+\frac{1}{3}\left(1\frac{1}{2}+4\frac{1}{2}\right)^2$$

Simplify each complex fraction as much as possible.

4. $\dfrac{\frac{2}{3}}{\frac{5}{9}}$

EXAMPLE 3   Simplify: $\dfrac{3}{5}+\dfrac{1}{2}\left(3\dfrac{2}{3}+4\dfrac{1}{3}\right)^2$

SOLUTION   We begin by combining the expressions inside the parentheses:

$$\frac{3}{5}+\frac{1}{2}\left(3\frac{2}{3}+4\frac{1}{3}\right)^2=\frac{3}{5}+\frac{1}{2}(8)^2 \qquad \text{The sum inside the parenthesis is 8}$$

$$=\frac{3}{5}+\frac{1}{2}(64) \qquad \text{The square of 8 is 64}$$

$$=\frac{3}{5}+32 \qquad \text{½ of 64 is 32}$$

$$=32\frac{3}{5} \qquad \text{The result is a mixed number} \qquad ■$$

Complex Fractions

DEFINITION

A **complex fraction** is a fraction in which the numerator and/or the denominator are themselves fractions or combinations of fractions.

Each of the following is a complex fraction:

$$\frac{\frac{3}{4}}{\frac{5}{6}}, \qquad \frac{3+\frac{1}{2}}{2-\frac{3}{4}}, \qquad \frac{\frac{1}{2}+\frac{2}{3}}{\frac{3}{4}-\frac{1}{6}}$$

EXAMPLE 4   Simplify: $\dfrac{\frac{3}{4}}{\frac{5}{6}}$

SOLUTION   This is actually the same as the problem $\frac{3}{4}\div\frac{5}{6}$, because the bar between $\frac{3}{4}$ and $\frac{5}{6}$ indicates division. Therefore, it must be true that

$$\frac{\frac{3}{4}}{\frac{5}{6}}=\frac{3}{4}\div\frac{5}{6}$$

$$=\frac{3}{4}\cdot\frac{6}{5}$$

$$=\frac{18}{20}$$

$$=\frac{9}{10} \qquad ■$$

Answers

3. $12\frac{3}{7}$ **4.** $1\frac{1}{5}$

EXAMPLE 5 Simplify: $\dfrac{\dfrac{1}{2} + \dfrac{2}{3}}{\dfrac{3}{4} - \dfrac{1}{6}}$

SOLUTION Let's decide to call the numerator of this complex fraction the *top* of the fraction and its denominator the *bottom* of the complex fraction. It will be less confusing if we name them this way. The LCD for all the denominators on the top and bottom is 12, so we can multiply the top and bottom of this complex fraction by 12 and be sure all the denominators will divide it exactly. This will leave us with only whole numbers on the top and bottom:

$$\frac{\dfrac{1}{2} + \dfrac{2}{3}}{\dfrac{3}{4} - \dfrac{1}{6}} = \frac{\mathbf{12}\left(\dfrac{1}{2} + \dfrac{2}{3}\right)}{\mathbf{12}\left(\dfrac{3}{4} - \dfrac{1}{6}\right)} \qquad \textbf{Multiply the top and bottom by the LCD}$$

$$= \frac{\mathbf{12} \cdot \dfrac{1}{2} + \mathbf{12} \cdot \dfrac{2}{3}}{\mathbf{12} \cdot \dfrac{3}{4} - \mathbf{12} \cdot \dfrac{1}{6}} \qquad \textbf{Distributive property}$$

$$= \frac{6 + 8}{9 - 2} \qquad \textbf{Multiply each fraction by 12}$$

$$= \frac{14}{7} \qquad \textbf{Add on top and subtract on bottom}$$

$$= 2 \qquad \textbf{Reduce to lowest terms}$$

The problem can be worked in another way also. We can simplify the top and bottom of the complex fraction separately. Simplifying the top, we have

$$\frac{1}{2} + \frac{2}{3} = \frac{1 \cdot \mathbf{3}}{2 \cdot \mathbf{3}} + \frac{2 \cdot \mathbf{2}}{3 \cdot \mathbf{2}} = \frac{3}{6} + \frac{4}{6} = \frac{7}{6}$$

Simplifying the bottom, we have

$$\frac{3}{4} - \frac{1}{6} = \frac{3 \cdot \mathbf{3}}{4 \cdot \mathbf{3}} - \frac{1 \cdot \mathbf{2}}{6 \cdot \mathbf{2}} = \frac{9}{12} - \frac{2}{12} = \frac{7}{12}$$

We now write the original complex fraction again using the simplified expressions for the top and bottom. Then we proceed as we did in Example 4.

$$\frac{\dfrac{1}{2} + \dfrac{2}{3}}{\dfrac{3}{4} - \dfrac{1}{6}} = \frac{\dfrac{7}{6}}{\dfrac{7}{12}}$$

$$= \frac{7}{6} \div \frac{7}{12} \qquad \textbf{The divisor is } \tfrac{7}{12}$$

$$= \frac{7}{6} \cdot \frac{12}{7} \qquad \textbf{Replace } \tfrac{7}{12} \textbf{ by its reciprocal and multiply}$$

$$= \frac{7 \cdot 2 \cdot \cancel{6}}{\cancel{6} \cdot \cancel{7}} \qquad \textbf{Divide out common factors}$$

$$= 2$$

In both cases the result is the same. We can use either method to simplify complex fractions. ∎

5. $\dfrac{\dfrac{1}{2} + \dfrac{3}{4}}{\dfrac{2}{3} - \dfrac{1}{4}}$

Note
We are going to simplify this complex fraction by two different methods. This is the first method.

Note
The fraction bar that separates the numerator of the complex fraction from its denominator works like parentheses. If we were to rewrite this problem without it, we would write it like this:

$$\left(\frac{1}{2} + \frac{2}{3}\right) \div \left(\frac{3}{4} - \frac{1}{6}\right)$$

That is why we simplify the top and bottom of the complex fraction separately and then divide.

Answer
5. 3

6. Simplify: $\dfrac{4 + \dfrac{2}{3}}{3 - \dfrac{1}{4}}$

EXAMPLE 6   Simplify: $\dfrac{3 + \dfrac{1}{2}}{2 - \dfrac{3}{4}}$

SOLUTION   The simplest approach here is to multiply both the top and bottom by the LCD for all fractions, which is 4:

$$\dfrac{3 + \dfrac{1}{2}}{2 - \dfrac{3}{4}} = \dfrac{\mathbf{4}\left(3 + \dfrac{1}{2}\right)}{\mathbf{4}\left(2 - \dfrac{3}{4}\right)} \qquad \textbf{Multiply the top and bottom by 4}$$

$$= \dfrac{\mathbf{4} \cdot 3 + \mathbf{4} \cdot \dfrac{1}{2}}{\mathbf{4} \cdot 2 - \mathbf{4} \cdot \dfrac{3}{4}} \qquad \textbf{Distributive property}$$

$$= \dfrac{12 + 2}{8 - 3} \qquad \textbf{Multiply each number by 4}$$

$$= \dfrac{14}{5} \qquad \textbf{Add on top and subtract on bottom}$$

$$= 2\dfrac{4}{5} \qquad \blacksquare$$

7. Simplify: $\dfrac{12\dfrac{1}{3}}{6\dfrac{2}{3}}$

EXAMPLE 7   Simplify: $\dfrac{10\dfrac{1}{3}}{8\dfrac{2}{3}}$

SOLUTION   The simplest way to simplify this complex fraction is to think of it as a division problem.

$$\dfrac{10\dfrac{1}{3}}{8\dfrac{2}{3}} = 10\dfrac{1}{3} \div 8\dfrac{2}{3} \qquad \textbf{Write with a} \div \textbf{symbol}$$

$$= \dfrac{31}{3} \div \dfrac{26}{3} \qquad \textbf{Change to improper fractions}$$

$$= \dfrac{31}{3} \cdot \dfrac{3}{26} \qquad \textbf{Write in terms of multiplication}$$

$$= \dfrac{31 \cdot \cancel{3}}{\cancel{3} \cdot 26} \qquad \textbf{Divide out the common factor 3}$$

$$= \dfrac{31}{26} \qquad \textbf{Answer as an improper fraction}$$

$$= 1\dfrac{5}{26} \qquad \textbf{Answer as a mixed number} \qquad \blacksquare$$

Answers

6. $1\dfrac{23}{33}$ **7.** $1\dfrac{17}{20}$

Problem Set 3.9

Use the rule for order of operations to simplify each of the following.

1. $3 + \left(1\frac{1}{2}\right)\left(2\frac{2}{3}\right)$

2. $7 - \left(1\frac{3}{5}\right)\left(2\frac{1}{2}\right)$

3. $8 - \left(\frac{6}{11}\right)\left(1\frac{5}{6}\right)$

4. $10 + \left(2\frac{4}{5}\right)\left(\frac{5}{7}\right)$

5. $\frac{2}{3}\left(1\frac{1}{2}\right) + \frac{3}{4}\left(1\frac{1}{3}\right)$

6. $\frac{2}{5}\left(2\frac{1}{2}\right) + \frac{5}{8}\left(3\frac{1}{5}\right)$

7. $2\left(1\frac{1}{2}\right) + 5\left(6\frac{2}{5}\right)$

8. $4\left(5\frac{3}{4}\right) + 6\left(3\frac{5}{6}\right)$

9. $\left(\frac{3}{5} + \frac{1}{10}\right)\left(\frac{1}{2} + \frac{3}{4}\right)$

10. $\left(\frac{2}{9} + \frac{1}{3}\right)\left(\frac{1}{5} + \frac{1}{10}\right)$

11. $\left(2 + \frac{2}{3}\right)\left(3 + \frac{1}{8}\right)$

12. $\left(3 - \frac{3}{4}\right)\left(3 + \frac{1}{3}\right)$

13. $\left(1 + \frac{5}{6}\right)\left(1 - \frac{5}{6}\right)$

14. $\left(2 - \frac{1}{4}\right)\left(2 + \frac{1}{4}\right)$

15. $\frac{2}{3} + \frac{1}{3}\left(2\frac{1}{2} + \frac{1}{2}\right)^2$

16. $\frac{3}{5} + \frac{1}{4}\left(2\frac{1}{2} - \frac{1}{2}\right)^3$

17. $2\frac{3}{8} + \frac{1}{2}\left(\frac{1}{3} + \frac{5}{3}\right)^3$

18. $8\frac{2}{3} + \frac{1}{3}\left(\frac{8}{5} + \frac{7}{5}\right)^2$

19. $2\left(\frac{1}{2} + \frac{1}{3}\right) + 3\left(\frac{2}{3} + \frac{1}{4}\right)$

20. $5\left(\frac{1}{5} + \frac{3}{10}\right) + 2\left(\frac{1}{10} + \frac{1}{2}\right)$

Simplify each complex fraction as much as possible.

21. $\dfrac{\frac{2}{3}}{\frac{3}{4}}$

22. $\dfrac{\frac{5}{6}}{\frac{3}{12}}$

23. $\dfrac{\frac{2}{3}}{\frac{4}{3}}$

24. $\dfrac{\frac{7}{9}}{\frac{5}{9}}$

25. $\dfrac{\frac{11}{20}}{\frac{5}{10}}$

26. $\dfrac{\frac{9}{16}}{\frac{3}{4}}$

27. $\dfrac{\frac{1}{2} + \frac{1}{3}}{\frac{1}{2} - \frac{1}{3}}$

28. $\dfrac{\frac{1}{4} + \frac{1}{5}}{\frac{1}{4} - \frac{1}{5}}$

29. $\dfrac{\frac{5}{8} - \frac{1}{4}}{\frac{1}{8} + \frac{1}{2}}$

30. $\dfrac{\frac{3}{4} + \frac{1}{3}}{\frac{2}{3} + \frac{1}{6}}$

31. $\dfrac{\frac{9}{20} - \frac{1}{10}}{\frac{1}{10} + \frac{9}{20}}$

32. $\dfrac{\frac{1}{2} + \frac{2}{3}}{\frac{3}{4} + \frac{5}{6}}$

33. $\dfrac{1 + \dfrac{2}{3}}{1 - \dfrac{2}{3}}$

34. $\dfrac{5 - \dfrac{3}{4}}{2 + \dfrac{3}{4}}$

35. $\dfrac{2 + \dfrac{5}{6}}{5 - \dfrac{1}{3}}$

36. $\dfrac{9 - \dfrac{11}{5}}{3 + \dfrac{13}{10}}$

37. $\dfrac{3 + \dfrac{5}{6}}{1 + \dfrac{5}{3}}$

38. $\dfrac{10 + \dfrac{9}{10}}{5 + \dfrac{4}{5}}$

39. $\dfrac{\dfrac{1}{3} + \dfrac{3}{4}}{2 - \dfrac{1}{6}}$

40. $\dfrac{3 + \dfrac{5}{2}}{\dfrac{5}{6} + \dfrac{1}{4}}$

41. $\dfrac{\dfrac{5}{6}}{3 + \dfrac{2}{3}}$

42. $\dfrac{9 - \dfrac{3}{2}}{\dfrac{7}{4}}$

Simplify each of the following complex fractions.

43. $\dfrac{2\dfrac{1}{2} + \dfrac{1}{2}}{3\dfrac{3}{5} - \dfrac{2}{5}}$

44. $\dfrac{5\dfrac{3}{8} + \dfrac{5}{8}}{4\dfrac{1}{4} + 1\dfrac{3}{4}}$

45. $\dfrac{2 + 1\dfrac{2}{3}}{3\dfrac{5}{6} - 1}$

46. $\dfrac{5 + 8\dfrac{3}{5}}{2\dfrac{3}{10} + 4}$

47. $\dfrac{3\dfrac{1}{4} - 2\dfrac{1}{2}}{5\dfrac{3}{4} + 1\dfrac{1}{2}}$

48. $\dfrac{9\dfrac{3}{8} + 2\dfrac{5}{8}}{6\dfrac{1}{2} + 7\dfrac{1}{2}}$

49. $\dfrac{3\dfrac{1}{4} + 5\dfrac{1}{6}}{2\dfrac{1}{3} + 3\dfrac{1}{4}}$

50. $\dfrac{8\dfrac{5}{6} + 1\dfrac{2}{3}}{7\dfrac{1}{3} + 2\dfrac{1}{4}}$

51. $\dfrac{6\dfrac{2}{3} + 7\dfrac{3}{4}}{8\dfrac{1}{2} + 9\dfrac{7}{8}}$

52. $\dfrac{3\dfrac{4}{5} - 1\dfrac{9}{10}}{6\dfrac{5}{6} - 2\dfrac{3}{4}}$

53. What is twice the sum of $2\dfrac{1}{5}$ and $\dfrac{3}{6}$?

54. Find 3 times the difference of $1\dfrac{7}{9}$ and $\dfrac{2}{9}$.

55. Add $5\dfrac{1}{4}$ to the sum of $\dfrac{3}{4}$ and 2.

56. Subtract $\dfrac{7}{8}$ from the product of 2 with $3\dfrac{1}{2}$.

Applying the Concepts

57. Manufacturing A dress manufacturer usually buys two rolls of cloth, one of $32\dfrac{1}{2}$ yards and the other of $25\dfrac{1}{3}$ yards, to fill his weekly orders. If his orders double one week, how much of the cloth should he order? (Give the total yardage.)

58. Body Temperature Suppose your normal body temperature is $98\dfrac{3}{5}°$ Fahrenheit. If your temperature goes up $3\dfrac{1}{5}°$ on Monday and then down $1\dfrac{4}{5}°$ on Tuesday, what is your temperature on Tuesday?

59. Stock Prices Suppose you want to buy 20 shares of a certain stock that is selling for $32\frac{1}{8}$ a share, but by the time you place your order the stock has gone up by $2\frac{3}{8}$ a share. How much will you have to pay for the 20 shares?

60. Stock Prices Suppose you buy 15 shares of a certain stock that is selling for $8\frac{1}{4}$ a share, but by the time you place your order the stock has gone down $2\frac{3}{4}$ a share. How much will you pay for the 15 shares?

61. Average Gain in Stock Price The table and bar chart below are from the introduction to this chapter. Each shows the amount of gain or loss each day of the week of March 6, 2000, for the stock price of eCollege.com, an Internet company specializing in distance learning for college students.

CHANGE IN STOCK PRICE	
DATE	**GAIN/LOSS**
Monday, March 6, 2000	$\frac{3}{4}$
Tuesday, March 7, 2000	$-\frac{9}{16}$
Wednesday, March 8, 2000	$\frac{3}{32}$
Thursday, March 9, 2000	$\frac{7}{32}$
Friday, March 10, 2000	$\frac{1}{16}$

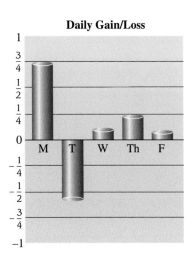

Find the mean gain or loss for the week for this stock.

62. Average Gain in Stock Price The table and bar chart below show the amount of gain or loss each day for the week of March 6, 2000, for the stock price of amazon.com, an online bookstore.

CHANGE IN STOCK PRICE	
DATE	**GAIN/LOSS**
Monday, March 6, 2000	$\frac{1}{16}$
Tuesday, March 7, 2000	$-1\frac{3}{8}$
Wednesday, March 8, 2000	$\frac{3}{8}$
Thursday, March 9, 2000	$5\frac{13}{16}$
Friday, March 10, 2000	$-\frac{3}{8}$

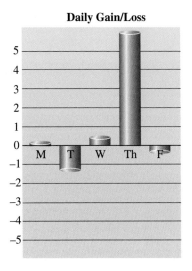

Find the mean gain/loss for the week.

63. Day Trading Jon is buying and selling stock from his home computer. He owns 150 shares of America Online (AOL) and 200 shares of Yahoo! (YHOO). On March 2, 2000, those stocks had the gain and loss shown in the table below. What was Jon's net gain or loss for the day on those two stocks?

STOCK	NUMBER OF SHARES	GAIN/LOSS
AOL	150	$+1\frac{3}{4}$
YHOO	200	$-3\frac{5}{8}$

64. Stock Gain/Loss Julie owns stock that she keeps in her retirement account. She owns 300 shares of General Electric (GE) and 150 shares of RealNetworks (RNWK). For the month of February 2000, those stocks had the gain and loss shown in the table below. What was Julie's net gain or loss for the month of February on those two stocks?

STOCK	NUMBER OF SHARES	GAIN/LOSS
GE	300	$-1\frac{7}{8}$
RNWK	150	$-16\frac{1}{16}$

Review Problems

These problems review the four basic operations with fractions from this chapter.

Perform the indicated operations.

65. $\dfrac{3}{4} \cdot \dfrac{8}{9}$

66. $-8 \cdot \dfrac{5}{6}$

67. $\dfrac{2}{3} \div (-4)$

68. $\dfrac{7}{8} \div \dfrac{14}{24}$

69. $-\dfrac{3}{7} + \dfrac{2}{7}$

70. $\dfrac{6}{7} - \left(-\dfrac{9}{14}\right)$

71. $10 - \dfrac{2}{9}$

72. $\dfrac{2}{3} - \dfrac{3}{5}$

73. $\dfrac{x}{3} + \dfrac{4}{5}$

74. $\dfrac{x}{2} - \dfrac{1}{3}$

75. $4 - \dfrac{2}{x}$

76. $3 + \dfrac{4}{x}$

77. $\dfrac{7}{a} + \dfrac{2}{3}$

78. $\dfrac{6}{a} + \dfrac{3}{4}$

Chapter 3 Summary

Definition of Fractions [3.1]

A fraction is any number that can be written in the form $\frac{a}{b}$, where a and b are integers and b is not 0. The number a is called the *numerator*, and the number b is called the *denominator*.

Properties of Fractions [3.1]

Multiplying the numerator and the denominator of a fraction by the same nonzero number will produce an equivalent fraction. The same is true for dividing the numerator and denominator by the same nonzero number. In symbols the properties look like this: If a, b, and c are integers and b and c are not 0, then

Property 1 $\quad \frac{a}{b} = \frac{a \cdot c}{b \cdot c} \qquad$ **Property 2** $\quad \frac{a}{b} = \frac{a \div c}{b \div c}$

Fractions and the Number 1 [3.1]

If a represents any number, then

$$\frac{a}{1} = a \quad \text{and} \quad \frac{a}{a} = 1 \quad \text{(where } a \text{ is not 0)}$$

Reducing Fractions to Lowest Terms [3.2]

To reduce a fraction to lowest terms, factor the numerator and the denominator, and then divide both the numerator and denominator by any factors they have in common.

Multiplying Fractions [3.3]

To multiply fractions, multiply numerators and multiply denominators.

The Area of a Triangle [3.3]

The formula for the area of a triangle with base b and height h is

$$A = \frac{1}{2}bh$$

Reciprocals [3.4]

Any two numbers whose product is 1 are called *reciprocals*. The numbers $\frac{2}{3}$ and $\frac{3}{2}$ are reciprocals, because their product is 1.

Division with Fractions [3.4]

To divide by a fraction, multiply by its reciprocal. That is, the quotient of two fractions is defined to be the product of the first fraction with the reciprocal of the second fraction (the divisor).

Examples

1. Each of the following is a fraction:
$$\frac{1}{2}, \frac{3}{4}, \frac{8}{1}, \frac{7}{3}$$

2. Change $\frac{3}{4}$ to an equivalent fraction with denominator 12.
$$\frac{3}{4} = \frac{3 \cdot 3}{4 \cdot 3} = \frac{9}{12}$$

3. $\frac{5}{1} = 5, \frac{5}{5} = 1$

4. $\frac{90}{588} = \frac{2 \cdot 3 \cdot 3 \cdot 5}{2 \cdot 2 \cdot 3 \cdot 7 \cdot 7}$
$$= \frac{3 \cdot 5}{2 \cdot 7 \cdot 7}$$
$$= \frac{15}{98}$$

5. $\frac{3}{5} \cdot \frac{4}{7} = \frac{3 \cdot 4}{5 \cdot 7} = \frac{12}{35}$

6. If the base of a triangle is 10 inches and the height is 7 inches, then the area is
$$A = \frac{1}{2}bh$$
$$= \frac{1}{2} \cdot 10 \cdot 7$$
$$= 5 \cdot 7$$
$$= 35 \text{ square inches}$$

7. $\frac{3}{8} \div \frac{1}{3} = \frac{3}{8} \cdot \frac{3}{1} = \frac{9}{8}$

Least Common Denominator (LCD) [3.5]

The *least common denominator* (LCD) for a set of denominators is the smallest number that is exactly divisible by each denominator.

8. $\dfrac{1}{8} + \dfrac{3}{8} = \dfrac{1+3}{8}$

$\phantom{\dfrac{1}{8} + \dfrac{3}{8}} = \dfrac{4}{8}$

$\phantom{\dfrac{1}{8} + \dfrac{3}{8}} = \dfrac{1}{2}$

Addition and Subtraction of Fractions [3.5]

To add (or subtract) two fractions with a common denominator, add (or subtract) numerators and use the common denominator. In symbols: If a, b, and c are integers with c not equal to 0, then

$$\frac{a}{c} + \frac{b}{c} = \frac{a+b}{c} \quad \text{and} \quad \frac{a}{c} - \frac{b}{c} = \frac{a-b}{c}$$

Additional Facts about Fractions

1. In some books fractions are called *rational numbers*.
2. Every whole number can be written as a fraction with a denominator of 1. [3.1]
3. The commutative, associative, and distributive properties are true for fractions.
4. The word *of* as used in the expression "$\frac{2}{3}$ *of* 12" indicates that we are to multiply $\frac{2}{3}$ and 12. [3.3]
5. Two fractions with the same value are called *equivalent* fractions. [3.1]

Mixed-Number Notation [3.6]

A mixed number is the sum of a whole number and a fraction. The + sign is not shown when we write mixed numbers; it is implied. The mixed number $4\frac{2}{3}$ is actually the sum $4 + \frac{2}{3}$.

9. $\underset{\text{Mixed number}}{4\frac{2}{3}} = \frac{3 \cdot 4 + 2}{3} = \underset{\text{Improper fraction}}{\frac{14}{3}}$

Changing Mixed Numbers to Improper Fractions [3.6]

To change a mixed number to an improper fraction, we write the mixed number showing the + sign and add as usual. The result is the same if we multiply the denominator of the fraction by the whole number and add what we get to the numerator of the fraction, putting this result over the denominator of the fraction.

10. Change $\dfrac{14}{3}$ to a mixed number.

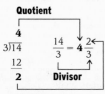

$$\dfrac{14}{3} = 4\dfrac{2}{3}$$

Changing an Improper Fraction to a Mixed Number [3.6]

To change an improper fraction to a mixed number, divide the denominator into the numerator. The quotient is the whole-number part of the mixed number. The fraction part is the remainder over the divisor.

11. $2\dfrac{1}{3} \cdot 1\dfrac{3}{4} = \dfrac{7}{3} \cdot \dfrac{7}{4} = \dfrac{49}{12} = 4\dfrac{1}{12}$

Multiplication and Division with Mixed Numbers [3.7]

To multiply or divide two mixed numbers, change each to an improper fraction and multiply or divide as usual.

12. $3\dfrac{4}{9} = 3\dfrac{4}{9} \phantom{= 3\dfrac{4}{9}} = 3\dfrac{4}{9}$

$+2\dfrac{2}{3} = 2\dfrac{2 \cdot 3}{3 \cdot 3} = 2\dfrac{6}{9}$

$\phantom{+2\dfrac{2}{3} = 2\dfrac{2 \cdot 3}{3 \cdot 3} =} 5\dfrac{10}{9} = 6\dfrac{1}{9}$

Common denominator · Add whole numbers · Add fractions

Addition and Subtraction with Mixed Numbers [3.8]

To add or subtract two mixed numbers, add or subtract the whole-number parts and the fraction parts separately. This is best done with the numbers written in columns.

Borrowing in Subtraction with Mixed Numbers [3.8]

It is sometimes necessary to borrow when doing subtraction with mixed numbers. We always change to a common denominator before we actually borrow.

$$
\begin{array}{r}
13. \quad 4\dfrac{1}{3} = 4\dfrac{2}{6} = 3\dfrac{8}{6} \\[4pt]
-1\dfrac{5}{6} = -1\dfrac{5}{6} = -1\dfrac{5}{6} \\[2pt]
\hline
2\dfrac{3}{6} = 2\dfrac{1}{2}
\end{array}
$$

COMMON MISTAKES

1. The most common mistake when working with fractions occurs when we try to add two fractions without using a common denominator. For example,

$$\frac{2}{3} + \frac{4}{5} \neq \frac{2+4}{3+5}$$

If the two fractions we are trying to add don't have the same denominators, then we *must* rewrite each one as an equivalent fraction with a common denominator. *We never add denominators when adding fractions.*

Note We do *not* need a common denominator when multiplying fractions.

2. A common mistake made with division of fractions occurs when we multiply by the reciprocal of the first fraction instead of the reciprocal of the divisor. For example,

$$\frac{2}{3} \div \frac{5}{6} \neq \frac{3}{2} \cdot \frac{5}{6}$$

Remember, we perform division by multiplying by the reciprocal of the divisor (the fraction to the right of the division symbol).

3. If the answer to a problem turns out to be a fraction, that fraction should always be written in lowest terms. It is a mistake not to reduce to lowest terms.

4. A common mistake when working with mixed numbers is to confuse mixed-number notation for multiplication of fractions. The notation $3\frac{2}{5}$ does *not* mean 3 *times* $\frac{2}{5}$. It means 3 *plus* $\frac{2}{5}$.

5. Another mistake occurs when multiplying mixed numbers. The mistake occurs when we don't change the mixed number to an improper fraction before multiplying and instead try to multiply the whole numbers and fractions separately. Like this:

$$2\frac{1}{2} \cdot 3\frac{1}{3} = (2 \cdot 3) + \left(\frac{1}{2} \cdot \frac{1}{3}\right) \quad \textbf{Mistake}$$

$$= 6 + \frac{1}{6}$$

$$= 6\frac{1}{6}$$

Remember, the correct way to multiply mixed numbers is to first change to improper fractions and then multiply numerators and multiply denominators. This is correct:

$$2\frac{1}{2} \cdot 3\frac{1}{3} = \frac{5}{2} \cdot \frac{10}{3} = \frac{50}{6} = 8\frac{2}{6} = 8\frac{1}{3} \quad \textbf{Correct}$$

Chapter **3** Review

Reduce each of the following fractions to lowest terms. [3.2]

1. $\dfrac{6}{8}$

2. $\dfrac{12}{36}$

3. $\dfrac{110a^3}{70a}$

4. $\dfrac{45xy}{75y}$

Multiply the following fractions. (That is, find the product in each case, and reduce to lowest terms.) [3.3]

5. $\dfrac{1}{5}(5x)$

6. $\dfrac{80}{27}\left(-\dfrac{3}{20}\right)$

7. $\dfrac{96}{25} \cdot \dfrac{15}{98} \cdot \dfrac{35}{54}$

8. $\dfrac{3}{5} \cdot 75 \cdot \dfrac{2}{3}$

Find the following quotients. (That is, divide and reduce to lowest terms.) [3.4]

9. $\dfrac{8}{9} \div \dfrac{4}{3}$

10. $\dfrac{9}{10} \div (-3)$

11. $\dfrac{a^3}{b} \div \dfrac{a}{b^2}$

12. $-\dfrac{18}{49} \div \left(-\dfrac{36}{28}\right)$

Perform the indicated operations. Reduce all answers to lowest terms. [3.5]

13. $\dfrac{6}{8} - \dfrac{2}{8}$

14. $\dfrac{9}{x} + \dfrac{11}{x}$

15. $-3 - \dfrac{1}{2}$

16. $\dfrac{3}{x} - \dfrac{5}{6}$

17. $\dfrac{11}{126} - \dfrac{5}{84}$

18. $\dfrac{3}{10} + \dfrac{7}{25} + \dfrac{3}{4}$

19. Change $3\dfrac{5}{8}$ to an improper fraction. [3.6]

20. Add: $3 + \dfrac{2}{x}$ [3.6]

21. Subtract: $x - \dfrac{3}{4}$ [3.6]

22. Change $\dfrac{110}{8}$ to a mixed number. [3.6]

Perform the indicated operations. [3.7, 3.8]

23. $2 \div 3\dfrac{1}{4}$

24. $4\dfrac{7}{8} \div 2\dfrac{3}{5}$

25. $6 \cdot 2\dfrac{1}{2} \cdot \dfrac{4}{5}$

26. $3\dfrac{1}{5} + 4\dfrac{2}{5}$

27. $8\dfrac{2}{3} + 9\dfrac{1}{4}$

28. $5\dfrac{1}{3} - 2\dfrac{8}{9}$

Simplify each of the following as much as possible. [3.9]

29. $3 + 2\left(4\dfrac{1}{3}\right)$

30. $\left(2\dfrac{1}{2} + \dfrac{3}{4}\right)\left(2\dfrac{1}{2} - \dfrac{3}{4}\right)$

Simplify each complex fraction as much as possible. [3.9]

31. $\dfrac{1 + \dfrac{2}{3}}{1 - \dfrac{2}{3}}$

32. $\dfrac{3 - \dfrac{3}{4}}{3 + \dfrac{3}{4}}$

33. $\dfrac{\dfrac{7}{8} - \dfrac{1}{2}}{\dfrac{1}{4} + \dfrac{1}{2}}$

34. $\dfrac{2\dfrac{1}{8} + 3\dfrac{1}{3}}{1\dfrac{1}{6} + 5\dfrac{1}{4}}$

35. Defective Parts If $\frac{1}{10}$ of the items in a shipment of 200 items are defective, how many are defective? [3.3]

36. Number of Students If 80 students took a math test and $\frac{3}{4}$ of them passed, then how many students passed the test? [3.3]

37. Translating What is 3 times the sum of $2\frac{1}{4}$ and $\frac{3}{4}$? [3.9]

38. Translating Subtract $\frac{5}{6}$ from the product of $1\frac{1}{2}$ with $\frac{2}{3}$. [3.9]

39. Cooking If a recipe that calls for $2\frac{1}{2}$ cups of flour will make 48 cookies, how much flour is needed to make 36 cookies? [3.7]

40. Length of Wood A piece of wood $10\frac{3}{4}$ inches long is divided into 6 equal pieces. How long is each piece? [3.7]

41. Cooking A recipe that calls for $3\frac{1}{2}$ tablespoons of oil is tripled. How much oil must be used in the tripled recipe? [3.7]

42. Sheep Feed A farmer fed his sheep $10\frac{1}{2}$ pounds of feed on Monday, $9\frac{3}{4}$ pounds on Tuesday, and $12\frac{1}{4}$ pounds on Wednesday. What is the total number of pounds of feed he used on these 3 days? [3.8]

43. Find the area and the perimeter of the triangle below. [3.7, 3.8]

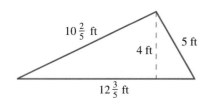

44. Comparing Area On April 3, 2000, *USA Today* changed the size of its paper. Previous to this date, each page of the paper was $13\frac{1}{2}$ inches wide and $22\frac{1}{4}$ inches long. The new paper size is $1\frac{1}{4}$ inches narrower and $\frac{1}{2}$ inch longer. [3.7, 3.8]

   a. What was the area of a page previous to April 3, 2000?

   b. What is the area of a page after April 3, 2000?

   c. What is the difference in the areas of the two page sizes?

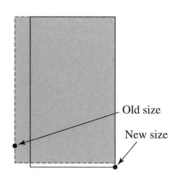

Old size

New size

Chapter **3** Test

1. Each circle below is divided into 8 equal parts. Shade each circle to represent the fraction below the circle.

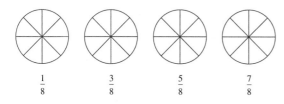

$$\frac{1}{8} \qquad \frac{3}{8} \qquad \frac{5}{8} \qquad \frac{7}{8}$$

2. Reduce each fraction to the lowest terms.

   a. $\dfrac{10}{15}$

   b. $\dfrac{130xy}{50x}$

Find each product, and reduce your answer to lowest terms.

3. $\dfrac{3}{5}(30x)$

4. $\dfrac{48}{49} \cdot \dfrac{35}{50} \cdot \dfrac{6}{18}$

Find each quotient, and reduce your answer to lowest terms.

5. $\dfrac{5}{18} \div \dfrac{15}{16}$

6. $\dfrac{4}{5} \div -8$

Perform the indicated operations. Reduce all answers to lowest terms.

7. $\dfrac{3}{10} + \dfrac{1}{10}$

8. $\dfrac{5}{x} - \dfrac{2}{x}$

9. $-4 - \dfrac{3}{5}$

10. $\dfrac{3}{x} + \dfrac{2}{5}$

11. $\dfrac{5}{6} + \dfrac{2}{9} + \dfrac{1}{4}$

12. Change $5\dfrac{2}{7}$ to an improper fraction.

13. Change $\dfrac{43}{5}$ to a mixed number.

14. Add: $5 + \dfrac{4}{x}$

Perform the indicated operations.

15. $6 \div 1\dfrac{1}{3}$

16. $7\dfrac{1}{3} + 2\dfrac{3}{8}$

17. $5\dfrac{1}{6} - 1\dfrac{1}{2}$

Simplify each of the following as much as possible.

18. $4 + 3\left(4\dfrac{1}{4}\right)$

19. $\left(2\dfrac{1}{3} + \dfrac{1}{2}\right)\left(3\dfrac{2}{3} - \dfrac{1}{6}\right)$

20. $\dfrac{\dfrac{11}{12} - \dfrac{2}{3}}{\dfrac{1}{6} + \dfrac{1}{3}}$

21. Number of Grapefruit If $\frac{1}{3}$ of a shipment of 120 grapefruit is spoiled, how many grapefruit are spoiled?

22. Sewing A dress that is $31\frac{1}{6}$ inches long is shortened by $3\frac{2}{3}$ inches. What is the length of the new dress?

23. Cooking A recipe that calls for $4\frac{2}{3}$ cups of sugar is doubled. How much sugar must be used in the doubled recipe?

24. Length of Rope A piece of rope $15\frac{2}{3}$ feet long is divided into 5 equal pieces. How long is each piece?

25. Find the area and the perimeter of the triangle below.

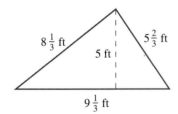

Chapters (1-3) Cumulative Review

Simplify.

1. $(6 + 2) + (3 + 6)$

2.
$$\begin{array}{r} 99 \\ 144 \\ 81 \\ +49 \\ \hline \end{array}$$

3. $1985 - 141$

4. $13 - (9 - 4)$

5. $13x - 19x$

6. $17\dfrac{13}{16} - 9\dfrac{5}{12}$

7. $(8t^3 - 4t - 6) - (5t^3 + 4t^2 - 9)$

8. $11(2n)$

9.
$$\begin{array}{r} 5280 \\ \times\ 26 \\ \hline \end{array}$$

10. $9050(373x)$

11. $2(-2)2 - 5(-3)$

12. $(13xyz)^2$

13. $(z^2)^3 \cdot (z^3)^4$

14. $(a^2b^3)^4 \cdot (a^3b^2)^5$

15. $(y + 6)(y - 6)$

16. $(x + 3)(x - 8)$

17. $(y - 3)^2$

18. $\left(\dfrac{3}{4}\right)^2 \cdot \left(-\dfrac{1}{2}\right)^3$

19. $\dfrac{11}{77}$

20. $\dfrac{104}{33}$

21. $121 \div 11 \div 11$

22. $\dfrac{4\frac{3}{8}}{5\frac{2}{3}}$

23. $\left(\dfrac{3}{5} + \dfrac{2}{25}\right) - \left(\dfrac{4}{125}\right)$

24. $6 + \dfrac{3}{11} \div \dfrac{1}{2}$

25. $2\dfrac{1}{3}\left(1\dfrac{1}{4} \div \dfrac{1}{5}\right)$

26. Round the following numbers to the nearest ten, then add.
$$\begin{array}{r} 747 \\ 116 \\ +222 \\ \hline \end{array}$$

27. Find the sum of $\dfrac{2}{3}$, $\dfrac{1}{9}$, and $\dfrac{3}{4}$.

28. Write the fraction $\frac{3}{13}$ as an equivalent fraction with a denominator of $39x$.

29. Find $\dfrac{7}{9}$ of $3\dfrac{1}{4}$.

30. Find the sum of 12 times 2 and 19 times 4.

31. Find the area of the figure:

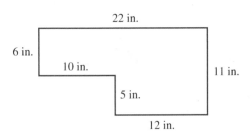

32. Place either $<$ or $>$ between the following numbers.

6    $|-7|$

33. Find the value of the polynomial below when $x = 3$.
$x^3 - 4x^2 + 4x - 9$

34. Medical Costs The table below shows the average yearly cost of visits to the doctor. Fill in the last column of the table by rounding each cost to the nearest hundred.

MEDICAL COSTS		
YEAR	AVERAGE ANNUAL COST	COST TO THE NEAREST HUNDRED
1990	$583	
1995	$739	
2000	$906	
2005	$1,172	

35. Reduce to lowest terms.
$$\frac{14x^2y}{49x^3y^4}$$

36. Find the area of the figure below:

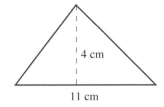

37. Add $-\dfrac{4}{5}$ to half of $\dfrac{4}{9}$.

38. Neptune's Diameter The planet Neptune has an equatorial diameter of about 30,760 miles. Write out Neptune's diameter in words and expanded form.

39. Photography A photographer buys one roll of 36-exposure 400-speed film and two rolls of 24-exposure 100-speed film. She uses $\frac{5}{6}$ of each type of film. How many pictures did she take?

40. Stopping Distances The table below shows how many feet it takes to stop a car traveling at different rates of speed, once the brakes are applied. Use the template to construct a line graph of the information in the table.

SPEED (MI/HR)	DISTANCE (FT)
0	0
20	22
30	49
40	88
50	137
60	198
70	269
80	352

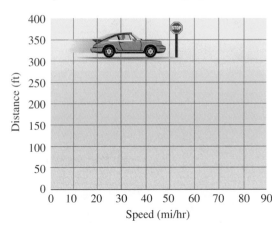

Solving Equations

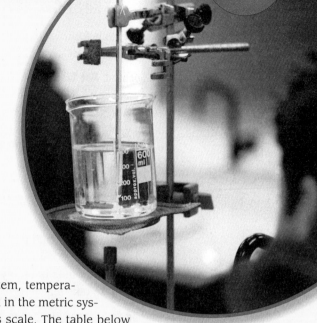

INTRODUCTION

As we mentioned previously, in the U.S. system, temperature is measured on the Fahrenheit scale, and in the metric system, temperature is measured on the Celsius scale. The table below gives some corresponding temperatures on both temperature scales. The line graph is constructed from the information in the table.

TEMPERATURE IN DEGREES CELSIUS	TEMPERATURE IN DEGREES FAHRENHEIT
0°C	32°F
25°C	77°F
50°C	122°F
75°C	167°F
100°C	212°F

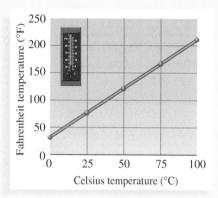

The information in the table is *numeric* in nature, whereas the information in the figure is *geometric*. In this chapter we will add a third category for presenting information and writing relationships. Information in this new category is *algebraic* in nature because it is presented using equations and formulas. The following equation is a formula that tells us how to convert from a temperature on the Celsius scale to the corresponding temperature on the Fahrenheit scale.

$$F = \frac{9}{5}C + 32$$

All three items above—the table, the line graph, and the equation—show the basic relationship between the two temperature scales. The differences among them are merely differences in form: The table is numeric, the line graph is geometric, and the equation is algebraic.

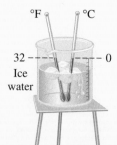

DEGREES CELSIUS	DEGREES FAHRENHEIT
0°C	32°F
25°C	77°F
50°C	122°F
75°C	167°F
100°C	212°F

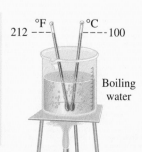

STUDY SKILLS

The study skills for this chapter are about attitude. They are points of view that point toward success.

1. Be Focused, Not Distracted. I have students who begin their assignments by asking themselves, "Why am I taking this class?" or, "When am I ever going to use this stuff?" If you are asking yourself similar questions, you may be distracting yourself from doing the things that will produce the results you want in this course. Don't dwell on questions and evaluations of the class that can be used as excuses for not doing well. If you want to succeed in this course, focus your energy and efforts toward success.

2. Be Resilient. Don't let setbacks keep you from your goals. You want to put yourself on the road to becoming someone who can succeed in this class or any class in college. Failing a test or quiz or having a difficult time on some topics is normal. No one goes through college without some setbacks. A low grade on a test or quiz is simply a signal that some reevaluation of your study habits needs to take place.

3. Intend to Succeed. I always have a few students who simply go through the motions of studying without intending to master the material. It is more important to them to look like they are studying than to actually study. You need to study with the intention of being successful in the course no matter what it takes.

4. Begin to Develop Confidence with Word Problems. The main difference between people who are good at working word problems and those who are not seems to be confidence. People with confidence know that no matter how long it takes them, they will eventually be able to solve the problem they are working on. Those without confidence begin by saying to themselves, "I'll never be able to work this problem." If you are in this second group, then instead of telling yourself that you can't do word problems, that you don't like them, or that they're not good for anything anyway, decide to do whatever it takes to master them.

4.1 The Distributive Property and Algebraic Expressions

Introduction . . .

A water bottling company charges $7 per month for their water dispenser and $2 for each gallon of water delivered. If you have g gallons of water delivered in a month, then the expression $7 + 2g$ gives the amount of your bill for that month. To find the monthly charge for using 10 gallons of water, you substitute 10 for g and simplify the result. This process is one of the topics we will study in this section.

We begin this section by extending the work we have done previously with the distributive property. Our first example is a review of one type of problem we covered in Chapter 2.

EXAMPLE 1 Apply the distributive property to the expression

$$5(x + 3)$$

SOLUTION Distributing the 5 over x and 3, we have

$$5(x + 3) = 5(x) + 5(3) \quad \textbf{Distributive property}$$
$$= 5x + 15 \quad \textbf{Multiplication}$$

Remember, $5x$ means "5 times x." ∎

The distributive property can be applied to more complicated expressions involving negative numbers.

EXAMPLE 2 Multiply: $-4(3x + 5)$

SOLUTION Multiplying both the $3x$ and the 5 by -4, we have

$$-4(3x + 5) = -4(3x) + (-4)5 \quad \textbf{Distributive property}$$
$$= -12x + (-20) \quad \textbf{Multiplication}$$
$$= -12x - 20 \quad \textbf{Definition of subtraction}$$

Notice, first of all, that when we apply the distributive property here, we multiply through by -4. It is important to include the sign with the number when we use the distributive property. Second, when we multiply -4 and $3x$, the result is $-12x$ because

$$-4(3x) = (-4 \cdot 3)x \quad \textbf{Associative property}$$
$$= -12x \quad \textbf{Multiplication}$$ ∎

Recall from Chapter 2 that we can also use the distributive property to simplify expressions like $4x + 3x$.

$$4x + 3x = (4 + 3)x \quad \textbf{Distributive property}$$
$$= 7x \quad \textbf{Addition}$$

SPECIAL INTRODUCTORY OFFER!
$7.00 per month plus $2.00 per gallon

AQUAPURE

Practice Problems

1. Apply the distributive property to the expression
   $6(x + 4)$

2. Multiply: $-3(2x + 4)$

3. Simplify: $6x - 2 + 3x + 8$

EXAMPLE 3   Simplify: $5x - 2 + 3x + 7$

SOLUTION   We begin by changing subtraction to addition of the opposite and applying the commutative property to rearrange the order of the terms. We want similar terms to be written next to each other.

$$5x - 2 + 3x + 7 = 5x + 3x + (-2) + 7 \qquad \textbf{Commutative property}$$
$$= (5 + 3)x + (-2) + 7 \qquad \textbf{Distributive property}$$
$$= 8x + 5 \qquad \textbf{Addition}$$

Notice that we take the negative sign in front of the 2 with the 2 when we rearrange terms. How do we justify doing this? ■

4. Simplify: $2(4x + 3) + 7$

EXAMPLE 4   Simplify: $3(4x + 5) + 6$

SOLUTION   We begin by distributing the 3 across the sum of $4x$ and 5. Then we combine similar terms.

$$3(4x + 5) + 6 = 12x + 15 + 6 \qquad \textbf{Distributive property}$$
$$= 12x + 21 \qquad \textbf{Add 15 and 6}$$ ■

5. Simplify: $3(2x + 1) + 5(4x - 3)$

EXAMPLE 5   Simplify: $2(3x + 1) + 4(2x - 5)$

SOLUTION   Again, we apply the distributive property first; then we combine similar terms. Here is the solution showing only the essential steps:

$$2(3x + 1) + 4(2x - 5) = 6x + 2 + 8x - 20 \qquad \textbf{Distributive property}$$
$$= 14x - 18 \qquad \textbf{Combine similar terms}$$ ■

The Value of an Algebraic Expression

An expression such as $3x + 5$ will take on different values depending on what x is. If we were to let x equal 2, the expression $3x + 5$ would become 11. On the other hand, if x is 10, the same expression has a value of 35:

	When	$x = 2$			When	$x = 10$
	the expression	$3x + 5$			the expression	$3x + 5$
	becomes	$3(2) + 5$			becomes	$3(10) + 5$
		$= 6 + 5$				$= 30 + 5$
		$= 11$				$= 35$

Table 1 lists some other algebraic expressions, along with specific values for the variables and the corresponding value of the expression after the variable has been replaced with the given number.

Table 1

ORIGINAL EXPRESSION	VALUE OF THE VARIABLE	VALUE OF THE EXPRESSION
$5x + 2$	$x = 4$	$5(4) + 2 = 20 + 2$ $= 22$
$3x - 9$	$x = 2$	$3(2) - 9 = 6 - 9$ $= -3$
$-2a + 7$	$a = 3$	$-2(3) + 7 = -6 + 7$ $= 1$
$6x + 3$	$x = -2$	$6(-2) + 3 = -12 + 3$ $= -9$
$-4y + 9$	$y = -1$	$-4(-1) + 9 = 4 + 9$ $= 13$

Answers
3. $9x + 6$ **4.** $8x + 13$
5. $26x - 12$

Facts from Geometry: Angles

An angle is formed by two rays with the same endpoint. The common endpoint is called the *vertex* of the angle, and the rays are called the *sides* of the angle.

In Figure 1, angle θ (theta) is formed by the two rays OA and OB. The vertex of θ is O. Angle θ is also denoted as angle AOB, where the letter associated with the vertex is always the middle letter in the three letters used to denote the angle.

Degree Measure The angle formed by rotating a ray through one complete revolution about its endpoint (Figure 2) has a measure of 360 degrees, which we write as 360°.

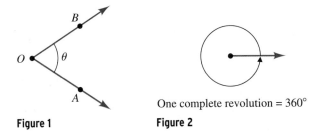

Figure 1   One complete revolution = 360°

Figure 2

One degree of angle measure, written 1°, is $\frac{1}{360}$ of a complete rotation of a ray about its endpoint; there are 360° in one full rotation. (The number 360 was decided upon by early civilizations because it was believed that Earth was at the center of the universe and the sun would rotate once around Earth every 360 days.) Similarly, 180° is half of a complete rotation, and 90° is a quarter of a full rotation. Angles that measure 90° are called *right angles*, and angles that measure 180° are called *straight angles*. If an angle measures between 0° and 90° it is called an *acute angle*, and an angle that measures between 90° and 180° is an *obtuse angle*. Figure 3 illustrates further.

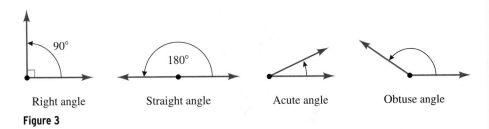

Right angle   Straight angle   Acute angle   Obtuse angle

Figure 3

Complementary Angles and Supplementary Angles If two angles add up to 90°, we call them *complementary angles*, and each is called the *complement* of the other. If two angles have a sum of 180°, we call them *supplementary angles*, and each is called the *supplement* of the other. Figure 4 illustrates the relationship between angles that are complementary and angles that are supplementary.

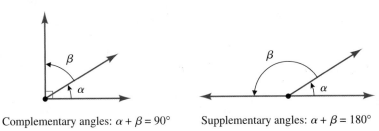

Complementary angles: $\alpha + \beta = 90°$ Supplementary angles: $\alpha + \beta = 180°$

Figure 4

6. Find *x* in each of the following diagrams.

a.

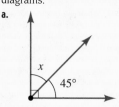

Complementary angles

b.

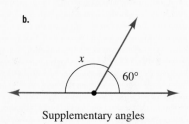

Supplementary angles

EXAMPLE 6 Find *x* in each of the following diagrams.

a.

b.

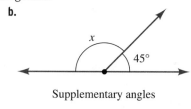

Complementary angles   Supplementary angles

SOLUTION   We use subtraction to find each angle.

a. Because the two angles are complementary, we can find *x* by subtracting 30° from 90°:

$$x = 90° − 30° = 60°$$

We say 30° and 60° are complementary angles. The complement of 30° is 60°.

b. The two angles in the diagram are supplementary. To find *x*, we subtract 45° from 180°:

$$x = 180° − 45° = 135°$$

We say 45° and 135° are supplementary angles. The supplement of 45° is 135°.

USING TECHNOLOGY

Protractors

When we think of technology, we think of computers and calculators. However, some simpler devices are also in the category of technology, because they help us do things that would be difficult to do without them. The protractor below can be used to draw and measure angles. In the diagram below, the protractor is being used to measure an angle of 120°. It can also be used to draw angles of any size.

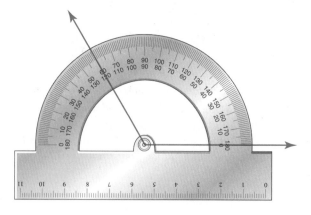

If you have a protractor, use it to draw the following angles: 30°, 45°, 60°, 120°, 135°, and 150°. Then imagine how you would draw these angles without a protractor.

Answer

6. a. 45°   **b.** 120°

Problem Set 4.1

For review, use the distributive property to combine each of the following pairs of similar terms.

1. $2x + 8x$

2. $3x + 7x$

3. $6a - 2a$

4. $9a - 3a$

5. $-4y + 5y$

6. $-3y + 10y$

7. $-6x - 2x$

8. $-9x - 4x$

9. $4a - a$

10. $9a - a$

11. $x - 6x$

12. $x - 9x$

Simplify the following expressions by combining similar terms. In some cases the order of the terms must be rearranged first by using the commutative property.

13. $4x + 2x + 3 + 8$

14. $7x + 5x + 2 + 9$

15. $7x - 5x + 6 - 4$

16. $10x - 7x + 9 - 6$

17. $-2a + a + 7 + 5$

18. $-8a + 3a + 12 + 1$

19. $6y - 2y - 5 + 1$

20. $4y - 3y - 7 + 2$

21. $4x + 2x - 8x + 4$

22. $6x + 5x - 12x + 6$

23. $9x - x - 5 - 1$

24. $2x - x - 3 - 8$

25. $2a + 4 + 3a + 5$

26. $9a + 1 + 2a + 6$

27. $3x + 2 - 4x + 1$

28. $7x + 5 - 2x + 6$

29. $12y + 3 + 5y$

30. $8y + 1 + 6y$

31. $4a - 3 - 5a + 2a$

32. $6a - 4 - 2a + 6a$

Simplify.

33. $2(3x + 4) + 8$

34. $2(5x + 1) + 10$

35. $5(2x - 3) + 4$

36. $6(4x - 2) + 7$

37. $8(2y + 4) + 3y$

38. $2(5y + 1) + 2y$

39. $6(4y - 3) + 6y$

40. $5(2y - 6) + 4y$

41. $2(x + 3) + 4(x + 2)$

42. $3(x + 1) + 2(x + 5)$

43. $5(2x - 3) + 2(x + 7)$

44. $4(3x - 1) + 5(x + 8)$

45. $3(2a + 4) + 7(3a - 1)$

46. $7(2a + 2) + 4(5a - 1)$

Find the value of each of the following expressions when $x = 5$.

47. $2x + 4$

48. $3x + 2$

49. $7x - 8$

50. $8x - 9$

51. $-4x + 1$

52. $-3x + 7$

53. $4 + 2x$

54. $9 + 3x$

55. $-8 + 3x$

56. $-7 + 2x$

Find the value of each of the following expressions when $a = -2$.

57. $2a + 5$ **58.** $3a + 4$ **59.** $a - 7$ **60.** $a - 8$

61. $-7a + 4$ **62.** $-9a + 3$ **63.** $-a + 10$ **64.** $-a + 8$

65. $-4 + 3a$ **66.** $-6 + 5a$

Find the value of each of the following expressions when $x = 3$. You may substitute 3 for x in each expression the way it is written, or you may simplify each expression first and then substitute 3 for x.

67. $3x + 5x + 4$ **68.** $6x + 8x + 7$ **69.** $9x + x + 3 + 7$ **70.** $5x + 3x + 2 + 4$

71. $4x + 3 + 2x + 5$ **72.** $7x + 6 + 2x + 9$ **73.** $3x - 8 + 2x - 3$ **74.** $7x - 2 + 4x - 1$

Applying the Concepts

75. Geometry Find the complement and supplement of 25°. Is 25° an acute angle or an obtuse angle?

76. Geometry Find the supplement of 125°. Is 125° an acute angle or an obtuse angle?

77. Temperature and Altitude On a certain day, the temperature on the ground is 72 degrees Fahrenheit, and the temperature at an altitude of A feet above the ground is found from the expression $72 - \frac{A}{300}$. Find the temperature at the following altitudes.
   a. 12,000 feet   **b.** 15,000 feet   **c.** 27,000 feet

78. Perimeter of a Rectangle As you know, the expression $2l + 2w$ gives the perimeter of a rectangle with length l and width w. The garden below has a width of $3\frac{1}{2}$ feet and a length of 8 feet. What is the length of the fence that surrounds the garden?

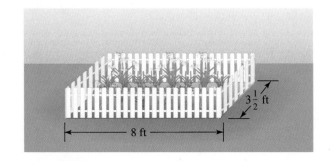

$3\frac{1}{2}$ ft

8 ft

79. Cost of Bottled Water A water bottling company charges $7 per month for their water dispenser and $2 for each gallon of water delivered. If you have g gallons of water delivered in a month, then the expression $7 + 2g$ gives the amount of your bill for that month. Find the monthly bill for each of the following deliveries.

  a. 10 gallons   **b.** 20 gallons

SPECIAL
INTRODUCTORY
OFFER!
$7.00 per month
plus
$2.00 per gallon

80. Cellular Phone Rates A cellular phone company charges $35 per month plus 25 cents for each minute, or fraction of a minute, that you use one of their cellular phones. The expression $\frac{3500 + 25t}{100}$ gives the amount of money, in dollars, you will pay for using one of their phones for t minutes a month. Find the monthly bill for using one of their phones:

  a. 20 minutes in a month   **b.** 40 minutes in a month

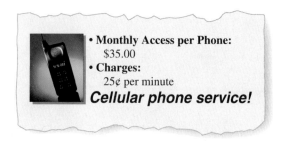

• **Monthly Access per Phone:**
  $35.00
• **Charges:**
  25¢ per minute
Cellular phone service!

Review Problems

Problems 81–84 review material we covered in Section 2.1. Give the opposite of each number.

81. 9   **82.** 12   **83.** −6   **84.** −5

Problems 85–90 review material we covered in Chapter 1. Match each statement on the left with the property that justifies it on the right.

85. $2(6 + 5) = 2(6) + 2(5)$

86. $3 + (4 + 1) = (3 + 4) + 1$

87. $x + 5 = 5 + x$

88. $(a + 3) + 2 = a + (3 + 2)$

89. $(x + 5) + 1 = 1 + (x + 5)$

90. $(a + 4) + 2 = (4 + 2) + a$

  a. Distributive property
  b. Associative property
  c. Commutative property
  d. Commutative and associative properties

4.2 The Addition Property of Equality

Introduction . . .

Previously we defined complementary angles as two angles whose sum is 90°. If A and B are complementary angles, then

$$A + B = 90°$$

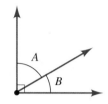

Complementary angles

If we know that $A = 30°$, then we can substitute 30° for A in the formula above to obtain the equation

$$30° + B = 90°$$

In this section we will learn how to solve equations like this one that involve addition and subtraction with the variable.

In general, solving an equation involves finding all replacements for the variable that make the equation a true statement.

> **DEFINITION**
>
> A **solution** for an equation is a number that when used in place of the variable makes the equation a true statement.

For example, the equation $x + 3 = 7$ has as its solution the number 4, because replacing x with 4 in the equation gives a true statement:

$$
\begin{aligned}
\text{When} \quad & x = 4 \\
\text{the equation} \quad & x + 3 = 7 \\
\text{becomes} \quad & 4 + 3 = 7 \\
\text{or} \quad & 7 = 7 \quad \textbf{A true statement}
\end{aligned}
$$

EXAMPLE 1 Is $x = 5$ the solution to the equation $3x + 2 = 17$?

SOLUTION To see if it is we replace x with 5 in the equation and see if the result is a true statement:

$$
\begin{aligned}
\text{When} \quad & x = 5 \\
\text{the equation} \quad & 3x + 2 = 17 \\
\text{becomes} \quad & 3(5) + 2 = 17 \\
& 15 + 2 = 17 \\
& 17 = 17 \quad \textbf{A true statement}
\end{aligned}
$$

Because the result is a true statement, we can conclude that $x = 5$ is the solution to $3x + 2 = 17$. ■

Note
Although an equation may have many solutions, the equations we work with in the first part of this chapter will always have a single solution.

Practice Problems
1. Show that $x = 3$ is a solution to the equation $5x - 4 = 11$.

Answer
1. See solutions section.

2. Is $a = -3$ the solution to the equation $6a - 3 = 2a + 4$?

EXAMPLE 2   Is $a = -2$ the solution to the equation $7a + 4 = 3a - 2$?

SOLUTION

$$
\begin{aligned}
\text{When} \qquad\qquad a &= -2 \\
\text{the equation} \qquad 7a + 4 &= 3a - 2 \\
\text{becomes} \quad 7(-2) + 4 &= 3(-2) - 2 \\
-14 + 4 &= -6 - 2 \\
-10 &= -8 \qquad \textbf{A false statement}
\end{aligned}
$$

Because the result is a false statement, we must conclude that $a = -2$ is *not* a solution to the equation $7a + 4 = 3a - 2$.   ■

We want to develop a process for solving equations with one variable. The most important property needed for solving the equations in this section is called the *addition property of equality*. The formal definition looks like this:

Addition Property of Equality

Let *A*, *B*, and *C* represent algebraic expressions.

$$
\begin{aligned}
\text{If} \qquad A &= B \\
\text{then} \quad A + C &= B + C
\end{aligned}
$$

In words: Adding the same quantity to both sides of an equation never changes the solution to the equation.

This property is extremely useful in solving equations. Our goal in solving equations is to isolate the variable on one side of the equation. We want to end up with an expression of the form

$$x = \text{A number}$$

To do so we use the addition property of equality.

3. Solve for x: $x + 5 = -2$

EXAMPLE 3   Solve for x: $x + 4 = -2$

SOLUTION   We want to isolate x on one side of the equation. If we add -4 to both sides, the left side will be $x + 4 + (-4)$, which is $x + 0$ or just x.

$$
\begin{aligned}
x + 4 &= -2 \\
x + 4 + \mathbf{(-4)} &= -2 + \mathbf{(-4)} \qquad \textbf{Add } -4 \textbf{ to both sides} \\
x + 0 &= -6 \qquad\qquad\quad \textbf{Addition} \\
x &= -6 \qquad\qquad\quad \boldsymbol{x + 0 = x}
\end{aligned}
$$

The solution is -6. We can check it if we want to by replacing x with -6 in the original equation:

$$
\begin{aligned}
\text{When} \qquad\qquad x &= -6 \\
\text{the equation} \qquad x + 4 &= -2 \\
\text{becomes} \quad -6 + 4 &= -2 \\
-2 &= -2 \qquad \textbf{A true statement}
\end{aligned}
$$
■

Note

With some of the equations in this section, you will be able to see the solution just by looking at the equation. But it is important that you show all the steps used to solve the equations anyway. The equations you come across in the future will not be as easy to solve, so you should learn the steps involved very well.

Answers

2. No   **3.** -7

EXAMPLE 4 Solve for a: $a - 3 = 5$

SOLUTION
$$a - 3 = 5$$
$$a - 3 + \mathbf{3} = 5 + \mathbf{3} \quad \text{Add 3 to both sides}$$
$$a + 0 = 8 \quad \text{Addition}$$
$$a = 8 \quad \mathbf{a + 0 = a}$$

The solution to $a - 3 = 5$ is $a = 8$. ■

4. Solve for a: $a - 2 = 7$

EXAMPLE 5 Solve for y: $y + 4 - 6 = 7 - 1$

SOLUTION Before we apply the addition property of equality, we must simplify each side of the equation as much as possible:

$$y + 4 - 6 = 7 - 1$$
$$y - 2 = 6 \quad \text{Simplify each side}$$
$$y - 2 + \mathbf{2} = 6 + \mathbf{2} \quad \text{Add 2 to both sides}$$
$$y + 0 = 8 \quad \text{Addition}$$
$$y = 8 \quad \mathbf{y + 0 = y}$$ ■

5. Solve for y: $y + 6 - 2 = 8 - 9$

EXAMPLE 6 Solve for x: $3x - 2 - 2x = 4 - 9$

SOLUTION Simplifying each side as much as possible, we have

$$3x - 2 - 2x = 4 - 9$$
$$x - 2 = -5 \quad \mathbf{3x - 2x = x}$$
$$x - 2 + \mathbf{2} = -5 + \mathbf{2} \quad \text{Add 2 to both sides}$$
$$x + 0 = -3 \quad \text{Addition}$$
$$x = -3 \quad \mathbf{x + 0 = x}$$ ■

6. Solve for x: $5x - 3 - 4x = 4 - 7$

EXAMPLE 7 Solve for x: $-3 - 6 = x + 4$

SOLUTION The variable appears on the right side of the equation in this problem. This makes no difference; we can isolate x on either side of the equation. We can leave it on the right side if we like:

$$-3 - 6 = x + 4$$
$$-9 = x + 4 \quad \text{Simplify the left side}$$
$$-9 + (\mathbf{-4}) = x + 4 + (\mathbf{-4}) \quad \text{Add } -4 \text{ to both sides}$$
$$-13 = x + 0 \quad \text{Addition}$$
$$-13 = x \quad \mathbf{x + 0 = x}$$

The statement $-13 = x$ is equivalent to the statement $x = -13$. In either case the solution to our equation is -13. ■

7. Solve for x: $-5 - 7 = x + 2$

EXAMPLE 8 Solve: $a - \dfrac{3}{4} = \dfrac{5}{8}$

SOLUTION To isolate a we add $\frac{3}{4}$ to each side:

$$a - \frac{3}{4} = \frac{5}{8}$$
$$a - \frac{3}{4} + \frac{\mathbf{3}}{\mathbf{4}} = \frac{5}{8} + \frac{\mathbf{3}}{\mathbf{4}}$$
$$a = \frac{11}{8}$$

When solving equations we will leave answers like $\frac{11}{8}$ as improper fractions, rather than change them to mixed numbers. ■

8. Solve: $a - \dfrac{2}{3} = \dfrac{5}{6}$

Answers
4. 9 **5.** −5 **6.** 0 **7.** −14
8. $\dfrac{3}{2}$

A Note on Subtraction

Although the addition property of equality is stated for addition only, we can subtract the same number from both sides of an equation as well. Because subtraction is defined as addition of the opposite, subtracting the same quantity from both sides of an equation will not change the solution. If we were to solve the equation in Example 3 using subtraction instead of addition, the steps would look like this:

$$x + 4 = -2 \qquad \text{Original equation}$$
$$x + 4 - \mathbf{4} = -2 - \mathbf{4} \qquad \text{Subtract 4 from each side}$$
$$x = -6 \qquad \text{Subtraction}$$

In my experience teaching algebra, I find that students make fewer mistakes if they think in terms of addition rather than subtraction. So, you are probably better off if you continue to use the addition property just the way we have used it in the examples in this section. But, if you are curious as to whether you can subtract the same number from both sides of an equation, the answer is yes.

Problem Set 4.2

Check to see if the number to the right of each of the following equations is the solution to the equation.

1. $2x + 1 = 5; 2$

2. $4x + 3 = 7; 1$

3. $3x + 4 = 19; 5$

4. $3x + 8 = 14; 2$

5. $2x - 4 = 2; 4$

6. $5x - 6 = 9; 3$

7. $2x + 1 = 3x + 3; -2$

8. $4x + 5 = 2x - 1; -6$

9. $x - 4 = 2x + 1; -4$

10. $x - 8 = 3x + 2; -5$

Solve each equation.

11. $x + 2 = 8$

12. $x + 3 = 5$

13. $x - 4 = 7$

14. $x - 6 = 2$

15. $a + 9 = -6$

16. $a + 3 = -1$

17. $x - 5 = -4$

18. $x - 8 = -3$

Solve each equation.

19. $y - 3 = -6$

20. $y - 5 = -1$

21. $a + \dfrac{1}{3} = -\dfrac{2}{3}$

22. $a + \dfrac{1}{4} = -\dfrac{3}{4}$

23. $x - \dfrac{3}{5} = \dfrac{4}{5}$

24. $x - \dfrac{7}{8} = \dfrac{3}{8}$

25. $a - \dfrac{5}{6} = \dfrac{1}{12}$

26. $a - \dfrac{3}{4} = \dfrac{1}{8}$

27. $\dfrac{12}{13} = a + \dfrac{1}{26}$

28. $\dfrac{14}{15} = a + \dfrac{1}{30}$

29. $x + 4 - 7 = 3 - 10$

30. $x + 6 - 2 = 5 - 12$

31. $x - 6 + 4 = -3 - 2$

32. $x - 8 + 2 = -7 - 1$

33. $3 - 5 = a - 4$

34. $2 - 6 = a - 1$

35. $3a + 7 - 2a = 1$

36. $5a + 6 - 4a = 4$

37. $6a - 2 - 5a = -9 + 1$

38. $7a - 6 - 6a = -3 + 1$

39. $8 - 5 = 3x - 2x + 4$

40. $10 - 6 = 8x - 7x + 6$

41. $4a + 1 - 3a = 3a + 2 - 3a$

42. $5a + 6 - 4a = 7a + 8 - 7a$

43. $4 - 7 - 2 = 5x - 8 - 4x$

44. $3 - 6 - 9 = 7x + 5 - 6x$

Applying the Concepts

45. Geometry Two angles are complementary angles. If one of the angles is 23°, then solving the equation $x + 23° = 90°$ will give you the other angle. Solve the equation.

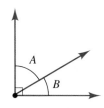

Complementary angles

46. Geometry Two angles are supplementary angles. If one of the angles is 23°, then solving the equation $x + 23° = 180°$ will give you the other angle. Solve the equation.

47. Theater Tickets The El Portal Center for the Arts in North Hollywood, California, holds a maximum of 400 people. The two balconies hold 86 and 89 people each; the rest of the seats are at the stage level. Solving the equation $x + 86 + 89 = 400$ will give you the number of seats on the stage level.

a. Solve the equation for x.

b. If tickets on the stage level are $30 each, and tickets in either balcony are $25 each, what is the maximum amount of money the theater can bring in for a show?

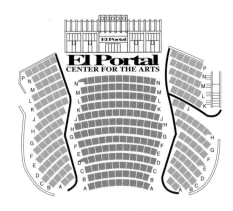

48. Geometry The sum of the angles in the triangle on the swing set is 180°. Use this fact to write an equation containing x. Then solve the equation.

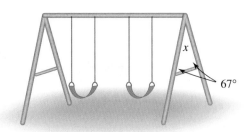

Translating Translate each of the following into an equation, and then solve the equation.

49. The sum of x and 12 is 30.

50. The difference of x and 12 is 30.

51. The difference of 8 and 5 is equal to the sum of x and 7.

52. The sum of 8 and 5 is equal to the difference of x and 7.

Review Problems

The problems below review material we covered in Sections 2.6 and 3.3. Apply the associative property and then simplify.

53. $4(5x)$

54. $3(7x)$

55. $-2(3y)$

56. $-6(8y)$

57. $4(-3a)$

58. $5(-2a)$

59. $\frac{1}{2}(2x)$

60. $\frac{1}{3}(3x)$

61. $\frac{1}{5}(5y)$

62. $\frac{1}{6}(6y)$

63. $3\left(\frac{1}{3}a\right)$

64. $4\left(\frac{1}{4}a\right)$

4.3 The Multiplication Property of Equality

In this section we will continue to solve equations in one variable. We will again use the addition property of equality, but we will also use another property—the *multiplication property of equality*—to solve the equations in this section. We state the multiplication property of equality and then see how it is used by looking at some examples.

Multiplication Property of Equality

Let A, B, and C represent algebraic expressions, with C not equal to 0.

$$\text{If} \quad A = B$$
$$\text{then} \quad AC = BC$$

In words: Multiplying both sides of an equation by the same nonzero quantity never changes the solution to the equation.

Now, because division is defined as multiplication by the reciprocal, we are also free to divide both sides of an equation by the same nonzero quantity and always be sure we have not changed the solution to the equation.

EXAMPLE 1 Solve for x: $\dfrac{1}{2}x = 3$

SOLUTION Our goal here is the same as it was in Section 4.2. We want to isolate x (that is, $1x$) on one side of the equation. We have $\frac{1}{2}x$ on the left side. If we multiply both sides by 2, we will have $1x$ on the left side. Here is how it looks:

$$\frac{1}{2}x = 3$$
$$2\left(\frac{1}{2}x\right) = 2(3) \quad \textbf{Multiply both sides by 2}$$
$$x = 6 \quad \textbf{Multiplication}$$

To see how $2(\frac{1}{2}x)$ is equivalent to x, we use the associative property:

$$2\left(\frac{1}{2}x\right) = \left(2 \cdot \frac{1}{2}\right)x \quad \textbf{Associative property}$$
$$= 1 \cdot x \quad \quad \mathbf{2 \cdot \frac{1}{2} = 1}$$
$$= x \quad \quad \mathbf{1 \cdot x = x}$$

Although we will not show this step when solving problems, it is implied. ∎

Practice Problems

1. Solve for x: $\dfrac{1}{3}x = 5$

Answer
1. 15

2. Solve for a: $\frac{1}{5}a + 3 = 7$

EXAMPLE 2   Solve for a: $\frac{1}{3}a + 2 = 7$

SOLUTION   We begin by adding -2 to both sides to get $\frac{1}{3}a$ by itself. We then multiply by 3 to solve for a.

$$\frac{1}{3}a + 2 = 7$$

$$\frac{1}{3}a + 2 + (-2) = 7 + (-2) \qquad \textbf{Add } -2 \textbf{ to both sides}$$

$$\frac{1}{3}a = 5 \qquad \textbf{Addition}$$

$$\textbf{3} \cdot \frac{1}{3}a = \textbf{3} \cdot 5 \qquad \textbf{Multiply both sides by 3}$$

$$a = 15 \qquad \textbf{Multiplication}$$

We can check our solution to see that it is correct:

When $\qquad a = 15$

the equation $\qquad \frac{1}{3}a + 2 = 7$

becomes $\quad \frac{1}{3}(15) + 2 = 7$

$$5 + 2 = 7$$

$$7 = 7 \qquad \textbf{A true statement} \quad \blacksquare$$

3. Solve: $\frac{3}{5}y = 6$

EXAMPLE 3   Solve for y: $\frac{2}{3}y = 12$

SOLUTION   In this case we multiply each side of the equation by the reciprocal of $\frac{2}{3}$, which is $\frac{3}{2}$.

$$\frac{2}{3}y = 12$$

$$\frac{\textbf{3}}{\textbf{2}}\left(\frac{2}{3}y\right) = \frac{\textbf{3}}{\textbf{2}}(12)$$

$$y = 18$$

The solution checks because $\frac{2}{3}$ of 18 is 12. $\qquad \blacksquare$

4. Solve: $-\frac{3}{4}x = \frac{6}{5}$

EXAMPLE 4   Solve: $-\frac{4}{5}x = \frac{8}{15}$

SOLUTION   The reciprocal of $-\frac{4}{5}$ is $-\frac{5}{4}$.

$$-\frac{4}{5}x = \frac{8}{15}$$

$$-\frac{\textbf{5}}{\textbf{4}}\left(-\frac{4}{5}x\right) = -\frac{\textbf{5}}{\textbf{4}}\left(\frac{8}{15}\right)$$

$$x = -\frac{2}{3}$$

$\blacksquare$

Answers

2. 20   **3.** 10   **4.** $-\frac{8}{5}$

Many times it is convenient to divide both sides by a nonzero number to solve an equation, as the next example shows.

EXAMPLE 5   Solve for x: $4x = -20$

SOLUTION   If we divide both sides by 4, the left side will be just x, which is what we want. It is okay to divide both sides by 4 because division by 4 is equivalent to multiplication by $\frac{1}{4}$, and the multiplication property of equality states that we can multiply both sides by any number so long as it isn't 0.

$$4x = -20$$
$$\frac{4x}{4} = \frac{-20}{4} \qquad \textbf{Divide both sides by 4}$$
$$x = -5 \qquad \textbf{Division}$$

Because $4x$ means "4 times x," the factors in the numerator of $\frac{4x}{4}$ are 4 and x.

Because the factor 4 is common to the numerator and the denominator, we divide it out to get just x. ∎

EXAMPLE 6   Solve for x: $-3x + 7 = -5$

SOLUTION   We begin by adding -7 to both sides to reduce the left side to just $-3x$.

$$-3x + 7 = -5$$
$$-3x + 7 + (-7) = -5 + (-7) \qquad \textbf{Add } -7 \textbf{ to both sides}$$
$$-3x = -12 \qquad \textbf{Addition}$$
$$\frac{-3x}{-3} = \frac{-12}{-3} \qquad \textbf{Divide both sides by } -3$$
$$x = 4 \qquad \textbf{Division}$$ ∎

With more complicated equations we simplify both sides first before we do any addition or multiplication on both sides. The examples below illustrate.

EXAMPLE 7   Solve for x: $5x - 8x + 3 = 4 - 10$

SOLUTION   We combine similar terms to simplify each side and then solve as usual.

$$5x - 8x + 3 = 4 - 10$$
$$-3x + 3 = -6 \qquad \textbf{Simplify each side}$$
$$-3x + 3 + (-3) = -6 + (-3) \qquad \textbf{Add } -3 \textbf{ to both sides}$$
$$-3x = -9 \qquad \textbf{Addition}$$
$$\frac{-3x}{-3} = \frac{-9}{-3} \qquad \textbf{Divide both sides by } -3$$
$$x = 3 \qquad \textbf{Division}$$ ∎

5. Solve for x: $6x = -42$

Note

If we multiply each side by $\frac{1}{4}$, the solution looks like this:

$$\frac{1}{4}(4x) = \frac{1}{4}(-20)$$
$$\left(\frac{1}{4} \cdot 4\right)x = -5$$
$$1x = -5$$
$$x = -5$$

6. Solve for x: $-5x + 6 = -14$

7. Solve for x: $3x - 7x + 5 = 3 - 18$

Answers

5. -7 **6.** 4   **7.** 5

8. Solve for x: $-5 + 4 = 2x - 11 + 3x$

EXAMPLE 8   Solve for x: $-8 + 11 = 4x - 11 + 3x$

SOLUTION   We begin by simplifying each side separately.

$$-8 + 11 = 4x - 11 + 3x$$

$$3 = 7x - 11 \qquad \textbf{Simplify both sides}$$

$$3 + \mathbf{11} = 7x - 11 + \mathbf{11} \qquad \textbf{Add 11 to both sides}$$

$$14 = 7x \qquad \textbf{Addition}$$

$$\frac{14}{7} = \frac{7x}{7} \qquad \textbf{Divide both sides by 7}$$

$$2 = x$$

$$x = 2$$

Again, it makes no difference which side of the equation x ends up on, so long as it is just one x.   ■

COMMON MISTAKE

Before we end this section we should mention a very common mistake made by students when they first begin to solve equations. It involves trying to subtract away the number in front of the variable—like this:

$$7x = 21$$

$$7x - \mathbf{7} = 21 - \mathbf{7} \qquad \textbf{Add } -7 \textbf{ to both sides}$$

$$x = 14 \longleftarrow \textbf{Mistake}$$

The mistake is not in trying to subtract 7 from both sides of the equation. The mistake occurs when we say $7x - 7 = x$. It just isn't true. We can add and subtract only similar terms. The numbers $7x$ and 7 are not similar, because one contains x and the other doesn't. The correct way to do the problem is like this:

$$7x = 21$$

$$\frac{7x}{7} = \frac{21}{7} \qquad \textbf{Divide both sides by 7}$$

$$x = 3 \qquad \textbf{Division}$$

Answer

8. 2

Problem Set 4.3

Use the multiplication property of equality to solve each of the following equations. In each case show all the steps.

1. $\frac{1}{4}x = 2$

2. $\frac{1}{3}x = 7$

3. $\frac{1}{2}x = -3$

4. $\frac{1}{5}x = -6$

5. $-\frac{1}{3}x = 2$

6. $-\frac{1}{3}x = 5$

7. $-\frac{1}{6}x = -1$

8. $-\frac{1}{2}x = -4$

9. $\frac{3}{4}y = 12$

10. $\frac{2}{3}y = 18$

11. $3a = 48$

12. $2a = 28$

13. $-\frac{3}{5}x = \frac{9}{10}$

14. $-\frac{4}{5}x = -\frac{8}{15}$

15. $5x = -35$

16. $7x = -35$

17. $-8y = 64$

18. $-9y = 27$

19. $-7x = -42$

20. $-6x = -42$

Using the addition property of equality first, solve each of the following equations.

21. $3x - 1 = 5$

22. $2x + 4 = 6$

23. $-4a + 3 = -9$

24. $-5a + 10 = 50$

25. $6x - 5 = 19$

26. $7x - 5 = 30$

27. $\frac{1}{3}a + 3 = -5$

28. $\frac{1}{2}a + 2 = -7$

29. $-\frac{1}{4}a + 5 = 2$

30. $-\frac{1}{5}a + 3 = 7$

31. $2x - 4 = -20$

32. $3x - 5 = -26$

33. $\frac{2}{3}x - 4 = 6$

34. $\frac{3}{4}x - 2 = 7$

35. $-11a + 4 = -29$

36. $-12a + 1 = -47$

37. $-3y - 2 = 1$

38. $-2y - 8 = 2$

39. $-2x - 5 = -7$

40. $-3x - 6 = -36$

Simplify each side of the following equations first, then solve.

41. $2x + 3x - 5 = 7 + 3$

42. $4x + 5x - 8 = 6 + 4$

43. $4x - 7 + 2x = 9 - 10$

44. $5x - 6 + 3x = -6 - 8$

45. $3a + 2a + a = 7 - 13$

46. $8a - 6a + a = 8 - 14$

47. $5x + 4x + 3x = 4 - 8$

48. $4x + 8x - 2x = 15 - 10$

49. $5 - 18 = 3y - 2y + 1$

50. $7 - 16 = 4y - 3y + 2$

Applying the Concepts

51. Basketball Kendra plays basketball for her high school. In one game she scored 21 points total, with a combination of free throws, field goals, and three-pointers. Each free throw is worth 1 point, each field goal is 2 points, and each three-pointer is worth 3 points. If she made 1 free throw and 4 field goals, then solving the equation

$$1 + 4(2) + 3x = 21$$

will give us the number of three-pointers she made. Solve the equation to find the number of three-point shots Kendra made.

52. Basketball Referring to Problem 51 list two other combinations of baskets Kendra could have made to give her a total of 21 points for the game. Note that this does not require any equations. There are a number of correct answers.

53. Stock Sales Suppose you purchase x shares of a stock at $25 per share, and then six months later you sell all your shares for $32 per share. Your broker charges you $18 for the transaction. If your profit is $2,082, then solving the equation

$$32x - 25x - 18 = 2,082$$

will tell you how many shares of the stock you purchased originally. Solve this equation to find the number of shares purchased originally.

54. Break-Even Point The El Portal Center for the Arts is showing a movie to raise money for a local charity. The cost to put on the event is $1,840, which includes rent on the Center and movie, insurance, and wages for the paid attendants. If tickets cost $8 each, then solving the equation $8x = 1,840$ gives the number of tickets they must sell in order to cover their costs. This number is called the break-even point. Solve the equation for x to find the break-even point.

Translations Translate each sentence below into an equation, then solve the equation.

55. The sum of $2x$ and 5 is 19.

56. The sum of 8 and $3x$ is 2.

57. The difference of $5x$ and 6 is -9.

58. The difference of 9 and $6x$ is 21.

Review Problems

The problems below review material we covered in Section 4.1.

Apply the distributive property to each of the following expressions.

59. $3(x + 4)$

60. $2(x + 6)$

61. $2(3a - 8)$

62. $4(2a - 5)$

63. $-3(5x - 1)$

64. $-2(7x - 3)$

Simplify each of the following expressions as much as possible.

65. $3(y - 5) + 6$

66. $5(y + 3) + 7$

67. $6(2x - 1) + 4x$

68. $8(3x - 2) + 4x$

4.4 Linear Equations in One Variable

Introduction . . .

The Rhind Papyrus is an ancient Egyptian document, created around 1650 BC, that contains some mathematical riddles. One problem on the Rhind Papyrus asked the reader to find a quantity such that when it is added to one-fourth of itself the sum is 15. The equation that describes this situation is

$$x + \frac{1}{4}x = 15$$

As you can see, this equation contains a fraction. One of the topics we will discuss in this section is how to solve equations that contain fractions.

In this chapter we have been solving what are called *linear equations in one variable.* They are equations that contain only one variable, and that variable is always raised to the first power and never appears in a denominator. Here are some examples of linear equations in one variable:

$$3x + 2 = 17, \qquad 7a + 4 = 3a - 2, \qquad 2(3y - 5) = 6$$

Because of the work we have done in the first three sections of this chapter, we are now able to solve any linear equation in one variable. The steps outlined below can be used as a guide to solving these equations.

Steps to Solve a Linear Equation in One Variable

Step 1 Simplify each side of the equation as much as possible. This step is done using the commutative, associative, and distributive properties.

Step 2 Use the addition property of equality to get all *variable terms* on one side of the equation and all *constant terms* on the other, then combine like terms. A *variable term* is any term that contains the variable. A *constant term* is any term that contains only a number.

Step 3 Use the multiplication property of equality to get the variable by itself on one side of the equation.

Step 4 Check your solution in the original equation if you think it is necessary.

Note
Once you have some practice at solving equations, these steps will seem almost automatic. Until that time, it is a good idea to pay close attention to these steps.

Practice Problems
1. Solve: $4(x + 3) = -8$

EXAMPLE 1   Solve: $3(x + 2) = -9$

SOLUTION   We begin by applying the distributive property to the left side:

Step 1 $\begin{cases} 3(x + 2) = -9 \\ 3x + 6 = -9 \end{cases}$ **Distributive property**

Step 2 $\begin{cases} 3x + 6 + \mathbf{(-6)} = -9 + \mathbf{(-6)} \\ 3x = -15 \end{cases}$ **Add −6 to both sides**   **Addition**

Step 3 $\begin{cases} \dfrac{3x}{3} = \dfrac{-15}{3} \\ x = -5 \end{cases}$ **Divide both sides by 3**   **Division** ■

This general method of solving linear equations involves using the two properties developed in Sections 4.2 and 4.3. We can add any number to both sides of an equation or multiply (or divide) both sides by the same nonzero number and always be sure we have not changed the solution to the equation. The equations may change in form, but the solution to the equation stays the same. Looking back to Example 1, we can see that each equation looks a little different from the preceding one. What is interesting, and useful, is that each of the equations says the same thing about x. They all say that x is −5. The last equation, of course, is the easiest to read. That is why our goal is to end up with x isolated on one side of the equation.

2. Solve $6a + 7 = 4a - 3$

EXAMPLE 2   Solve: $4a + 5 = 2a - 7$

SOLUTION   Neither side can be simplified any further. What we have to do is get the variable terms ($4a$ and $2a$) on the same side of the equation. We can eliminate the variable term from the right side by adding $-2a$ to both sides:

Step 2 $\begin{cases} 4a + 5 = 2a - 7 \\ 4a + \mathbf{(-2a)} + 5 = 2a + \mathbf{(-2a)} - 7 \\ 2a + 5 = -7 \\ \\ 2a + 5 + \mathbf{(-5)} = -7 + \mathbf{(-5)} \\ 2a = -12 \end{cases}$ **Add −2a to both sides**   **Addition**   **Add −5 to both sides**   **Addition**

Step 3 $\begin{cases} \dfrac{2a}{2} = \dfrac{-12}{2} \\ a = -6 \end{cases}$ **Divide by 2**   **Division** ■

3. Solve: $5(x - 2) + 3 = -12$

EXAMPLE 3   Solve: $2(x - 4) + 5 = -11$

SOLUTION   We begin by applying the distributive property to multiply 2 and $x - 4$:

Step 1 $\begin{cases} 2(x - 4) + 5 = -11 \\ 2x - 8 + 5 = -11 \\ 2x - 3 = -11 \end{cases}$ **Distributive property**   **Addition**

Step 2 $\begin{cases} 2x - 3 + \mathbf{3} = -11 + \mathbf{3} \\ 2x = -8 \end{cases}$ **Add 3 to both sides**   **Addition**

Step 3 $\begin{cases} \dfrac{2x}{2} = \dfrac{-8}{2} \\ x = -4 \end{cases}$ **Divide by 2**   **Division** ■

Answers
1. −5   **2.** −5   **3.** −1

EXAMPLE 4 Solve: $5(2x - 4) + 3 = 4x - 5$

4. Solve: $3(4x - 5) + 6 = 3x + 9$

SOLUTION   We apply the distributive property to multiply 5 and $2x - 4$. We then combine similar terms and solve as usual:

Step 1

$$5(2x - 4) + 3 = 4x - 5$$
$$10x - 20 + 3 = 4x - 5 \qquad \text{Distributive property}$$
$$10x - 17 = 4x - 5 \qquad \text{Simplify the left side}$$

Step 2

$$10x + (-4x) - 17 = 4x + (-4x) - 5 \qquad \text{Add } -4x \text{ to both sides}$$
$$6x - 17 = -5 \qquad \text{Addition}$$
$$6x - 17 + 17 = -5 + 17 \qquad \text{Add 17 to both sides}$$
$$6x = 12 \qquad \text{Addition}$$

Step 3

$$\frac{6x}{6} = \frac{12}{6} \qquad \text{Divide by 6}$$
$$x = 2 \qquad \text{Division} \qquad ■$$

Equations Involving Fractions

In the rest of this section we will solve some equations that involve fractions. Because integers are usually easier to work with than fractions, we will begin each problem by clearing the equation we are trying to solve of all fractions. To do this we will use the multiplication property of equality to multiply each side of the equation by the LCD for all fractions appearing in the equation. Here is an example.

EXAMPLE 5 Solve the equation $\dfrac{x}{2} + \dfrac{x}{6} = 8$.

5. Solve: $\dfrac{x}{3} + \dfrac{x}{6} = 9$

SOLUTION   The LCD for the fractions $x/2$ and $x/6$ is 6. It has the property that both 2 and 6 divide it evenly. Therefore, if we multiply both sides of the equation by 6, we will be left with an equation that does not involve fractions.

$$6\left(\frac{x}{2} + \frac{x}{6}\right) = 6(8) \qquad \text{Multiply each side by 6}$$
$$6\left(\frac{x}{2}\right) + 6\left(\frac{x}{6}\right) = 6(8) \qquad \text{Apply the distributive property}$$
$$3x + x = 48 \qquad \text{Multiplication}$$
$$4x = 48 \qquad \text{Combine similar terms}$$
$$x = 12 \qquad \text{Divide each side by 4}$$

We could check our solution by substituting 12 for x in the original equation. If we do so, the result is a true statement. The solution is 12.   ■

As you can see from Example 5, the most important step in solving an equation that involves fractions is the first step. In that first step we multiply both sides of the equation by the LCD for all the fractions in the equation. After we have done so, the equation is clear of fractions because the LCD has the property that all the denominators divide it evenly.

Answers
4. 2   **5.** 18

6. Solve: $3x + \dfrac{1}{4} = \dfrac{5}{8}$

EXAMPLE 6 Solve the equation $2x + \dfrac{1}{2} = \dfrac{3}{4}$.

SOLUTION   This time the LCD is 4. We begin by multiplying both sides of the equation by 4 to clear the equation of fractions.

$$4\left(2x + \frac{1}{2}\right) = 4\left(\frac{3}{4}\right) \qquad \textbf{Multiply each side by the LCD, 4}$$

$$4(2x) + 4\left(\frac{1}{2}\right) = 4\left(\frac{3}{4}\right) \qquad \textbf{Apply the distributive property}$$

$$8x + 2 = 3 \qquad \textbf{Multiplication}$$

$$8x = 1 \qquad \textbf{Add } -2 \textbf{ to each side}$$

$$x = \frac{1}{8} \qquad \textbf{Divide each side by 8}$$

7. Solve: $\dfrac{4}{x} + 3 = \dfrac{11}{5}$

EXAMPLE 7 Solve for x: $\dfrac{3}{x} + 2 = \dfrac{1}{2}$ (Assume x is not 0.)

SOLUTION   This time the LCD is $2x$. Following the steps we used in Examples 5 and 6, we have

$$2x\left(\frac{3}{x} + 2\right) = 2x\left(\frac{1}{2}\right) \qquad \textbf{Multiply through by the LCD, } 2x$$

$$2x\left(\frac{3}{x}\right) + 2x(2) = 2x\left(\frac{1}{2}\right) \qquad \textbf{Distributive property}$$

$$6 + 4x = x \qquad \textbf{Multiplication}$$

$$6 = -3x \qquad \textbf{Add } -4x \textbf{ to each side}$$

$$-2 = x \qquad \textbf{Divide each side by } -3$$

Answers

6. $\dfrac{1}{8}$ **7.** -5

Problem Set 4.4

Solve each equation using the methods shown in this section.

1. $5(x + 1) = 20$

2. $4(x + 2) = 24$

3. $6(x - 3) = -6$

4. $7(x - 2) = -7$

5. $2x + 4 = 3x + 7$

6. $5x + 3 = 2x + (-3)$

7. $7y - 3 = 4y - 15$

8. $3y + 5 = 9y + 8$

9. $12x + 3 = -2x + 17$

10. $15x + 1 = -4x + 20$

11. $6x - 8 = -x - 8$

12. $7x - 5 = -x - 5$

13. $7(a - 1) + 4 = 11$

14. $3(a - 2) + 1 = 4$

15. $8(x + 5) - 6 = 18$

16. $7(x + 8) - 4 = 10$

Solve each equation.

17. $2(3x - 6) + 1 = 7$

18. $5(2x - 4) + 8 = 38$

19. $10(y + 1) + 4 = 3y + 7$

20. $12(y + 2) + 5 = 2y - 1$

21. $4(x - 6) + 1 = 2x - 9$

22. $7(x - 4) + 3 = 5x - 9$

23. $2(3x + 1) = 4(x - 1)$

24. $7(x - 8) = 2(x - 13)$

25. $3a + 4 = 2(a - 5) + 15$

26. $10a + 3 = 4(a - 1) + 1$

27. $9x - 6 = -3(x + 2) - 24$

28. $8x - 10 = -4(x + 3) + 2$

29. $3x - 5 = 11 + 2(x - 6)$

30. $5x - 7 = -7 + 2(x + 3)$

Solve each equation by first finding the LCD for the fractions in the equation and then multiplying both sides of the equation by it. (Assume x is not 0 in Problems 39–46.)

31. $\dfrac{x}{3} + \dfrac{x}{6} = 5$

32. $\dfrac{x}{2} - \dfrac{x}{4} = 3$

33. $\dfrac{x}{5} - x = 4$

34. $\dfrac{x}{3} + x = 8$

35. $3x + \dfrac{1}{2} = \dfrac{1}{4}$

36. $3x - \dfrac{1}{3} = \dfrac{1}{6}$

37. $\dfrac{x}{3} + \dfrac{1}{2} = -\dfrac{1}{2}$

38. $\dfrac{x}{2} + \dfrac{4}{3} = -\dfrac{2}{3}$

39. $\dfrac{4}{x} = \dfrac{1}{5}$

40. $\dfrac{2}{3} = \dfrac{6}{x}$

41. $\dfrac{3}{x} + 1 = \dfrac{2}{x}$

42. $\dfrac{4}{x} + 3 = \dfrac{1}{x}$

43. $\dfrac{3}{x} - \dfrac{2}{x} = \dfrac{1}{5}$

44. $\dfrac{7}{x} + \dfrac{1}{x} = 2$

45. $\dfrac{1}{x} - \dfrac{1}{2} = -\dfrac{1}{4}$

46. $\dfrac{3}{x} - \dfrac{4}{5} = -\dfrac{1}{5}$

Applying the Concepts

The problems below contain a variety of the equations we have solved in this chapter.

47. Geometry The figure below shows part of a room. From a point on the floor, the angle of elevation to the top of the window is 45°, while the angle of elevation to the ceiling above the window is 58°. Solving either of the equations $58 - x = 45$ or $45 + x = 58$ will give us the number of degrees in the angle labeled $x°$. Solve both equations.

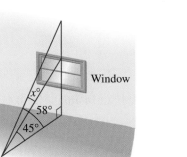

Window

48. Stock Market On the stock market, one share of Yahoo.com rises from $177\frac{1}{8}$ at the beginning of the week to a high of $183\frac{1}{4}$ at the end of the week. Solving the equation $177\frac{1}{8} + x = 183\frac{1}{4}$ for x will yield the increase in the stock price.

a. Solve the equation for x.

b. If you own 240 shares of this stock, how much would your investment have increased during the week?

49. Rhind Papyrus As we mentioned in the introduction to this section, the Rhind Papyrus was created around 1650 BC and contains the riddle "What quantity when added to one-fourth of itself becomes 15?" This riddle can be solved by finding x in the equation below. Solve this equation.

$$x + \frac{1}{4}x = 15$$

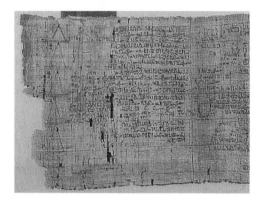

50. Test Average Justin earns scores of 75, 83, 77, and 81 on four tests in his math class. The final exam has twice the weight of a test. Justin wants to end the semester with an average of 80 for the course. Solving the equation below will give the lowest score Justin can earn on the final exam and still end the course with an 80 average. Solve the equation.

$$\frac{75 + 83 + 77 + 81 + 2x}{6} = 80$$

Review Problems

The problems below review material we covered in Chapter 1.

Write the mathematical expression that is equivalent to each of the following English phrases.

51. Twice the sum of a number and 6

52. Three times the sum of a number and 8

53. The difference of x and 4

54. The difference of 4 and x

55. The sum of twice a number and 5

56. The sum of three times a number and 4

4.5 Applications

Introduction . . .

As you begin reading through the examples in this section, you may find yourself asking why some of these problems seem so contrived. The title of the section is "Applications," but many of the problems here don't seem to have much to do with real life. You are right about that. Example 5 is what we refer to as an "age problem." Realistically, it is not the kind of problem you would expect to find if you choose a career in which you use algebra. However, solving age problems is good practice for someone with little experience with application problems, because the solution process has a form that can be applied to all similar age problems.

To begin this section we list the steps used in solving application problems. We call this strategy *Blueprint for Problem Solving*. It is an outline that will overlay the solution process we use on all application problems.

Blueprint for Problem Solving

Step 1 **Read** the problem, and then mentally **list** the items that are known and the items that are unknown.

Step 2 **Assign a variable** to one of the unknown items. (In most cases this will amount to letting x equal the item that is asked for in the problem.) Then **translate** the other **information** in the problem to expressions involving the variable.

Step 3 **Reread** the problem, and then **write an equation,** using the items and variables listed in Steps 1 and 2, that describes the situation.

Step 4 **Solve the equation** found in Step 3.

Step 5 **Write** your **answer** using a complete sentence.

Step 6 **Reread** the problem, and **check** your solution with the original words in the problem.

There are a number of substeps within each of the steps in our blueprint. For instance, with Steps 1 and 2 it is always a good idea to draw a diagram or picture if it helps you to visualize the relationship between the items in the problem.

Number Problems

EXAMPLE 1 The sum of a number and 2 is 8. Find the number.

SOLUTION Using our blueprint for problem solving as an outline, we solve the problem as follows:

Step 1 *Read* the problem, and then mentally *list* the items that are known and the items that are unknown.

> *Known items:* The numbers 2 and 8
> *Unknown item:* The number in question

Practice Problems

1. The sum of a number and 3 is 10. Find the number.

Answer

1. 7

Step 2 *Assign a variable* to one of the unknown items. Then *translate* the other *information* in the problem to expressions involving the variable.

> Let x = the number asked for in the problem.
> Then "The sum of a number and 2" translates to $x + 2$.

Step 3 *Reread* the problem, and then *write an equation,* using the items and variables listed in Steps 1 and 2, that describes the situation.
With all word problems, the word "is" translates to =.

> The sum of x and 2 is 8
> $$x + 2 = 8$$

Step 4 *Solve the equation* found in Step 3.

$$x + 2 = 8$$
$$x + 2 + (-2) = 8 + (-2) \quad \textbf{Add } -\textbf{2 to each side}$$
$$x = 6$$

Step 5 *Write* your *answer* using a complete sentence.

> The number is 6.

Step 6 *Reread* the problem, and *check* your solution with the original words in the problem.

> The sum of **6** and 2 is 8.   **A true statement**   ■

To help with other problems of the type shown in Example 1, here are some common English words and phrases and their mathematical translations.

ENGLISH	ALGEBRA
The sum of a and b	$a + b$
The difference of a and b	$a - b$
The product of a and b	$a \cdot b$
The quotient of a and b	$\dfrac{a}{b}$
Of	$\cdot$ (multiply)
Is	= (equals)
A number	x
4 more than x	$x + 4$
4 times x	$4x$
4 less than x	$x - 4$

You may find some examples and problems in this section and the problem set that follows that you can solve without using algebra or our blueprint. It is very important that you solve those problems using the methods we are showing here. The purpose behind these problems is to give you experience using the blueprint as a guide to solving problems written in words. Your answers are much less important than the work that you show in obtaining your answer.

EXAMPLE 2 If 5 is added to the sum of twice a number and three times the number, the result is 25. Find the number.

SOLUTION

Step 1 *Read and list.*

> *Known items:* The numbers 5 and 25, twice a number, and three times a number
> *Unknown item:* The number in question

2. If 4 is added to the sum of twice a number and three times the number, the result is 34. Find the number.

Answer

2. 6

Step 2 *Assign a variable and translate the information.*

Let x = the number asked for in the problem.
Then "The sum of twice a number and three times the number"
translates to $2x + 3x$.

Step 3 *Reread and write an equation.*

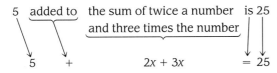

$$5 + 2x + 3x = 25$$

Step 4 *Solve the equation.*

$$5 + 2x + 3x = 25$$
$$5x + 5 = 25 \qquad \textbf{Simplify the left side}$$
$$5x + 5 + (\mathbf{-5}) = 25 + (\mathbf{-5}) \qquad \textbf{Add } \mathbf{-5} \textbf{ to both sides}$$
$$5x = 20 \qquad \textbf{Addition}$$
$$\frac{5x}{5} = \frac{20}{5} \qquad \textbf{Divide by 5}$$
$$x = 4 \qquad \textbf{Division}$$

Step 5 *Write your answer.*

The number is 4.

Step 6 *Reread and check.*

Twice **4** is 8, and three times **4** is 12. Their sum is $8 + 12 = 20$. Five added to this is 25. Therefore, 5 added to the sum of twice **4** and three times **4** is 25.   ■

Geometry Problems

EXAMPLE 3 The length of a rectangle is three times the width. The perimeter is 72 centimeters. Find the width and the length.

SOLUTION

Step 1 *Read and list.*

Known items: The length is three times the width.
                  The perimeter is 72 centimeters.
Unknown items: The length and the width

Step 2 *Assign a variable, and translate the information.* We let x = the width. Because the length is three times the width, the length must be $3x$. A picture will help.

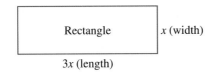

Figure 1

3. The length of a rectangle is twice the width. The perimeter is 42 centimeters. Find the length and the width.

Answer
3. Width 7 cm, length 14 cm

Step 3 *Reread and write an equation.* Because the perimeter is the sum of the sides, it must be $x + x + 3x + 3x$ (the sum of the four sides). But the perimeter is also given as 72 centimeters. Hence,

$$x + x + 3x + 3x = 72$$

Step 4 *Solve the equation.*

$$x + x + 3x + 3x = 72$$
$$8x = 72$$
$$x = 9$$

Step 5 *Write your answer.* The width, x, is 9 centimeters. The length, $3x$, must be 27 centimeters.

Step 6 *Reread and check.* From the diagram below, we see that these solutions check:

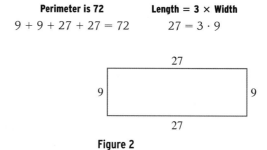

Perimeter is 72   **Length = 3 × Width**

$$9 + 9 + 27 + 27 = 72 \qquad 27 = 3 \cdot 9$$

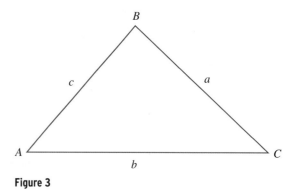

Figure 2

FACTS FROM GEOMETRY: Labeling Triangles and the Sum of the Angles in a Triangle

One way to label the important parts of a triangle is to label the vertices with capital letters and the sides with small letters, as shown in Figure 1.

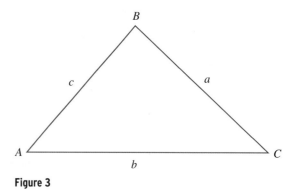

Figure 3

In Figure 3, notice that side a is opposite vertex A, side b is opposite vertex B, and side c is opposite vertex C. Also, because each vertex is the vertex of one of the angles of the triangle, we refer to the three interior angles as A, B, and C.

In any triangle, the sum of the interior angles is 180°. For the triangle shown in Figure 3, the relationship is written

$$A + B + C = 180°$$

EXAMPLE 4 The angles in a triangle are such that one angle is twice the smallest angle, while the third angle is three times as large as the smallest angle. Find the measure of all three angles.

SOLUTION

Step 1 *Read and list.*

Known items: The sum of all three angles is 180°, one angle is twice the smallest angle, and the largest angle is three times the smallest angle.

Unknown items: The measure of each angle

Step 2 *Assign a variable and translate information.* Let *x* be the smallest angle, then 2*x* will be the measure of another angle, and 3*x* will be the measure of the largest angle.

Step 3 *Reread and write an equation.* When working with geometric objects, drawing a generic diagram will sometimes help us visualize what it is that we are asked to find. In Figure 4, we draw a triangle with angles *A*, *B*, and *C*.

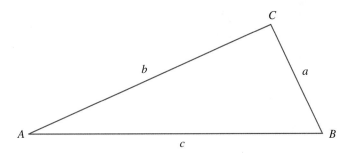

Figure 4

We can let the value of $A = x$, the value of $B = 2x$, and the value of $C = 3x$. We know that the sum of angles *A*, *B*, and *C* will be 180°, so our equation becomes

$$x + 2x + 3x = 180°$$

Step 4 *Solve the equation.*

$$x + 2x + 3x = 180°$$
$$6x = 180°$$
$$x = 30°$$

Step 5 *Write the answer.*

The smallest angle *A* measures 30°
Angle *B* measures 2*x*, or 2(30°) = 60°
Angle *C* measures 3*x*, or 3(30°) = 90°

Step 6 *Reread and check.* The angles must add to 180°:

$$A + B + C = 180°$$
$$30° + 60° + 90° = 180°$$
$$180° = 180° \quad \textbf{Our answers check}$$

4. The angles in a triangle are such that one angle is three times the smallest angle, while the largest angle is five times the smallest angle. Find the measure of all three angles.

Answer
4. 20°, 60°, and 100°

5. Joyce is 21 years older than her son Travis. In six years the sum of their ages will be 49. How old are they now?

Age Problem

EXAMPLE 5 Jo Ann is 22 years older than her daughter Stacey. In six years the sum of their ages will be 42. How old are they now?

SOLUTION

Step 1 *Read and list.*

> *Known items:* Jo Ann is 22 years older than Stacey. Six years from now their ages will add to 42.
> *Unknown items:* Their ages now

Step 2 *Assign a variable and translate the information.* Let x = Stacey's age now. Because Jo Ann is 22 years older than Stacey, her age is $x + 22$.

Step 3 *Reread and write an equation.* As an aid in writing the equation we use the following table:

	NOW	IN SIX YEARS	
STACEY	x	$x + 6$	Their ages in six years
JO ANN	$x + 22$	$x + 28$	will be their ages now plus 6.

Because the sum of their ages six years from now is 42, we write the equation as

$$(x + 6) + (x + 28) = 42$$

> Stacey's age in 6 years   Jo Ann's age in 6 years

Step 4 *Solve the equation.*

$$x + 6 + x + 28 = 42$$
$$2x + 34 = 42$$
$$2x = 8$$
$$x = 4$$

Step 5 *Write your answer.* Stacey is now 4 years old, and Jo Ann is $4 + 22 = 26$ years old.

Step 6 *Reread and check.* To check, we see that in six years, Stacey will be 10, and Jo Ann will be 32. The sum of 10 and 32 is 42, which checks. ∎

Answer
5. Travis is 8; Joyce is 29.

Problem Set 4.5

Write each of the following English phrases in symbols using the variable x.

1. The sum of x and 3

2. The difference of x and 2

3. The sum of twice x and 1

4. The sum of three times x and 4

5. Five x decreased by 6

6. Twice the sum of x and 5

7. Three times the sum of x and 1

8. Four times the sum of twice x and 1

9. Five times the sum of three x and 4

10. Three x added to the sum of twice x and 1

Applying the Concepts

Use the six steps in the "Blueprint for Problem Solving" to solve the following word problems. You may recognize the solution to some of them by just reading the problem. In all cases, be sure to assign a variable and write the equation used to describe the problem. Write your answer using a complete sentence.

Number Problems

11. The sum of a number and 3 is 5. Find the number.

12. If 2 is subtracted from a number, the result is 4. Find the number.

13. The sum of twice a number and 1 is −3. Find the number.

14. If three times a number is increased by 4, the result is −8. Find the number.

15. When 6 is subtracted from fives times a number, the result is 9. Find the number.

16. Twice the sum of a number and 5 is 4. Find the number.

17. Three times the sum of a number and 1 is 18. Find the number.

18. Four times the sum of twice a number and 6 is −8. Find the number.

19. Five times the sum of three times a number and 4 is −10. Find the number.

20. If the sum of three times a number and two times the same number is increased by 1, the result is 16. Find the number.

Geometry Problems

21. The length of a rectangle is twice its width. The perimeter is 30 meters. Find the length and the width.

22. The width of a rectangle is 3 feet less than its length. If the perimeter is 22 feet, what is the width?

23. The perimeter of a square is 32 centimeters. What is the length of one side?

24. Two sides of a triangle are equal in length, and the third side is 10 inches. If the perimeter is 26 inches, how long are the two equal sides?

25. Two angles in a triangle are equal, and their sum is equal to the third angle in the triangle. What are the measures of each of the three interior angles?

26. One angle in a triangle measures twice the smallest angle, while the largest angle is six times the smallest angle. Find the measures of all three angles.

27. The smallest angle in a triangle is $\frac{1}{3}$ as large as the largest angle. The third angle is twice the smallest angle. Find the three angles.

28. One angle in a triangle is half the largest angle, but three times the smallest. Find all three angles.

Age Problems

29. Pat is 20 years older than his son Patrick. In two years, the sum of their ages will be 90. How old are they now?

	NOW	IN 2 YEARS
PATRICK	x	
PAT		

30. Diane is 23 years older than her daughter Amy. In 5 years, the sum of their ages will be 91. How old are they now?

	NOW	IN 5 YEARS
AMY	x	
DIANE		

31. Dale is 4 years older than Sue. Five years ago the sum of their ages was 64. How old are they now?

32. Pat is 2 years younger than his wife, Wynn. Ten years ago the sum of their ages was 48. How old are they now?

Miscellaneous Problems

33. Magic Square The sum of the numbers in each row, each column, and each diagonal of the magic square below is 15. Use this fact, along with the information in the first column of the magic square, to write an equation containing the variable x, then solve the equation to find x. Next, write and solve equations that will give you y and z.

x	1	y
3	5	7
4	z	2

34. Magic Square The sum of the numbers in each row, each column, and each diagonal of the magic square below is 3. Use this fact, along with the information in the second row of the magic square, to write an equation containing the variable a, then solve the equation to find a. Next, write and solve an equation that will allow you to find the value of b. Next, write and solve equations that will give you c and d.

4	d	b
a	1	3
0	c	−2

35. Wages JoAnn works in the publicity office at the state university. She is paid $14 an hour for the first 35 hours she works each week and $21 an hour for every hour after that. If she makes $574 one week, how many hours did she work?

36. Ticket Sales Stacey is selling tickets to the school play. The tickets are $6 for adults and $4 for children. She sells twice as many adult tickets as children's tickets and brings in a total of $112. How many of each kind of ticket did she sell?

Dance Lessons Ike and Nancy Lara give western dance lessons at the Elks Lodge on Sunday nights. The lessons cost $3 for members of the lodge and $5 for nonmembers. Half of the money collected for the lessons is paid to Ike and Nancy. The Elks Lodge keeps the other half. One Sunday night Ike counts 36 people in the dance lesson. Use this information to work Problems 37 through 40.

37. What is the least amount of money Ike and Nancy will make?

38. What is the largest amount of money Ike and Nancy will make?

39. At the end of the evening, the Elks Lodge gives Ike and Nancy a check for $80 to cover half of the receipts. Can this amount be correct?

40. Besides the number of people in the dance lesson, what additional information does Ike need to know in order to always be sure he is being paid the correct amount?

Review Problems

The problems below review some of the work we did with fractions in Chapter 3.

Perform the indicated operations.

41. $\dfrac{3}{4} + \dfrac{2}{3}$

42. $\dfrac{3}{4} - \dfrac{2}{3}$

43. $\dfrac{3}{4}\left(\dfrac{2}{3}\right)$

44. $\dfrac{3}{4} \div \dfrac{2}{3}$

45. $2 + \dfrac{4}{5}$

46. $2 - \dfrac{4}{5}$

47. $2\left(\dfrac{4}{5}\right)$

48. $2 \div \dfrac{4}{5}$

49. $3\dfrac{1}{3} + 2\dfrac{1}{6}$

50. $3\dfrac{1}{3} - 2\dfrac{1}{6}$

51. $3\dfrac{1}{3} \cdot 2\dfrac{1}{6}$

52. $3\dfrac{1}{3} \div 2\dfrac{1}{6}$

4.6 Evaluating Formulas

Introduction . . .

In mathematics a formula is an equation that contains more than one variable. The equation $P = 2w + 2l$ is an example of a formula. This formula tells us the relationship between the perimeter P of a rectangle, its length l, and its width w.

There are many formulas with which you may be familiar already. Perhaps you have used the formula $d = r \cdot t$ to find out how far you would go if you traveled at 50 miles an hour for 3 hours. If you take a chemistry class while you are in college, you will certainly use the formula that gives the relationship between the two temperature scales, Fahrenheit and Celsius, that we mentioned at the beginning of this chapter.

$$F = \frac{9}{5}C + 32$$

Although there are many kinds of problems we can work using formulas, we will limit ourselves to those that require only substitutions. The examples that follow illustrate this type of problem.

EXAMPLE 1   The perimeter P of a rectangular livestock pen is 40 feet. If the width w is 6 feet, find the length.

SOLUTION   First we substitute 40 for P and 6 for w in the formula $P = 2l + 2w$. Then we solve for l:

$$\begin{array}{rll} \text{When} & P = 40 \text{ and } w = 6 & \\ \text{the formula} & P = 2l + 2w & \\ \text{becomes} & 40 = 2l + 2(6) & \\ \text{or} & 40 = 2l + 12 & \textbf{Multiply 2 and 6} \\ & 28 = 2l & \textbf{Add } -12 \textbf{ to each side} \\ & 14 = l & \textbf{Multiply each side by } \frac{1}{2} \end{array}$$

To summarize our results, if a rectangular pen has a perimeter of 40 feet and a width of 6 feet, then the length must be 14 feet.   ■

EXAMPLE 2   Use the formula $C = \frac{5}{9}(F - 32)$ to find C when F is 95 degrees.

SOLUTION   Substituting 95 for F in the formula gives us the following.

$$\begin{array}{rl} \text{When} & F = 95 \\ \text{the formula} & C = \frac{5}{9}(F - 32) \\ \text{becomes} & C = \frac{5}{9}(95 - 32) \\ & = \frac{5}{9}(63) \\ & = \frac{5}{9} \cdot \frac{63}{1} \\ & = \frac{315}{9} \\ & = 35 \end{array}$$

A temperature of 95 degrees Fahrenheit is the same as a temperature of 35 degrees Celsius.   ■

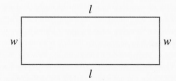

Figure 1

3. Use the formula in Example 3 to find y when x is 0.

EXAMPLE 3   Use the formula $y = 2x + 6$ to find y when x is -2.

SOLUTION   Proceeding as we have in the previous examples gives us the following.

$$\begin{aligned}
\text{When} \quad x &= -2 \\
\text{the formula} \quad y &= 2x + 6 \\
\text{becomes} \quad y &= 2(-2) + 6 \\
&= -4 + 6 \\
&= 2
\end{aligned}$$ ∎

In some cases evaluating a formula also involves solving an equation, as the next example illustrates.

4. Use the formula in Example 4 to find y when x is -3.

EXAMPLE 4   Find y when x is 3 in the formula $2x + 3y = 4$.

SOLUTION   First we substitute 3 for x; then we solve the resulting equation for y.

$$\begin{aligned}
\text{When} \quad x &= 3 \\
\text{the equation} \quad 2x + 3y &= 4 \\
\text{becomes} \quad 2(3) + 3y &= 4 \\
6 + 3y &= 4 \\
3y &= -2 \qquad \textbf{Add } -6 \textbf{ to each side} \\
y &= -\frac{2}{3} \qquad \textbf{Divide each side by 3}
\end{aligned}$$ ∎

Facts From Geometry

Earlier we defined complementary angles as angles that add to 90°. That is, if x and y are complementary angles, then

$$x + y = 90°$$

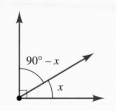

Complementary angles

If we solve this formula for y, we obtain a formula equivalent to our original formula:

$$y = 90° - x$$

Because y is the complement of x, we can generalize by saying that the complement of angle x is the angle $90° - x$. By a similar reasoning process, we can say that the supplement of angle x is the angle $180° - x$. To summarize, if x is an angle, then

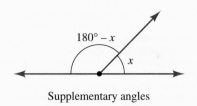

Supplementary angles

the complement of x is $90° - x$, and
the supplement of x is $180° - x$

If you go on to take a trigonometry class, you will see these formulas again.

5. Find the complement and the supplement of 35°.

EXAMPLE 5   Find the complement and the supplement of 25°.

SOLUTION   We can use the formulas above with $x = 25°$.

The complement of 25° is $90° - 25° = 65°$.
The supplement of 25° is $180° - 25° = 155°$. ∎

Answers

3. 6   **4.** $\dfrac{10}{3}$

5. Complement = 55°; supplement = 145°

Problem Set 4.6

The formula for the area A of a rectangle with length l and width w is $A = l \cdot w$. Find A if:

1. $l = 32$ feet and $w = 22$ feet

2. $l = 22$ feet and $w = 12$ feet

3. $l = \dfrac{3}{2}$ inch and $w = \dfrac{3}{4}$ inch

4. $l = \dfrac{3}{5}$ inch and $w = \dfrac{3}{10}$ inch

The formula $G = H \cdot R$ tells us how much gross pay G a person receives for working H hours at an hourly rate of pay R. In Problems 5–8, find G.

5. $H = 40$ hours and $R = \$6$

6. $H = 36$ hours and $R = \$8$

7. $H = 30$ hours and $R = \$9\dfrac{1}{2}$

8. $H = 20$ hours and $R = \$6\dfrac{3}{4}$

Because there are 3 feet in every yard, the formula $F = 3 \cdot Y$ will convert Y yards into F feet. In Problems 9–12, find F.

9. $Y = 4$ yards

10. $Y = 8$ yards

11. $Y = 2\dfrac{2}{3}$ yards

12. $Y = 6\dfrac{1}{3}$ yards

If you invest P dollars (P is for *principal*) at simple interest rate R for T years, the amount of interest you will earn is given by the formula $I = P \cdot R \cdot T$. In Problems 13 and 14, find I.

13. $P = \$1{,}000$, $R = \dfrac{7}{100}$, and $T = 2$ years

14. $P = \$2{,}000$, $R = \dfrac{6}{100}$, and $T = 2\dfrac{1}{2}$ years

In Problems 15–18, use the formula $P = 2W + 2L$ to find P.

15. $W = 10$ inches and $L = 19$ inches

16. $W = 12$ inches and $L = 22$ inches

17. $W = \dfrac{3}{4}$ foot and $L = \dfrac{7}{8}$ foot

18. $W = \dfrac{1}{2}$ foot and $L = \dfrac{3}{2}$ foot

We have mentioned the two temperature scales, Fahrenheit and Celsius, a number of times in the first four chapters. Table 1 is intended to give you a more intuitive idea of the relationship between the two temperature scales.

Table 1

COMPARING TWO TEMPERATURE SCALES

SITUATION	TEMPERATURE FAHRENHEIT	TEMPERATURE CELSIUS
Water freezes	32°F	0°C
Room temperature	68°F	20°C
Normal body temperature	$98\frac{3}{5}$°F	37°C
Water boils	212°F	100°C
Bake cookies	365°F	185°C

Table 2 gives the formulas, in both symbols and words, that are used to convert between the two scales.

Table 2

FORMULAS FOR CONVERTING BETWEEN TEMPERATURE SCALES

TO CONVERT FROM	FORMULA IN SYMBOLS	FORMULA IN WORDS
Fahrenheit to Celsius	$C = \dfrac{5}{9}(F - 32)$	Subtract 32, then multiply by $\dfrac{5}{9}$.
Celsius to Fahrenheit	$F = \dfrac{9}{5}C + 32$	Multiply by $\dfrac{9}{5}$, then add 32.

19. Let $F = 212$ in the formula $C = \frac{5}{9}(F - 32)$, and solve for C. Does the value of C agree with the information in Table 1?

20. Let $C = 100$ in the formula $F = \frac{9}{5}C + 32$, and solve for F. Does the value of F agree with the information in Table 1?

21. Let $F = 68$ in the formula $C = \frac{5}{9}(F - 32)$, and solve for C. Does the value of C agree with the information in Table 1?

22. Let $C = 37$ in the formula $F = \frac{9}{5}C + 32$, and solve for F. Does the value of F agree with the information in Table 1?

23. Find C when F is 32°.

24. Find C when F is −4°.

25. Find F when C is −15°.

26. Find F when C is 35°.

The circumference C (the distance around the outside of the circle) can be found from the formula $C = 2\pi r$. In Problems 27–30, find C if $\pi = \frac{22}{7}$ and:

27. $r = 7$ inches

28. $r = 14$ inches

29. $r = \dfrac{7}{2}$ inches

30. $r = \dfrac{7}{11}$ inch

The area A of a circle is found from the formula $A = \pi r^2$. In Problems 31–34, find A if $\pi = \frac{22}{7}$ and:

31. $r = 7$ feet

32. $r = 5$ feet

33. $r = \dfrac{1}{2}$ foot

34. $r = \dfrac{3}{4}$ foot

As you know, the volume V enclosed by a rectangular solid with length l, width w, and height h is $V = l \cdot w \cdot h$. In Problems 35–38, find V if:

35. $l = 6$ inches, $w = 12$ inches, and $h = 5$ inches

36. $l = 16$ inches, $w = 22$ inches, and $h = 15$ inches

37. $l = 6$ yards, $w = \dfrac{1}{2}$ yard, and $h = \dfrac{1}{3}$ yard

38. $l = 30$ yards, $w = \dfrac{5}{2}$ yards, and $h = \dfrac{5}{3}$ yards

Suppose $y = 3x - 2$. In Problems 39–44, find y if:

39. $x = 3$ **40.** $x = -5$ **41.** $x = -\dfrac{1}{3}$ **42.** $x = \dfrac{2}{3}$ **43.** $x = 0$ **44.** $x = 5$

Suppose $x + y = 5$. In Problems 45–50, find x if:

45. $y = 2$ **46.** $y = -2$ **47.** $y = 0$ **48.** $y = 5$ **49.** $y = -3$ **50.** $y = 3$

Suppose $x + y = 3$. In Problems 51–56, find y if:

51. $x = 2$ **52.** $x = -2$ **53.** $x = 0$ **54.** $x = 3$ **55.** $x = \dfrac{1}{2}$ **56.** $x = -\dfrac{1}{2}$

Suppose $4x + 3y = 12$. In Problems 57–62, find y if:

57. $x = 3$ **58.** $x = -5$ **59.** $x = -\dfrac{1}{4}$ **60.** $x = \dfrac{3}{2}$ **61.** $x = 0$ **62.** $x = -3$

Suppose $4x + 3y = 12$. In Problems 63–68, find x if:

63. $y = 4$ **64.** $y = -4$ **65.** $y = -\dfrac{1}{3}$ **66.** $y = \dfrac{5}{3}$ **67.** $y = 0$ **68.** $y = -3$

Find the complement and supplement of each angle.

69. $45°$ **70.** $75°$ **71.** $31°$ **72.** $59°$

Baseball In Problem Set 2.4, we worked a problem in which we found Rolaids points earned in 1998 for some relief pitchers. We are going to work the same problems again, but this time using a formula. The formula $P = 3s + 2w - 2l - 2b$, where s = saves, w = wins, l = losses, and b = blown saves, gives the Rolaids points a pitcher earns. Use this formula to complete the following tables.

73.

NATIONAL LEAGUE					
PITCHER, TEAM	W	L	S	BS	PTS
Trevor Hoffman, S.D.	3	1	39	1	
Robb Nen, S.F.	7	3	31	3	
Rod Beck, Chi.	1	2	36	6	
Jeff Shaw, L.A.	2	5	36	6	
Ugueth Urbina, Mon.	4	2	26	4	
Billy Wagner, Hou.	3	3	23	2	
Kerry Ligtenberg, Atl.	3	2	20	2	

Source: Reprinted with permission of The Associated Press.

74.

AMERICAN LEAGUE					
PITCHER, TEAM	W	L	S	BS	PTS
Tom Gordon, Bos.	6	3	34	1	
Mariano Rivera, N.Y.	2	0	32	3	
John Wetteland, Texas	3	1	32	4	
Troy Percival, Ana.	2	5	33	4	
Mike Jackson, Cle.	1	1	29	5	
Randy Myers, Tor.	3	4	28	5	
Rick Aguilera, Min.	3	7	30	8	

Source: Reprinted with permission of The Associated Press.

Estimating Vehicle Weight If you can measure the area that the tires on your car contact the ground, and you know the air pressure in the tires, then you can estimate the weight of your car with the following formula:

$$W = \frac{APN}{2,000}$$

where W is the vehicle's weight in tons, A is the average tire contact area with a hard surface in square inches, P is the air pressure in the tires in pounds per square inch (psi, or lb/in^2), and N is the number of tires.

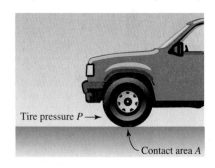

Tire pressure $P \longrightarrow$

Contact area A

75. What is the approximate weight of a car if the average tire contact area is a rectangle 6 inches by 5 inches and if the air pressure in the tires is 30 psi?

76. What is the approximate weight of a car if the average tire contact area is a rectangle 5 inches by 4 inches, and the tire pressure is 30 psi?

Review Problems

The problems that follow review material we covered in Section 3.9 on complex fractions.

Simplify.

77. $\dfrac{\dfrac{3}{5}}{\dfrac{4}{5}}$

78. $\dfrac{\dfrac{5}{7}}{\dfrac{6}{7}}$

79. $\dfrac{1 + \dfrac{1}{2}}{1 - \dfrac{1}{2}}$

80. $\dfrac{1 + \dfrac{1}{3}}{1 - \dfrac{1}{3}}$

81. $\dfrac{\dfrac{1}{2} + \dfrac{1}{4}}{\dfrac{1}{4} + \dfrac{1}{8}}$

82. $\dfrac{\dfrac{1}{2} - \dfrac{1}{4}}{\dfrac{1}{4} - \dfrac{1}{8}}$

83. $\dfrac{\dfrac{3}{5} + \dfrac{3}{7}}{\dfrac{3}{5} - \dfrac{3}{7}}$

84. $\dfrac{\dfrac{5}{7} + \dfrac{5}{8}}{\dfrac{5}{7} - \dfrac{5}{8}}$

4.7 Paired Data and Equations in Two Variables

Introduction . . .

The introduction to this chapter contained a table and a line graph that showed the relationship between the Fahrenheit and Celsius temperature scales. We repeat them below.

Table 1

COMPARING TEMPERATURES ON TWO SCALES (NUMERIC)

TEMPERATURE IN DEGREES CELSIUS	TEMPERATURE IN DEGREES FAHRENHEIT
0°C	32°F
25°C	77°F
50°C	122°F
75°C	167°F
100°C	212°F

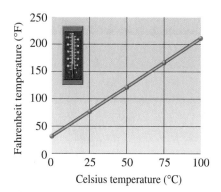

Figure 1   A line graph of the data in the table (geometric)

The data in Table 1 are called *paired data* because it is organized so that each number in the first column is paired with a specific number in the second column: A Celsius temperature of 100°C (first column) corresponds to a Fahrenheit temperature of 212°F (second column). The information in Figure 1 is also paired data because each dot on the line graph comes from one of the pairs of numbers in the table: The upper rightmost dot on the line graph corresponds to 100°C and 212°F.

We considered paired data again in Section 4.6, when we used formulas to produce pairs of numbers: When we substitute $C = 100$ into the formula below, we obtain $F = 212$, the same pair of numbers shown in Table 1 and again in the line graph in Figure 1.

$$F = \frac{9}{5}C + 32$$

The concept of paired data is very important in mathematics, so we want to continue our study of it. We start with a more detailed look at equations that contain two variables.

Recall that a solution to an equation in one variable is a single number that, when replaced for the variable in the equation, turns the equation into a true statement. For example, $x = 5$ is a solution to the equation $2x - 4 = 6$, because replacing x with 5 turns the equation into $6 = 6$, a true statement.

Next, consider the equation $x + y = 5$. It has two variables instead of one. Therefore, a solution to $x + y = 5$ will have to consist of two numbers, one for x and one for y, that together make the equation a true statement. One pair of numbers is $x = 2$ and $y = 3$, because when we substitute 2 for x and 3 for y into the equation $x + y = 5$, the result is a true statement. That is,

$$\begin{array}{rl}\text{When} & x = 2 \text{ and } y = 3 \\ \text{the equation} & x + y = 5 \\ \text{becomes} & 2 + 3 = 5 \\ & 5 = 5 \quad \textbf{A true statement}\end{array}$$

To simplify our work, we write the pair of numbers $x = 2$, $y = 3$ in the shorthand form $(2, 3)$. The expression $(2, 3)$ is called an *ordered pair* of numbers. Here is the formal definition:

> ## DEFINITION
>
> A pair of numbers enclosed in parentheses and separated by a comma, such as $(2, 3)$, is called an **ordered pair** of numbers. The first number in the pair is called the **x-coordinate** of the ordered pair, while the second number is called the **y-coordinate**. For the ordered pair $(2, 3)$, the x-coordinate is 2 and the y-coordinate is 3.

In the equation $x + y = 5$, we find that $(2, 3)$ is not the only solution. Another solution is $(0, 5)$, because when $x = 0$ and $y = 5$, then

$$0 + 5 = 5 \quad \textbf{A true statement}$$

Still another solution is the ordered pair $(-7, 12)$, because

$$\begin{array}{rl}\text{When} & x = -7 \text{ and } y = 12 \\ \text{the equation} & x + y = 5 \\ \text{becomes} & -7 + 12 = 5 \\ & 5 = 5 \quad \textbf{A true statement}\end{array}$$

As you can imagine, there are many more ordered pairs that are solutions to the equation $x + y = 5$. As a matter of fact, for any number we choose for x, there is another number we can use for y that will make the equation a true statement. There is an infinite number of ordered pairs that are solutions to the equation $x + y = 5$.

Practice Problems

1. Fill in the ordered pairs $(0, \quad)$, $(\quad, 0)$, and $(-5, \quad)$ so they are solutions to the equation $3x + 5y = 15$.

EXAMPLE 1   Fill in the ordered pairs $(0, \quad)$, $(\quad, -2)$, and $(3, \quad)$ so they are solutions to the equation $2x + 3y = 6$.

SOLUTION   To complete the ordered pair $(0, \quad)$, we substitute 0 for x in the equation and then solve for y:

$$\begin{array}{rl}\text{When} & x = 0 \\ \text{the equation} & 2x + 3y = 6 \\ \text{becomes} & 2 \cdot 0 + 3y = 6 \\ & 3y = 6 \\ & y = 2\end{array}$$

Therefore, the ordered pair $(0, 2)$ is a solution to $2x + 3y = 6$.

To complete the ordered pair $(\quad, -2)$, we substitute -2 for y and then solve for x.

$$\begin{array}{rl}\text{When} & y = -2 \\ \text{the equation} & 2x + 3y = 6 \\ \text{becomes} & 2x + 3(-2) = 6 \\ & 2x - 6 = 6 \\ & 2x = 12 \\ & x = 6\end{array}$$

Answer

1. $(0, 3)$, $(5, 0)$, and $(-5, 6)$

Therefore, the ordered pair $(6, -2)$ is another solution to our equation.

Finally, to complete the ordered pair (3,   ), we substitute 3 for x and then solve for y. The result is $y = 0$. The ordered pair $(3, 0)$ is a third solution to our equation.

∎

EXAMPLE 2   Use the equation $5x - 2y = 20$ to complete the table below.

x	y
2	
0	
	5
	0

2. Use the equation $5x + 2y = 20$ to complete the table below.

x	y
2	
0	
	5
	0

SOLUTION   Filling in the table is equivalent to completing the following ordered pairs: (2,), (0,), (, 5), and   . We proceed as in Example 1.

When $x = 2$, we have
$$5 \cdot 2 - 2y = 20$$
$$10 - 2y = 20$$
$$-2y = 10$$
$$y = -5$$

When $x = 0$, we have
$$5 \cdot 0 - 2y = 20$$
$$0 - 2y = 20$$
$$-2y = 20$$
$$y = -10$$

When $y = 5$, we have
$$5x - 2 \cdot 5 = 20$$
$$5x - 10 = 20$$
$$5x = 30$$
$$x = 6$$

When $y = 0$, we have
$$5x - 2 \cdot 0 = 20$$
$$5x - 0 = 20$$
$$5x = 20$$
$$x = 4$$

Using these results, we complete our table.

x	y
2	−5
0	−10
6	5
4	0

∎

EXAMPLE 3   Complete the table below for the equation $y = 2x + 1$.

x	y
0	
5	
	7
	3

3. Complete the table below for the equation $y = \frac{1}{2}x + 1$.

x	y
0	
4	
	7
	−3

SOLUTION   When $x = 0$, we have
$$y = 2 \cdot 0 + 1$$
$$y = 0 + 1$$
$$y = 1$$

When $x = 5$, we have
$$y = 2 \cdot 5 + 1$$
$$y = 10 + 1$$
$$y = 11$$

When $y = 7$, we have
$$7 = 2x + 1$$
$$6 = 2x$$
$$3 = x$$

When $y = 3$, we have
$$3 = 2x + 1$$
$$2 = 2x$$
$$1 = x$$

Answers

2.

x	y
2	5
0	10
2	5
4	0

3.

x	y
0	1
4	3
12	7
−8	−3

The completed table looks like this:

x	y
0	1
5	11
3	7
1	3

■

4. Which of the ordered pairs $(1, 5)$ and $(2, 4)$ are solutions to the equation $y = 5x - 6$?

EXAMPLE 4 Which of the ordered pairs $(1, 5)$ and $(2, 4)$ are solutions to the equation $y = 3x + 2$?

SOLUTION   If an ordered pair is a solution to an equation, then it must yield a true statement when the coordinates of the ordered pair are substituted for x and y in the equation.

First, we try $(1, 5)$ in the equation $y = 3x + 2$:

$$5 = 3 \cdot 1 + 2$$
$$5 = 3 + 2$$
$$5 = 5 \qquad \textbf{A true statement}$$

Next, we try $(2, 4)$ in the equation:

$$4 = 3 \cdot 2 + 2$$
$$4 = 6 + 2$$
$$4 = 8 \qquad \textbf{A false statement}$$

The ordered pair $(1, 5)$ is a solution to the equation $y = 3x + 2$, but $(2, 4)$ is not a solution to the equation.   ■

Answer

4. $(2, 4)$

Problem Set 4.7

For each equation, complete the given ordered pairs.

1. $x + y = 4$ (0,), (3,), (−2,)

2. $x + y = 6$ (0,), (2,), (,2)

3. $x + 2y = 6$ (0,), (2,), (,−6)

4. $3x + y = 5$ (0,), (1,), (,5)

5. $4x + 3y = 12$ (0,), (,0), (−3,)

6. $5x + 5y = 20$ (0,), (,−2), (1,)

7. $y = 4x − 3$ (1,), (,−3), (5,)

8. $y = 3x − 5$ (,13), (0,), (−2,)

9. $y = 2x + 3$ (0,), (2,), (−2,)

10. $y = 3x + 2$ (0,), (2,), (−2,)

11. $y = 7x$ (2,), (,6), (0,)

12. $y = 8x$ (3,), (,0), (,−6)

13. $y = -2x$ (0,),(-2,),(2,)

14. $y = -5x$ (0,),(-1,),(1,)

15. $y = \dfrac{1}{2}x$ (0,),(2,),(4,)

16. $y = \dfrac{1}{3}x$ (-3,),(0,),(3,)

17. $y = -\dfrac{1}{2}x + 2$ (-2,),(0,),(2,)

18. $y = -\dfrac{1}{3}x + 1$ (-3,),(0,),(3,)

For each of the following equations, complete the given table.

19. $x + y = 5$

x	y
2	
3	
	4
	5

20. $x - y = 2$

x	y
3	
-2	
	2
	6

21. $y = -4x$

x	y
0	
	4
	8
1	

22. $y = 5x$

x	y
	5
	0
-1	
2	

23. $3x - 2y = 6$

x	y
-2	
	0
4	
	-3

24. $2x - y = 6$

x	y
1	
	6
-6	
	-6

25. $y = 6x - 1$

x	y
-1	
	5
	-13
0	

26. $y = 3x + 1$

x	y
-2	
	-2
	4
0	

In Problems 27–34, indicate which of the given ordered pairs are solutions for each equation.

27. $2x - 5y = 10$ $(2, 3), (0, -2), \left(\dfrac{5}{2}, 1\right)$

28. $3x + 7y = 21$ $(0, 3), (7, 0), (1, 2)$

29. $y = 2x + 3$ $(0, 3), (5, 4), (2, 0)$

30. $y = -3x + 2$ $(0, -3), (0, 2), (-3, 0)$

31. $x + y = 0$ $(0, 0), (5, -5), (-3, 3)$

32. $x - y = 0$ $(0, 0), (5, -5), (3, 3)$

33. $y = 5x$ $(0, 5), (1, 5), \left(2, \dfrac{5}{2}\right)$

34. $y = -3x$ $(0, 0), (-3, 0), (-1, 3)$

Applying the Concepts

35. Complementary Angles The diagram below shows sunlight hitting the ground. Angle α (alpha) is called the angle of inclination and angle θ (theta) is called the angle of incidence. As the sun moves across the sky, the values of these angles change, but the two angles are always complementary, meaning that $\alpha + \theta = 90°$. Use this information to fill in the table below.

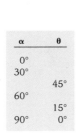

α	θ
0°	
30°	
	45°
60°	
	15°
90°	0°

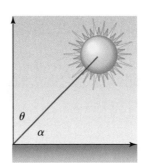

36. Temperature and Altitude On a certain day, the temperature on the ground is 72° Fahrenheit, and the temperature T at an altitude of A feet above the ground is given by the equation $T = 72 - \frac{1}{300}A$. Use this equation to fill in the following table.

ALTITUDE (FEET) A	TEMPERATURE (°F) T
0	
300	
1200	
3000	
	52
	-28

37. Comparing Tables Which of the tables below could be produced from the equation $y = 2x - 6$?

(a)

x	y
0	6
1	4
2	2
3	0

(b)

x	y
0	-6
1	-4
2	-2
3	0

(c)

x	y
0	-6
1	-5
2	-4
3	-3

38. Cost of Bottled Water A water bottling company charges $7 per month for their water dispenser and $2 for each gallon of water delivered. If you have g gallons of water delivered in a month, then the equation $y = 7 + 2g$ gives the amount of your bill for that month. Use this equation to fill in the following table.

SPECIAL
INTRODUCTORY
OFFER!
$7.00 per month
plus
$2.00 per gallon

WATER (GALLONS)	MONTHLY BILL (DOLLARS)
g	y
0	
1	
2	
	13
	15
5	

Review Problems

The problems that follow review material we covered in Sections 1.3 and 3.3.

Find the area and the perimeter of each triangle.

39.

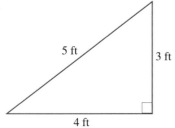

5 ft   3 ft

4 ft

40.

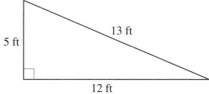

5 ft   13 ft

12 ft

41.

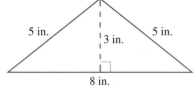

5 in.   5 in.

3 in.

8 in.

42.

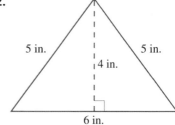

5 in.   5 in.

4 in.

6 in.

4.8   The Rectangular Coordinate System

Introduction . . .

The table and line graph below are similar to those that we have used previously in this chapter to discuss temperature. Note that we have extended both the table and the line graph to show some temperatures below zero on both scales.

Table 1

**COMPARING TEMPERATURES
ON TWO SCALES**

TEMPERATURE IN DEGREES CELSIUS	TEMPERATURE IN DEGREES FAHRENHEIT
−100°C	−148°F
−75°C	−103°F
−50°C	−58°F
−25°C	−13°F
0°C	32°F
25°C	77°F
50°C	122°F
75°C	167°F
100°C	212°F

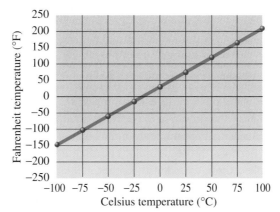

Figure 1   A line graph of the data in Table 1

We know from the previous section that the information in both the table and the line graph is paired data. We also know that solutions to equations in two variables consist of pairs of numbers that together satisfy the equations. What we want to do next is look at solutions to equations in two variables from a visual perspective. In order to do so, we need to standardize the way in which we present paired data visually. This is accomplished with the *rectangular coordinate system.*

The Rectangular Coordinate System

The rectangular coordinate system can be used to plot (or graph) pairs of numbers (see Figure 2 on the following page). It consists of two number lines, called *axes,* which intersect at right angles. (A right angle is a 90° angle.) The point at which the axes intersect is called the *origin.*

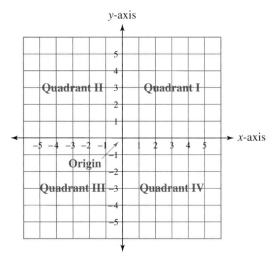

Figure 2

The horizontal number line is exactly the same as the real number line and is called the *x-axis*. The vertical number line is also the same as the real number line with the positive direction up and the negative direction down. It is called the *y-axis*. As you can see, the axes divide the plane into four regions, called *quadrants*, which are numbered I through IV in a counterclockwise direction.

Because the rectangular coordinate system consists of two number lines, one called the *x*-axis and the other called the *y*-axis, we can plot pairs of numbers such as $x = 2$ and $y = 3$. As a matter of fact, each point in the rectangular coordinate system is named by exactly one pair of numbers. We call the pair of numbers that name a point the *coordinates* of that point. To find the point that is associated with the pair of numbers $x = 2$ and $y = 3$, we start at the origin and move 2 units horizontally to the right and then 3 units vertically up (see Figure 3). The place where we end up is the point named by the pair of numbers $x = 2$, $y = 3$, which we write in shorthand form as the ordered pair $(2, 3)$.

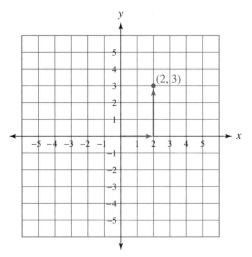

Figure 3

In general, to graph an ordered pair (a, b) on the rectangular coordinate system, we start at the origin and move a units right or left (right if a is positive, left if a is negative). From there we move b units up or down (up if b is positive, down if b is negative). The point were we end up is the graph of the ordered pair (a, b).

EXAMPLE 1 Plot (graph) the ordered pairs $(2, 3), (-2, 3), (-2, -3)$, and $(2, -3)$.

SOLUTION   To graph the ordered pair $(2, 3)$, we start at the origin and move 2 units to the right, then 3 units up. We are now at the point whose coordinates are $(2, 3)$. We plot the other three ordered pairs in the same manner (Figure 4).

Practice Problems
1. Plot the ordered pairs $(3, 2)$, $(3, -2), (-3, -2)$, and $(-3, 2)$ on the coordinate system used in Example 1.

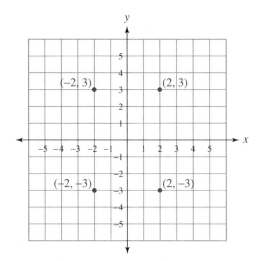

Figure 4

Note Looking at Example 1, we see that any point in quadrant I must have positive x- and y-coordinates $(+, +)$. In quadrant II, x-coordinates are negative and y-coordinates are positive, $(-, +)$. In quadrant III, both coordinates are negative $(-, -)$. Finally, in quadrant IV, all ordered pairs must have the form $(+, -)$.

EXAMPLE 2 Plot the ordered pairs $(1, -4), \left(\frac{1}{2}, 3\right), (2, 0), (0, -2)$, and $(-3, 0)$.

SOLUTION

2. Plot the ordered pairs $(-1, -4)$, $\left(-\frac{1}{2}, 3\right), (0, 2), (5, 0), (0, -5)$, and $(-1, 0)$ on the coordinate system used in Example 2.

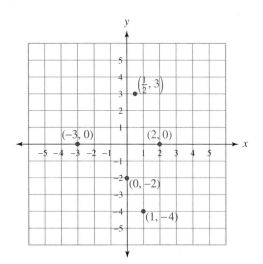

Answers
1. See solutions section.
2. See solutions section.

3. Give the coordinates of each point in the figure below.

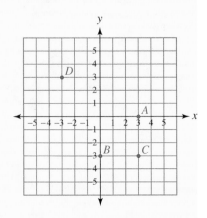

EXAMPLE 3 Give the coordinates of each point in Figure 5.

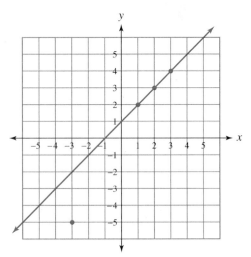

Figure 5

SOLUTION   *A* is named by the ordered pair $(2, 3)$. *B* is named by the ordered pair $(-3, -2)$. *C* is named by the ordered pair $(4, -3)$. *D* is named by the ordered pair $(-2, 5)$.   ■

4. The points $(2, 0)$, $(-3, 0)$, $(\frac{5}{2}, 0)$, and $(-8, 0)$ all lie on which axis?

EXAMPLE 4 Where are all the points that have coordinates of the form $(x, 0)$?

SOLUTION   Because the *y*-coordinate is 0, these points must lie on the *x*-axis. Remember, the *y*-coordinate tells us how far up or down we move to find the point in question. If the *y*-coordinate is 0, then we don't move up or down at all. Therefore, we must stay on the *x*-axis.   ■

5. a. Does the point $(-3, -2)$ lie on the line shown in Example 5?
   b. Does the point $(3, 2)$ lie on the line shown in Example 5?

EXAMPLE 5 Graph the points $(1, 2)$ and $(3, 4)$, and draw a line through them. Then use your result to answer these questions.
   a. Does the graph of $(2, 3)$ lie on this line?
   b. Does the graph of $(-3, -5)$ lie on this line?

SOLUTION   Figure 6 shows the graphs of $(1, 2)$ and $(3, 4)$ and the line that connects them. The line does not pass through the point $(-3, -5)$ but does pass through $(2, 3)$

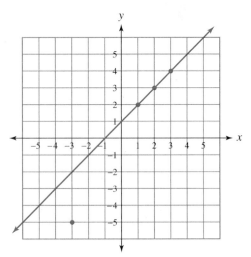

Answers
3. *A* is $(3, 0)$; *B* is $(0, -3)$; *C* is $(3, -3)$; *D* is $(-3, 3)$.
4. The *x*-axis
5. a. Yes   **b.** No

Figure 6   ■

Problem Set 4.8

Graph each of the following ordered pairs.

1. $(4, 2)$　　**2.** $(4, -2)$　　**3.** $(-4, 2)$　　**4.** $(-4, -2)$　　**5.** $(3, 4)$　　**6.** $(-3, 4)$

7. $(-3, -4)$　　**8.** $(3, -4)$　　**9.** $(4, 3)$　　**10.** $(-4, 3)$　　**11.** $\left(5, \dfrac{1}{2}\right)$　　**12.** $\left(-5, -\dfrac{1}{2}\right)$

13. $\left(1, -\dfrac{3}{2}\right)$　　**14.** $\left(-\dfrac{3}{2}, 1\right)$　　**15.** $(2, 0)$　　**16.** $(0, -5)$　　**17.** $(-2, 0)$　　**18.** $(0, 5)$

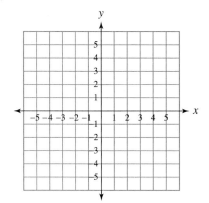

USE FOR ODD-NUMBERED PROBLEMS

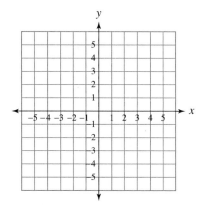

USE FOR EVEN-NUMBERED PROBLEMS

19.-25. Give the coordinates of each point in the figure below.

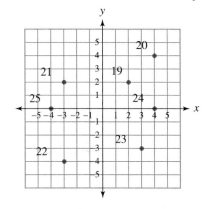

26. Where will you find all the ordered pairs of the form $(0, y)$?

Graph the points $(1, 3)$ and $(2, 4)$, and draw a line through them. Use that graph to answer Problems 27–30.

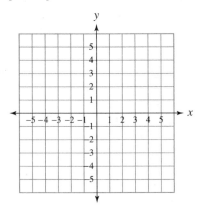

27. Does the graph of $(-1, 1)$ lie on this line?

28. Does the graph of $(3, 4)$ lie on this line?

29. Does the graph of $(-3, -5)$ lie on this line?

30. Does the graph of $(3, 5)$ lie on this line?

Graph the points $(-3, 2)$ and $(1, 3)$, and draw a line through them. Use that graph to answer Problems 31–34.

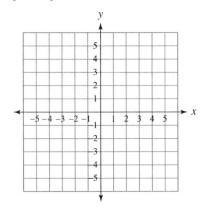

31. Does the graph of $(5, 4)$ lie on this line?

32. Does the graph of $(3, -1)$ lie on this line?

33. Does the graph of $(4, 5)$ lie on this line?

34. Does the graph of $(-7, 1)$ lie on this line?

Applying the Concepts

35. Hourly Wages Jane is considering a job at Gigi's Boutique. This job pays $6 per hour plus $50 per week in commissions. The following line graph shows how much Jane will make for working from 0 to 40 hours in a week at Gigi's. List three ordered pairs that lie on the line graph.

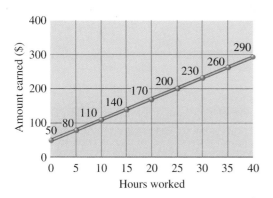

36. Hourly Wage Jane is also considering a job at Marcy's department store. The job pays $8 per hour. The following line graph shows how much Jane will make for working from 0 to 40 hours in a week. List three ordered pairs that lie on the line graph.

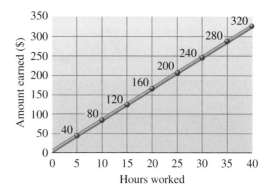

37. Digital Camera Sales The table and bar chart from Problem 55 in Problem Set 1.5 are shown here. Each shows the sales of digital cameras from 1996 to 1999. Use the information from the table and chart to construct a line graph on the template below.

YEAR	SALES (IN MILLIONS)
1996	$386
1997	$518
1998	$573
1999	$819

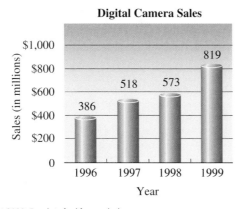

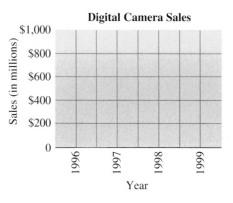

Source: Data from *USA TODAY*, copyright 2000. Reprinted with permission.

38. Wireless Phone Costs The table and bar chart from Problem 56 in Problem Set 1.5 are shown here. Each shows the projected cost of wireless phone use. Use the information from the table and chart to construct a line graph on the template below.

YEAR	CENTS PER MINUTE
1998	33
1999	28
2000	25
2001	23
2002	22
2003	20

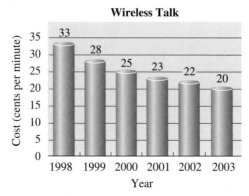

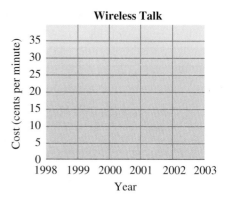

Source: Data from *USA TODAY*, copyright 2000. Reprinted with permission.

Review Problems

The problems below review material we covered in Chapter 3.

Multiply.

39. $\dfrac{x^2}{y} \cdot \dfrac{y^3}{x}$

40. $\dfrac{a^3}{15} \cdot \dfrac{12}{a^2}$

Divide.

41. $\dfrac{x^2}{y^3} \div \dfrac{x^3}{y^2}$

42. $\dfrac{a^2b}{c^3} \div \dfrac{ab^2}{c^3}$

Add.

43. $\dfrac{x}{5} + \dfrac{3}{4}$

44. $5 + \dfrac{2}{x}$

Subtract.

45. $3 - \dfrac{1}{x}$

46. $\dfrac{x}{8} - \dfrac{5}{6}$

4.9 Graphing Straight Lines

In this section we use what we have learned in the previous two sections to graph straight lines.

EXAMPLE 1   Graph the solution set for $x + y = 4$.

SOLUTION   We know from our work in the previous section that there are an infinite number of solutions to this equation, and that each solution is an ordered pair of numbers. Some of the ordered pairs that satisfy the equation $x + y = 4$ are $(0, 4), (2, 2), (4, 0)$, and $(5, -1)$. If we plot these ordered pairs, we have the points shown in Figure 1.

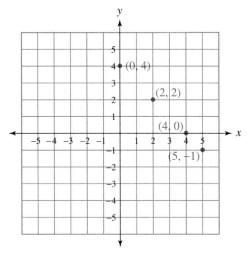

Figure 1   The graph of four solutions to $x + y = 4$

Notice that all four points lie in a straight line. If we were to find other solutions to the equation $x + y = 4$, we would find that they too would line up with the points shown in Figure 1. In fact, every solution to $x + y = 4$ is a point that lies in line with the points shown in Figure 1. Therefore, to graph the solution set to $x + y = 4$, we simply draw a line through the points in Figure 1, to obtain the graph shown in Figure 2.

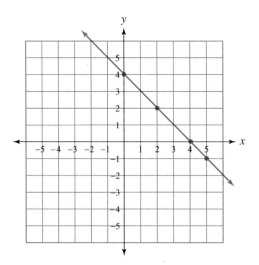

Figure 2   The graph of the line $x + y = 4$

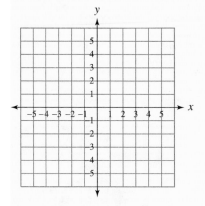

> **To Graph a Straight Line**
>
> **Step 1**   Find any three ordered pairs that satisfy the equation. This is usually accomplished by substituting a number for one of the variables in the equation, and then solving for the other variable.
>
> **Step 2**   Plot the three ordered pairs found in Step 1. (Actually, we need only two points to graph a straight line. The third point serves as a check. If all three points don't line up, then we have made a mistake in our work.)
>
> **Step 3**   Draw a line through the three points you plotted in Step 2.

2. Graph the equation $y = -2x + 1$ by completing the ordered pairs $(1, \ )$, $(0, \ )$, and $(-1, \ )$.

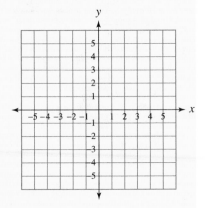

EXAMPLE 2   Graph the equation $y = 2x + 1$ by completing the ordered pairs $(1, \ )$, $(0, \ )$, and $(-1, \ )$.

SOLUTION   To complete the ordered pair $(1, \ )$, we let $x = 1$ in the equation $y = 2x + 1$:

$$y = 2 \cdot 1 + 1$$
$$y = 2 + 1$$
$$y = 3$$

To complete the ordered pair $(0, \ )$, we let $x = 0$ in the equation:

$$y = 2 \cdot 0 + 1$$
$$y = 0 + 1$$
$$y = 1$$

To complete the ordered pair $(-1, \ )$, we let $x = -1$ in the equation:

$$y = 2(-1) + 1$$
$$y = -2 + 1$$
$$y = -1$$

The ordered pairs $(1, 3)$, $(0, 1)$, and $(-1, -1)$ each satisfy the equation $y = 2x + 1$. Graphing each ordered pair and then drawing a line through them, we have the graph shown in Figure 3.

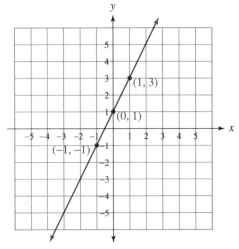

Figure 3   **The graph of $y = 2x + 1$**

EXAMPLE 3 Graph the equation $y = 3x - 2$.

SOLUTION   We need to find three solutions to the equation. We can do so by choosing some numbers to use for x and then seeing what values of y they yield. Since the choice of the numbers we will use for x is up to us, let's make it easy on ourselves and use numbers like $x = -1, 0$, and 2, that are easy to work with.

$$\text{Let } x = -1: \quad y = 3(-1) - 2$$
$$y = -3 - 2$$
$$y = -5 \quad (-1, -5) \text{ is one solution.}$$
$$\text{Let } x = 0: \quad y = 3 \cdot 0 - 2$$
$$y = 0 - 2$$
$$y = -2 \quad (0, -2) \text{ is another solution.}$$
$$\text{Let } x = 2: \quad y = 3 \cdot 2 - 2$$
$$y = 6 - 2$$
$$y = 4 \quad (2, 4) \text{ is a third solution.}$$

Now we plot the solutions we found above and then draw a line through them.

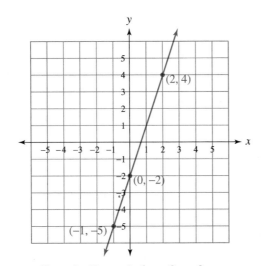

Figure 4   The graph of $y = 3x - 2$

EXAMPLE 4 Graph the line $3x + 2y = 6$.

SOLUTION   This time, let's substitute values for both x and y to find three solutions we need to draw the graph.

$$\text{Let } x = 0: \quad 3 \cdot 0 + 2y = 6$$
$$0 + 2y = 6$$
$$2y = 6$$
$$y = 3 \quad (0, 3) \text{ is one solution.}$$
$$\text{Let } y = 0: \quad 3x + 2 \cdot 0 = 6$$
$$3x + 0 = 6$$
$$3x = 6$$
$$x = 2 \quad (2, 0) \text{ is a second solution.}$$
$$\text{Let } y = -3: \quad 3x + 2(-3) = 6$$
$$3x - 6 = 6$$
$$3x = 12$$
$$x = 4 \quad (4, -3) \text{ is a third solution.}$$

Plotting these three solutions and drawing a line through them, we have the graph shown in Figure 5.

3. Graph the equation $y = 2x - 3$.

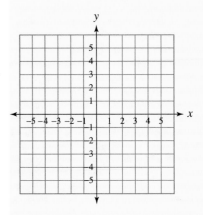

Note
Every point on the line shown in Figure 4 has coordinates that satisfy the equation $y = 3x - 2$. Likewise, every ordered pair that satisfies the equation $y = 3x - 2$ has its graph on the line shown in Figure 4. We say that there is a one-to-one correspondence between points on the graph of $y = 3x - 2$ and ordered pairs that satisfy the equation $y = 3x - 2$.

4. Graph the line $3x - 2y = 6$.

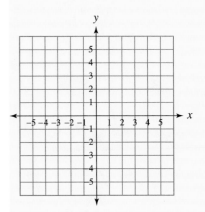

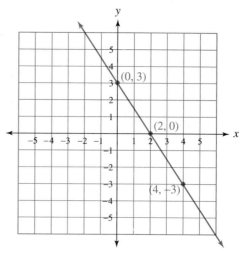

Figure 5   The graph of $3x + 2y = 6$ ■

5. Graph the line $x = -3$.

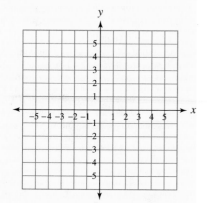

EXAMPLE 5   Graph the line $x = 3$.

SOLUTION   When we encounter an equation like $x = 3$, we assume that it looks like $x + 0y = 3$, so that no matter what we use for y, x will always be 3. A solution for $x = 3$ is any ordered pair that has an x-coordinate of 3; y can be any number, so long as x is 3. Some ordered pairs with this characteristic are $(3, 0)$, $(3, 1), (3, -1), (3, 4)$, and $(3, -4)$, to name just a few. Plotting each of these, we see that they line up vertically in a straight line (Figure 6).

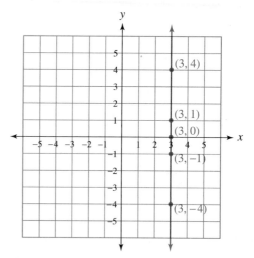

Figure 6   The graph of $x = 3$ ■

As you can see from Example 5, the graph of $x = 3$ is a vertical line. If we take a similar approach to graphing the equation $y = -4$, we find that the graph is a *horizontal* line, because any solution to the equation $y = -4$ will be an ordered pair with a y-coordinate of -4. For example, the points $(0, -4), (3, -4), (-2, -4)$, and $(4, -4)$ are all solutions to the equation $y = -4$. If we were to graph these points, they would line up in a horizontal line 4 units below the x-axis. Plot these points on the coordinate system in Figure 6 and see for yourself.

Graphing Calculators—Graphing with Trace and Zoom

All graphing calculators have the ability to graph a function and then trace over the points on the graph, giving their coordinates. Furthermore, all graphing calculators can zoom in and out on a graph that has been drawn. To graph a linear equation on a graphing calculator, we first set the graph window. Most calculators label the smallest value of x as Xmin and the largest value of x as Xmax. The counterpart values of y are Ymin and Ymax.

Set your calculator with the following window:

$$\text{Xmin} = -10 \quad \text{Ymin} = -10$$
$$\text{Xmax} = 10 \quad \text{Ymax} = 10$$

Graph the equation $y = -x + 8$. On the TI-82/83, you use the ☐ key to enter the equation; you enter a negative sign with the ☐ key, and a subtraction sign with the ☐ key. The graph will be similar to the one shown below.

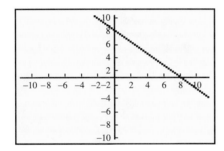

Use the Trace feature of your calculator to name three points on the graph. Next, use the Zoom feature of your calculator to zoom out so your window is twice as large.

Solving for y First To graph the equation $2x + 3y = 6$ on a graphing calculator, you must first solve it for y. When you do so, you will get $y = -\frac{2}{3}x + 2$, which you enter into your calculator as Y = -(2/3)X + 2. Graph this equation.

$$\text{Xmin} = -6 \quad \text{Ymin} = -6$$
$$\text{Xmax} = 6 \quad \text{Ymax} = 6$$

Hint on Tracing If you are going to use the Trace feature and you want the x-coordinates to be exact numbers, set your window so that the range of X inputs is a multiple of the number of horizontal pixels on your calculator screen. On the TI-82/83, the screen has 94 pixels across. Here are a few convenient trace windows:

X from −4.7 to 4.7	To trace to the nearest tenth
X from −47 to 47	To trace to the nearest integer
X from 0 to 9.4	To trace to the nearest tenth
X from 0 to 94	To trace to the nearest integer
X from −94 to 94	To trace to the nearest even integer

Problem Set 4.9

For the equations in Problems 1–16, complete the given ordered pairs, and use the results to graph the equation.

1. $x + y = 2$ $(0, \ ),(2, \ ),(-1, \ )$

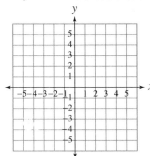

2. $x - y = 2$ $(1, \ ),(2, \ ),(\ ,2)$

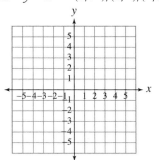

3. $y = 2x - 4$ $(0, \ ),(1, \ ),(2, \ )$

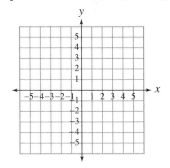

4. $y = -2x + 4$ $(0, \ ),(1, \ ),(2, \ )$

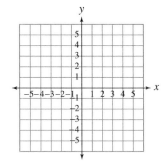

5. $3x + y = 3$ $(0, \ ),(\ ,0),(3, \ )$

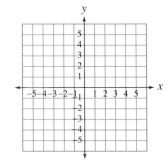

6. $x + 3y = 3$ $(0, \ ),(\ ,0),(-3, \ )$

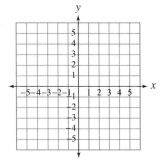

7. $3x + 4y = 12$ $(0, \ ),(\ ,0),(-4, \ )$

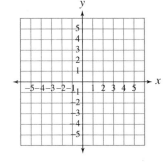

8. $4x + 3y = 12$ $(0, \ ),(\ ,0),(\ ,-4)$

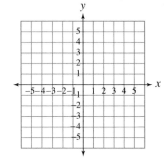

9. $3x - 4y = 12$ $(0,\ \ ), (\ \ , 0), (-4,\ \ )$

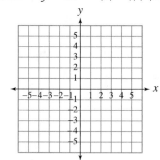

10. $4x - 3y = 12$ $(0,\ \ ), (\ \ , 0), (\ \ , -4)$

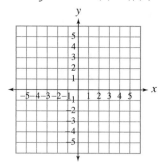

11. $y = \dfrac{1}{2}x$ $(0,\ \ ), (2,\ \ ), (-2,\ \ )$

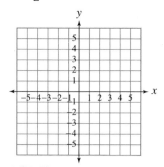

12. $y = -\dfrac{1}{2}x$ $(0,\ \ ), (2,\ \ ), (-2,\ \ )$

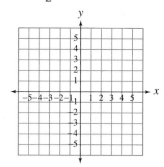

13. $y = \dfrac{1}{2}x + 2$ $(-2,\ \ ), (0,\ \ ), (2,\ \ )$

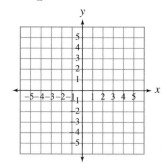

14. $y = \dfrac{1}{3}x + 1$ $(-3,\ \ ), (0,\ \ ), (3,\ \ )$

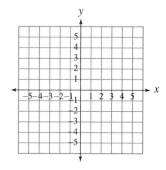

15. $x = 2$ $(\ \ , 0), (\ \ , 3), (2,\ \ \ \ \ \ )$

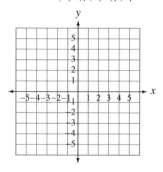

16. $y = 2$ $(0,\ \ ), (3,\ \ ), (\ \ \ \ \ \ , 2)$

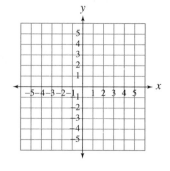

Find three solutions to each of the equations in Problems 17–34, and use them to draw the graph.

17. $x + y = 5$

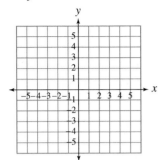

18. $x - y = 4$

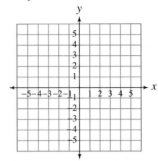

19. $x - y = 5$

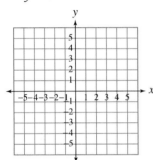

20. $x + y = 3$

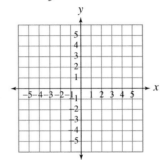

21. $2x - 4y = 4$

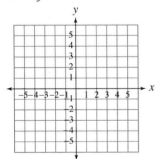

22. $4x + 2y = 4$

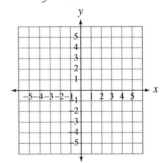

23. $2x + 4y = 4$

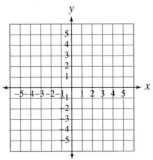

24. $4x - 2y = 4$

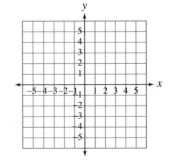

25. $y = 2x + 3$

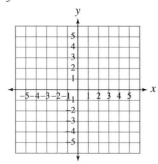

26. $y = 3x - 4$

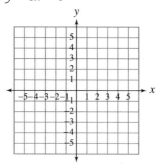

27. $y = 2x$

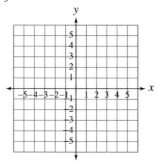

28. $y = -2x$

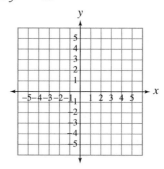

29. $y = \dfrac{1}{3}x$

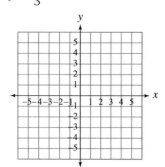

30. $y = -\dfrac{1}{3}x$

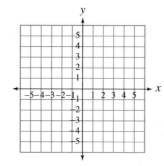

31. $x = 4$

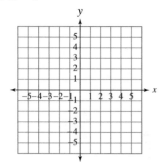

32. $x = -3$

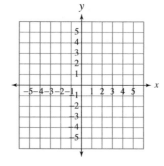

33. $y = -2$

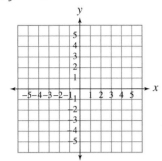

34. $y = 3$

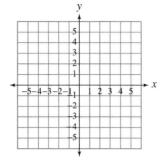

35. The graph shown here is the graph of which of the following equations?
   a. $3x - 2y = 6$
   b. $2x - 3y = 6$
   c. $2x + 3y = 6$

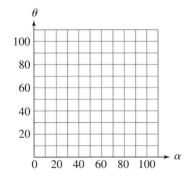

36. The graph shown here is the graph of which of the following equations?
   a. $3x - 2y = 8$
   b. $2x - 3y = 8$
   c. $2x + 3y = 8$

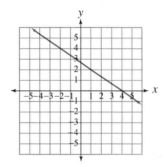

Applying the Concepts

37. Complementary Angles The diagram below shows sunlight hitting the ground. Angle α (alpha) is called the angle of inclination, and angle θ (theta) is called the angle of incidence. As the sun moves across the sky, the values of these angles change, but the two angles are always complementary, meaning that $\alpha + \theta = 90°$. Graph this equation on the template below. (The graph is restricted to the first quadrant so that we don't have to define negative angles.)

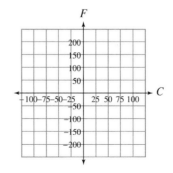

38. Temperature Scales As you know from the introduction to this chapter, the relationship between the Fahrenheit and Celsius temperature scales is given by the equation $F = \frac{9}{5}C + 32$. Graph this equation on the template below.

39. Cost of Bottled Water A water bottling company charges $7 per month for their water dispenser and $2 for each gallon of water delivered. If you have g gallons of water delivered in a month, then the equation $y = 7 + 2g$ gives the amount of your bill for that month. Graph this equation on a rectangular coordinate system below. Note that we are restricting our graph to positive values of g only. Do you know why?

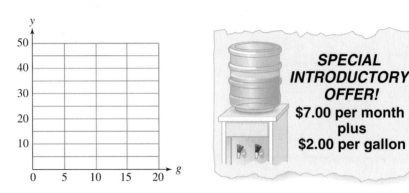

40. Temperature and Altitude On a certain day, the temperature on the ground is 72° Fahrenheit, and the temperature T at an altitude of A feet above the ground is given by the equation $T = 72 - \frac{1}{300}A$. Graph this equation on the rectangular coordinate system below. Note that we have restricted our graph to positive values of A only.

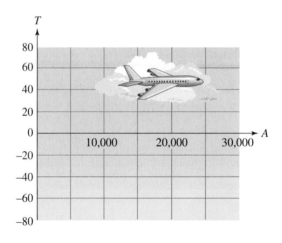

Review Problems

Problems 41–46 review material we covered in Section 3.9.

Simplify each expression.

41. $\dfrac{3}{5}\left(2\dfrac{1}{5} - 1\dfrac{1}{10}\right)$

42. $\dfrac{7}{8}\left(5\dfrac{3}{4} - 2\dfrac{1}{2}\right)$

43. $\left(\dfrac{3}{5} + \dfrac{1}{3}\right)\left(\dfrac{3}{5} - \dfrac{1}{3}\right)$

44. $\left(1 + \dfrac{3}{8}\right)\left(1 - \dfrac{3}{8}\right)$

45. $\left(\dfrac{7}{15} - \dfrac{11}{30}\right)^2$

46. $\left(\dfrac{13}{21} - \dfrac{13}{35}\right)^2$

Chapter ④ Summary

Combining Similar Terms [4.1]

Two terms are similar terms if they have the same variable part. The expressions $7x$ and $2x$ are similar because the variable part in each is the same. Similar terms are combined by using the distributive property.

1. $7x + 2x = (7 + 2)x$
$\qquad = 9x$

Finding the Value of an Algebraic Expression [4.1]

An algebraic expression is a mathematical expression that contains numbers and variables. Expressions that contain a variable will take on different values depending on the value of the variable.

2. When $x = 5$, the expression $2x + 7$ becomes
$2(5) + 7 = 10 + 7 = 17$

The Solution to an Equation [4.2]

A solution to an equation is a number that, when used in place of the variable, makes the equation a true statement.

The Addition Property of Equality [4.2]

Let A, B, and C represent algebraic expressions.

$$\text{If} \quad A = B$$
$$\text{then} \quad A + C = B + C$$

In words: Adding the same quantity to both sides of an equation will not change the solution.

3. We solve $x - 4 = 9$ by adding 4 to each side.
$$x - 4 = 9$$
$$x - 4 + \mathbf{4} = 9 + \mathbf{4}$$
$$x + 0 = 13$$
$$x = 13$$

The Multiplication Property of Equality [4.3]

Let A, B, and C represent algebraic expressions with C not equal to 0.

$$\text{If} \quad A = B$$
$$\text{then} \quad AC = BC$$

In words: Multiplying both sides of an equation by the same nonzero number will not change the solution to the equation. This property holds for division as well.

4. Solve $\frac{1}{3}x = 5$.
$$\frac{1}{3}x = 5$$
$$\mathbf{3} \cdot \frac{1}{3}x = \mathbf{3} \cdot 5$$
$$x = 15$$

Steps Used to Solve a Linear Equation in One Variable [4.4]

Step 1 Simplify each side of the equation.

Step 2 Use the addition property of equality to get all variable terms on one side and all constant terms on the other side.

Step 3 Use the multiplication property of equality to get just one x isolated on either side of the equation.

Step 4 Check the solution in the original equation if necessary.

If the original equation contains fractions, you can begin by multiplying each side by the LCD for all fractions in the equation.

5. $2(x - 4) + 5 = -11$
$$2x - 8 + 5 = -11$$
$$2x - 3 = -11$$
$$2x - 3 + \mathbf{3} = -11 + \mathbf{3}$$
$$2x = -8$$
$$\frac{2x}{\mathbf{2}} = \frac{-8}{\mathbf{2}}$$
$$x = -4$$

Evaluating Formulas [4.6]

In mathematics, a formula is an equation that contains more than one variable. For example, the formula for the perimeter of a rectangle is $P = 2l + 2w$. We evaluate a formula by substituting values for all but one of the variables and then solving the resulting equation for that variable.

6. When $w = 8$ and $l = 13$
the formula $P = 2w + 2l$
becomes $P = 2 \cdot 8 + 2 \cdot 13$
$\qquad = 16 + 26$
$\qquad = 42$

The Rectangular Coordinate System [4.8]

The rectangular coordinate system consists of two number lines, called axes, which intersect at right angles. The point at which the axes intersect is called the origin.

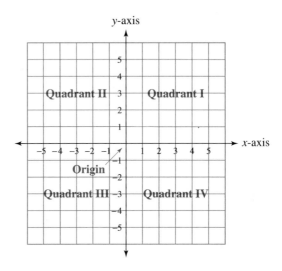

Graphing Ordered Pairs [4.8]

To graph an ordered pair (a, b) on the rectangular coordinate system, we start at the origin and move a units right or left (right if a is positive, left if a is negative). From there we move b units up or down (up if b is positive, down if b is negative). The point where we end up is the graph of the ordered pair (a, b).

To Graph a Straight Line [4.9]

Step 1   Find any three ordered pairs that satisfy the equation. This is usually accomplished by substituting a number for one of the variables in the equation and then solving for the other variable.

Step 2   Plot the three ordered pairs found in Step 1. (Actually, we need only two points to graph a straight line. The third point serves as a check. If all three points don't line up, then we have made a mistake in our work.)

Step 3   Draw a line through the three points you plotted in Step 2.

7.   The graph of $3x + 2y = 6$

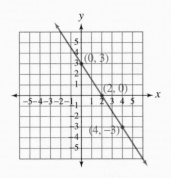

Chapter **4** Review

Simplify the expressions by combining similar terms. [4.1]

1. $10x + 7x$ 　　　　**2.** $8x - 12x$ 　　　　**3.** $2a + 9a + 3 - 6$ 　　　　**4.** $4y - 7y + 8 - 10$

5. $6x - x + 4$ 　　　　**6.** $-5a + a + 4 - 3$ 　　　　**7.** $2a - 6 + 8a + 2$ 　　　　**8.** $12y - 4 + 3y - 9$

Find the value of each expression when x is 4. [4.1]

9. $10x + 2$ 　　　　**10.** $5x - 12$ 　　　　**11.** $-2x + 9$ 　　　　**12.** $-x + 8$

13. Is $x = -3$ a solution to $5x - 2 = -17$? [4.2] 　　　　**14.** Is $x = 4$ a solution to $3x - 2 = 2x + 1$? [4.2]

Solve the equations. [4.2, 4.3, 4.4]

15. $x - 5 = 4$ 　　　　**16.** $-x + 3 + 2x = 6 - 7$ 　　　　**17.** $2x + 1 = 7$ 　　　　**18.** $3x - 5 = 1$

19. $2x + 4 = 3x - 5$ 　　　　**20.** $4x + 8 = 2x - 10$ 　　　　**21.** $3(x - 2) = 9$ 　　　　**22.** $4(x - 3) = -20$

23. $3(2x + 1) - 4 = -7$ 　　　　**24.** $4(3x + 1) = -2(5x - 2)$ 　　　　**25.** $5x + \dfrac{3}{8} = -\dfrac{1}{4}$ 　　　　**26.** $\dfrac{7}{x} - \dfrac{2}{5} = 1$

27. Number Problem The sum of a number and -3 is -5. Find the number. [4.5]

28. Number Problem If twice a number is added to 3, the result is 7. Find the number. [4.5]

29. Number Problem Three times the sum of a number and 2 is -6. Find the number. [4.5]

30. Number Problem If 7 is subtracted from twice a number, the result is 5. Find the number. [4.5]

31. Geometry The length of a rectangle is twice its width. If the perimeter is 42 meters, find the length and the width. [4.5]

32. Age Problem Patrick is 3 years older than Amy. In 5 years the sum of their ages will be 31. How old are they now? [4.5]

In Problems 33–36, use the equation $3x + 2y = 6$ to find y. [4.6, 4.7]

33. $x = -2$ 　　　　**34.** $x = 6$ 　　　　**35.** $x = 0$ 　　　　**36.** $x = \dfrac{1}{3}$

In Problems 37–39, use the equation $3x + 2y = 6$ to find x. [4.6, 4.7]

37. $y = 3$ **38.** $y = -3$ **39.** $y = 0$

40. Medical Costs The table below shows the average yearly cost of visits to the doctor as reported in 1999.

MEDICAL COSTS	
YEAR	**AVERAGE ANNUAL COST**
1990	$583
1995	$739
2000	$906
2005	$1,172

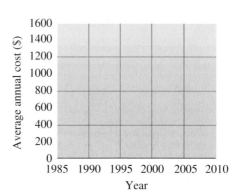

a. Use the template to construct a line graph of the information in the table.
b. From the line graph, estimate the annual cost of seeing a doctor in 2010.

In Problems 41–48, plot the points. (Put 41, 43, 45, and 47 on one graph; put 42, 44, 46, and 48 on the other.) [4.8]

41. $(4, 2)$ **42.** $(4, -2)$

43. $(-4, 2)$ **44.** $(-4, -2)$

45. $(4, 0)$ **46.** $(0, 4)$

47. $(0, -4)$ **48.** $(-4, 0)$

In Problems 49–58, graph each line. [4.9]

49. $x + y = 2$ **50.** $x - y = 2$

51. $3x + 2y = 6$ **52.** $2x + 3y = 6$

53. $y = \dfrac{1}{2}x$ **54.** $y = 2x$

55. $y = 2x - 3$ **56.** $y = \dfrac{1}{3}x + 1$

57. $x = -2$ **58.** $y = 3$

Chapter **4** Test

Simplify each expression by combining similar terms.

1. $9x - 3x + 7 - 12$

2. $4b - 1 - b - 3$

Find the value of each expression when $x = 3$.

3. $3x - 12$

4. $-x + 12$

5. Is $x = -1$ a solution to $4x - 3 = -7$?

Solve each equation.

6. $x - 7 = -3$

7. $\dfrac{2}{3}y = 18$

8. $3x - 7 = 5x + 1$

9. $2(x - 5) = -8$

10. $3(2x + 3) = -3(x - 5)$

11. $6(3x - 2) - 8 = 4x - 6$

12. Number Problem Twice the sum of a number and 3 is −10. Find the number.

13. Number Problem If 8 is subtracted from three times a number, the result is 1. Find the number.

14. Geometry The length of a rectangle is 4 centimeters longer than its width. If the perimeter is 28 centimeters, find the length and the width.

15. Age Problem Karen is 5 years younger than Susan. Three years ago, the sum of their ages was 11. How old are they now?

16. Use the equation $4x + 3y = 12$ to find y when $x = -3$.

17. Plot the following points: $(3, 2), (-3, 2), (3, 0), (0, -2)$.

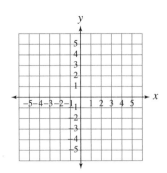

18. Graph $2x + y = 4$.

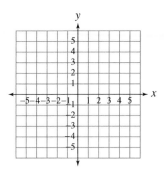

19. Graph $y = 2x - 1$.

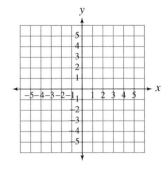

20. Graph $x = 2$.

Chapters **1-4** Cumulative Review

Simplify.

1. $\begin{array}{r} 7520 \\ 599 \\ +8640 \\ \hline \end{array}$

2. $\begin{array}{r} 6000 \\ -3999 \\ \hline \end{array}$

3. $156 \div 13$

4. $9(7 \cdot 2)$

5. $64\overline{)31{,}362}$

6. 2^8

7. $12 + 81 \div 3^2$

8. $\dfrac{329}{47}$

9. $-25 + (-13)$

10. $(-10 + 4) + (212 - 100)$

11. $\dfrac{-39}{-3}$

12. $(15z)(2z)^2$

13. $(2x^2y)^2(3x^2y)^3$

14. $(9x - 3y) - (2x + 4y - 2)$

15. $(5x + 5)(7x + 3)$

16. $\left(\dfrac{1}{4}\right)^3\left(-\dfrac{1}{2}\right)^2$

17. $17 \div \left(\dfrac{1}{3}\right)^2$

18. $\dfrac{t}{25} + \left(-\dfrac{7}{50}\right)$

19. $\left(16 \div 1\dfrac{1}{4}\right) \div 2$

20. $15 - 3\dfrac{1}{2}$

21. $3(z + 2) - 2(3z - 1)$

22. $-2a - 3 + a + 7$

23. $13 + \dfrac{3}{14} \div \dfrac{5}{42}$

Solve.

24. $4b + 3 + 2b = 12b - 6$

25. $5a - 3a + 6a = -8 + (-8)$

26. Graph the equation $2x + y = 6$, and name 3 ordered pairs on the graph.

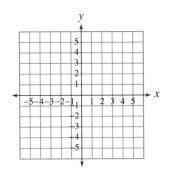

27. Find the perimeter and area of the figure below.

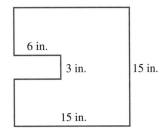

28. Find the perimeter of the figure below.

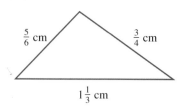

$\frac{5}{6}$ cm $\frac{3}{4}$ cm

$1\frac{1}{3}$ cm

29. Find the difference between -15 and -62.

30. Find the value of $(5m^2n^4)^2$ when $m = 2$ and $n = 1$.

31. Find the volume of a cube with sides 21 cm.

32. Factor 126 into a product of prime factors.

33. Find $\frac{2}{3}$ of the product of 7 and 9.

34. Use the equation $4x + y = 26$ to find y when $x = 3$.

35. Use the equation $2x - 2y = 12$ to complete the following ordered pairs. $(10, \quad), (\quad , 2), (4, \quad)$

36. On the graph of $x = -2$, when $y = 4$, what is the value of x?

37. Hammerhead Shark The largest hammerhead shark ever recorded was $18\frac{1}{3}$ feet long. Multiply the shark's length by 12 to find its length in inches.

38. Running Over two days, James ran 18 miles. He ran three times farther on the first day than he did on the second day. What was the distance of James's shorter run?

39. Volume What is the volume of a rectangular solid that is 36 inches long, 24 inches wide, and 12 inches deep?

40. NYC Marathon The table below shows the number of people who finished the New York City Marathons in certain years.

NEW YORK CITY MARATHON		
YEAR	NUMBER OF FINISHERS	ROUNDED TO THE NEAREST HUNDRED
1975	339	
1980	12,512	
1985	15,881	
1990	23,774	
1995	26,754	

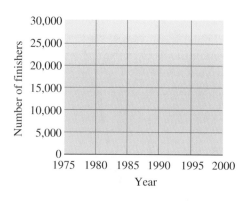

a. Fill in the last column of the table by rounding each number in the second column to the nearest hundred.

b. Use the template and the rounded numbers from Part a to construct a line graph of the information in the table.

c. From the line graph, estimate the number of finishers for the 2000 NYC Marathon.

Decimals

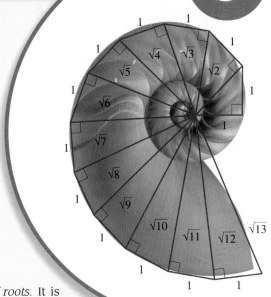

INTRODUCTION

The diagram shown here is called the *spiral of roots.* It is constructed using the Pythagorean theorem, which is one of the topics we will work with in this chapter. The spiral of roots gives us a way to visualize positive square roots, another of the topics we will cover in this chapter. The table below gives us the decimal equivalents (some of which are approximations) of the first 10 square roots in the spiral. The line graph can be constructed from the table or from the spiral.

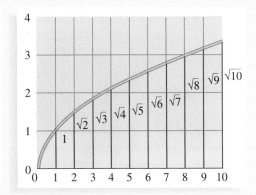

APPROXIMATE LENGTH OF DIAGONALS

NUMBER	POSITIVE SQUARE ROOT
1	1
2	1.41
3	1.73
4	2
5	2.24
6	2.45
7	2.65
8	2.83
9	3
10	3.16

Figure 1

As you can see from the diagrams above, there is an attractive visual compo- nent to square roots. If you think about it, spirals like the one above can be found in a number of places in our everyday world. With square roots, some of these spirals are easy to model with mathematics.

STUDY SKILLS

This is the last chapter in which we will mention study skills. You know by now what works best for you, and what you have to do to achieve your goals for this course. From now on it is simply a matter of continuing to use the skills that work for you and avoiding those that do not. No one can do this for you: It is up to you to maintain the skills that will get you where you want to be in the course.

If you intend to take more classes in mathematics, and you want to ensure your success in those classes, then you can work toward this goal: *Become the type of student who can learn mathematics on his or her own.* Most people who have degrees in mathematics were students who could learn mathematics on their own. This doesn't mean that you have to learn it all on your own; it simply means that if you do have to, you can. Attaining this goal gives you independence and puts you in control of your success in any mathematics class you take.

5.1   Decimal Notation and Place Value

Introduction . . .

In this chapter we will focus our attention on *decimals*. Anyone who has used money in the United States has worked with decimals already. For example, if you have been paid an hourly wage, such as

$6.25 per hour

↑
└── **Decimal point**

you have had experience with decimals. What is interesting and useful about decimals is their relationship to fractions and to powers of ten. The work we have done up to now—especially our work with fractions—can be used to develop the properties of decimal numbers.

In Chapter 1 we developed the idea of place value for the digits in a whole number. At that time we gave the name and the place value of each of the first seven columns in our number system, as follows:

MILLIONS COLUMN	HUNDRED THOUSANDS COLUMN	TEN THOUSANDS COLUMN	THOUSANDS COLUMN	HUNDREDS COLUMN	TENS COLUMN	ONES COLUMN
1,000,000	100,000	10,000	1,000	100	10	1

As we move from right to left, we multiply by 10 each time. The value of each column is 10 times the value of the column on its right, with the rightmost column being 1. Up until now we have always looked at place value as increasing by a factor of 10 each time we move one column to the left:

TEN THOUSANDS	THOUSANDS	HUNDREDS	TENS	ONES
10,000 ←	1,000 ←	100 ←	10 ←	1
Multiply by 10	Multiply by 10	Multiply by 10	Multiply by 10	

To understand the idea behind decimal numbers, we notice that moving in the opposite direction, from left to right, we *divide* by 10 each time:

TEN THOUSANDS	THOUSANDS	HUNDREDS	TENS	ONES
10,000 →	1,000 →	100 →	10 →	1
Divide by 10	Divide by 10	Divide by 10	Divide by 10	

If we keep going to the right, the next column will have to be

$$1 \div 10 = \frac{1}{10}$$ **Tenths**

The next one after that will be

$$\frac{1}{10} \div 10 = \frac{1}{10} \cdot \frac{1}{10} = \frac{1}{100} \quad \textbf{Hundredths}$$

After that, we have

$$\frac{1}{100} \div 10 = \frac{1}{100} \cdot \frac{1}{10} = \frac{1}{1,000} \quad \textbf{Thousandths}$$

We could continue this pattern as long as we wanted. We simply divide by 10 to move one column to the right. (And remember, dividing by 10 gives the same result as multiplying by $\frac{1}{10}$.)

To show where the ones column is, we use a *decimal point* between the ones column and the tenths column.

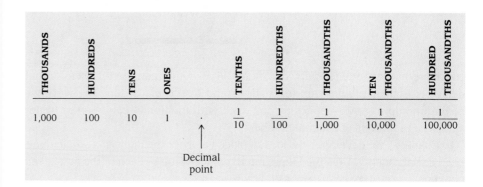

The ones column can be thought of as the middle column, with columns larger than 1 to the left and columns smaller than 1 to the right. The first column to the right of the ones column is the tenths column, the next column to the right is the hundredths column, the next is the thousandths column, and so on. The decimal point is always written between the ones column and the tenths column.

We can use the place value of decimal fractions to write them in expanded form.

EXAMPLE 1   Write 423.576 in expanded form.

SOLUTION $423.576 = 400 + 20 + 3 + \dfrac{5}{10} + \dfrac{7}{100} + \dfrac{6}{1,000}$ ∎

EXAMPLE 2   Write each number in words.
  a. 0.4
  b. 0.04
  c. 0.004

SOLUTION
  a. 0.4 is "four tenths."
  b. 0.04 is "four hundredths."
  c. 0.004 is "four thousandths."   ∎

When a decimal fraction contains digits to the left of the decimal point, we use the word "and" to indicate where the decimal point is when writing the number in words.

Practice Problems
1. Write 785.462 in expanded form.

2. Write in words.
  a. 0.06
  b. 0.7
  c. 0.008

Answers
1-2. See solutions section.

EXAMPLE 3 Write each number in words.

    a. 5.4

    b. 5.04

    c. 5.004

SOLUTION   **a.** 5.4 is "five and four tenths."

    b. 5.04 is "five and four hundredths."

    c. 5.004 is "five and four thousandths."

EXAMPLE 4 Write 3.64 in words.

SOLUTION   The number 3.64 is read "three and sixty-four hundredths." The place values of the digits are as follows:

    3   ·   6   4

3 ones   6 tenths   4 hundredths

We read the decimal part as "sixty-four hundredths" because

$$6 \text{ tenths} + 4 \text{ hundredths} = \frac{6}{10} + \frac{4}{100} = \frac{60}{100} + \frac{4}{100} = \frac{64}{100}$$

EXAMPLE 5 Write 25.4936 in words.

SOLUTION   Using the idea given in Example 4, we write 25.4936 in words as "twenty-five and four thousand, nine hundred thirty-six ten thousandths."

    In order to understand addition and subtraction of decimals in the next section, we need to be able to convert decimal numbers to fractions or mixed numbers.

EXAMPLE 6 Write each number as a fraction or a mixed number. Do not reduce to lowest terms.

    a. 0.004   **b.** 3.64   **c.** 25.4936

SOLUTION   **a.** Because 0.004 is 4 thousandths, we write

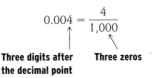

$$0.004 = \frac{4}{1,000}$$

Three digits after   **Three zeros**
the decimal point

    b. Looking over the work in Example 4, we can write

$$3.64 = 3\frac{64}{100}$$

Two digits after   **Two zeros**
the decimal point

    c. From the way in which we wrote 25.4936 in words in Example 5, we have

$$25.4936 = 25\frac{4936}{10,000}$$

Four digits after   **Four zeros**
the decimal point

---

3. Write in words.

    a. 5.06

    b. 4.7

    c. 3.008

Note

Sometimes we name decimal fractions by simply reading the digits from left to right and using the word "point" to indicate where the decimal point is. For example, using this method the number 5.04 is read "five point zero four."

4. Write 5.98 in words.

5. Write 305.406 in words.

6. Write each number as a fraction or a mixed number. Do not reduce to lowest terms.

    a. 0.06

    b. 5.98

    c. 305.406

Answers

3-5. See solutions section.

6. a. $\dfrac{6}{100}$ **b.** $5\dfrac{98}{100}$

    c. $305\dfrac{406}{1,000}$

Rounding Decimal Numbers

The rule for rounding decimal numbers is similar to the rule for rounding whole numbers. If the digit in the column to the right of the one we are rounding to is 5 or more, we add 1 to the digit in the column we are rounding to; otherwise, we leave it alone. We then replace all digits to the right of the column we are rounding to with zeros if they are to the left of the decimal point; otherwise, we simply delete them. Table 1 illustrates the procedure.

Table 1

NUMBER	ROUNDED TO THE NEAREST		
	WHOLE NUMBER	TENTH	HUNDREDTH
24.785	25	24.8	24.79
2.3914	2	2.4	2.39
0.98243	1	1.0	0.98
14.0942	14	14.1	14.09
0.545	1	0.5	0.55

7. Round 8,935.042 to the nearest hundred.

EXAMPLE 7   Round 9,235.492 to the nearest hundred.

SOLUTION   The number next to the hundreds column is 3, which is less than 5. We change all digits to the right to 0, and we can drop all digits to the right of the decimal point, so we write

9,200   ■

8. Round 0.05067 to the nearest ten thousandth.

EXAMPLE 8   Round 0.00346 to the nearest ten thousandth.

SOLUTION   Because the number to the right of the ten thousandths column is more than 5, we add 1 to the 4 and get

0.0035   ■

9. Round each number in the table to the nearest tenth of a dollar.

EXAMPLE 9   The bar chart below shows some ticket prices for major league baseball for the 1999 season. Round each ticket price to the nearest dollar.

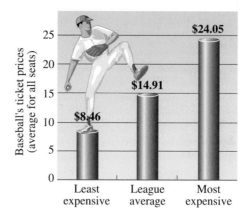

SOLUTION   Using our rule for rounding decimal numbers, we have the following results:

Least expensive: $8.46 rounds to $8
League average: $14.91 rounds to $15
Most expensive: $24.05 rounds to $24   ■

Problem Set ⬤5.1

Write out the name of each number in words.

1. 0.3

2. 0.03

3. 0.015

4. 0.0015

5. 3.4

6. 2.04

7. 52.7

8. 46.8

Write each number as a fraction or a mixed number. Do not reduce your answers.

9. 405.36

10. 362.78

11. 9.009

12. 60.06

13. 1.234

14. 12.045

15. 0.00305

16. 2.00106

Give the place value of the 5 in each of the following numbers.

17. 458.327

18. 327.458

19. 29.52

20. 25.92

21. 0.00375

22. 0.00532

23. 275.01

24. 0.356

25. 539.76

26. 0.123456

Write each of the following as a decimal number.

27. Fifty-five hundredths

28. Two hundred thirty-five ten thousandths

29. Six and nine tenths

30. Forty-five thousand and six hundred twenty-one thousandths

31. Eleven and eleven hundredths

32. Twenty-six thousand, two hundred forty-five and sixteen hundredths

33. One hundred and two hundredths

34. Seventy-five and seventy-five hundred thousandths

35. Three thousand and three thousandths

36. One thousand, one hundred eleven and one hundred eleven thousandths

Complete the following table.

NUMBER	ROUNDED TO THE NEAREST			
	WHOLE NUMBER	TENTH	HUNDREDTH	THOUSANDTH
37. 47.5479	_____	_____	_____	_____
38. 100.9256	_____	_____	_____	_____
39. 0.8175	_____	_____	_____	_____
40. 29.9876	_____	_____	_____	_____
41. 0.1562	_____	_____	_____	_____
42. 128.9115	_____	_____	_____	_____
43. 2,789.3241	_____	_____	_____	_____
44. 0.8743	_____	_____	_____	_____
45. 99.9999	_____	_____	_____	_____
46. 71.7634	_____	_____	_____	_____

Applying the Concepts

47. Penny Weight If you have a penny dated anytime from 1959 through 1982, its original weight was 3.11 grams. If the penny has a date of 1983 or later, the original weight was 2.5 grams. Write the two weights in words.

48. Savings Rates A local savings and loan company advertises savings notes that give a 7.78 percent interest rate. Round 7.78 to the nearest tenth.

49. Speed of Light The speed of light is 186,282.3976 miles per second. Round this number to the nearest hundredth.

50. Halley's Comet Halley's comet was seen from the earth during 1986. It will be another 76.1 years before it returns. Write 76.1 in words.

51. Nutrition A 50-gram egg contains 0.15 milligram of riboflavin. Write 0.15 in words.

52. Nutrition One medium banana contains 0.64 milligram of B$_6$. Write 0.64 in words.

53. Gasoline Prices The bar chart below was created from a survey by the U.S. Department of Energy's Energy Information Administration during the month of February 2000. It gives the average price of regular gasoline for the state of California on each Monday of the month. Use the information in the chart to fill in the table.

PRICE OF 1 GALLON OF REGULAR GASOLINE

DATE	PRICE (DOLLARS)
2/7/00	
2/14/00	
	1.444
	1.518

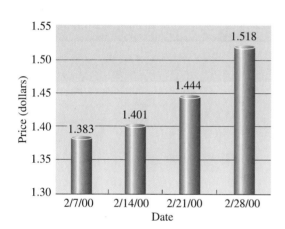

54. Speed and Time The bar chart below was created from data given by *Car and Driver* magazine. It gives the minimum time in seconds for a Toyota Echo to reach various speeds from a complete stop. Use the information in the chart to fill in the table.

SPEED (MILES PER HOUR)	TIME (SECONDS)
30	
40	
50	
	8.5
	11.6
80	
90	

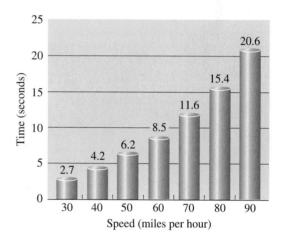

55. Write the following numbers in order from smallest to largest.

0.02 0.05 0.025 0.052 0.005 0.002

56. Write the following numbers in order from smallest to largest.

0.2 0.02 0.4 0.04 0.42 0.24

57. Which of the following numbers will round to 7.5?

7.451 7.449 7.54 7.56

58. Which of the following numbers will round to 3.2?

3.14999 3.24999 3.279 3.16111

Change each decimal to a fraction, and then reduce to lowest terms.

59. 0.25

60. 0.75

61. 0.35

62. 0.65

63. 0.125

64. 0.375

65. 0.625

66. 0.0625

67. 0.875

68. 0.1875

Estimating

For each pair of numbers, choose the number that is closest to 10.

69. 9.9 and 9.99

70. 8.5 and 8.05

71. 10.5 and 10.05

72. 10.9 and 10.99

For each pair of numbers, choose the number that is closest to 0.

73. 0.5 and 0.05

74. 0.10 and 0.05

75. 0.01 and 0.02

76. 0.1 and 0.01

Review Problems

In the next section we will do addition and subtraction with decimals. To understand the process of addition and subtraction, we need to understand the process of addition and subtraction with mixed numbers from Section 3.8.

Find each of the following sums and differences. (Add and subtract.)

77. $4\frac{3}{10} + 2\frac{1}{100}$

78. $5\frac{35}{100} + 2\frac{3}{10}$

79. $8\frac{5}{10} - 2\frac{4}{100}$

80. $6\frac{3}{100} - 2\frac{125}{1,000}$

81. $5\frac{1}{10} + 6\frac{2}{100} + 7\frac{3}{1,000}$

82. $4\frac{27}{100} + 6\frac{3}{10} + 7\frac{123}{1,000}$

83. $2 + \frac{3}{x}$

84. $3 + \frac{2}{x}$

85. $x - \frac{3}{4}$

86. $x - \frac{5}{6}$

Extending the Concepts

87. List all the three-digit numbers that round to 2.9.

88. List all the three-digit numbers that round to 2.5.

89. List all the two-digit numbers that round to 3.

90. List all the two-digit numbers that round to 5.

91. In your own words, give a step-by-step explanation of how you would round the number 3.456 to the nearest hundredth.

92. In your own words, give a step-by-step explanation of how you would round the number 3.456 to the nearest whole number.

5.2 Addition and Subtraction with Decimals

Introduction . . .

If you are earning $8.50 per hour and you get a raise of $1.25 per hour, then your new hourly rate of pay is

$$\begin{array}{r} \$8.50 \\ +\$1.25 \\ \hline \$9.75 \end{array}$$

To add the two rates of pay, we align the decimal points, and then add in columns.

To see why this is true in general, we can use mixed-number notation:

$$8.50 = 8\frac{50}{100}$$

$$+1.25 = 1\frac{25}{100}$$

$$\rule{2cm}{0.4pt}$$

$$9\frac{75}{100} = 9.75$$

We can visualize the mathematics above by thinking in terms of money:

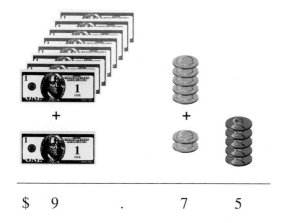

$$\$\quad 9\quad .\quad 7\quad 5$$

EXAMPLE 1 Add by first changing to fractions: $25.43 + 2.897 + 379.6$

SOLUTION We first change each decimal to a mixed number. We then write each fraction using the least common denominator and add as usual:

$$\begin{aligned} 25.43 &= 25\frac{43}{100} = 25\frac{430}{1,000} \\ 2.897 &= 2\frac{897}{1,000} = 2\frac{897}{1,000} \\ +379.6 &= 379\frac{6}{10} = 379\frac{600}{1,000} \\ \hline & \quad\quad 406\frac{1,927}{1,000} = 407\frac{927}{1,000} = 407.927 \end{aligned}$$

Again, the result is the same if we just line up the decimal points and add as if we were adding whole numbers:

$$\begin{array}{r} 25.430 \\ 2.897 \\ +379.600 \\ \hline 407.927 \end{array}$$

Notice that we can fill in zeros on the right to help keep the numbers in the correct columns. Doing this does not change the value of any of the numbers.

Note: **The decimal point in the answer is directly below the decimal points in the problem.**

The same thing would happen if we were to subtract two decimal numbers. We can use these facts to write a rule for addition and subtraction of decimal numbers.

RULE

To add (or subtract) decimal numbers, we line up the decimal points and add (or subtract) as usual. The decimal point in the result is written directly below the decimal points in the problem.

We will use this rule for the rest of the examples in this section.

2. Subtract: $78.674 - 23.431$

EXAMPLE 2 Subtract: $39.812 - 14.236$

SOLUTION   We write the numbers vertically, with the decimal points lined up, and subtract as usual.

$$\begin{array}{r} 39.812 \\ -14.236 \\ \hline 25.576 \end{array}$$

3. Add: $16 + 0.033 + 4.6 + 0.08$

EXAMPLE 3 Add: $8 + 0.002 + 3.1 + 0.04$

SOLUTION   To make sure we keep the digits in the correct columns, we can write zeros to the right of the rightmost digits:

$$\begin{aligned} 8 &= 8.000 \\ 3.1 &= 3.100 \\ 0.04 &= 0.040 \end{aligned}$$

Writing the extra zeros here is really equivalent to finding a common denominator for the fractional parts of the original four numbers—now we have a thousandths column in all the numbers

This doesn't change the value of any of the numbers, and it makes our task easier. Now we have

$$\begin{array}{r} 8.000 \\ 0.002 \\ 3.100 \\ +\ 0.040 \\ \hline 11.142 \end{array}$$

4. Subtract: $6.7 - 2.0563$

EXAMPLE 4 Subtract: $5.9 - 3.0814$

SOLUTION   In this case it is very helpful to write 5.9 as 5.9000, since we will have to borrow in order to subtract.

$$\begin{array}{r} 5.9000 \\ -3.0814 \\ \hline 2.8186 \end{array}$$

Answers
2. 55.243   **3.** 20.713
4. 4.6437

EXAMPLE 5 Subtract 3.09 from the sum of 9 and 5.472.

SOLUTION   Writing the problem in symbols, we have

$$(9 + 5.472) - 3.09 = 14.472 - 3.09$$
$$= 11.382$$

5. Subtract 5.89 from the sum of 7 and 3.567.

EXAMPLE 6 Add: $2.89 + (-5.93)$

SOLUTION   Recall from Chapter 2 that to add two numbers with different signs, we subtract the smaller absolute value from the larger. The sign of the answer is the same as the sign of the number with the larger absolute value.

$$2.89 + (-5.93) = -3.04$$

The answer has the sign of the number with the larger absolute value

6. Add: $4.93 + (-7.85)$

EXAMPLE 7 Subtract: $-8 - 1.37$

SOLUTION   Because subtraction can be thought of as addition of the opposite, instead of subtracting 1.37 we can add -1.37.

$$-8 - 1.37 = -8 + (-1.37)$$

Now, to add two numbers with the same sign, we add their absolute values. The answer has the same sign as the original numbers.

$$-8 + (-1.37) = -9.37$$

The answer has the same sign as the original two numbers

$$\begin{array}{r} 8.00 \\ +1.37 \\ \hline 9.37 \end{array}$$

Add the absolute values of the two numbers

7. Subtract: $-4.09 - 3$

EXAMPLE 8 While I was writing this section of the book, I stopped to have lunch with a friend at a coffee shop near my office. The bill for lunch was $15.64. I gave the person at the cash register a $20 bill. For change, I received four $1 bills, a quarter, a nickel, and a penny. Was my change correct?

SOLUTION   To find the total amount of money I received in change, we add:

Four $1 bills = $4.00
One quarter = 0.25
One nickel   = 0.05
One penney   = 0.01
Total   = $4.31

To find out if this is the correct amount, we subtract the amount of the bill from $20.00.

$$\begin{array}{r} \$20.00 \\ -15.64 \\ \hline \$ \ 4.36 \end{array}$$

The change was not correct. It is off by 5 cents. Instead of the nickel, I should have been given a dime.

8. If you pay for a purchase of $9.56 with a $10 bill, how much money should you receive in change? What will you do if the change that is given to you is one quarter, two dimes, and four pennies?

Answers
5. 4.677   **6.** −2.92   **7.** −7.09
8. $0.44; Tell the clerk that you have been given too much change. Instead of two dimes, you should have received one dime and one nickel.

9. Below are two more cars and the time it takes each to accelerate from 0 miles per hour to 60 miles per hour. Extend the histogram in Figure 1 to include these two cars.

CAR	TIME (IN SECONDS)
Chevrolet Corvette LT1	5.7
Honda Prelude	7.1

EXAMPLE 9   Table 1 lists the number of seconds it takes various cars to accelerate from 0 miles per hour to 60 miles per hour. Construct a bar chart from the information in the table. Then find the differences in elapsed time between the slowest car and the fastest car.

Table 1

ELAPSED TIME FROM 0 MPH TO 60 MPH

CAR	TIME (IN SECONDS)
BMW 325i	7.4
Ford Mustang GT	6.7
Mazda Miata	8.8
Porsche 911 Carrera	5.2
Toyota Supra Turbo	5.3
Volkswagen GTI	7.5

SOLUTION   We construct a bar chart (Figure 1) with the car names on the horizontal axis. The numbers on the vertical axis represent the time, in seconds.

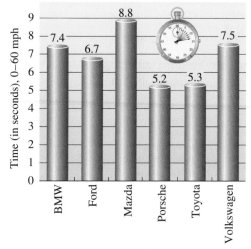

Figure 1   A bar chart of the data in Table 1

The slowest car is the Mazda at 8.8 seconds, whereas the fastest car is the Porsche at 5.2 seconds. The difference between these two times is

$$8.8 - 5.2 = 3.6 \text{ seconds}$$

If the two cars started at the same time, the Porsche would reach a speed of 60 miles per hour 3.6 seconds before the Mazda. ■

Answer

9. See solutions section.

Problem Set 5.2

Find each of the following sums. (Add.)

1. 2.91 + 3.28

2. 8.97 + 2.04

3. 0.04 + 0.31 + 0.78

4. 0.06 + 0.92 + 0.65

5. 3.89 + (−2.4)

6. 7.65 + (−3.8)

7. 4.532 + 1.81 + 2.7

8. 9.679 + 3.49 + 6.5

9. 0.081 + (−5) + 2.94

10. 0.396 + (−7) + 3.96

11. 5.0003 + 6.78 + 0.004

12. 27.0179 + 7.89 + 0.009

13. 7.123
8.12
9.1

14. 5.432
4.32
3.2

15. 9.001
8.01
7.1

16. 6.003
5.02
4.1

17. 89.7854
3.4
65.35
100.006

18. 57.4698
9.89
32.032
572.0079

19. 543.21
123.45

20. 987.654
456.789

Find each of the following differences. (Subtract.)

21. 99.34 − 88.23

22. 47.69 − 36.58

23. 2.4 − 5.97

24. 1.04 − 9.87

25. 6.3 − 2.08

26. 7.5 − 3.04

27. 28.96 − 149.37

28. 32.68 − 796.45

29. 45 − 0.067

30. 48 − 0.075

31. −8 − 0.327

32. −12 − 0.962

33. 765.432 − 234.567

34. 654.321 − 123.456

Subtract.

35. 34.07
      − 6.18

36. 25.008
      − 3.119

37. 40.04
      − 4.4

38. 50.05
      − 5.5

39. 768.436
      −356.998

40. 495.237
      −247.668

Add and subtract as indicated.

41. (7.8 − 4.3) + 2.5

42. (8.3 − 1.2) + 3.4

43. 7.8 − (4.3 + 2.5)

44. 8.3 − (1.2 + 3.4)

45. (9.7 − 5.2) − 1.4

46. (7.8 − 3.2) − 1.5

47. 9.7 − (5.2 − 1.4)

48. 7.8 − (3.2 − 1.5)

49. Subtract 5 from the sum of 8.2 and 0.072.

50. Subtract 8 from the sum of 9.37 and 2.5.

Applying the Concepts

51. Gasoline Prices The table and bar chart below are from Problem 53 in Problem Set 5.1. Use the template below to create a line graph of the information in the table and bar chart.

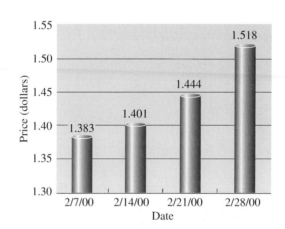

PRICE OF ONE GALLON OF REGULAR GASOLINE	
DATE	**PRICE** (dollars)
2/7/00	1.383
2/14/00	1.401
2/21/00	1.444
2/28/00	1.518

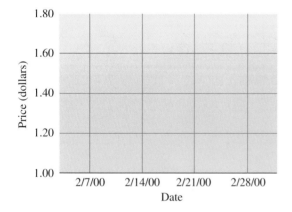

52. Speed and Time The table and bar chart below are from Problem 54 in Problem Set 5.1. Use the template below to create a line graph of the information in the table and bar chart.

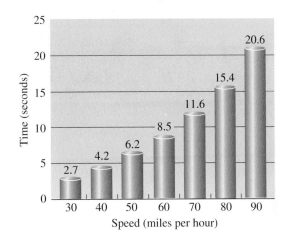

SPEED (miles per hour)	TIME (seconds)
30	2.7
40	4.2
50	6.2
60	8.5
70	11.6
80	15.4
90	20.6

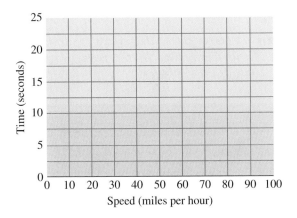

53. Shopping A family buying school clothes for their two children spends $25.37 at one store, $39.41 at another, and $52.04 at a third store. What is the total amount spent at the three stores?

54. Fund Raising A 4-H Club member is raising a lamb to take to the county fair. If she spent $75 for the lamb, $25.60 for feed, and $35.89 for shearing tools, what was the total cost of the project?

55. Wages A waiter making $5.75 per hour is given a raise to $6.04 per hour. How much is the raise?

56. Wages A person making $5.19 per hour is given a raise to $5.72 per hour. How much is the raise?

57. Take-Home Pay A college professor making $2,105.96 per month has deducted from her check $311.93 for federal income tax, $158.21 for retirement, and $64.72 for state income tax. How much does the professor take home after the deductions have been taken from her monthly income?

58. Take-Home Pay A cook making $1,504.75 a month has deductions of $157.32 for federal income tax, $58.52 for Social Security, and $45.12 for state income tax. How much does the cook take home after the deductions have been taken from his check?

59. Change A person buys $4.57 worth of candy. If he pays for the candy with a $10 bill, how much change should he receive?

60. Change If a person buys $15.37 worth of groceries and pays for them with a $20 bill, how much change should she receive?

61. Sale Price A three-piece suit that usually sells for $179.95 is marked down by $25.83. What is the new price of the suit?

62. Sale Price A dress that usually sells for $49.95 is marked down by $9.99. What is the new price of the dress?

Reduced!!!
Top quality suit
Was $179.95
NOW ONLY
¢

63. Checking Account A checking account contains $42.35. If checks are written for $9.25, $3.77, and $24.50, how much money is left in the account?

64. Checking Account A checking account contains $342.38. If checks are written for $25.04, $36.71, and $210, how much money is left in the account?

		RECORD ALL CHARGES OR CREDITS THAT AFFECT YOUR ACCOUNT							BALANCE	
NUMBER	DATE	DESCRIPTION OF TRANSACTION	PAYMENT/DEBIT (-)	√ T	FEE (IF ANY) (+)	DEPOSIT/CREDIT (+)	$			
	2/8	Deposit	$		$	$ 342 38		342 38		
1457	2/8	Woolworth's	25 04							
1458	2/9	Walgreen's	36 71							
1459	2/11	Electric Company	210 00				?			

65. Car Speeds The table below gives the number of seconds it takes various cars to accelerate from 0 miles per hour to 60 miles per hour. Construct a bar chart from the information in the table. Then find the difference in elapsed time between the slowest car and the fastest car.

CAR	TIME (IN SECONDS)
BMW M3	5.4
Chrysler Cirrus LXi	9.4
Ford Contour SE	9.5
Honda Civic EX	8.8
Nissan Sentra GLE	11.0
Toyota Corolla DX	10.2

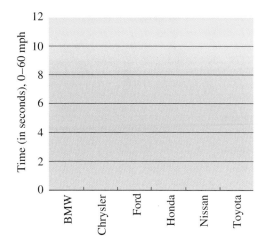

66. Car Speeds The table below gives the number of seconds it takes various sport-utility vehicles to accelerate from 0 miles per hour to 60 miles per hour. Construct a bar chart from the information in the table. Then find the difference in elapsed time between the slowest vehicle and the fastest vehicle.

CAR	TIME (IN SECONDS)
Chevrolet Blazer	9.1
Ford Explorer	10.7
Isuzu Trooper	10.9
Jeep Cherokee	9.7
Nissan Pathfinder	12.3
Toyota 4Runner	15.7

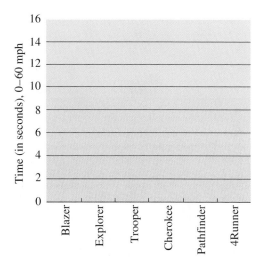

Estimating

67. Which number below is closest to $0.005 + 0.09$?
   a. 1.0   **b.** 0.1   **c.** 0.01   **d.** 0.001

68. Which number below is closest to $0.05 + 0.009$?
   a. 1.0   **b.** 0.1   **c.** 0.01   **d.** 0.001

Calculator Problems

Work the following problems on your calculator.

69. $39.0715 + 6.5498 + 173.4629$

70. $5.00073 - 3.98995$

71. $(28.40962 + 77.35674) - 36.42099$

72. $(37.929394 - 16.50732) - 18.79765$

73. Add: 279.8024
          429.07
           22.9009
        $+304.021$

74. Add: 46.35772
           4.0404
          32.6262
        $+\ 0.0457$

Review Problems

To understand how to multiply decimals, we need to understand multiplication with whole numbers and fractions. The following problems review these two concepts from Sections 1.7, 2.4, and 3.3.

Multiply.

75. $\dfrac{1}{10} \cdot \dfrac{3}{10}$

76. $-\dfrac{5}{10} \cdot \dfrac{6}{10}$

77. $\dfrac{3}{100} \cdot \dfrac{17}{100}$

78. $\dfrac{7}{100} \cdot \dfrac{31}{100}$

79. $-5\left(-\dfrac{3}{10}\right)$

80. $7 \cdot \dfrac{7}{10}$

81. $56 \cdot 25$

82. $39(-48)$

Extending the Concepts

83. Geometry A rectangle has a perimeter of 9.5 inches. If the length is 2.75 inches, find the width.

84. Geometry A rectangle has a perimeter of 11 inches. If the width is 2.5 inches, find the length.

85. Change Suppose you eat dinner in a restaurant and the bill comes to $16.76. If you give the cashier a $20 bill and a penny, how much change should you receive? List the bills and coins you should receive for change.

86. Change Suppose you buy some tools at the hardware store and the bill comes to $37.87. If you give the cashier two $20 bills and 2 pennies, how much change should you receive? List the bills and coins you should receive for change.

Find the next number in each sequence.

87. 2.5, 2.75, 3,...

88. 3.125, 3.375, 3.625,...

5.3 Multiplication with Decimals; Circumference and Area of a Circle

Introduction . . .

During a half-price sale a calendar that usually sells for $6.42 is priced at $3.21. Therefore it must be true that

$$\frac{1}{2} \text{ of } 6.42 \text{ is } 3.21$$

But, because $\frac{1}{2}$ can be written as 0.5 and *of* translates to *multiply*, we can write this problem again as

$$0.5 \times 6.42 = 3.21$$

If we were to ignore the decimal points in this problem and simply multiply 5 and 642, the result would be 3,210. So, multiplication with decimal numbers is similar to multiplication with whole numbers. The difference lies in deciding where to place the decimal point in the answer. To find out how this is done, we can use fraction notation.

EXAMPLE 1 Change each decimal to a fraction and multiply:

0.5×0.3 — **To indicate multiplication we are using a × sign here instead of a dot so we won't confuse the decimal points with the multiplication symbol.**

SOLUTION Changing each decimal to a fraction and multiplying, we have

$$0.5 \times 0.3 = \frac{5}{10} \times \frac{3}{10} \quad \textbf{Change to fractions}$$

$$= \frac{15}{100} \quad \textbf{Multiply numerators and multiply denominators}$$

$$= 0.15 \quad \textbf{Write the answer in decimal form}$$

The result is 0.15, which has two digits to the right of the decimal point. ■

What we want to do now is find a shortcut that will allow us to multiply decimals without first having to change each decimal number to a fraction. Let's look at another example.

EXAMPLE 2 Change each decimal to a fraction and multiply: 0.05×0.003

SOLUTION
$$0.05 \times 0.003 = \frac{5}{100} \times \frac{3}{1,000} \quad \textbf{Change to fractions}$$

$$= \frac{15}{100,000} \quad \textbf{Multiply numerators and multiply denominators}$$

$$= 0.00015 \quad \textbf{Write the answer in decimal form}$$

The result is 0.00015, which has a total of five digits to the right of the decimal point. ■

Looking over these first two examples, we can see that the digits in the result are just what we would get if we simply forgot about the decimal points and

Practice Problems

1. Change each decimal to a fraction and multiply:
   0.4×0.6
   Write your answer as a decimal.

2. Change each decimal to a fraction and multiply:
   0.5×0.007
   Write your answer as a decimal.

Answers

1. 0.24   2. 0.0035

multiplied; that is, $3 \times 5 = 15$. The decimal point in the result is placed so that the total number of digits to its right is the same as the total number of digits to the right of both decimal points in the original two numbers. The reason this is true becomes clear when we look at the denominators after we have changed from decimals to fractions.

3. Change to fractions and multiply:
3.5×0.04

EXAMPLE 3   Multiply: 2.1×0.07

SOLUTION $2.1 \times 0.07 = 2\dfrac{1}{10} \times \dfrac{7}{100}$ **Change to fractions**

$\qquad\qquad\qquad = \dfrac{21}{10} \times \dfrac{7}{100}$

$\qquad\qquad\qquad = \dfrac{147}{1,000}$ **Multiply numerators and multiply denominators**

$\qquad\qquad\qquad = 0.147$ **Write the answer as a decimal**

Again, the digits in the answer come from multiplying $21 \times 7 = 147$. The decimal point is placed so that there are three digits to its right, because that is the total number of digits to the right of the decimal points in 2.1 and 0.07.   ■

We summarize this discussion with a rule.

RULE

To multiply two decimal numbers:

1. Multiply as you would if the decimal points were not there.
2. Place the decimal point in the answer so that the number of digits to its right is equal to the total number of digits to the right of the decimal points in the original two numbers in the problem.

4. How many digits will be to the right of the decimal point in the following product?
3.705×55.88

EXAMPLE 4   How many digits will be to the right of the decimal point in the following product?

$$2.987 \times 24.82$$

SOLUTION   There are three digits to the right of the decimal point in 2.987 and two digits to the right in 24.82. Therefore, there will be $3 + 2 = 5$ digits to the right of the decimal point in their product.   ■

5. Multiply: 4.03×5.22

EXAMPLE 5   Multiply: 3.05×4.36

SOLUTION   We can set this up as if it were a multiplication problem with whole numbers. We multiply and then place the decimal point in the correct position in the answer.

$$
\begin{array}{r}
3.05 \leftarrow \textbf{2 digits to the right of decimal point} \\
\times 4.36 \leftarrow \textbf{2 digits to the right of decimal point} \\
\hline
1830 \\
915 \\
12\,20 \\
\hline
13.2980
\end{array}
$$

↑
└──**The decimal point is placed so that there are 2 + 2 = 4 digits to its right**   ■

Answers
3. 0.14   **4.** 5   **5.** 21.0366

As you can see, multiplying decimal numbers is just like multiplying whole numbers, except that we must place the decimal point in the result in the correct position.

EXAMPLE 6   Multiply: $-1.3(-5.6)$

SOLUTION   In this case we are multiplying two negative numbers. To do so we simply multiply their absolute values. The answer is positive, because the original two numbers had the same sign.

$$-1.3(-5.6) = 7.28$$ ∎

EXAMPLE 7   Multiply: $4.56(-100)$

SOLUTION   The product of two numbers with different signs is negative.

$$4.56(-100) = -456$$ ∎

Estimating

Look back to Example 5. We could have placed the decimal point in the answer by rounding the two numbers to the nearest whole number and then multiplying them. Because 3.05 rounds to 3 and 4.36 rounds to 4, and the product of 3 and 4 is 12, we estimate that the answer to 3.05×4.36 will be close to 12. We then place the decimal point in the product 132980 between the 3 and the 2 in order to make it into a number close to 12.

EXAMPLE 8   Estimate the answer to each of the following products.
  a. 29.4×8.2 **b.** 68.5×172 **c.** $(6.32)^2$

SOLUTION   **a.** Because 29.4 is approximately 30 and 8.2 is approximately 8, we estimate this product to be about $30 \times 8 = 240$. (If we were to multiply 29.4 and 8.2, we would find the product to be exactly 241.08.)
  b. Rounding 68.5 to 70 and 172 to 170, we estimate this product to be $70 \times 170 = 11,900$. (The exact answer is 11,782.) Note here that we do not always round the numbers to the nearest whole number when making estimates. The idea is to round to numbers that will be easy to multiply.
  c. Because 6.32 is approximately 6 and $6^2 = 36$, we estimate our answer to be close to 36. (The actual answer is 39.9424.)   ∎

Combined Operations

We can use the rule for order of operations to simplify expressions involving decimal numbers and addition, subtraction, and multiplication.

EXAMPLE 9   Perform the indicated operations: $0.05(4.2 + 0.03)$

SOLUTION   We begin by adding inside the parentheses:

$$0.05(4.2 + 0.03) = 0.05(4.23) \quad \textbf{Add}$$
$$= 0.2115 \quad \textbf{Multiply}$$

6. Multiply: $1.3(-5.6)$

7. Multiply: $-4.56(-100)$

8. Estimate the answer to each product.
  a. 82.3×5.8
  b. 37.5×178
  c. $(8.21)^2$

9. Perform the indicated operations: $0.03(5.5 + 0.02)$

Answers
6. -7.28 **7.** 456
8. a. 480   **b.** 7,200   **c.** 64
9. 0.1656

Notice that we could also have used the distributive property first, and the result would be unchanged:

$$0.05(4.2 + 0.03) = 0.05(4.2) + 0.05(0.03) \quad \textbf{Distributive property}$$
$$= 0.210 + 0.0015 \quad \textbf{Multiply}$$
$$= 0.2115 \quad \textbf{Add} \quad ∎$$

10. Simplify: $5.7 + 14(2.4)^2$

EXAMPLE 10 Simplify: $4.8 + 12(3.2)^2$

SOLUTION According to the rule for order of operations, we must first evaluate the number with an exponent, then multiply, and finally add.

$$4.8 + 12(3.2)^2 = 4.8 + 12(10.24) \quad \textbf{(3.2)}^2 = \textbf{10.24}$$
$$= 4.8 + 122.88 \quad \textbf{Multiply}$$
$$= 127.68 \quad \textbf{Add} \quad ∎$$

Applications

11. How much will the phone company in Example 11 charge for a 20-minute call?

EXAMPLE 11 Suppose a phone company charges $0.43 for the first minute and then $0.32 for each additional minute for a long-distance phone call. How much will it cost for a 10-minute call?

SOLUTION For this 10-minute call, the first minute will cost $0.43 and the next 9 minutes will cost $0.32 each. The total cost of the call will be

Cost of the first minute   Cost of the next 9 minutes

$$0.43 + 9(0.32) = 0.43 + 2.88$$
$$= 3.31$$

The call will cost $3.31.   ∎

12. How much will Sally make if she works 50 hours in one week?

EXAMPLE 12 Sally earns $6.32 for each of the first 36 hours she works in one week and $9.48 in overtime pay for each additional hour she works in the same week. How much money will she make if she works 42 hours in one week?

SOLUTION The difference between 42 and 36 is 6 hours of overtime pay. The total amount of money she will make is

Pay for the first 36 hours   Pay for the next 6 hours

$$6.32(36) + 9.48(6) = 227.52 + 56.88$$
$$= 284.40$$

Note
To estimate the answer to Example 12 before doing the actual calculations, we would do the following:
$6(40) + 9(6) = 240 + 54 = 294$

She will make $284.40 for working 42 hours in one week.   ∎

Facts from Geometry: The Circumference of a Circle

The *circumference* of a circle is the distance around the outside, just as the perimeter of a polygon is the distance around the outside. The circumference of a circle can be found by measuring its radius or diameter and then using the appropriate formula. The *radius* of a circle is the distance from the center of the circle to the circle itself. The radius is denoted by the letter r. The *diameter* of a circle is the distance from one side to the other, through the center. The diameter is denoted by the letter d. In Figure 1 we can see that the diameter is twice the radius, or

$$d = 2r$$

Answers
10. 86.34   **11.** $6.51
12. $360.24

The relationship between the circumference and the diameter or radius is not as obvious. As a matter of fact, it takes some fairly complicated mathematics to show just what the relationship between the circumference and the diameter is.

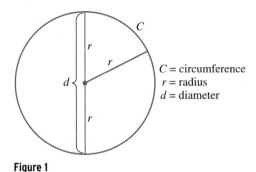

C = circumference
r = radius
d = diameter

Figure 1

If you took a string and actually measured the circumference of a circle by wrapping the string around the circle and then measured the diameter of the same circle, you would find that the ratio of the circumference to the diameter, C/d, would be approximately equal to 3.14. The actual ratio of C to d in any circle is an irrational number. It can't be written in decimal form. We use the symbol π (Greek pi) to represent this ratio. In symbols the relationship between the circumference and the diameter in any circle is

$$\frac{C}{d} = \pi$$

Knowing what we do about the relationship between division and multiplication, we can rewrite this formula as

$$C = \pi d$$

This is the formula for the circumference of a circle. When we do the actual calculations, we will use the approximation 3.14 for π.

Because $d = 2r$, the same formula written in terms of the radius is

$$C = 2\pi r$$

Here are some examples that show how we use the formulas given above to find the circumference of a circle.

EXAMPLE 13 Find the circumference of a circle with a diameter of 5 feet.

SOLUTION   Substituting 5 for d in the formula $C = \pi d$, and using 3.14 for π, we have

$$C = 3.14(5)$$
$$= 15.7 \text{ feet}$$ ∎

EXAMPLE 14 Find the circumference of a circle with a radius of 8 centimeters.

SOLUTION   We use the formula $C = 2\pi r$ this time and substitute 8 for r to get

$$C = 2(3.14)8 = 50.24 \text{ centimeters}$$ ∎

13. Find the circumference of a circle with a diameter of 3 centimeters.

14. Find the circumference of a circle with a radius of 5 feet.

Answers
13. 9.42 cm   **14.** 31.4 ft

Facts from Geometry: Other Formulas Involving π

Two figures are presented here, along with some important formulas that are associated with each figure. As you can see, each of the formulas contains the number π.

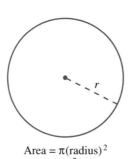

Area = π(radius)^2
$A = \pi r^2$

Figure 2   Circle

Volume = π(radius)^2(height)
$V = \pi r^2 h$

Figure 3   Right circular cylinder

15. Find the area of a circle with a diameter of 20 feet.

EXAMPLE 15 Find the area of a circle with a diameter of 10 feet.

SOLUTION   The formula for the area of a circle is $A = \pi r^2$. Because the radius r is half the diameter and the diameter is 10 feet, the radius is 5 feet. Therefore,

$$A = \pi r^2 = (3.14)(5)^2 = (3.14)(25) = 78.5 \text{ ft}^2 \qquad \blacksquare$$

16. Find the volume of the straw in Example 16, if the radius is doubled. Round your answer to the nearest thousandth.

EXAMPLE 16 The drinking straw shown in Figure 4 has a radius of 0.125 inch and a length of 6 inches. To the nearest thousandth, find the volume of liquid that it will hold.

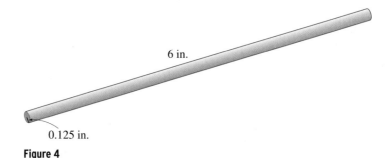

6 in.

0.125 in.

Figure 4

SOLUTION   The total volume is found from the formula for the volume of a right circular cylinder. In this case, the radius is $r = 0.125$, and the height is $h = 6$. We approximate π with 3.14.

$$\begin{aligned}
V &= \pi r^2 h \\
&= (3.14)(0.125)^2(6) \\
&= (3.14)(0.015625)(6) \\
&= 0.294 \text{ in}^3 \text{ to the nearest thousandth} \qquad \blacksquare
\end{aligned}$$

Problem Set 5.3

Find each of the following products. (Multiply.)

1. 0.7
×0.4

2. 0.8
×0.3

3. 0.07
× 0.4

4. 0.8
×0.03

5. 0.03
×0.09

6. 0.07
×0.002

7. $-2.6(0.3)$

8. $-8.9(0.2)$

9. 0.9
×0.88

10. 0.8
×0.99

11. 3.12
×0.005

12. 4.69
×0.006

13. 4.003
× 6.07

14. 7.0001
× 3.04

15. $-5(0.006)$

16. $-7(0.005)$

17. 75.14
× 2.5

18. 963.8
× 0.24

19. 0.1
×0.02

20. 0.3
×0.02

21. $2.796(-10)$

22. $97.531(-100)$

23. 0.0043
× 100

24. 12.345
× 1,000

25. 49.94
×1,000

26. 157.02
×10,000

27. 987.654
× 10,000

28. 1.23
×100,000

Perform the following operations according to the rule for order of operations.

29. $2.1(3.5 - 2.6)$

30. $5.4(9.9 - 6.6)$

31. $-0.05(0.02 + 0.03)$

32. $-0.04(0.07 + 0.09)$

33. $2.02(0.03 - 2.5)$

34. $4.04(0.05 - 6.6)$

35. $(2.1 + 0.03)(3.4 + 0.05)$

36. $(9.2 + 0.01)(3.5 + 0.03)$

37. $(2.1 - 0.1)(2.1 + 0.1)$

38. $(9.6 - 0.5)(9.6 + 0.5)$

39. $3.08 - 0.2(5 + 0.03)$

40. $4.09 + 0.5(6 + 0.02)$

41. $4.23 - 5(0.04 + 0.09)$

42. $7.89 - 2(0.31 + 0.76)$

43. $2.5 + 10(4.3)^2$

44. $3.6 + 15(2.1)^2$

45. $100(1 + 0.08)^2$

46. $500(1 + 0.12)^2$

47. $(1.5)^2 + (2.5)^2 + (3.5)^2$

48. $(1.1)^2 + (2.1)^2 + (3.1)^2$

Applying the Concepts

Solve each of the following word problems. Note that not all of the problems are solved by simply multiplying the numbers in the problems. Many of the problems involve addition and subtraction as well as multiplication.

49. Number Problem What is the product of 6 and the sum of 0.001 and 0.02?

50. Number Problem Find the product of 8 and the sum of 0.03 and 0.002.

51. Number Problem What does multiplying a decimal number by 100 do to the decimal point?

52. Number Problem What does multiplying a decimal number by 1,000 do to the decimal point?

53. Price of Beef Ground beef is priced at $1.98 per pound. How much will it cost for 3.5 pounds?

54. Caffeine Content If 1 cup of regular coffee contains 105 milligrams of caffeine, how much caffeine is contained in 3.5 cups of coffee?

55. Price of Gasoline If gasoline costs 119.9¢ per gallon, how much does it cost to fill an 11-gallon tank?

56. Price of Gasoline If gasoline costs $1.37 per gallon when you pay with a credit card, but $0.06 per gallon less if you pay with cash, how much do you save by filling up a 12-gallon tank and paying for it with cash?

57. Home Mortgage On a certain home mortgage, there is a monthly payment of $9.66 for every $1,000 that is borrowed. What is the monthly payment on this type of loan if $43,000 is borrowed?

58. Home Mortgage If the loan mentioned in Problem 57 is for $53,000, how much more will the payments be each month?

59. Long-Distance Charges If a phone company charges $0.45 for the first minute and $0.35 for each additional minute for a long-distance call, how much will a 20-minute long-distance call cost?

60. Long-Distance Charges Suppose a phone company charges $0.53 for the first minute and $0.27 for each additional minute for a long-distance call. How much will a 15-minute long-distance call cost?

61. Car Rental Suppose it costs $15 per day and $0.12 per mile to rent a car. What is the total bill if a car is rented for 2 days and is driven 120 miles?

62. Car Rental Suppose it costs $20 per day and $0.08 per mile to rent a car. What is the total bill if the car is rented for 2 days and is driven 120 miles?

63. Wages A man earns $5.92 for each of the first 36 hours he works in one week and $8.88 in overtime pay for each additional hour he works in the same week. How much money will he make if he works 45 hours in one week?

64. Wages A student earns $8.56 for each of the first 40 hours she works in one week and $12.84 in overtime pay for each additional hour she works in the same week. How much money will she make if she works 44 hours in one week?

Circumference Find the circumference and the area of each circle. Use 3.14 for π.

65.

4 in.

66.

2 in.

67. Circumference A dinner plate has a radius of 6 inches. Find the circumference and the area.

68. Circumference A salad plate has a radius of 3 inches. Find the circumference and the area.

69. Circumference The radius of the earth is approximately 3,900 miles. Find the circumference of the earth at the equator. (The equator is a circle around the earth that divides the earth into two equal halves.)

70. Circumference The radius of the moon is approximately 1,100 miles. Find the circumference of the moon around its equator.

Volume Find the volume of each cylinder.

71.

8 ft

2 ft

72.

8 ft

4 ft

73.

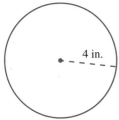

4 ft

2 ft

74.

4 ft

4 ft

Calculator Problems

Find each of the following products by using a calculator.

75. $(429.5)(246.7)$

76. $(0.0799)(8.437)$

77. $(6.59)(7.32)(4.2)$

78. $(0.04)(0.35)(0.172)$

79. $(0.75)^3$

80. $(7.5)^3$

81. $(1.11)^2$

82. $(0.111)^2$

Review Problems

To get ready for the next section, which covers division with decimals, we will review division with whole numbers and fractions from Sections 1.8 and 3.4.

Perform each of the following divisions. (Find the quotients.)

83. $3{,}758 \div 2$

84. $9{,}900 \div 22$

85. $50{,}032 \div 33$

86. $90{,}902 \div 5$

87. $\dfrac{3}{10} \div \dfrac{4}{10}$

88. $\dfrac{6}{10} \div \dfrac{3}{100}$

89. $-\dfrac{6}{100} \div \left(-\dfrac{3}{10}\right)$

90. $\dfrac{27}{100} \div \left(-\dfrac{9}{10}\right)$

91. $\dfrac{45}{100} \div \dfrac{5}{1{,}000}$

92. $\dfrac{225}{1{,}000} \div \dfrac{25}{1{,}000}$

93. $\dfrac{x^3}{y^5} \div \dfrac{x}{y^8}$

94. $\dfrac{x^4}{y^3} \div \dfrac{x}{y^9}$

95. $\dfrac{a^2b^3}{c^4} \div \dfrac{a^2b}{c^7}$

96. $\dfrac{a^3b^2}{c} \div \dfrac{a^3b}{c^4}$

97. $\dfrac{10x^5}{21y^3} \div \dfrac{5x^3}{42y^5}$

98. $\dfrac{15x^7}{14y^3} \div \dfrac{3x^5}{28y^8}$

Extending the Concepts: Number Sequences

Each sequence below is either arithmetic or geometric. Find the next number in each sequence, and then give a written description of the sequence. For example, the sequence 0.5, 0.6, 0.7, . . . can be described as follows: "An arithmetic sequence, starting with 0.5, in which each term is found by adding 0.1 to the term before it."

99. 0.1, 0.2, 0.3, . . .

100. 5.2, 6.3, 7.4, . . .

101. 0.1, 0.01, 0.001, . . .

102. 50, 5, 0.5, . . .

A sequence of numbers is referred to as *increasing* if each term is larger than the term that came before it. Likewise, a sequence is *decreasing* if each term is smaller than the term that came before it. Simplify each term in the following sequences, and then identify the sequence as increasing or decreasing.

103. $(0.2)^2, (0.2)^3, (0.2)^4, \ldots$

104. $(1.2)^2, (1.2)^3, (1.2)^4, \ldots$

105. $1.1, (1.1)^2, (1.1)^3, \ldots$

106. $0.1, (0.1)^2, (0.1)^3, \ldots$

5.4 Division with Decimals

Introduction . . .

Suppose three friends go out for lunch and their total bill comes to $18.75. If they decide to split the bill equally, how much does each person owe? To find out, they will have to divide 18.75 by 3. If you think about the dollars and cents separately, you will see that each person owes $6.25. Therefore,

$$\$18.75 \div 3 = \$6.25$$

In this section we will find out how to do division with any combination of decimal numbers.

EXAMPLE 1 Divide: $5,974 \div 20$

SOLUTION

```
       298
  20)5,974
     4 0
     1 97
     1 80
       174
       160
        14
```

In the past we have written this answer as $298\frac{14}{20}$ or, after reducing the fraction, $298\frac{7}{10}$. Because $\frac{7}{10}$ can be written as 0.7, we could also write our answer as 298.7. This last form of our answer is exactly the same result we obtain if we write 5,974 as 5,974.0 and continue the division until we have no remainder. Here is how it looks:

```
       298.7
  20)5,974.0
     4 0
     1 97
     1 80
       174
       160
        14 0
        14 0
           0
```

Notice that we place the decimal point in the answer directly above the decimal point in the problem

∎

Let's try another division problem. This time one of the numbers in the problem will be a decimal.

Practice Problems
1. Divide: $4,626 \div 30$

Note
We can estimate the answer to Example 1 by rounding 5,974 to 6,000 and dividing by 20:
$$\frac{6,000}{20} = 300$$

Answer
1. 154.2

2. Divide: $33.5 \div 5$

Note

We never need to make a mistake with division, because we can always check our results with multiplication.

EXAMPLE 2   Divide: $34.8 \div 4$

SOLUTION   We can use the ideas from Example 1 and divide as usual. The decimal point in the answer will be placed directly above the decimal point in the problem.

$$
\begin{array}{r}
8.7 \\
4\overline{)34.8} \\
\underline{32} \\
2\,8 \\
\underline{2\,8} \\
0
\end{array}
$$

Check:
$$
\begin{array}{r}
8.7 \\
\times\ \ \ 4 \\
\hline
34.8
\end{array}
$$

The answer is 8.7.   ■

We can use these facts to write a rule for dividing decimal numbers.

> **RULE**
>
> To divide a decimal by a whole number, we do the usual long division as if there were no decimal point involved. The decimal point in the answer is placed directly above the decimal point in the problem.

Here are some more examples to illustrate the procedure.

3. Divide: $47.448 \div 18$

EXAMPLE 3   Divide: $49.896 \div 27$

SOLUTION
$$
\begin{array}{r}
1.848 \\
27\overline{)49.896} \\
\underline{27} \\
22\,8 \\
\underline{21\,6} \\
1\,29 \\
\underline{1\,08} \\
216 \\
\underline{216} \\
0
\end{array}
$$

Check this result by multiplication:

$$
\begin{array}{r}
1.848 \\
\times\ \ \ 27 \\
\hline
12\,936 \\
36\,96 \\
\hline
49.896
\end{array}
$$

■

We can write as many zeros as we chose after the rightmost digit in a decimal number without changing the value of the number. For example,

$$6.91 = 6.910 = 6.9100 = 6.91000$$

There are times when this can be very useful, as Example 4 shows.

EXAMPLE 4 Divide: 1,138.9 ÷ 35

SOLUTION

$$
\begin{array}{r}
32.54 \\
35{\overline{\smash{\big)}\,1{,}138.90}} \\
\underline{1\ 05} \\
88 \\
\underline{70} \\
18\ 9 \\
\underline{17\ 5} \\
1\ 40 \\
\underline{1\ 40} \\
0
\end{array}
$$

Write 0 after the 9. It doesn't change the original number, but it gives us another digit to bring down.

Check:

$$
\begin{array}{r}
32.54 \\
\times\ \ \ \ 35 \\
\hline
162\ 70 \\
976\ 2 \\
\hline
1{,}138.90
\end{array}
$$

■

4. Divide: 1,138.5 ÷ 25

Until now we have considered only division by whole numbers. Extending division to include division by decimal numbers is a matter of knowing what to do about the decimal point in the divisor.

EXAMPLE 5 Divide: 31.35 ÷ 3.8

SOLUTION   In fraction form, this problem is equivalent to

$$\frac{31.35}{3.8}$$

If we want to write the divisor as a whole number, we can multiply the numerator and the denominator of this fraction by 10:

$$\frac{31.35 \times \mathbf{10}}{3.8 \times \mathbf{10}} = \frac{313.5}{38}$$

So, since this fraction is equivalent to the original fraction, our original division problem is equivalent to

$$
\begin{array}{r}
8.25 \\
38{\overline{\smash{\big)}\,313.50}} \\
\underline{304} \\
9\ 5 \\
\underline{7\ 6} \\
1\ 90 \\
\underline{1\ 90} \\
0
\end{array}
$$

Put 0 after the last digit

■

5. Divide: 13.23 ÷ 4.2

Note
We do not always use the rules for rounding numbers to make estimates. For example, to estimate the answer to Example 5, 31.35 ÷ 3.8, we can get a rough estimate of the answer by reasoning that 3.8 is close to 4 and 31.35 is close to 32. Therefore, our answer will be approximately 32 ÷ 4 = 8.

We can summarize division with decimal numbers by listing the following points, as illustrated by the first five examples.

Summary of Division with Decimals

1. We divide decimal numbers by the same process used in Chapter 1 to divide whole numbers. The decimal point in the answer is placed directly above the decimal point in the dividend.
2. We are free to write as many zeros after the last digit in a decimal number as we need.
3. If the divisor is a decimal, we can change it to a whole number by moving the decimal point to the right as many places as necessary so long as we move the decimal point in the dividend the same number of places.

Answers
4. 45.54 **5.** 3.15

6. Divide, and round your answer to
the nearest hundredth:
0.4553 ÷ 0.32

Note

Moving the decimal point two places
in both the divisor and the dividend is
justified like this:

$$\frac{0.3778 \times \mathbf{100}}{0.25 \times \mathbf{100}} = \frac{37.78}{25}$$

EXAMPLE 6   Divide, and round the answer to the nearest hundredth:

0.3778 ÷ 0.25

SOLUTION   First, we move the decimal point two places to the right:

0.25.‾)‾3̲7̲.̲7̲8̲

Then we divide, using long division:

```
        1.5112
  25)37.7800
     25
     ───
     12 7
     12 5
     ────
        28
        25
        ──
        30
        25
        ──
        50
        50
        ──
         0
```

Rounding to the nearest hundredth, we have 1.51. We actually did not need to
have this many digits to round to the hundredths column. We could have stopped
at the thousandths column and rounded off.   ■

7. Divide, and round to the nearest
tenth:
19 ÷ 0.06

EXAMPLE 7   Divide, and round to the nearest tenth: 17 ÷ 0.03

SOLUTION   Because we are rounding to the nearest tenth, we will continue dividing until we have a digit in the hundredths column. We don't have to go any
further to round to the tenths column.

```
        5 66.66
  0.03.)17.00.00
        15
        ──
         2 0
         1 8
         ───
           20
           18
           ──
            2 0
            1 8
            ───
              20
              18
              ──
               2
```

Rounding to the nearest tenth, we have 566.7.   ■

Answers
6. 1.42   **7.** 316.7

EXAMPLE 8 If a man earning $5.26 an hour receives a paycheck for $170.95, how many hours did he work?

SOLUTION To find the number of hours the man worked, we divide $170.95 by $5.26.

$$
\begin{array}{r}
32.5 \\
5.26.\overline{)170.95.0} \\
157\ 8 \\
\overline{13\ 15} \\
10\ 52 \\
\overline{2\ 63\ 0} \\
2\ 63\ 0 \\
\overline{0}
\end{array}
$$

The man worked 32.5 hours. ∎

EXAMPLE 9 A telephone company charges $0.43 for the first minute and then $0.33 for each additional minute for a long-distance call. If a long-distance call costs $3.07, how many minutes was the call?

SOLUTION To solve this problem we need to find the number of additional minutes for the call. To do so, we first subtract the cost of the first minute from the total cost, and then we divide the result by the cost of each additional minute. Without showing the actual arithmetic involved, the solution looks like this:

Total cost **Cost of the**
of the call **first minute**

$$
\text{The number of additional minutes} = \frac{3.07 - 0.43}{0.33} = \frac{2.64}{0.33} = 8
$$

Cost of each
additional minute

The call was 9 minutes long. (The number 8 is the number of additional minutes past the first minute.) ∎

Descriptive Statistics: Grade Point Average

I have always been surprised by the number of my students who have difficulty calculating their grade point average (GPA). During her first semester in college, my daughter, Amy, earned the following grades:

CLASS	UNITS	GRADE
Algebra	5	B
Chemistry	4	C
English	3	A
History	3	B

When her grades arrived in the mail, she told me she had a 3.0 grade point average, because the A and C grades averaged to a B. I told her that her GPA was a little less than a 3.0. What do you think? Can you calculate her GPA? If not, you will be able to after you finish this section.

When you calculate your grade point average (GPA), you are calculating what is called a *weighted average*. To calculate your grade point average, you must first calculate the number of grade points you have earned in each class that you have

8. A woman earning $6.54 an hour receives a paycheck for $186.39. How many hours did the woman work?

9. If the phone company in Example 9 charged $4.39 for a call, how long was the call?

Answers
8. 28.5 hours **9.** 13 minutes

completed. The number of grade points for a class is the product of the number of units the class is worth times the value of the grade received. The table below shows the value that is assigned to each grade.

GRADE	VALUE
A	4
B	3
C	2
D	1
F	0

If you earn a B in a 4-unit class, you earn $4 \times 3 = 12$ grade points. A grade of C in the same class gives you $4 \times 2 = 8$ grade points. To find your grade point average for one term (a semester or quarter), you must add your grade points and divide that total by the number of units. Round your answer to the nearest hundredth.

10. If Amy had earned a B in chemistry, instead of a C, what grade point average would she have?

EXAMPLE 10 Calculate Amy's grade point average using the information above.

SOLUTION   We begin by writing in two more columns, one for the value of each grade (4 for an A, 3 for a B, 2 for a C, 1 for a D, and 0 for an F), and another for the grade points earned for each class. To fill in the grade points column, we multiply the number of units by the value of the grade:

CLASS	UNITS	GRADE	VALUE	GRADE POINTS
Algebra	5	B	3	$5 \times 3 = 15$
Chemistry	4	C	2	$4 \times 2 = 8$
English	3	A	4	$3 \times 4 = 12$
History	3	B	3	$3 \times 3 = 9$
Total Units:	15			Total Grade Points: 44

To find her grade point average, we divide 44 by 15 and round (if necessary) to the nearest hundredth:

$$\text{Grade point average} = \frac{44}{15} = 2.93$$

■

Problem Set 5.4

Perform each of the following divisions.

1. $394 \div 20$

2. $486 \div 30$

3. $248 \div 40$

4. $372 \div 80$

5. $5\overline{)26}$

6. $8\overline{)36}$

7. $25\overline{)276}$

8. $50\overline{)276}$

9. $28.8 \div (-6)$

10. $15.5 \div (-5)$

11. $-77.6 \div (-8)$

12. $-31.48 \div (-4)$

13. $35\overline{)92.05}$

14. $26\overline{)146.38}$

15. $45\overline{)190.8}$

16. $55\overline{)342.1}$

17. $86.7 \div 34$

18. $411.4 \div 44$

19. $-29.7 \div 22$

20. $-488.4 \div 88$

21. $4.5\overline{)29.25}$

22. $3.3\overline{)21.978}$

23. $0.11\overline{)1.089}$

24. $0.75\overline{)2.40}$

25. $2.3\overline{)0.115}$

26. $6.6\overline{)0.198}$

27. $0.012\overline{)1.068}$

28. $0.052\overline{)0.23712}$

29. $1.1\overline{)2.42}$

30. $2.2\overline{)7.26}$

Carry out each of the following divisions only so far as needed to round the results to the nearest hundredth.

31. $26\overline{)35}$　　　　**32.** $18\overline{)47}$　　　　**33.** $3.3\overline{)56}$　　　　**34.** $4.4\overline{)75}$

35. $0.1234 \div 0.5$　　**36.** $0.543 \div 2.1$　　**37.** $19 \div 7$　　　　**38.** $16 \div 6$

39. $0.059\overline{)0.69}$　　　**40.** $0.048\overline{)0.49}$

In Problems 41–44, indicate whether the statements are *True* or *False*.

41. $6.99 \div 0.3 = 6.99 \div 3$　　**42.** $3.5 \div 7 = 35 \div 70$　　**43.** $42 \div 0.06 = 420 \div 0.6$　　**44.** $75 \div 0.25 = 7.5 \div 25$

Applying the Concepts

45. If a six-pack of soft drinks costs $1.98, how many six-packs were purchased for $9.90?

46. Feed Bill Patrick owns 15 sheep. If his monthly feed bill is $103.50, how much does it cost him to feed one sheep for a month? If he buys two more sheep, how much will his monthly feed bill increase?

47. Women's Golf The table below gives the top five money earners for the Ladies' Professional Golf Association (LPGA) in 1999. Fill in the last column of the table by finding the average earning per tournament for each golfer. Round your answers to the nearest dollar.

PLAYER	NUMBER OF TOURNAMENTS	TOTAL EARNINGS	AVERAGE EARNINGS PER TOURNAMENT
Karrie Web	25	$1,591,959	
Juli Inkster	24	$1,337,256	
Si Re Pak	27	$959,926	
Annika Sorenstam	22	$863,816	
Lorie Kane	30	$757,844	

48. Men's Golf The table below gives the top five money earners for the men's Professional Golf Association (PGA) in 1999. Fill in the last column of the table by finding the average earnings per tournament for each golfer. Round your answers to the nearest dollar.

RANK	PLAYER	EVENTS	MONEY	AVERAGE PER TOURNAMENT
1	Tiger Woods	21	$6,616,585	
2	David Duval	21	$3,641,906	
3	Davis Love III	23	$2,475,328	
4	Vijay Singh	29	$2,283,233	
5	Chris Perry	31	$2,145,707	

49. Wages If a woman earns $33.90 for working 6 hours, how much does she earn per hour?

50. Wages How many hours does a person making $6.78 per hour have to work in order to earn $257.64?

51. Gas Mileage If a car travels 336 miles on 15 gallons of gas, how far will the car travel on 1 gallon of gas?

52. Gas Mileage If a car travels 392 miles on 16 gallons of gas, how far will the car travel on 1 gallon of gas?

53. Wages Suppose a woman earns $6.78 an hour for the first 36 hours she works in a week and then $10.17 an hour in overtime pay for each additional hour she works in the same week. If she makes $294.93 in one week, how many hours did she work overtime?

54. Wages Suppose a woman makes $286.08 in one week. If she is paid $5.96 an hour for the first 36 hours she works and then $8.94 an hour in overtime pay for each additional hour she works in the same week, how many hours did she work overtime that week?

55. Phone Bill Suppose a telephone company charges $0.41 for the first minute and then $0.32 for each additional minute for a long-distance call. If a long-distance call costs $2.33, how many minutes was the call?

56. Phone Bill Suppose a telephone company charges $0.45 for the first three minutes and then $0.29 for each additional minute for a long-distance call. If a long-distance call costs $2.77, how many minutes was the call?

Grade Point Average The following grades were earned by Steve during his first term in college. Use these data to answer Problems 57–60.

CLASS	UNITS	GRADE
Basic mathematics	3	A
Health	2	B
History	3	B
English	3	C
Chemistry	4	C

57. Calculate Steve's GPA.

58. If his grade in chemistry had been a B instead of a C, by how much would his GPA have increased?

59. If his grade in health had been a C instead of a B, by how much would his grade point average have dropped?

60. If his grades in both English and chemistry had been B's, what would his GPA have been?

Calculator Problems

Work each of the following problems on your calculator. If rounding is necessary, round to the nearest hundred thousandth.

61. $7 \div 9$

62. $11 \div 13$

63. $243 \div 0.791$

64. $67.8 \div 37.92$

65. $0.0503 \div 0.0709$

66. $429.87 \div 16.925$

Review Problems

In the next section we will consider the relationship between fractions and decimals in more detail. The problems below review some of the material from Sections 3.1 and 3.2 that is necessary to make a successful start in the next section.

Reduce to lowest terms.

67. $\dfrac{75}{100}$

68. $\dfrac{220}{1,000}$

69. $\dfrac{12x}{18xy}$

70. $\dfrac{15xy}{30x}$

71. $\dfrac{75x^2y^3}{100xy^2}$

72. $\dfrac{220x^3y^2}{1,000x^2y}$

Write each fraction as an equivalent fraction with denominator 10.

73. $\dfrac{3}{5}$

74. $\dfrac{1}{2}$

Write each fraction as an equivalent fraction with denominator 100.

75. $\dfrac{3}{5}$

76. $\dfrac{17}{20}$

Write each fraction as an equivalent fraction with denominator $15x$.

77. $\dfrac{4}{5}$

78. $\dfrac{2}{3}$

79. $\dfrac{4}{x}$

80. $\dfrac{2}{x}$

81. $\dfrac{6}{5x}$

82. $\dfrac{7}{3x}$

5.5 Fractions and Decimals and the Volume of a Sphere

Introduction . . .

If you are shopping for clothes and a store has a sale advertising $\frac{1}{3}$ off the regular price, how much can you expect to pay for a pair of pants that normally sells for $31.95? If the sale price of the pants is $22.30, have they really been marked down by $\frac{1}{3}$? To answer questions like these, we need to know how to solve problems that involve fractions and decimals together.

We begin this section by showing how to convert back and forth between fractions and decimals.

Converting Fractions to Decimals

You may recall that the notation we use for fractions can be interpreted as implying division. That is, the fraction $\frac{3}{4}$ can be thought of as meaning "3 divided by 4." We can use this idea to convert fractions to decimals.

EXAMPLE 1 Write $\frac{3}{4}$ as a decimal.

SOLUTION Dividing 3 by 4, we have

```
    .75
4)3.00
  2 8
   20
   20
    0
```

The fraction $\frac{3}{4}$ is equal to the decimal 0.75. ∎

EXAMPLE 2 Write $\frac{7}{12}$ as a decimal correct to the thousandths column.

SOLUTION Because we want the decimal to be rounded to the thousandths column, we divide to the ten thousandths column and round off to the thousandths column:

```
      .5833
12)7.0000
   6 0
   1 00
     96
     40
     36
     40
     36
      4
```

Practice Problems

1. Write $\frac{3}{5}$ as a decimal.

2. Write $\frac{11}{12}$ as a decimal correct to the thousandths column.

Answers
1. 0.6 2. 0.917

Rounding off to the thousandths column, we have 0.583. Because $\frac{7}{12}$ is not exactly the same as 0.583, we write

$$\frac{7}{12} \approx 0.583$$

where the symbol $\approx$ is read "is approximately." ■

If we wrote more zeros after 7.0000 in Example 2, the pattern of 3's would continue for as many places as we could want. When we get a sequence of digits that repeat like this, 0.58333 . . ., we can indicate the repetition by writing

$0.58\overline{3}$ **The bar over the 3 indicates that the 3 repeats from there on**

3. Write $\frac{5}{11}$ as a decimal.

EXAMPLE 3   Write $\frac{3}{11}$ as a decimal.

SOLUTION   Dividing 3 by 11, we have

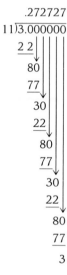

No matter how long we continue the division, the remainder will never be 0, and the pattern will continue. We write the decimal form of $\frac{3}{11}$ as $0.\overline{27}$, where

$$0.\overline{27} = 0.272727 . . .   \text{\textbf{The dots mean "and so on"}}$$ ■

Note
The bar over the 2 and the 7 in $0.\overline{27}$ is used to indicate that the pattern repeats itself indefinitely.

Converting Decimals to Fractions

To convert decimals to fractions, we take advantage of the place values we assigned to the digits to the right of the decimal point.

4. Write 0.48 as a fraction in lowest terms.

EXAMPLE 4   Write 0.38 as a fraction in lowest terms.

SOLUTION   0.38 is 38 hundredths, or

$$0.38 = \frac{38}{100}$$

$$= \frac{19}{50}   \text{\textbf{Divide the numerator and the denominator by 2 to reduce to lowest terms}}$$

The decimal 0.38 is equal to the fraction $\frac{19}{50}$.

We could check our work here by converting $\frac{19}{50}$ back to a decimal. We do this by dividing 19 by 50. That is,

$$
\begin{array}{r}
.38 \\
50\overline{)19.00} \\
\underline{15\ 0} \\
4\ 00 \\
\underline{4\ 00} \\
0
\end{array}
$$

■

EXAMPLE 5   Convert 0.075 to a fraction.

SOLUTION   We have 75 thousandths, or

$$0.075 = \frac{75}{1,000}$$

$$= \frac{3}{40} \qquad \textbf{Divide the numerator and the denominator}$$
$$\textbf{by 25 to reduce to lowest terms}$$

■

5.   Convert 0.025 to a fraction.

EXAMPLE 6   Write 15.6 as a mixed number.

SOLUTION   Converting 0.6 to a fraction, we have

$$0.6 = \frac{6}{10} = \frac{3}{5} \qquad \textbf{Reduce to lowest terms}$$

Since $0.6 = \frac{3}{5}$, we have $15.6 = 15\frac{3}{5}$.   ■

6.   Write 12.8 as a mixed number.

Problems Containing Both Fractions and Decimals

We will end this section by working some problems that involve both fractions and decimals.

EXAMPLE 7   Simplify: $\frac{19}{50}(1.32 + 0.48)$

SOLUTION   In Example 4, we found that $0.38 = \frac{19}{50}$. Therefore we can rewrite the problem as

$$\frac{19}{50}(1.32 + 0.48) = 0.38(1.32 + 0.48) \qquad \textbf{Convert all numbers to decimals}$$
$$= 0.38(1.80) \qquad\qquad\qquad \textbf{Add: 1.32 + 0.48}$$
$$= 0.684 \qquad\qquad\qquad\quad \textbf{Multiply: 0.38 × 1.80}$$

■

7.   Simplify: $\frac{14}{25}(2.43 + 0.27)$

EXAMPLE 8   Simplify: $\frac{1}{2} + (0.75)\left(\frac{2}{5}\right)$

SOLUTION   We could do this problem one of two different ways. First, we could convert all fractions to decimals and then simplify:

$$\frac{1}{2} + (0.75)\left(\frac{2}{5}\right) = 0.5 + 0.75(0.4) \qquad \textbf{Convert to decimals}$$
$$= 0.5 + 0.300 \qquad\qquad \textbf{Multiply: 0.75 × 0.4}$$
$$= 0.8 \qquad\qquad\qquad\quad \textbf{Add}$$

8.   Simplify: $\frac{1}{4} + 0.25\left(\frac{3}{5}\right)$

Answers

5. $\frac{1}{40}$ **6.** $12\frac{4}{5}$ **7.** 1.512

8. $\frac{2}{5}$, or 0.4

Or, we could convert 0.75 to $\frac{3}{4}$ and then simplify:

$$\frac{1}{2} + 0.75\left(\frac{2}{5}\right) = \frac{1}{2} + \frac{3}{4}\left(\frac{2}{5}\right) \qquad \textbf{Convert decimals to fractions}$$

$$= \frac{1}{2} + \frac{3}{10} \qquad \textbf{Multiply: } \frac{3}{4} \times \frac{2}{5}$$

$$= \frac{5}{10} + \frac{3}{10} \qquad \textbf{The common denominator is 10}$$

$$= \frac{8}{10} \qquad \textbf{Add numerators}$$

$$= \frac{4}{5} \qquad \textbf{Reduce to lowest terms}$$

The answers are equivalent. That is, $0.8 = \frac{8}{10} = \frac{4}{5}$. Either method can be used with problems of this type. ■

9. Simplify: $\left(\frac{1}{3}\right)^3 (5.4) + \left(\frac{1}{5}\right)^2 (2.5)$

EXAMPLE 9 Simplify: $\left(\frac{1}{2}\right)^3 (2.4) + \left(\frac{1}{4}\right)^2 (3.2)$

SOLUTION This expression can be simplified without any conversions between fractions and decimals. To begin, we evaluate all numbers that contain exponents. Then we multiply. After that, we add.

$$\left(\frac{1}{2}\right)^3 (2.4) + \left(\frac{1}{4}\right)^2 (3.2) = \frac{1}{8}(2.4) + \frac{1}{16}(3.2) \qquad \textbf{Evaluate exponents}$$

$$= 0.3 + 0.2 \qquad \textbf{Multiply by } \frac{1}{8} \textbf{ and } \frac{1}{16}$$

$$= 0.5 \qquad \textbf{Add}$$ ■

10. A shirt that normally sells for $35.50 is on sale for $\frac{1}{4}$ off. What is the sale price of the shirt? (Round to the nearest cent.)

EXAMPLE 10 If a shirt that normally sells for $27.99 is on sale for $\frac{1}{3}$ off, what is the sale price of the shirt?

SOLUTION To find out how much the shirt is marked down, we must find $\frac{1}{3}$ of 27.99. That is, we multiply $\frac{1}{3}$ and 27.99, which is the same as dividing 27.99 by 3.

$$\frac{1}{3}(27.99) = \frac{27.99}{3} = 9.33$$

The shirt is marked down $9.33. The sale price is the original price less the amount it is marked down:

$$\text{Sale price} = 27.99 - 9.33 = 18.66$$

The sale price is $18.66. We also could have solved this problem by simply multiplying the original price by $\frac{2}{3}$, since, if the shirt is marked $\frac{1}{3}$ off, then the sale price must be $\frac{2}{3}$ of the original price. Multiplying by $\frac{2}{3}$ is the same as dividing by 3 and then multiplying by 2. The answer would be the same. ■

Facts from Geometry: The Volume of a Sphere

Figure 1 shows a sphere and the formula for its volume. Because the formula contains both the fraction $\frac{4}{3}$ and the number π, and we have been using 3.14 for π, we can think of the formula as containing both a fraction and a decimal.

Answers
9. 0.3 **10.** $26.63

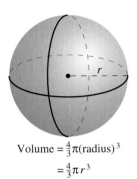

$$\text{Volume} = \tfrac{4}{3}\pi(\text{radius})^3$$
$$= \tfrac{4}{3}\pi r^3$$

Figure 1 Sphere

EXAMPLE 11 Figure 2 is composed of a right circular cylinder with half a sphere on top. (A half-sphere is called a *hemisphere*.) To the nearest tenth, find the total volume enclosed by the figure.

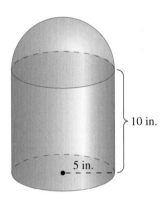

10 in.

5 in.

Figure 2

SOLUTION The total volume is found by adding the volume of the cylinder to the volume of the hemisphere.

$$V = \text{volume of cylinder} + \text{volume of hemisphere}$$

$$= \pi r^2 h + \frac{1}{2} \cdot \frac{4}{3} \pi r^3$$

$$= (3.14)(5)^2(10) + \frac{1}{2} \cdot \frac{4}{3}(3.14)(5)^3$$

$$= (3.14)(25)(10) + \frac{1}{2} \cdot \frac{4}{3}(3.14)(125)$$

$$= 785 + \frac{2}{3}(392.5) \quad \textbf{Multiply: } \frac{1}{2} \cdot \frac{4}{3} = \frac{4}{6} = \frac{2}{3}$$

$$= 785 + \frac{785}{3} \quad \textbf{Multiply: 2(392.5) = 785}$$

$$= 785 + 261.7 \quad \textbf{Divide 785 by 3, and round to the nearest tenth}$$

$$= 1{,}046.7 \text{ in}^3 \qquad\qquad\blacksquare$$

11. If the radius in Figure 2 is doubled so that it becomes 10 inches instead of 5 inches, what is the new volume of the figure? Round your answer to the nearest tenth.

Answer
11. 5,233.3 in³

Problem Set 5.5

Each circle below is divided into 8 equal parts. The number below each circle indicates what fraction of the circle is shaded. Convert each fraction to a decimal.

1.

$$\frac{1}{8}$$

2.

$$\frac{3}{8}$$

3.

$$\frac{5}{8}$$

4.

$$\frac{7}{8}$$

Complete the following tables by converting each fraction to a decimal.

5.

Fraction	$\frac{1}{4}$	$\frac{2}{4}$	$\frac{3}{4}$	$\frac{4}{4}$
Decimal				

6.

Fraction	$\frac{1}{5}$	$\frac{2}{5}$	$\frac{3}{5}$	$\frac{4}{5}$	$\frac{5}{5}$
Decimal					

7.

Fraction	$\frac{1}{6}$	$\frac{2}{6}$	$\frac{3}{6}$	$\frac{4}{6}$	$\frac{5}{6}$	$\frac{6}{6}$
Decimal						

Convert each of the following fractions to a decimal.

8. $\dfrac{1}{2}$ **9.** $\dfrac{12}{25}$ **10.** $\dfrac{14}{25}$ **11.** $\dfrac{14}{32}$ **12.** $\dfrac{18}{32}$

Write each fraction as a decimal correct to the hundredths column.

13. $\dfrac{12}{13}$ **14.** $\dfrac{17}{19}$ **15.** $\dfrac{3}{11}$ **16.** $\dfrac{5}{11}$

17. $\dfrac{2}{23}$ **18.** $\dfrac{3}{28}$ **19.** $\dfrac{12}{43}$ **20.** $\dfrac{15}{51}$

Complete the following table by converting each decimal to a fraction.

21.

Decimal	0.125	0.250	0.375	0.500	0.625	0.750	0.875
Fraction							

22.

Decimal	0.1	0.2	0.3	0.4	0.5	0.6	0.7	0.8	0.9
Fraction									

Write each decimal as a fraction in lowest terms.

23. 0.15

24. 0.45

25. 0.08

26. 0.06

27. 0.375

28. 0.475

Write each decimal as a mixed number.

29. 5.6

30. 8.4

31. 5.06

32. 8.04

33. 17.26

34. 39.35

35. 1.22

36. 2.11

Simplify each of the following as much as possible, and write all answers as decimals.

37. $\frac{1}{2}(2.3 + 2.5)$

38. $\frac{3}{4}(1.8 + 7.6)$

39. $\frac{3}{8}(4.7)$

40. $\frac{5}{8}(1.2)$

41. $3.4 - \frac{1}{2}(0.76)$

42. $6.7 - \frac{1}{5}(0.45)$

43. $\frac{2}{5}(0.3) + \frac{3}{5}(0.3)$

44. $\frac{1}{8}(0.7) + \frac{3}{8}(0.7)$

45. $6\left(\frac{3}{5}\right)(0.02)$

46. $8\left(\frac{4}{5}\right)(0.03)$

47. $\frac{5}{8} + 0.35\left(\frac{1}{2}\right)$

48. $\frac{7}{8} + 0.45\left(\frac{3}{4}\right)$

49. $\left(\frac{1}{3}\right)^2(5.4) + \left(\frac{1}{2}\right)^3(3.2)$

50. $\left(\frac{1}{5}\right)^2(7.5) + \left(\frac{1}{4}\right)^2(6.4)$

51. $\left(\frac{1}{2}\right)^2\left(\frac{2}{5}\right)^3(12.5)$

52. $\left(\frac{1}{4}\right)^2\left(\frac{2}{3}\right)^3(10.8)$

53. $(0.25)^2 + \left(\frac{1}{4}\right)^2(3)$

54. $(0.75)^2 + \left(\frac{1}{4}\right)^2(7)$

Applying the Concepts

55. Price of Beef If each pound of beef costs $2.59, how much does $3\frac{1}{4}$ pounds cost?

56. Price of Gasoline What does it cost to fill a $15\frac{1}{2}$-gallon gas tank if the gasoline is priced at 129.9¢ per gallon?

57. Filling a Container How many $6\frac{1}{2}$-ounce glasses can be filled from a 32-ounce container of orange juice?

58. Filling a Container How many $8\frac{1}{2}$-ounce glasses can be filled from a 32-ounce container of orange juice?

59. Sale Price A dress that costs $57.99 is on sale for $\frac{1}{3}$ off. What is the sale price of the dress?

60. Sale Price A suit that normally sells for $121 is on sale for $\frac{1}{4}$ off. What is the sale price of the suit?

61. Average Gain in Stock Price The table below is from the introduction to Chapter 3. It shows the amount of gain or loss each day of the week of March 6, 2000, for the price of eCollege.com, an Internet company specializing in distance learning for college students. Complete the table by converting each fraction to a decimal, rounding to the nearest hundredth if necessary.

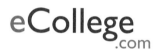

CHANGE IN STOCK PRICE		
DATE	GAIN/LOSS ($)	AS A DECIMAL ($) (TO THE NEAREST HUNDREDTH)
Monday, March 6, 2000	$\frac{3}{4}$	
Tuesday, March 7, 2000	$-\frac{9}{16}$	
Wednesday, March 8, 2000	$\frac{3}{32}$	
Thursday, March 9, 2000	$\frac{7}{32}$	
Friday, March 10, 2000	$\frac{1}{16}$	

62. Average Gain in Stock Price The table below shows the amount of gain or loss each day for the week of March 6, 2000, for the stock price of amazon.com, an online bookstore. Complete the table by converting each fraction to a decimal, rounding to the nearest hundredth, if necessary.

amazon.com.

CHANGE IN STOCK PRICE		
DATE	GAIN/LOSS	AS A DECIMAL ($) (TO THE NEAREST HUNDREDTH)
Monday, March 6, 2000	$\frac{1}{16}$	
Tuesday, March 7, 2000	$-1\frac{3}{8}$	
Wednesday, March 8, 2000	$\frac{3}{8}$	
Thursday, March 9, 2000	$5\frac{13}{16}$	
Friday, March 10, 2000	$-\frac{3}{8}$	

63. Nutrition If 1 ounce of ground beef contains 50.75 calories and 1 ounce of halibut contains 27.5 calories, what is the difference in calories between a $4\frac{1}{2}$-ounce serving of ground beef and a $4\frac{1}{2}$-ounce serving of halibut?

64. Nutrition If a 1-ounce serving of baked potato contains 48.3 calories and a 1-ounce serving of chicken contains 24.6 calories, how many calories are in a meal of $5\frac{1}{4}$ ounces of chicken and a $3\frac{1}{3}$-ounce baked potato?

Volume Find the volume of each sphere. Round to the nearest hundredth.

65.

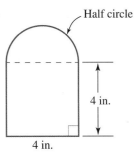

2 mi

66.

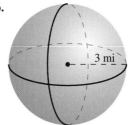

3 mi

67. Volume The radius of a sphere is 3.9 inches. Find the volume to the nearest hundredth.

68. Volume The radius of a sphere is 1.1 inches. Find the volume to the nearest hundredth.

Area Find the total area enclosed by each figure below.

69.

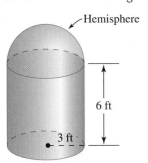

Half circle

4 in.

4 in.

70.

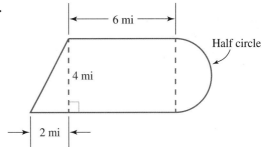

6 mi

4 mi

2 mi

Half circle

Find the volume of each figure. Round your answers to the nearest tenth.

71.

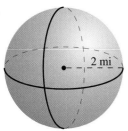

Hemisphere

6 ft

3 ft

72.

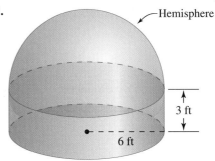

Hemisphere

3 ft

6 ft

Calculator Problems

Work each of the following problems using a calculator.

73. Write $\dfrac{7}{9}$ as a decimal.

74. Write $\dfrac{12}{99}$ as a decimal.

75. Write $\dfrac{123}{999}$ as a decimal.

76. Write $\dfrac{73}{99}$ as a decimal.

Taxi Ride Recently, the Texas Junior College Teachers Association annual conference was held in Austin. At that time a taxi ride in Austin was $1.25 for the first $\frac{1}{5}$ of a mile and $0.25 for each additional $\frac{1}{5}$ of a mile. The charge for a taxi to wait is $12.00 per hour. Use this information for Problems 77 through 80.

77. If the distance from one of the convention hotels to the airport is 7.5 miles, how much will it cost to take a taxi from that hotel to the airport?

78. If you were to tip the driver of the taxi in Problem 77 $1.50, how much would it cost to take a taxi from the hotel to the airport?

79. Suppose the distance from one of the hotels to one of the western dance clubs in Austin is 12.4 miles. If the fare meter in the taxi gives the charge for that trip as $16.50, is the meter working correctly?

80. Suppose that the distance from a hotel to the airport is 8.2 miles, and the ride takes 20 minutes. Is it more expensive to take a taxi to the airport or to just sit in the taxi?

Review Problems

The problems below review some of the material on solving equations that we covered in Chapter 4. Reviewing these problems will help you with the next section.

81. $x + 3 = -6$

82. $-4y = 28$

83. $\dfrac{1}{7}a = -7$

84. $\dfrac{1}{5}a = -3$

85. $5n + 4 = -26$

86. $6n - 2 = 40$

87. $5x + 8 = 3x + 2$

88. $7x - 3 = 5x + 9$

89. $2(x + 3) = 10$

90. $3(x - 2) = 6$

91. $3(y - 4) + 5 = -4$

92. $5(y - 1) + 6 = -9$

Extending the Concepts: Number Sequences

Give the next two numbers in each sequence.

93. $5.125, 6\dfrac{3}{8}, 7.625, 8\dfrac{7}{8}, \ldots$

94. $3.2, 5\dfrac{3}{5}, 8, 10\dfrac{2}{5}, \ldots$

95. $0.1, \dfrac{1}{100}, 0.001, \ldots$

96. $5.5, \dfrac{11}{20}, 0.055, \ldots$

5.6 Equations Containing Decimals

In this section we will continue our work with equations by considering some equations that involve decimals. We will also look at some application problems whose solutions come from equations with decimals.

EXAMPLE 1 Solve the equation $x + 8.2 = 5.7$.

SOLUTION   We use the addition property of equality to add -8.2 to each side of the equation.

$$x + 8.2 = 5.7$$
$$x + 8.2 + (\mathbf{-8.2}) = 5.7 + (\mathbf{-8.2}) \quad \textbf{Add } -\textbf{8.2 to each side}$$
$$x + 0 = -2.5 \quad \textbf{Simplify each side}$$
$$x = -2.5 \qquad ■$$

EXAMPLE 2 Solve: $3y = 2.73$

SOLUTION   To isolate y on the left side, we divide each side by 3.

$$3y = 2.73$$
$$\frac{3y}{\mathbf{3}} = \frac{2.73}{\mathbf{3}} \quad \textbf{Divide each side by 3}$$
$$y = 0.91 \quad \textbf{Division} \qquad ■$$

EXAMPLE 3 Solve: $\frac{1}{2}x - 3.78 = 2.52$

SOLUTION   We begin by adding 3.78 to each side of the equation. Then we multiply each side by 2.

$$\frac{1}{2}x - 3.78 = 2.52$$
$$\frac{1}{2}x - 3.78 + \mathbf{3.78} = 2.52 + \mathbf{3.78} \quad \textbf{Add 3.78 to each side}$$
$$\frac{1}{2}x = 6.30$$
$$\mathbf{2}\left(\frac{1}{2}x\right) = \mathbf{2}(6.30) \quad \textbf{Multiply each side by 2}$$
$$x = 12.6 \qquad ■$$

EXAMPLE 4 Solve: $5a - 0.42 = -3a + 0.98$

SOLUTION   We can isolate a on the left side of the equation by adding $3a$ to each side.

$$5a + \mathbf{3a} - 0.42 = -3a + \mathbf{3a} + 0.98 \quad \textbf{Add 3a to each side}$$
$$8a - 0.42 = 0.98$$
$$8a - 0.42 + \mathbf{0.42} = 0.98 + \mathbf{0.42} \quad \textbf{Add 0.42 to each side}$$
$$8a = 1.40$$
$$\frac{8a}{\mathbf{8}} = \frac{1.40}{\mathbf{8}} \quad \textbf{Divide each side by 8}$$
$$a = 0.175 \qquad ■$$

Practice Problems

1. Solve: $x - 3.4 = 6.7$

2. Solve: $4y = 3.48$

3. Solve: $\frac{1}{5}x - 2.4 = 8.3$

4. Solve: $7a - 0.18 = 2a + 0.77$

Answers
1. 10.1   **2.** 0.87   **3.** 53.5
4. 0.19

5. If a car was rented from the company in Example 5 for 2 days and the total charge was $40.88, how many miles was the car driven?

EXAMPLE 5   A car rental company charges $11 per day and 16 cents per mile for their cars. If a car was rented for 1 day and the charge was $25.40, how many miles was it driven?

SOLUTION   We use our Blueprint for Problem Solving as a guide to solving this application problem.

Step 1 *Read and list*
Known items: Charges are $11 per day and 16 cents per mile; car was rented for 1 day for a total cost of $25.40.
Unknown item: Number of miles it was driven.

Step 2 *Assign a variable and translate information*
If we let x = the number of miles driven, then the charge for the miles driven will be $0.16x$, the cost per mile times the number of miles.

Step 3 *Reread and write an equation*

$$\underbrace{\frac{\$11 \text{ per}}{\text{day}}}_{11} + \underbrace{\frac{16 \text{ cents}}{\text{per mile}}}_{0.16x} = \underbrace{\text{Total cost}}_{25.40}$$

Step 4 *Solve the equation*
To solve the equation, we add -11 to each side and then divide each side by 0.16.

$$11 + (-11) + 0.16x = 25.40 + (-11) \qquad \textbf{Add } -\textbf{11 to each side}$$
$$0.16x = 14.40$$
$$\frac{0.16x}{\textbf{0.16}} = \frac{14.40}{\textbf{0.16}} \qquad \textbf{Divide each side by 0.16}$$
$$x = 90 \qquad \textbf{14.40} \div \textbf{0.16} = \textbf{90}$$

Step 5 *Write the answer*
The car was driven 90 miles.

Step 6 *Reread and check*

$$\text{The 1-day charge} = \$11.00$$
$$\underline{\text{The mileage charge is } \$.16(90) = \$14.40}$$
$$\text{Total charge} = \$25.40$$

■

EXAMPLE 6 Diane has $1.60 in dimes and nickels. If she has 7 more dimes than nickels, how many of each coin does she have?

SOLUTION We use our Blueprint for Problem Solving as a guide to solving this application problem.

Step 1 *Read and list*

Known items: We have dimes and nickels, seven more dimes than nickels.
Unknown items: Number of dimes and the number of nickels.

Step 2 *Assign a variable and translate information*

If we let $x =$ the number of nickels, then the number of dimes must be $x + 7$, because Diane has 7 more dimes than nickels. Since each nickel is worth 5 cents, the amount of money she has in nickels is $0.05x$. Similarly, since each dime is worth 10 cents, the amount of money she has in dimes is $0.10(x + 7)$. Here is a table that summarizes what we have so far:

	NICKELS	DIMES
Number of	x	$x + 7$
Value of	$0.05x$	$0.10(x + 7)$

Step 3 *Reread and write an equation*

Because the total value of all the coins is $1.60, the equation that describes this situation is

$$\underbrace{\text{Amount of money in nickels}}_{0.05x} + \underbrace{\text{Amount of money in dimes}}_{0.10(x + 7)} = \underbrace{\text{Total amount of money}}_{1.60}$$

Step 4 *Solve the equation*

This time, let's show only the essential steps in the solution.

$0.05x + 0.10x + 0.70 = 1.60$	**Distributive property**
$0.15x + 0.70 = 1.60$	**Add $0.05x$ and $0.10x$ to get $0.15x$**
$0.15x = 0.90$	**Add -0.70 to each side**
$x = 6$	**Divide each side by 0.15**

Step 5 *Write the answer*

Because $x = 6$, Diane has 6 nickels. To find the number of dimes, we add 7 to the number of nickels (she has 7 more dimes than nickels). The number of dimes is $6 + 7 = 13$.

Step 6 *Reread and check*

$$\begin{aligned}
6 \text{ nickels are worth } 6(\$0.05) &= \$0.30 \\
\underline{13 \text{ dimes are worth } 13(\$0.10)} &= \underline{\$1.30} \\
\text{The total value is} \qquad\qquad &\quad \$1.60
\end{aligned}$$

■

6. Amy has $1.75 in dimes and quarters. If she has 7 more dimes than quarters, how many of each coin does she have?

Problem Set 5.6

Solve each equation.

1. $x + 3.7 = 2.2$

2. $x + 4.8 = 9.1$

3. $x - 0.45 = 0.32$

4. $x - 23.3 = -4.5$

5. $8a = 1.2$

6. $6a = 18.6$

7. $-4y = 1.4$

8. $-7y = -0.63$

9. $0.5n = -0.4$

10. $0.6n = -0.12$

11. $4x - 4.7 = 3.5$

12. $2x + 3.8 = -7.7$

13. $0.02 + 5y = -0.3$

14. $0.8 + 10y = -0.7$

15. $\frac{1}{3}x - 2.99 = 1.02$

16. $\frac{1}{7}x + 2.87 = -3.01$

17. $7n - 0.32 = 5n + 0.56$

18. $6n + 0.88 = 2n - 0.77$

19. $3a + 4.6 = 7a + 5.3$

20. $2a - 3.3 = 7a - 5.2$

21. $0.5x + 0.1(x + 20) = 3.2$

22. $0.1x + 0.5(x + 8) = 7$

23. $0.08x + 0.09(x + 2000) = 690$

24. $0.11x + 0.12(x + 4000) = 940$

Applying the Concepts

25. Car Rental A car rental company charges $10 a day and 16 cents per mile to rent their cars. If a car was rented for 1 day for a total charge of $23.92, how many miles was it driven?

26. Car Rental A car rental company charges $12 a day and 18 cents per mile to rent their cars. If the total charge for a 1-day rental was $33.78, how many miles was the car driven?

27. Car Rental A car rental company charges $9 per day and 15 cents a mile to rent their cars. If a car was rented for 2 days for a total charge of $40.05, how many miles was it driven?

28. Car Rental A car rental company charges $11 a day and 18 cents per mile to rent their cars. If the total charge for a 2-day rental was $61.60, how many miles was it driven?

29. Coin Problem Mary has $2.20 in dimes and nickels. If she has 10 more dimes than nickels, how many of each coin does she have?

30. Coin Problem Bob has $1.65 in dimes and nickels. If he has 9 more nickels than dimes, how many of each coin does he have?

31. Coin Problem Suppose you have $9.60 in dimes and quarters. How many of each coin do you have if you have twice as many quarters as dimes?

32. Coin Problem A collection of dimes and quarters has a total value of $2.75. If there are three times as many dimes as quarters, how many of each coin is in the collection?

33. Long-Distance Charges The cost of a long-distance phone call is $0.41 for the first minute and $0.32 for each additional minute. If the total charge for a long-distance call is $5.21, how many minutes was the call?

34. Long-Distance Charges Danny, who is 1 year old, is playing with the telephone when he accidentally presses one of the buttons his mother has programmed to dial her friend Sue's number. Sue answers the phone and realizes Danny is on the other end. She talks to Danny, trying to get him to hang up. The cost for a call is $0.23 for the first minute and $0.14 for every minute after that. If the total charge for the call is $3.73, how long did it take Sue to convince Danny to hang up the phone?

35. Coin Problem Katie has a collection of nickels, dimes, and quarters with a total value of $4.35. There are 3 more dimes than nickels and 5 more quarters than nickels. How many of each coin is in her collection? (*Hint:* Let x = the number of nickels.)

36. Coin Problem Mary Jo has $3.90 worth of nickels, dimes, and quarters. The number of nickels is 3 more than the number of dimes. The number of quarters is 7 more than the number of dimes. How many of each coin does she have? (*Hint:* Let x = the number of dimes.)

Review Problems

The problems below review the material on exponents we have covered previously.

Expand and simplify.

37. 5^3

38. 2^5

39. $(-3)^2$

40. $(-2)^3$

41. $\left(\dfrac{1}{3}\right)^4$

42. $\left(\dfrac{3}{4}\right)^3$

43. $\left(-\dfrac{5}{6}\right)^2$

44. $\left(-\dfrac{3}{5}\right)^3$

45. $(0.5)^2$

46. $(0.1)^3$

47. $(-1.2)^2$

48. $(-2.1)^2$

Use the properties of exponents from Section 2.7 to simplify each expression.

49. $x^5 \cdot x^8$

50. $x^3 \cdot x^5 \cdot x^7$

51. $5a^3 \cdot 9a^4$

52. $8a^5 \cdot 6a^4$

53. $(15x^5y^3)^2$

54. $(6x^3y^2)^3$

5.7 Square Roots and the Pythagorean Theorem

Introduction . . .

Figure 1 shows the front view of the roof of a tool shed. How do we find the length d of the diagonal part of the roof? (Imagine that you are drawing the plans for the shed. Since the shed hasn't been built yet, you can't just measure the diagonal, but you need to know how long it will be so you can buy the correct amount of material to build the shed.)

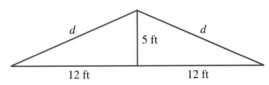

Figure 1

There is a formula from geometry that gives the length d:

$$d = \sqrt{12^2 + 5^2}$$

where $\sqrt{}$ is called the *square root symbol.* If we simplify what is under the square root symbol, we have this:

$$d = \sqrt{144 + 25}$$
$$= \sqrt{169}$$

The expression $\sqrt{169}$ stands for the number we *square* to get 169. Because $13 \cdot 13 = 169$, that number is 13. Therefore the length d in our original diagram is 13 feet.

Here is a more detailed discussion of square roots. In Chapter 1, we did some work with exponents. In particular, we spent some time finding squares of numbers. For example, we considered expressions like this:

$$5^2 = 5 \cdot 5 = 25$$
$$7^2 = 7 \cdot 7 = 49$$
$$x^2 = x \cdot x$$

We say that "the square of 5 is 25" and "the square of 7 is 49." To square a number, we multiply it by itself. When we ask for the *square root* of a given number, we want to know what number we *square* in order to obtain the given number. We say that the square root of 49 is 7, because 7 is the number we square to get 49. Likewise, the square root of 25 is 5, because $5^2 = 25$. The symbol we use to denote square root is $\sqrt{}$, which is also called a *radical sign.* Here is the precise definition of square root.

DEFINITION

The **square root** of a positive number a, written $\sqrt{a}$, is the number we square to get a. *In symbols:*

If $\sqrt{a} = b$ then $b^2 = a$.

We list some common square roots in Table 1.

Note

The square root we are describing here is actually the principal square root. There is another square root that is a negative number. We won't see it in this book, but, if you go on to take an algebra course, you will see it there.

Table 1

STATEMENT	IN WORDS	REASON
$\sqrt{0} = 0$	The square root of 0 is 0	Because $0^2 = 0$
$\sqrt{1} = 1$	The square root of 1 is 1	Because $1^2 = 1$
$\sqrt{4} = 2$	The square root of 4 is 2	Because $2^2 = 4$
$\sqrt{9} = 3$	The square root of 9 is 3	Because $3^2 = 9$
$\sqrt{16} = 4$	The square root of 16 is 4	Because $4^2 = 16$
$\sqrt{25} = 5$	The square root of 25 is 5	Because $5^2 = 25$

Numbers like 1, 9, and 25, whose square roots are whole numbers, are called *perfect squares.* To find the square root of a perfect square, we look for the whole number that is squared to get the perfect square. The following examples involve square roots of perfect squares.

EXAMPLE 1   Simplify: $7\sqrt{64}$

SOLUTION   The expression $7\sqrt{64}$ means 7 times $\sqrt{64}$. To simplify this expression, we write $\sqrt{64}$ as 8 and multiply:

$$7\sqrt{64} = 7 \cdot 8 = 56$$

We know $\sqrt{64} = 8$, because $8^2 = 64$.   ∎

EXAMPLE 2   Simplify: $\sqrt{9} + \sqrt{16}$

SOLUTION   We write $\sqrt{9}$ as 3 and $\sqrt{16}$ as 4. Then we add:

$$\sqrt{9} + \sqrt{16} = 3 + 4 = 7$$

       ∎

EXAMPLE 3   Simplify: $\sqrt{\dfrac{25}{81}}$

SOLUTION   We are looking for the number we square (multiply times itself) to get $\frac{25}{81}$. We know that when we multiply two fractions, we multiply the numerators and multiply the denominators. Because $5 \cdot 5 = 25$ and $9 \cdot 9 = 81$, the square root of $\frac{25}{81}$ must be $\frac{5}{9}$:

$$\sqrt{\frac{25}{81}} = \frac{5}{9} \quad \text{because} \quad \left(\frac{5}{9}\right)^2 = \frac{5}{9} \cdot \frac{5}{9} = \frac{25}{81}$$

       ∎

In Examples 4–6, we simplify each expression as much as possible.

EXAMPLE 4   Simplify: $12\sqrt{25} = 12 \cdot 5 = 60$ ∎

EXAMPLE 5   Simplify: $\sqrt{100} - \sqrt{36} = 10 - 6 = 4$ ∎

EXAMPLE 6   Simplify: $\sqrt{\dfrac{49}{121}} = \dfrac{7}{11}$ because $\left(\dfrac{7}{11}\right)^2 = \dfrac{7}{11} \cdot \dfrac{7}{11} = \dfrac{49}{121}$ ∎

So far in this section we have been concerned only with square roots of perfect squares. The next question is, "What about square roots of numbers that are not perfect squares, like $\sqrt{7}$, for example?" We know that

$$\sqrt{4} = 2 \quad \text{and} \quad \sqrt{9} = 3$$

Practice Problems

1. Simplify: $4\sqrt{25}$

2. Simplify: $\sqrt{36} + \sqrt{4}$

3. Simplify: $\sqrt{\dfrac{36}{100}}$

Simplify each expression as much as possible.

4. $14\sqrt{36}$

5. $\sqrt{81} - \sqrt{25}$

6. $\sqrt{\dfrac{64}{121}}$

Answers

1. 20   2. 8   3. $\dfrac{3}{5}$ 4. 84

5. 4   6. $\dfrac{8}{11}$

And because 7 is between 4 and 9, $\sqrt{7}$ should be between $\sqrt{4}$ and $\sqrt{9}$. That is, $\sqrt{7}$ should be between 2 and 3. But what is it exactly? The answer is, we cannot write it exactly in decimal or fraction form. Because of this, it is called an *irrational number*. We can approximate it with a decimal, but we can never write it exactly with a decimal. Table 2 gives some decimal approximations for $\sqrt{7}$. The decimal approximations were obtained by using a calculator. We could continue the list to any accuracy we desired. However, we would never reach a number in decimal form whose square was exactly 7.

Table 2

APPROXIMATIONS FOR THE SQUARE ROOT OF 7		
ACCURATE TO THE NEAREST	THE SQUARE ROOT OF 7 IS	CHECK BY SQUARING
Tenth	$\sqrt{7} = 2.6$	$(2.6)^2 = 6.76$
Hundredth	$\sqrt{7} = 2.65$	$(2.65)^2 = 7.0225$
Thousandth	$\sqrt{7} = 2.646$	$(2.646)^2 = 7.001316$
Ten thousandth	$\sqrt{7} = 2.6458$	$(2.6458)^2 = 7.00025764$

EXAMPLE 7   Give a decimal approximation for the expression $5\sqrt{12}$ that is accurate to the nearest ten thousandth.

SOLUTION   Let's agree not to round to the nearest ten thousandth until we have first done all the calculations. Using a calculator, we find $\sqrt{12} \approx 3.4641016$. Therefore,

$$5\sqrt{12} \approx 5(3.4641016) \qquad \sqrt{12} \text{ on calculator}$$
$$= 17.320508 \qquad \text{Multiplication}$$
$$= 17.3205 \qquad \text{To the nearest ten thousandth} \quad \blacksquare$$

EXAMPLE 8   Approximate $\sqrt{301} + \sqrt{137}$ to the nearest hundredth.

SOLUTION   Using a calculator to approximate the square roots, we have

$$\sqrt{301} + \sqrt{137} \approx 17.349352 + 11.704700 = 29.054052$$

To the nearest hundredth, the answer is 29.05. $\quad \blacksquare$

EXAMPLE 9   Approximate $\sqrt{\dfrac{7}{11}}$ to the nearest thousandth.

SOLUTION   Because we are using calculators, we first change $\frac{7}{11}$ to a decimal and then find the square root:

$$\sqrt{\frac{7}{11}} \approx \sqrt{0.6363636} \approx 0.7977240$$

To the nearest thousandth, the answer is 0.798. $\quad \blacksquare$

Facts from Geometry: The Pythagorean Theorem

Next, we will work some problems involving the Pythagorean theorem, which we mentioned in the introduction to this section. It may interest you to know that Pythagoras formed a secret society around the year 540 BC. Known as the Pythagoreans, members kept no written record of their work; everything was handed down by spoken word. They influenced not only mathematics, but religion, science, medicine, and music as well. Among other things, they discovered

7. Give a decimal approximation for the expression $5\sqrt{14}$ that is accurate to the nearest ten thousandth.

8. Approximate $\sqrt{405} + \sqrt{147}$ to the nearest hundredth.

9. Approximate $\sqrt{\dfrac{7}{12}}$ to the nearest thousandth.

Answers
7. 18.7083   **8.** 32.25   **9.** 0.764

DECIMALS

the correlation between musical notes and the reciprocals of counting numbers, $\frac{1}{2}, \frac{1}{3}, \frac{1}{4}$, and so on. In their daily lives, they followed strict dietary and moral rules to achieve a higher rank in future lives.

At the beginning of this section we mentioned a formula from geometry that we used to find the length of the diagonal of the roof of a tool shed. That formula is called the Pythagorean theorem, and it applies to right triangles. A *right triangle* is a triangle that contains a 90° (or right) angle. The longest side in a right triangle is called the *hypotenuse,* and we use the letter c to denote it. The two shorter sides are denoted by the letters a and b. The Pythagorean theorem states that the hypotenuse is the square root of the sum of the squares of the two shorter sides. In symbols:

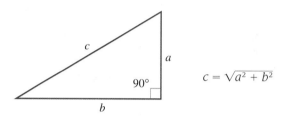

10. Find the length of the hypotenuse in each right triangle.

a

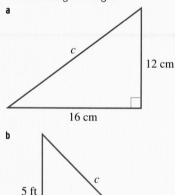

EXAMPLE 10 Find the length of the hypotenuse in each right triangle.

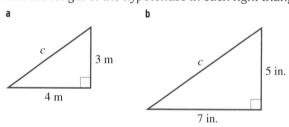

SOLUTION   We apply the formula given above.

a When $a = 3$ and $b = 4$:
$$c = \sqrt{3^2 + 4^2}$$
$$= \sqrt{9 + 16}$$
$$= \sqrt{25}$$
$$c = 5 \text{ meters}$$

b When $a = 5$ and $b = 7$:
$$c = \sqrt{5^2 + 7^2}$$
$$= \sqrt{25 + 49}$$
$$= \sqrt{74}$$
$$c \approx 8.60 \text{ inches}$$

In part (a), the solution is a whole number, whereas in part (b), we must use a calculator to get 8.60 as an approximation to $\sqrt{74}$. ■

11. A wire from the top of a 12-foot pole is fastened to the ground by a stake that is 5 feet from the bottom of the pole. What is the length of the wire?

EXAMPLE 11 A ladder is leaning against the top of a 6-foot wall. If the bottom of the ladder is 8 feet from the wall, how long is the ladder?

SOLUTION   A picture of the situation is shown in Figure 2. We let c denote the length of the ladder. Applying the Pythagorean theorem, we have

$$c = \sqrt{6^2 + 8^2}$$
$$= \sqrt{36 + 64}$$
$$= \sqrt{100}$$
$$= 10 \text{ feet}$$

The ladder is 10 feet long.

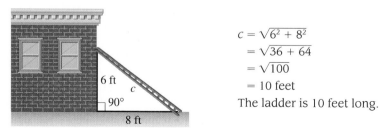

Figure 2

Answers
10. a 20 cm   **b** 7.07 ft
11. 13 ft

Problem Set 5.7

Find each of the following square roots without using a calculator.

1. $\sqrt{64}$　　　　**2.** $\sqrt{100}$　　　　**3.** $\sqrt{81}$　　　　**4.** $\sqrt{49}$

5. $\sqrt{36}$　　　　**6.** $\sqrt{144}$　　　　**7.** $\sqrt{25}$　　　　**8.** $\sqrt{169}$

Simplify each of the following expressions without using a calculator.

9. $3\sqrt{25}$　　　　**10.** $9\sqrt{49}$　　　　**11.** $6\sqrt{64}$　　　　**12.** $11\sqrt{100}$

13. $15\sqrt{9}$　　　　**14.** $8\sqrt{36}$　　　　**15.** $16\sqrt{9}$　　　　**16.** $9\sqrt{16}$

17. $\sqrt{49} + \sqrt{64}$　　　**18.** $\sqrt{1} + \sqrt{0}$　　　**19.** $\sqrt{16} - \sqrt{9}$　　　**20.** $\sqrt{25} - \sqrt{4}$

21. $3\sqrt{25} + 9\sqrt{49}$　　**22.** $6\sqrt{64} + 11\sqrt{100}$　　**23.** $15\sqrt{9} - 9\sqrt{16}$　　**24.** $7\sqrt{49} - 2\sqrt{4}$

25. $\sqrt{\dfrac{16}{49}}$　　　**26.** $\sqrt{\dfrac{100}{121}}$　　　**27.** $\sqrt{\dfrac{36}{64}}$　　　**28.** $\sqrt{\dfrac{81}{144}}$

Indicate whether each of the expressions in Problems 29–32 is *True* or *False*.

29. $\sqrt{4} + \sqrt{9} = \sqrt{4 + 9}$　　　　　　**30.** $\sqrt{\dfrac{16}{25}} = \dfrac{\sqrt{16}}{\sqrt{25}}$

31. $\sqrt{25 \cdot 9} = \sqrt{25} \cdot \sqrt{9}$　　　　　　**32.** $\sqrt{100} - \sqrt{36} = \sqrt{100 - 36}$

Find the length of the hypotenuse in each right triangle. Round to the nearest hundredth, if rounding is necessary.

33.

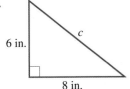

6 in.　　*c*　　8 in.

34.

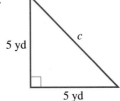

5 yd　　*c*　　5 yd

35.

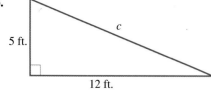

5 ft.

12 ft.

c

36.

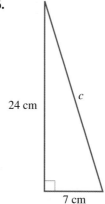

24 cm

7 cm

c

37.

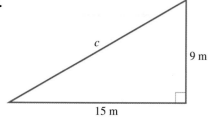

c

4 in.

5 in.

38.

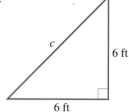

c

6 ft

6 ft

39.

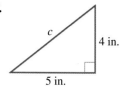

c

9 m

15 m

40.

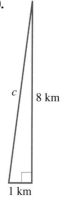

c

8 km

1 km

Applying the Concepts

41. Geometry One end of a wire is attached to the top of a 24-foot pole; the other end of the wire is anchored to the ground 18 feet from the bottom of the pole. If the pole makes an angle of 90° with the ground, find the length of the wire.

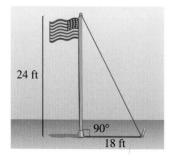

24 ft

90°

18 ft

42. Geometry Two children are trying to cross a stream. They want to use a log that goes from one bank to the other. If the left bank is 5 feet higher than the right bank and the stream is 12 feet wide, how long must a log be to just barely reach?

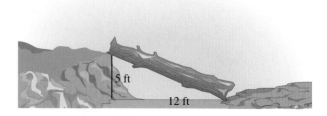

5 ft

12 ft

43. Geometry A ladder is leaning against the top of a 15-foot wall. If the bottom of the ladder is 20 feet from the wall, how long is the ladder?

44. Geometry A wire from the top of a 24-foot pole is fastened to the ground by a stake that is 10 feet from the bottom of the pole. How long is the wire?

Calculator Problems

Use a calculator to work each of the following problems.

Approximate each of the following square roots to the nearest ten thousandth.

45. $\sqrt{12}$ **46.** $\sqrt{18}$ **47.** $\sqrt{125}$ **48.** $\sqrt{75}$ **49.** $\sqrt{324}$ **50.** $\sqrt{1296}$

Approximate each of the following expressions to the nearest hundredth.

51. $2\sqrt{3}$ **52.** $3\sqrt{2}$ **53.** $5\sqrt{5}$ **54.** $5\sqrt{3}$

55. $\dfrac{\sqrt{3}}{3}$ **56.** $\dfrac{\sqrt{2}}{2}$ **57.** $\sqrt{\dfrac{1}{3}}$ **58.** $\sqrt{\dfrac{1}{2}}$

Approximate each of the following expressions to the nearest thousandth.

59. $\sqrt{12} + \sqrt{75}$ **60.** $\sqrt{18} + \sqrt{50}$ **61.** $\sqrt{87}$ **62.** $\sqrt{68}$

63. $2\sqrt{3} + 5\sqrt{3}$ **64.** $3\sqrt{2} + 5\sqrt{2}$ **65.** $7\sqrt{3}$ **66.** $8\sqrt{2}$

67. Lighthouse Problem The higher you are above the ground, the farther you can see. If your view is unobstructed, then the distance in miles that you can see from h feet above the ground is given by the formula

$$d = \sqrt{\dfrac{3h}{2}}$$

The following figure shows a lighthouse with a door and windows at various heights. The preceding formula can be used to find the distance to the ocean horizon from these heights. Use the formula and a calculator to complete the following table. Round your answers to the nearest whole number.

HEIGHT h(feet)	DISTANCE d(miles)
10	
50	
90	
130	
170	
190	

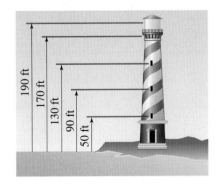

68. Pendulum Problem The time (in seconds) it takes for the pendulum on a clock to swing through one complete cycle is given by the formula

$$T = \frac{11}{7} \sqrt{\frac{L}{2}}$$

where L is the length (in feet) of the pendulum. Use this formula and a calculator to complete the following table. Round your answers to the nearest hundredth.

LENGTH L (feet)	TIME T (seconds)
1	
2	
3	
4	
5	
6	

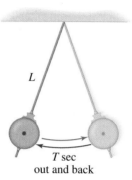

L

T sec
out and back

Review Problems

The problems below review some of the material we covered in Chapter 3, involving fractions and mixed numbers. Perform the indicated operations. Write your answers as whole numbers, proper fractions, or mixed numbers.

69. $\dfrac{5}{7} \cdot \dfrac{14}{25}$

70. $1\dfrac{1}{4} \div 2\dfrac{1}{8}$

71. $4\dfrac{3}{10} + 5\dfrac{2}{100}$

72. $8\dfrac{1}{5} + 1\dfrac{1}{10}$

73. $3\dfrac{2}{10} \cdot 2\dfrac{5}{10}$

74. $6\dfrac{9}{10} \div 2\dfrac{3}{10}$

75. $7\dfrac{1}{10} - 4\dfrac{3}{10}$

76. $3\dfrac{7}{10} - 1\dfrac{97}{100}$

Extending the Concepts

77. Geometry The hypotenuse of a right triangle is 10 inches long. If one of the sides is 8 inches long, find the length of the other side and the area of the triangle.

78. Geometry The hypotenuse of a right triangle is 13 inches long. If one of the sides is 5 inches long, find the length of the other side and the area of the triangle.

79. Geometry The area of a right triangle is 6 square feet. If one of the shorter sides is 3 feet long, how long is the hypotenuse?

80. Geometry The area of a right triangle is 54 square feet. If one of the shorter sides is 12 feet long, how long is the hypotenuse?

5.8 Simplifying Square Roots

Do you know that $\sqrt{50}$ and $5 \cdot \sqrt{2}$ are the same number? One way to convince yourself that this is true is with a calculator. On a scientific calculator, to find a decimal approximation to the first expression, we enter 50 and then press the $\sqrt{}$ key:

Note
On a graphing calculator, the keys are
$\boxed{\sqrt{}}$ 50 $\boxed{\text{ENTER}}$
and
$_5$ $\boxed{\times}$ $\boxed{\sqrt{}}$ $_2$ $\boxed{\text{ENTER}}$

$$50 \boxed{\sqrt{}} \qquad \text{The calculator shows 7.0710678}$$

To find a decimal approximation to the second expression, we multiply 5 and $\sqrt{2}$:

$$5 \boxed{\times} 2 \boxed{\sqrt{}} \boxed{=} \qquad \text{The calculator shows 7.0710678}$$

Although a calculator will give the same result for both $\sqrt{50}$ and $5 \cdot \sqrt{2}$, it does not tell us *why* the answers are the same. The discussion below shows why the results are the same.

First, notice that the expressions $\sqrt{4 \cdot 9}$ and $\sqrt{4} \cdot \sqrt{9}$ have the same value:

$$\sqrt{4 \cdot 9} = \sqrt{36} = 6 \quad \text{and} \quad \sqrt{4} \cdot \sqrt{9} = 2 \cdot 3 = 6$$

Both are equal to 6. When we are multiplying and taking square roots, we can either multiply first and then take the square root of what we get, or we can take square roots first and then multiply. In symbols, we write it this way:

Multiplication Property for Square Roots

If a and b are positive numbers, then

$$\sqrt{a \cdot b} = \sqrt{a} \cdot \sqrt{b}$$

In words: The square root of a product is the product of the square roots.

Second, when a number occurs twice as a factor in a square root, then the square root simplifies to just that number. For example, $\sqrt{5 \cdot 5}$ is really $\sqrt{25}$, which is the same as just 5. Therefore, $\sqrt{5 \cdot 5} = 5$.

Repeated Factor Property for Square Roots

If a is a positive number, then

$$\sqrt{a \cdot a} = a$$

But how do these two properties help us simplify expressions such as $\sqrt{50}$? To see the answer to this question, we must factor 50 into the product of its prime factors:

$$\sqrt{50} = \sqrt{5 \cdot 5 \cdot 2}$$

The factor 5 occurs twice, meaning that we have a perfect square ($5 \cdot 5 = 25$) under the radical. Writing this as two separate square roots, we have

$$\sqrt{5 \cdot 5 \cdot 2} = \sqrt{5 \cdot 5} \cdot \sqrt{2}$$
$$= 5 \cdot \sqrt{2}$$

> **RULE**
>
> When the number under a square root is factored completely, any factor that occurs twice can be taken out from under the square root symbol.

Practice Problems

1. Simplify: $\sqrt{63}$

EXAMPLE 1   Simplify: $\sqrt{45}$

SOLUTION   To begin we factor 45 into the product of prime factors:

$$
\begin{aligned}
\sqrt{45} &= \sqrt{3 \cdot 3 \cdot 5} && \textbf{Factor} \\
&= \sqrt{3 \cdot 3} \cdot \sqrt{5} && \textbf{Multiplication property} \\
&= 3 \cdot \sqrt{5} && \textbf{Repeated factor property}
\end{aligned}
$$

The expressions $\sqrt{45}$ and $3 \cdot \sqrt{5}$ are equivalent. The expression $3 \cdot \sqrt{5}$ is said to be in *simplified form* because the number under the radical is as small as possible.   ■

Note

When we work with square roots, the expressions $3\sqrt{5}$ and $3 \cdot \sqrt{5}$ are the same. Both of them represent the product of 3 and $\sqrt{5}$. For simplicity, we usually omit the multiplication dot.

In our next example, a variable appears under the square root symbol. We will assume that all variables that appear under a radical represent positive numbers.

2. Simplify: $\sqrt{45x^2}$

EXAMPLE 2   Simplify: $\sqrt{18x^2}$

SOLUTION   We factor $18x^2$ into $3 \cdot 3 \cdot 2 \cdot x \cdot x$. Because the factor 3 occurs twice, it can be taken out from under the radical. Likewise, because the factor x appears twice, it also can be taken out from under the radical. Therefore,

$$
\begin{aligned}
\sqrt{18x^2} &= \sqrt{3 \cdot 3 \cdot 2 \cdot x \cdot x} \\
&= 3 \cdot x \cdot \sqrt{2} \\
&= 3x\sqrt{2}
\end{aligned}
$$

You may be wondering if we can check this answer on a calculator. The answer is yes, but we need to substitute a value for x first. Suppose x is 5. Then

$$
\sqrt{18x^2} = \sqrt{18 \cdot 25} = \sqrt{450} \approx 21.213203
$$

and

$$
3x\sqrt{2} = 3 \cdot 5\sqrt{2} = 15\sqrt{2} \approx 15(1.4142136) = 21.213203 \quad ■
$$

3. Simplify: $\sqrt{300}$

EXAMPLE 3   Simplify: $\sqrt{180}$

SOLUTION   We factor and then look for factors occurring twice.

$$
\begin{aligned}
\sqrt{180} &= \sqrt{2 \cdot 2 \cdot 3 \cdot 3 \cdot 5} \\
&= 2 \cdot 3 \cdot \sqrt{5} \\
&= 6\sqrt{5}
\end{aligned}
$$

■

4. Simplify: $\sqrt{50x^3}$

EXAMPLE 4   Simplify: $\sqrt{48x^3}$

SOLUTION $\begin{aligned}[t]
\sqrt{48x^3} &= \sqrt{2 \cdot 2 \cdot 2 \cdot 2 \cdot 3 \cdot x \cdot x \cdot x} \\
&= 2 \cdot 2 \cdot x \cdot \sqrt{3 \cdot x} \\
&= 4x\sqrt{3x}
\end{aligned}$

■

Answers

1. $3\sqrt{7}$ **2.** $3x\sqrt{5}$ **3.** $10\sqrt{3}$
4. $5x\sqrt{2x}$

Facts from Geometry: The Spiral of Roots

To visualize the square roots of the positive integers, we can construct the spiral of roots, which we mentioned in the introduction to this chapter. To begin, we draw two line segments, each of length 1, at right angles to each other. Then we use the Pythagorean theorem to find the length of the diagonal. Figure 1 illustrates:

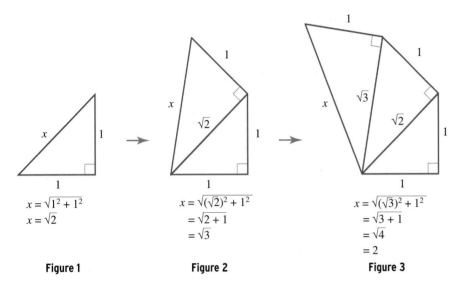

$$x = \sqrt{1^2 + 1^2}$$
$$x = \sqrt{2}$$

Figure 1

$$x = \sqrt{(\sqrt{2})^2 + 1^2}$$
$$= \sqrt{2 + 1}$$
$$= \sqrt{3}$$

Figure 2

$$x = \sqrt{(\sqrt{3})^2 + 1^2}$$
$$= \sqrt{3 + 1}$$
$$= \sqrt{4}$$
$$= 2$$

Figure 3

Note
We can simplify $(\sqrt{2})^2$ like this:
$$(\sqrt{2})^2 = \sqrt{2} \cdot \sqrt{2}$$
$$= \sqrt{2 \cdot 2}$$ **Multiplication property**
$$= 2$$ **Repeated factor property**

Next, we construct a second triangle by connecting a line segment of length 1 to the end of the first diagonal so that the angle formed is a right angle. We find the length of the second diagonal using the Pythagorean theorem. Figure 2 illustrates this procedure. As we continue to draw new triangles by connecting line segments of length 1 to the end of each previous diagonal, so that the angle formed is a right angle, the spiral of roots begins to appear (see Figure 3).

Problem Set 5.8

Simplify each expression by taking as much out from under the radical as possible. You may assume that all variables represent positive numbers.

1. $\sqrt{12}$

2. $\sqrt{18}$

3. $\sqrt{20}$

4. $\sqrt{27}$

5. $\sqrt{72}$

6. $\sqrt{48}$

7. $\sqrt{98}$

8. $\sqrt{75}$

9. $\sqrt{28}$

10. $\sqrt{44}$

11. $\sqrt{200}$

12. $\sqrt{300}$

13. $\sqrt{12x^2}$

14. $\sqrt{18x^2}$

15. $\sqrt{50x^2}$

16. $\sqrt{45x^2}$

17. $\sqrt{75x^3}$

18. $\sqrt{8x^3}$

19. $\sqrt{50x^3}$

20. $\sqrt{45x^3}$

21. $\sqrt{32x^2y^3}$

22. $\sqrt{90x^2y^3}$

23. $\sqrt{243x^4}$

24. $\sqrt{288x^4}$

25. $\sqrt{72x^2y^4}$

26. $\sqrt{72x^4y^2}$

27. $\sqrt{12x^3y^3}$

28. $\sqrt{20x^3y^3}$

The triangles below are called *isosceles* right triangles because the two shorter sides are the same length. In each case, use the Pythagorean theorem to find the length of the hypotenuse. Simplify your answers, but do not use a calculator to approximate them.

29.

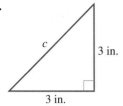

30.

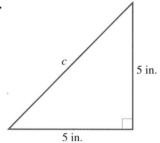

31.

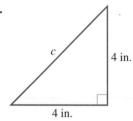

32.

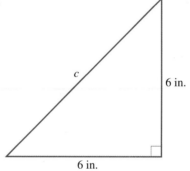

33.

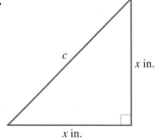

34.

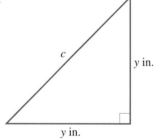

Applying the Concepts

35. Falling Time The formula that gives the number of seconds t it takes for an object to reach the ground when dropped from the top of a building h feet high is

$$t = \sqrt{\frac{h}{16}}$$

If a rock is dropped from the top of a building 25 feet high, how long will it take for the rock to hit the ground?

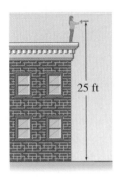

25 ft

36. Falling Time Using the formula given in Problem 35, how long will it take for an object dropped from a building 100 feet high to reach the ground?

37. Spiral of Roots Construct your own spiral of roots by using a ruler. Draw the first triangle by using two 1-inch lines. The first diagonal will have a length of $\sqrt{2}$ inches. Each new triangle will be formed by drawing a 1-inch line segment at the end of the previous diagonal so that the angle formed is 90°. Draw your spiral until you have at least six right triangles.

38. Spiral of Roots Construct a spiral of roots by using line segments of length 2 inches. The length of the first diagonal will be $2\sqrt{2}$ inches. The length of the second diagonal will be $2\sqrt{3}$ inches. What will be the length of the third diagonal?

Calculator Problems

Use a calculator to find decimal approximations for each of the following numbers.

39. $\sqrt{72}$ and $6\sqrt{2}$ **40.** $\sqrt{75}$ and $5\sqrt{3}$ **41.** $\sqrt{24}$ and $2\sqrt{6}$ **42.** $\sqrt{45}$ and $3\sqrt{5}$

Substitute $x = 5$ into each of the following expressions, and then use a calculator to obtain a decimal approximation to each.

43. $x\sqrt{x}$ **44.** $\sqrt{x^3}$ **45.** $x^2\sqrt{x}$ **46.** $\sqrt{x^5}$

Use a calculator to help complete the following tables. If an answer needs rounding, round to the nearest thousandth.

47.

x	$\sqrt{x}$	$2\sqrt{x}$	$\sqrt{4x}$
1			
2			
3			
4			

48.

x	$\sqrt{x}$	$2\sqrt{x}$	$\sqrt{4x}$
1			
4			
9			
16			

49.

x	$\sqrt{x}$	$3\sqrt{x}$	$\sqrt{9x}$
1			
2			
3			
4			

50.

x	$\sqrt{x}$	$3\sqrt{x}$	$\sqrt{9x}$
1			
4			
9			
16			

Review Problems

The problems below review material we covered in Section 4.6.

In Problems 51–56, use the formula $y = \frac{1}{2}x - 3$ to find y if:

51. $x = 0$ **52.** $x = 1$ **53.** $x = -4$ **54.** $x = 4$

55. $x = 2$ **56.** $x = -2$

In Problems 57–60, use the formula $2x + 5y = 10$ to find y if:

57. $x = 0$ **58.** $x = -5$ **59.** $x = 5$ **60.** $x = \dfrac{5}{2}$

In Problems 61–64, use the formula $2x + 5y = 10$ to find x if:

61. $y = 0$ **62.** $y = 2$ **63.** $y = -2$ **64.** $y = \dfrac{2}{5}$

5.9 Adding and Subtracting Roots

In the past we have combined similar terms by applying the distributive property. Here are some examples that will remind you of that procedure. The middle step in each example shows the distributive property.

$$7x + 3x = (7 + 3)x = 10x$$
$$5a^2 - 2a^2 = (5 - 2)a^2 = 3a^2$$
$$6xy + 3xy = (6 + 3)xy = 9xy$$
$$4x + 3x - 2x = (4 + 3 - 2)x = 5x$$

The distributive property is the only property that allows us to combine similar terms as we have done above. In order for the distributive property to be applied, the variable parts in each expression must be the same.

We add radical expressions in the same way we add similar terms, that is, by applying the distributive property. Here is how we use the distributive property to add $7\sqrt{2}$ and $3\sqrt{2}$:

$$7\sqrt{2} + 3\sqrt{2} = (7 + 3)\sqrt{2} \quad \textbf{Distributive property}$$
$$= 10\sqrt{2} \quad \textbf{Add 7 and 3}$$

To understand the steps shown here, you must remember that $7\sqrt{2}$ means 7 times $\sqrt{2}$; the 7 and the $\sqrt{2}$ are not "stuck together." Here are some additional examples. Compare them with the problems we did at the beginning of this section. Combine using the distributive property. (Assume all variables represent positive numbers.)

EXAMPLE 1 $5\sqrt{6} - 2\sqrt{6} = (5 - 2)\sqrt{6} = 3\sqrt{6}$ ▓

EXAMPLE 2 $6\sqrt{x} + 3\sqrt{x} = (6 + 3)\sqrt{x} = 9\sqrt{x}$ ▓

EXAMPLE 3 $4\sqrt{3} + 3\sqrt{3} - 2\sqrt{3} = (4 + 3 - 2)\sqrt{3} = 5\sqrt{3}$ ▓

As you can see, it is easy to combine radical expressions when each term contains the same square root.

Next, suppose we try to add $\sqrt{12}$ and $\sqrt{75}$. How should we go about it? You may think we should add 12 and 75 to get $\sqrt{87}$, but notice that we have not done that in any of the examples above. In fact, in Examples 1–3, the square root in the answer is the same square root we started with; we never added the number under the square roots!

A calculator can help us decide if $\sqrt{12} + \sqrt{75}$ is the same as $\sqrt{87}$. Here are the decimal approximations a calculator will give us:

$$\sqrt{12} + \sqrt{75} \approx 3.4641016 + 8.6602540 = 12.1243556$$
$$\sqrt{12 + 75} = \sqrt{87} \approx 9.3273791$$

As you can see, the two results are quite different, so we can assume that it would be a mistake to add the numbers under the square roots. That is:

$$\sqrt{12} + \sqrt{75} \neq \sqrt{12 + 75}$$

Practice Problems
Use the distributive property to combine each of the following.
1. $5\sqrt{3} - 2\sqrt{3}$

2. $7\sqrt{y} + 3\sqrt{y}$

3. $8\sqrt{5} - 2\sqrt{5} + 9\sqrt{5}$

Answers
1. $3\sqrt{3}$ 2. $10\sqrt{y}$ 3. $15\sqrt{5}$

The correct way to add $\sqrt{12}$ and $\sqrt{75}$ is to simplify each expression by taking as much out from under each square root as possible. Then, if the square roots in the resulting expressions are the same, we can add using the distributive property. Here is the way the problem is done correctly:

$$\sqrt{12} + \sqrt{75} = \sqrt{2 \cdot 2 \cdot 3} + \sqrt{5 \cdot 5 \cdot 3} \quad \text{\textbf{Simplify each square root}}$$
$$= 2\sqrt{3} + 5\sqrt{3}$$
$$= (2 + 5)\sqrt{3} \quad \text{\textbf{Distributive property}}$$
$$= 7\sqrt{3} \quad \text{\textbf{Add 2 and 5}}$$

On a calculator, $7\sqrt{3} \approx 7(1.7320508) = 12.1243556$, which matches the approximation a calculator gives for $\sqrt{12} + \sqrt{75}$.

Note You may be thinking "Why did he show us the wrong way to do the problem first?" The reason is simple: Many people will try to add $\sqrt{12}$ and $\sqrt{75}$ by adding 12 and 75—it is a natural thing to want to do. The reason we don't is that it gives us the wrong answer every time! One of the things you need to know about learning algebra is that your intuition may lead you to a mistake.

4. Combine, if possible:
$\sqrt{27} + \sqrt{75} - \sqrt{12}$

EXAMPLE 4 Combine, if possible: $\sqrt{18} + \sqrt{50} - \sqrt{8}$

SOLUTION   First we simplify each term by taking as much out from under the square root as possible. Then we use the distributive property to combine terms if they contain the same square root.

$$\sqrt{18} + \sqrt{50} - \sqrt{8} = \sqrt{3 \cdot 3 \cdot 2} + \sqrt{5 \cdot 5 \cdot 2} - \sqrt{2 \cdot 2 \cdot 2}$$
$$= 3\sqrt{2} + 5\sqrt{2} - 2\sqrt{2}$$
$$= (3 + 5 - 2)\sqrt{2}$$
$$= 6\sqrt{2} \quad \blacksquare$$

5. Subtract: $5\sqrt{45} - 3\sqrt{20}$

EXAMPLE 5 Subtract: $5\sqrt{54} - 3\sqrt{24}$

SOLUTION   Proceeding as we did in the previous example, we simplify each term first; then we subtract by applying the distributive property.

$$5\sqrt{54} - 3\sqrt{24} = 5\sqrt{3 \cdot 3 \cdot 6} - 3\sqrt{2 \cdot 2 \cdot 6}$$
$$= 5 \cdot 3\sqrt{6} - 3 \cdot 2\sqrt{6}$$
$$= 15\sqrt{6} - 6\sqrt{6}$$
$$= (15 - 6)\sqrt{6}$$
$$= 9\sqrt{6} \quad \blacksquare$$

6. Combine, if possible:
$7\sqrt{18x^3} - 3\sqrt{50x^3}$

EXAMPLE 6 Assume x is a positive number and combine, if possible:

$$5\sqrt{12x^3} - 3\sqrt{75x^3}$$

SOLUTION   We simplify each square root, and then we subtract.

$$5\sqrt{12x^3} - 3\sqrt{75x^3} = 5\sqrt{2 \cdot 2 \cdot 3 \cdot x \cdot x \cdot x} - 3\sqrt{5 \cdot 5 \cdot 3 \cdot x \cdot x \cdot x}$$
$$= 5 \cdot 2 \cdot x\sqrt{3x} - 3 \cdot 5 \cdot x\sqrt{3x}$$
$$= 10x\sqrt{3x} - 15x\sqrt{3x}$$
$$= -5x\sqrt{3x} \quad \blacksquare$$

Answers
4. $6\sqrt{3}$ **5.** $9\sqrt{5}$ **6.** $6x\sqrt{2x}$

Problem Set 5.9

Combine by applying the distributive property. Assume all variables represent positive numbers.

1. $2\sqrt{3} + 8\sqrt{3}$

2. $2\sqrt{3} - 8\sqrt{3}$

3. $7\sqrt{5} - 3\sqrt{5}$

4. $7\sqrt{5} + 3\sqrt{5}$

5. $9\sqrt{x} + 3\sqrt{x} - 5\sqrt{x}$

6. $6\sqrt{x} + 10\sqrt{x} - 3\sqrt{}$

7. $8\sqrt{7} + \sqrt{7}$

8. $9\sqrt{7} + \sqrt{7}$

9. $2\sqrt{y} + \sqrt{y} + 3\sqrt{y}$

10. $7\sqrt{y} + \sqrt{y} + 2\sqrt{y}$

Simplify each square root, and then combine if possible. Assume all variables represent positive numbers.

11. $\sqrt{18} + \sqrt{32}$

12. $\sqrt{12} + \sqrt{27}$

13. $\sqrt{75} + \sqrt{27}$

14. $\sqrt{50} + \sqrt{8}$

15. $2\sqrt{75} - 4\sqrt{27}$

16. $4\sqrt{50} - 5\sqrt{8}$

17. $2\sqrt{90} + 3\sqrt{40} - 4\sqrt{10}$

18. $5\sqrt{40} - 2\sqrt{90} + 3\sqrt{10}$

19. $\sqrt{72x^2} - \sqrt{50x^2}$

20. $\sqrt{98x^2} - \sqrt{72x^2}$

21. $4\sqrt{20x^3} + 3\sqrt{45x^3}$

22. $8\sqrt{48x^3} + 2\sqrt{12x^3}$

In each diagram below, find the distance from *A* to *B*. Simplify your answers, but do not use a calculator.

23.

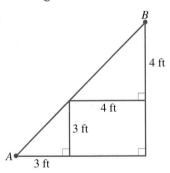

24.

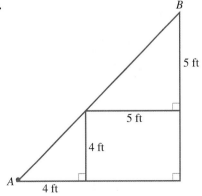

Calculator Problems

25. Use a calculator to show that $\sqrt{2} + \sqrt{3}$ is not the same as $\sqrt{5}$.

26. Use a calculator to show that $\sqrt{5} - \sqrt{2}$ is not the same as $\sqrt{3}$.

Use a calculator to help complete the following tables. If an answer needs rounding, round to the nearest thousandth.

27.

x	$\sqrt{x^2 + 9}$	$x + 3$
1		
2		
3		
4		
5		
6		

28.

x	$\sqrt{x^2 + 16}$	$x + 4$
1		
2		
3		
4		
5		
6		

29.

x	$\sqrt{x + 3}$	$\sqrt{x} + \sqrt{3}$
1		
2		
3		
4		
5		
6		

30.

x	$\sqrt{x + 4}$	$\sqrt{x} + 2$
1		
2		
3		
4		
5		
6		

Review Problems

The problems below review material we covered in Section 4.9.

Graph each equation.

31. $x + y = 3$

32. $x - y = 3$

33. $y = \dfrac{1}{2}x - 3$

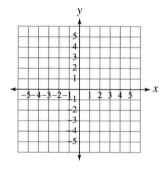
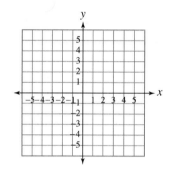
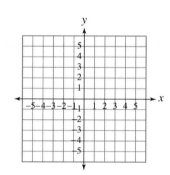

34. $y = \dfrac{1}{2}x + 3$

35. $2x + 5y = 10$

36. $5x + 2y = 10$

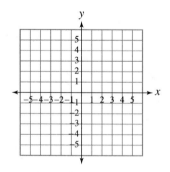
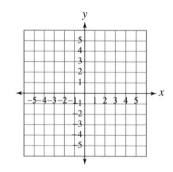
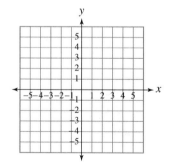

Chapter ⑤ Summary

Place Value [5.1]

The place values for the first five places to the right of the decimal point are

DECIMAL POINT	TENTHS	HUNDREDTHS	THOUSANDTHS	TEN THOUSANDTHS	HUNDRED THOUSANDTHS
.	$\dfrac{1}{10}$	$\dfrac{1}{100}$	$\dfrac{1}{1,000}$	$\dfrac{1}{10,000}$	$\dfrac{1}{100,000}$

Rounding Decimals [5.1]

If the digit in the column to the right of the one we are rounding to is 5 or more, we add 1 to the digit in the column we are rounding to; otherwise, we leave it alone. We then replace all digits to the right of the column we are rounding to with zeros if they are to the left of the decimal point; otherwise, we simply delete them.

Addition and Subtraction with Decimals [5.2]

To add (or subtract) decimal numbers, we align the decimal points and add (or subtract) as if we were adding (or subtracting) whole numbers. The decimal point in the answer goes directly below the decimal points in the problem.

Multiplication with Decimals [5.3]

To multiply two decimal numbers, we multiply as if the decimal points were not there. The decimal point in the product has as many digits to the right as there are total digits to the right of the decimal points in the two original numbers.

Division with Decimals [5.4]

To begin a division problem with decimals, we make sure that the divisor is a whole number. If it is not, we move the decimal point in the divisor to the right as many places as it takes to make it a whole number. We must then be sure to move the decimal point in the dividend the same number of places to the right. Once the divisor is a whole number, we divide as usual. The decimal point in the answer is placed directly above the decimal point in the dividend.

Changing Fractions to Decimals [5.5]

To change a fraction to a decimal, we divide the numerator by the denominator.

Changing Decimals to Fractions [5.5]

To change a decimal to a fraction, we write the digits to the right of the decimal point over the appropriate power of 10.

Examples

1. The number 4.123 in words is "four and one hundred twenty-three thousandths."

2. 357.753 rounded to the nearest
Tenth: 357.8
Ten: 360

3.
$$\begin{array}{r} 3.400 \\ 25.060 \\ +\ 0.347 \\ \hline 28.807 \end{array}$$

4. If we multiply 3.49×5.863, there will be a total of $2 + 3 = 5$ digits to the right of the decimal point in the answer.

5.
$$\begin{array}{r} 1.39 \\ 2.5\overline{)3.4.75} \\ \underline{2\ 5} \\ 9\ 7 \\ \underline{7\ 5} \\ 2\ 25 \\ \underline{2\ 25} \\ 0 \end{array}$$

6. $\dfrac{4}{15} = 0.2\overline{6}$ because
$$\begin{array}{r} .266 \\ 15\overline{)4.000} \\ \underline{3\ 0} \\ 1\ 00 \\ \underline{90} \\ 100 \\ \underline{90} \\ 10 \end{array}$$

7. $0.781 = \dfrac{781}{1,000}$

8.
$$\frac{1}{2}x - 3.78 = 2.52$$
$$\frac{1}{2}x - 3.78 + \mathbf{3.78} = 2.52 + \mathbf{3.78}$$
$$\frac{1}{2}x = 6.30$$
$$\mathbf{2}\left(\frac{1}{2}x\right) = \mathbf{2}(6.30)$$
$$x = 12.6$$

9. $\sqrt{49} = 7$ because
$7^2 = 7 \cdot 7 = 49$

10.
$$\sqrt{50} = \sqrt{5 \cdot 5 \cdot 2} \qquad \textbf{Factor 50}$$
$$= \sqrt{5 \cdot 5} \cdot \sqrt{2} \qquad \textbf{Multiplication property}$$
$$= 5 \cdot \sqrt{2} \qquad \textbf{Repeated factor property}$$

11. $\sqrt{12} + \sqrt{75}$
$$= \sqrt{2 \cdot 2 \cdot 3} + \sqrt{5 \cdot 5 \cdot 3}$$
$$= 2\sqrt{3} + 5\sqrt{3}$$
$$= (2 + 5)\sqrt{3}$$
$$= 7\sqrt{3}$$

Equations Containing Decimals [5.6]

We solve equations that contain decimals by applying the addition property of equality and the multiplication property of equality, as we did with the equations in Chapter 4.

Square Roots [5.7]

The square root of a positive number a, written $\sqrt{a}$, is the number we square to get a.

Pythagorean Theorem [5.7]

In any right triangle, the length of the longest side (the hypotenuse) is equal to the square root of the sum of the squares of the two shorter sides.

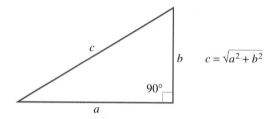

Multiplication Property for Square Roots [5.8]

If a and b are positive numbers, then

$$\sqrt{a \cdot b} = \sqrt{a} \cdot \sqrt{b}$$

In words: The square root of a product is the product of the square roots.

Repeated Factor Property for Square Roots [5.8]

If a is a positive number, then

$$\sqrt{a \cdot a} = a$$

Simplifying Square Roots [5.8]

To simplify a square root, we factor the number under the square root symbol into the product of prime factors. Then we use the two properties of square roots shown above to take as much out from under the square root symbol as possible.

Adding and Subtracting Roots [5.9]

We add expressions containing square roots by first simplifying each square root and then, if the square roots in the resulting terms are the same, applying the distributive property.

Chapter **5** Review

Give the place value of the 7 in each of the following numbers. [5.1]

1. 36.007

2. 121.379

Write each of the following with a decimal number. [5.1]

3. Thirty-seven and forty-two ten thousandths

4. One hundred and two hundred two hundred thousandths

5. Round 98.7654 to the nearest hundredth. [5.1]

Perform the following operations. [5.2, 5.3, 5.4]

6. $3.78 + 2.036$

7. $11.076 - 3.297$

8. 6.7×5.43

9. $0.89(-24.24)$

10. $-29.07 \div (-3.8)$

11. $0.7134 \div 0.58$

12. Write $\dfrac{7}{8}$ as a decimal. [5.5]

13. Write 0.705 as a fraction in lowest terms. [5.5]

14. Write 14.125 as a mixed number. [5.5]

Simplify each of the following expressions as much as possible. [5.5]

15. $3.3 - 4(0.22)$

16. $54.987 - 2(3.05 + 0.151)$

17. $125\left(\dfrac{3}{5}\right) + 4$

18. $\dfrac{3}{5}(0.9) + \dfrac{2}{5}(0.4)$

Solve each equation. [5.6]

19. $x + 9.8 = 3.9$

20. $5x = 23.4$

21. $0.5y - 0.2 = 3$

22. $5x - 7.2 = 3x + 3.8$

Simplify each expression as much as possible. [5.7]

23. $3\sqrt{25}$

24. $\sqrt{64} - \sqrt{36}$

25. $4\sqrt{25} + 3\sqrt{81}$

26. $\sqrt{\dfrac{16}{49}}$

27. A student has bills of $19.48 for heating and electricity, $6.72 for the telephone, and $241.50 for rent each month. What is the total of these three bills? [5.2]

28. A person purchases $7.23 worth of goods at a drugstore. If a $10 bill is used to pay for the purchases, how much change is received? [5.2]

29. What is the product of $\frac{3}{4}$ and the sum of 2.8 and 3.7? [5.3]

30. If a person earns $223.60 for working 40 hours, what is the person's hourly wage? [5.4]

31. Jayma earns the following grades during the spring term of 2001. [5.4]

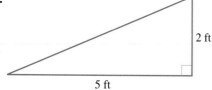

CLASS	UNITS	GRADE
College algebra	4	C
Speech	3	A
Accounting	3	B
Marketing	3	B
Real estate	2	C

a. Calculate her GPA.

b. If her grade in college algebra had been a B instead of a C, by how much would her GPA have increased?

32. Comparing Area On April 3, 2000, *USA Today* changed the size of its paper. Previous to this date, each page of the paper was 13.5 inches wide and 22.25 inches long. The new paper size is 1.25 inches narrower and 0.5 inches longer. (After you have finished, compare your answers with the answers to Problem 44 in the Chapter 3 Review.) [5.2, 5.3]

a. What is the area of a page previous to April 3, 2000?

b. What is the area of a page after April 3, 2000?

c. What is the difference in the areas of the two page sizes?

Find the length of the hypotenuse in each right triangle. Round your answers to the nearest tenth. [5.7]

33.

2 ft

5 ft

34.

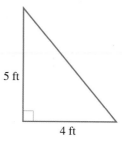

5 ft

4 ft

35. Find the perimeter (to the nearest tenth) and the area of the triangle in Problem 33. [3.3, 5.2]

36. Find the perimeter (to the nearest tenth) and the area of the triangle in Problem 34. [3.3, 5.2]

Simplify each expression by taking as much out from under the radical as possible. [5.8]

37. $\sqrt{12}$

38. $\sqrt{50}$

39. $\sqrt{32}$

40. $\sqrt{24}$

41. $\sqrt{20x^2}$

42. $\sqrt{20x^3}$

43. $\sqrt{18x^2y^3}$

44. $\sqrt{75x^3y^2}$

Combine. [5.9]

45. $8\sqrt{3} + 2\sqrt{3}$

46. $7\sqrt{6} + 10\sqrt{6}$

47. $4\sqrt{3} + \sqrt{3}$

48. $9\sqrt{2} + \sqrt{2}$

Simplify each square root, and then combine if possible. [5.9]

49. $\sqrt{24} + \sqrt{54}$

50. $\sqrt{18} + \sqrt{8}$

51. $3\sqrt{75} - 8\sqrt{27}$

52. $7\sqrt{20} - 2\sqrt{45}$

Chapter **5** Test

1. Write the decimal number 5.053 in words.

2. Give the place value of the 4 in the number 53.0543.

3. Write seventeen and four hundred six ten thousandths as a decimal number.

4. Round 46.7549 to the nearest hundredth.

Perform the following operations.

5. $7 + 0.6 + 0.58$

6. $12.032 - 5.976$

7. $5.7(6.24)$

8. $-22.672 \div (-2.6)$

9. Write $\dfrac{23}{25}$ as a decimal.

10. Write 0.56 as a fraction in lowest terms.

Simplify each expression as much as possible.

11. $5.2(2.8 + 0.02)$

12. $5.2 - 3(0.17)$

13. $23.852 - 3(2.01 + 0.231)$

14. $\dfrac{3}{5}(0.6) - \dfrac{2}{3}(0.15)$

15. Solve: $6a - 0.18 = a + 0.77$

Simplify each expression as much as possible.

16. $2\sqrt{36} + 3\sqrt{64}$

17. $\sqrt{\dfrac{25}{81}}$

18. A person purchases $8.47 worth of goods at a drug-store. If a $20 bill is used to pay for the purchases, how much change is received?

19. If coffee sells for $5.44 per pound, how much will 3.5 pounds of coffee cost?

20. If a person earns $262 for working 40 hours, what is the person's hourly wage?

21. Find the length of the hypotenuse of the right triangle below.

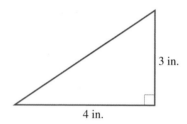

Simplify by taking as much out from under the radical as possible.

22. $\sqrt{72}$

23. $\sqrt{48x^2y}$

24. Combine $5\sqrt{7} + 3\sqrt{7}$.

25. Simplify each square root, and then combine, if possible: $6\sqrt{12} - 5\sqrt{48}$

Chapters (1-5) Cumulative Review

Simplify.

1. $6 + \dfrac{5}{x}$

2. $(3x^2 - 5x + 7) - (x^2 - 3x - 8)$

3. $56(287)$

4. $(x - 6)^2$

5. $5a^3(3a^2 - 2)$

6. $(3x - 2)(4x + 5)$

7. $-4.3(-12.96)$

8. $1,292 \div 17$

9. $\dfrac{5}{14} \div \dfrac{15}{21}$

10. Round 463,612 to the nearest thousand.

11. Change $\dfrac{63}{4}$ to a mixed number.

12. Give the opposite and absolute value of 11.

13. Is $x = -2$ a solution to $3x - 5 = 1$?

14. Change each decimal into a fraction.

Decimal	0.125	0.250	0.375	0.500	0.625	0.750	0.875	1
Fraction								

15. Give the quotient of 72 and -8.

16. Identify the property or properties used in the following: $2 \cdot (x \cdot 3) = (2 \cdot 3) \cdot x$

17. Translate into symbols, then simplify: Three times the sum of 13 and 4 is 51.

18. Reduce $\dfrac{120ab}{70b}$

19. Use the equation $5x - 10y = 15$ to find x when $y = -3$.

Simplify.

20. $6(4)^2 - 8(2)^3$

21. $\dfrac{-6 + 2(-4)}{8 - 10}$

22. $\sqrt{72x^3y^2}$

23. $\dfrac{2}{3}(0.45) - \dfrac{4}{5}(0.8)$

24. $2b + 6 - 4b - 5$

25. $\left(3\dfrac{1}{3} - \dfrac{1}{2}\right)\left(4\dfrac{1}{2} + \dfrac{3}{4}\right)$

26. $3\sqrt{24} - 5\sqrt{54}$

Solve.

27. $\dfrac{3}{4}x = 21$

28. $8b - 0.22 = 2b + 0.68$

29. $2(x - 6) = -10$

30. $5(3x + 4) - 9 = -2(x - 8)$

31. $a - \dfrac{3}{4} = \dfrac{3}{8}$

32. Gambling A gambler loses $15 playing poker one night, wins $25 the next night, and then loses $30 the last night. How much did the gambler win or lose overall?

33. Average Score Lorena has scores of 83, 85, 79, 93, and 80 on her first five math tests. What is her average score for these five tests?

34. Geometry Find the length of the hypotenuse of a right triangle with shorter sides of 6 in. and 8 in.

35. Age Carrie is seven years younger than Martin. Two years ago, the sum of their ages was 13. How old are they now?

36. Recipe A muffin recipe calls for $2\frac{3}{4}$ cups of flour. If the recipe is tripled, how many cups of flour will be needed?

37. Volume Find the volume of a rectangular solid with length 5 in., width 3 in., and height 7 in.

38. Hourly Wage If you earn $384 for working 40 hours, what is your hourly wage?

39. Test Taking If 40 students take a Spanish test and $\frac{4}{5}$ of them pass, how many students passed?

40. Going Out to Eat The table below shows the annual amount of money an average household spent eating out, as reported by the Bureau of Labor Statistics, for the years 1994 through 1998.

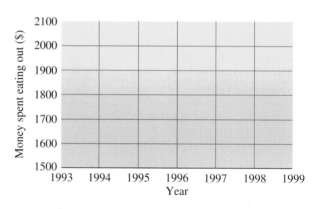

| | GOING OUT TO EAT | |
YEAR	ANNUAL AMOUNT SPENT BY AVERAGE FAMILY ($)	ROUNDED TO THE NEAREST HUNDRED
1994	1,698	
1995	1,702	
1996	1,823	
1997	1,921	
1998	2,030	

a. Fill in the last column of the table by rounding each number in the second column to the nearest hundred.
b. Use the template and the rounded numbers from part (a) to construct a line graph of the information in the table.
c. From the line graph, estimate the annual amount of money spent eating out by an average family in 1999.

Ratio, Proportion, and Unit Analysis

INTRODUCTION

Model railroads continue to be as popular today as they ever have been. One of the first things model railroaders ask each other is what scale they work with. The scale of a model train indicates its size relative to a full-size train. Each scale is associated with a ratio and a fraction, as shown in the table and bar chart below. An HO scale model train has a ratio of 1 to 87, meaning it is $\frac{1}{87}$ as large as an actual train.

SCALE	RATIO	AS A FRACTION
LGB	1 to 22.5	$\frac{1}{22.5}$
#1	1 to 32	$\frac{1}{32}$
O	1 to 43.5	$\frac{1}{43.5}$
S	1 to 64	$\frac{1}{64}$
HO	1 to 87	$\frac{1}{87}$
TT	1 to 120	$\frac{1}{120}$

Model trains

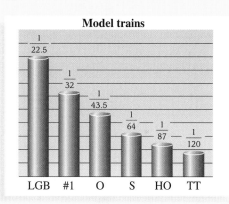

How long is an actual box car that has an HO scale model 5 inches long? In this chapter we will solve this problem.

6.1 Ratios

The *ratio* of two numbers is a way of comparing them. If we say that the ratio of two numbers is 2 to 1, then the first number is twice as large as the second number. For example, if there are 10 men and 5 women enrolled in a math class, then the ratio of men to women is 10 to 5. Because 10 is twice as large as 5, we can also say that the ratio of men to women is 2 to 1.

We can define the ratio of two numbers in terms of fractions.

DEFINITION

The **ratio** of two numbers can be expressed as a fraction, where the first number in the ratio is the numerator and the second number in the ratio is the denominator. *In symbols:*

If a and b are any two numbers,

then the ratio of a to b is $\dfrac{a}{b}$ $(b \neq 0)$

We handle ratios the same way we handle fractions. For example, when we said that the ratio of 10 men to 5 women was the same as the ratio 2 to 1, we were actually saying

$$\frac{10}{5} = \frac{2}{1} \quad \textbf{Reducing to lowest terms}$$

Because we have already studied fractions in detail, much of the introductory material on ratio will seem like review.

EXAMPLE 1   Express the ratio of 16 to 48 as a fraction in lowest terms.

SOLUTION   Because the ratio is 16 to 48, the numerator of the fraction is 16 and the denominator is 48:

$$\frac{16}{48} = \frac{1}{3} \quad \textbf{In lowest terms}$$

Notice that the first number in the ratio becomes the numerator of the fraction, and the second number in the ratio becomes the denominator. ■

EXAMPLE 2   Give the ratio of $\frac{2}{3}$ to $\frac{4}{9}$ as a fraction in lowest terms.

SOLUTION   We begin by writing the ratio of $\frac{2}{3}$ to $\frac{4}{9}$ as a complex fraction. The numerator is $\frac{2}{3}$, and the denominator is $\frac{4}{9}$. Then we simplify.

$$\frac{\frac{2}{3}}{\frac{4}{9}} = \frac{2}{3} \cdot \frac{9}{4} \quad \textbf{Division by } \tfrac{4}{9} \textbf{ is the same as multiplication by } \tfrac{9}{4}$$

$$= \frac{18}{12} \quad \textbf{Multiply}$$

$$= \frac{3}{2} \quad \textbf{Reduce to lowest terms}$$

■

Practice Problems

1. Express the ratio of 32 to 48 as a fraction in lowest terms.

2. Give the ratio of $\frac{3}{5}$ to $\frac{9}{10}$ as a fraction in lowest terms.

Answers

1. $\dfrac{2}{3}$ 2. $\dfrac{2}{3}$

3. Write the ratio of 0.06 to 0.12 as a fraction in lowest terms.

Note

Another symbol used to denote ratio is the colon (:). The ratio of, say, 5 to 4 can be written as 5:4. Although we will not use it here, this notation is fairly common.

EXAMPLE 3 Write the ratio of 0.08 to 0.12 as a fraction in lowest terms.

SOLUTION   When the ratio is in reduced form, it is customary to write it with whole numbers and not decimals. For this reason we multiply the numerator and the denominator of the ratio by 100 to clear it of decimals. Then we reduce to lowest terms.

$$\frac{0.08}{0.12} = \frac{0.08 \times 100}{0.12 \times 100}$$ **Multiply the numerator and the denominator by 100 to clear the ratio of decimals**

$$= \frac{8}{12}$$ **Multiply**

$$= \frac{2}{3}$$ **Reduce to lowest terms**   ■

Table 1 shows several more ratios and their fractional equivalents. Notice that in each case the fraction has been reduced to lowest terms. Also, the ratio that contains decimals has been rewritten as a fraction that does not contain decimals.

Table 1

RATIO	FRACTION	FRACTION IN LOWEST TERMS
25 to 35	$\frac{25}{35}$	$\frac{5}{7}$
35 to 25	$\frac{35}{25}$	$\frac{7}{5}$
8 to 2	$\frac{8}{2}$	$\frac{4}{1}$ We can also write this as just 4.
$\frac{1}{4}$ to $\frac{3}{4}$	$\frac{\frac{1}{4}}{\frac{3}{4}}$	$\frac{1}{3}$ because $\frac{\frac{1}{4}}{\frac{3}{4}} = \frac{1}{4} \cdot \frac{4}{3} = \frac{1}{3}$
0.6 to 1.7	$\frac{0.6}{1.7}$	$\frac{6}{17}$ because $\frac{0.6 \times 10}{1.7 \times 10} = \frac{6}{17}$

4. Suppose the basketball player in Example 4 makes 12 out of 16 free throws. Write the ratio again using these new numbers.

EXAMPLE 4 During a game, a basketball player makes 12 out of the 18 free throws he attempts. Write the ratio of the number of free throws he makes to the number of free throws he attempts as a fraction in lowest terms.

SOLUTION   Because he makes 12 out of 18, we want the ratio 12 to 18, or

$$\frac{12}{18} = \frac{2}{3}$$

Because the ratio is 2 to 3, we can say that, in this particular game, he made 2 out of every 3 free throws he attempted.   ■

Answers

3. $\frac{1}{2}$ **4.** $\frac{3}{4}$

EXAMPLE 5   A solution of alcohol and water contains 15 milliliters of water and 5 milliliters of alcohol. Find the ratio of alcohol to water, water to alcohol, water to total solution, and alcohol to total solution. Write each ratio as a fraction and reduce to lowest terms.

SOLUTION   There are 5 milliliters of alcohol and 15 milliliters of water, so there is 20 milliliters of solution (alcohol + water). The ratios are as follows:

The ratio of alcohol to water is 5 to 15, or

$$\frac{5}{15} = \frac{1}{3}$$ **In lowest terms**

— 15 ml
— 5 ml

The ratio of water to alcohol is 15 to 5, or

$$\frac{15}{5} = \frac{3}{1}$$ **In lowest terms**

The ratio of water to total solution is 15 to 20, or

$$\frac{15}{20} = \frac{3}{4}$$ **In lowest terms**

The ratio of alcohol to total solution is 5 to 20, or

$$\frac{5}{20} = \frac{1}{4}$$ **In lowest terms**   ∎

Rates

Whenever a ratio compares two quantities that have different units (and neither unit can be converted to the other), then the ratio is called a *rate*. For example, if we were to travel 120 miles in 3 hours, then our average rate of speed expressed as the ratio of miles to hours would be

$$\frac{120 \text{ miles}}{3 \text{ hours}} = \frac{40 \text{ miles}}{1 \text{ hour}}$$ **Divide the numerator and the denominator by 3 to reduce to lowest terms**

The ratio $\frac{40 \text{ miles}}{1 \text{ hour}}$ can be expressed as

$$40 \frac{\text{miles}}{\text{hour}} \quad \text{or} \quad 40 \text{ miles/hour} \quad \text{or} \quad 40 \text{ miles per hour}$$

A rate is expressed in simplest form when the numerical part of the denominator is 1. To accomplish this we use division.

EXAMPLE 6   A train travels 125 miles in 2 hours. What is the train's rate in miles per hour?

SOLUTION   The ratio of miles to hours is

$$\frac{125 \text{ miles}}{2 \text{ hours}} = 62.5 \frac{\text{miles}}{\text{hour}}$$ **Divide 125 by 2**

$$= 62.5 \text{ miles per hour}$$

If the train travels 125 miles in 2 hours, then its average rate of speed is 62.5 miles per hour.   ∎

5. A solution of alcohol and water contains 12 milliliters of water and 4 milliliters of alcohol. Find the ratio of alcohol to water, water to alcohol, and water to total solution. Write each ratio as a fraction and reduce to lowest terms.

6. A car travels 107 miles in 2 hours. What is the car's rate in miles per hour?

Answers

5. $\frac{1}{3}, \frac{3}{1}, \frac{3}{4}$ **6.** 53.5 miles/hour

7. A car travels 192 miles on 6 gallons of gas. Give the ratio of miles to gallons as a rate in miles per gallon.

EXAMPLE 7   A car travels 90 miles on 5 gallons of gas. Give the ratio of miles to gallons as a rate in miles per gallon.

SOLUTION   The ratio of miles to gallons is

$$\frac{90 \text{ miles}}{5 \text{ gallons}} = 18 \frac{\text{miles}}{\text{gallon}} \qquad \textbf{Divide 90 by 5}$$

$$= 18 \text{ miles/gallon}$$

The gas mileage of the car is 18 miles per gallon. ■

Unit Pricing

One kind of rate that is very common is *unit pricing*. Unit pricing is the ratio of price to quantity. Suppose a 1-liter bottle of a certain soft drink costs $1.19, whereas a 2-liter bottle of the same drink costs $1.39. Which is the better buy? That is, which has the lower price per liter?

$$\frac{\$1.19}{1 \text{ liter}} = \$1.19 \text{ per liter}$$

$$\frac{\$1.39}{2 \text{ liters}} = \$0.695 \text{ per liter}$$

The unit price for the 1-liter bottle is $1.19 per liter, whereas the unit price for the 2-liter bottle is 69.5¢ per liter. The 2-liter bottle is a better buy.

8. A supermarket sells vegetable juice in three different containers at the following prices:
  5.5 ounces, 48¢
  11.5 ounces, 75¢
  46 ounces, $2.19
Give the unit price in cents per ounce for each one. Round to the nearest tenth of a cent, if necessary.

EXAMPLE 8   A supermarket sells low-fat milk in three different containers at the following prices:

1 gallon $3.59
½ gallon $1.99
1 quart $1.29   **(1 quart = ¼ gallon)**

$3.59 $1.99 $1.29

Give the unit price in dollars per gallon for each one.

SOLUTION   Because 1 quart = ¼ gallon, we have

1-gallon container $\dfrac{\$3.59}{1 \text{ gallon}} = \dfrac{\$3.59}{1 \text{ gallon}} = \$3.59 \text{ per gallon}$

½-gallon container $\dfrac{\$1.99}{\frac{1}{2} \text{ gallon}} = \dfrac{\$1.99}{0.5 \text{ gallon}} = \$3.98 \text{ per gallon}$

1-quart container $\dfrac{\$1.29}{1 \text{ quart}} = \dfrac{\$1.29}{0.25 \text{ gallon}} = \$5.16 \text{ per gallon}$

The 1-gallon container has the lowest unit price, whereas the 1-quart container has the highest unit price. ■

Answers
7. 32 miles/gallon
8. 8.7¢/ounce, 6.5¢/ounce, 4.8¢/ounce

Problem Set 6.1

Write each of the following ratios as a fraction in lowest terms. None of the answers should contain decimals.

1. 8 to 6   **2.** 6 to 8   **3.** 64 to 12   **4.** 12 to 64

5. 100 to 250   **6.** 250 to 100   **7.** 13 to 26   **8.** 36 to 18

9. $\frac{3}{4}$ to $\frac{1}{4}$ **10.** $\frac{5}{8}$ to $\frac{3}{8}$ **11.** $\frac{7}{3}$ to $\frac{6}{3}$ **12.** $\frac{9}{5}$ to $\frac{11}{5}$

13. $\frac{6}{5}$ to $\frac{6}{7}$ **14.** $\frac{5}{3}$ to $\frac{1}{3}$ **15.** $2\frac{1}{2}$ to $3\frac{1}{2}$ **16.** $5\frac{1}{4}$ to $1\frac{3}{4}$

17. $2\frac{2}{3}$ to $\frac{5}{3}$ **18.** $\frac{1}{2}$ to $3\frac{1}{2}$ **19.** 0.05 to 0.15   **20.** 0.21 to 0.03

21. 0.3 to 3   **22.** 0.5 to 10   **23.** 1.2 to 10   **24.** 6.4 to 0.8

25. $\frac{1}{2}$ to 1.5   **26.** $\frac{1}{4}$ to 0.75

Applying the Concepts

27. Coffee Prices The table below shows the unit price of some popular coffees. Use the template below to construct a bar chart of the information in the table.

COFFEE	PRICE (cents/cup)
Green Mountain Kona Estate	32
Gevalia Kaffe Select	17
Dunkin' Donuts Original	11
Brothers Gourmet	10
Trader Joe's Colombian	7
Yuban Colombian	6
Folgers Classic	5

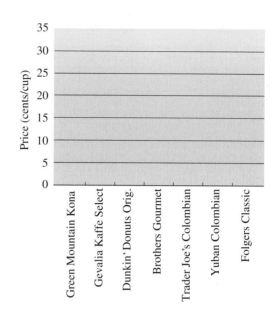

28. Geometry Regarding the diagram below, *AC* represents the length of the line segment that starts at *A* and ends at *C*. From the diagram we see that *AC* = 8.

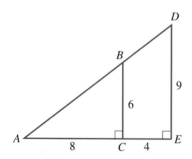

a. Find the ratio of *BC* to *AC*.

b. What is the length *AE*?

c. Find the ratio of *DE* to *AE*.

29. Family Budget A family of four budgeted the following amounts for some of their monthly bills:

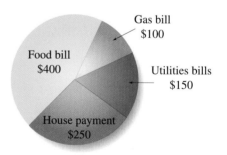

a. What is the ratio of the house payment to the food bill?

b. What is the ratio of the gas bill to the food bill?

c. What is the ratio of the utilities bills to the food bill?

d. What is the ratio of the house payment to the utilities bills?

30. Nutrition One cup of breakfast cereal was found to contain the following nutrients:

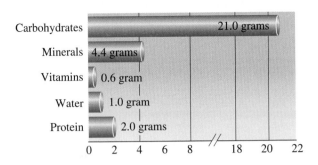

a. Find the ratio of water to protein.

b. Find the ratio of carbohydrates to protein.

c. Find the ratio of vitamins to minerals.

d. Find the ratio of protein to vitamins and minerals.

Rates

31. Miles/Hour A car travels 220 miles in 4 hours. What is the rate of the car in miles per hour?

32. Miles/Hour A train travels 360 miles in 5 hours. What is the rate of the train in miles per hour?

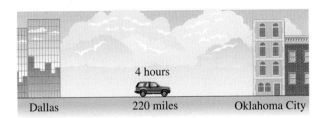

33. Kilometers/Hour It takes a car 3 hours to travel 252 kilometers. What is the rate in kilometers per hour?

34. Kilometers/Hour In 6 hours an airplane travels 4,200 kilometers. What is the rate of the airplane in kilometers per hour?

35. Gallons/Second The flow of water from a water faucet can fill a 3-gallon container in 15 seconds. Give the ratio of gallons to seconds as a rate in gallons per second.

36. Gallons/Minute A 225-gallon drum is filled in 3 minutes. What is the rate in gallons per minute?

37. Liters/Minute It takes 4 minutes to fill a 56-liter gas tank. What is the rate in liters per minute?

38. Liters/Hour The gas tank on a car holds 60 liters of gas. At the beginning of a 6-hour trip, the tank is full. At the end of the trip, it contains only 12 liters. What is the rate at which the car uses gas in liters per hour?

39. Miles/Gallon A car travels 95 miles on 5 gallons of gas. Give the ratio of miles to gallons as a rate in miles per gallon.

40. Miles/Gallon On a 384-mile trip, an economy car uses 8 gallons of gas. Give this as a rate in miles per gallon.

41. Miles/Liter The gas tank on a car has a capacity of 75 liters. On a full tank of gas, the car travels 325 miles. What is the gas mileage in miles per liter?

42. Miles/Liter A car pulling a trailer can travel 105 miles on 70 liters of gas. What is the gas mileage in miles per liter?

Unit Pricing

43. Cents/Ounce A 6-ounce can of frozen orange juice costs 96¢. Give the unit price in cents per ounce.

44. Cents/Liter A 2-liter bottle of root beer costs $1.25. Give the unit price in cents per liter.

45. Cents/Ounce A 20-ounce package of frozen peas is priced at 99¢. Give the unit price in cents per ounce.

46. Cents/Pound A 4-pound bag of cat food costs $8.12. Give the unit price in dollars per pound.

Calculator Problems

Write each of the following ratios as a fraction, and then use a calculator to change the fraction to a decimal. Round all answers to the nearest hundredth. Do not reduce fractions.

47. Number of Students The total number of students attending a community college in the Midwest is 4,722. Of these students, 2,314 are male and 2,408 are female.

  a. Give the ratio of males to females as a fraction and as a decimal.

  b. Give the ratio of females to males as a fraction and as a decimal.

  c. Give the ratio of males to total number of students as a fraction and as a decimal.

  d. Give the ratio of total number of students to females as a fraction and as a decimal.

48. Stock Market One method of comparing stocks on the stock market is the price to earnings ratio, or P/E.

$$P/E = \frac{\text{Current Stock Price}}{\text{Earnings per Share}}$$

Most stocks have a P/E between 25 and 40. A stock with a P/E of less than 25 may be undervalued, while a stock with a P/E greater than 40 may be overvalued. Fill in the P/E for each stock listed in the table below. Round to the nearest whole number.

STOCK	PRICE	EARNINGS PER SHARE	P/E
IBM	149	$6.35	
AOL	$139\frac{1}{2}$	$0.61	
DIS	$30\frac{3}{16}$	$0.91	
KM	$15\frac{15}{16}$	$0.68	
GE	$90\frac{13}{16}$	$2.75	
TOY	$20\frac{1}{2}$	$1.66	

49. Miles/Hour A car travels 675.4 miles in $12\frac{1}{2}$ hours. Give the rate in miles per hour to the nearest hundredth.

50. Miles/Hour At the beginning of a trip, the odometer on a car read 32,567.2 miles. At the end of the trip, it read 32,741.8 miles. If the trip took $4\frac{1}{4}$ hours, what was the rate of the car in miles per hour to the nearest tenth?

51. Miles/Gallon If a truck travels 128.4 miles on 13.8 gallons of gas, what is the gas mileage in miles per gallon? (Round to the nearest tenth.)

52. Cents/Ounce If a 15-day supply of vitamins costs $1.62, what is the price in cents per day?

Review Problems

The problems that follow review multiplication with fractions from Section 3.3, along with some material covered in Section 5.5. Reviewing these problems will help you with the next section.

Multiply.

53. $8 \cdot \dfrac{1}{3}$

54. $9 \cdot \dfrac{1}{3}$

55. $25 \cdot \dfrac{1}{1,000}$

56. $25 \cdot \dfrac{1}{100}$

57. $36.5 \cdot \dfrac{1}{100} \cdot 10$

58. $36.5 \cdot \dfrac{1}{1,000} \cdot 100$

59. $248 \cdot \dfrac{1}{10} \cdot \dfrac{1}{10}$

60. $969 \cdot \dfrac{1}{10} \cdot \dfrac{1}{10}$

61. $48 \cdot \dfrac{1}{12} \cdot \dfrac{1}{3}$

62. $56 \cdot \dfrac{1}{12} \cdot \dfrac{1}{2}$

Extending the Concepts

63. Geometry Regarding the diagram below, AC represents the length of the line segment that starts at A and ends at C. From the diagram we see that $AC = 8$.

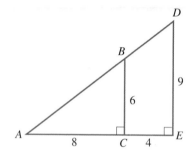

a.   Use the Pythagorean theorem to find AB.

b.   Use the Pythagorean theorem to find AD.

c.   Find BD.

d.   Show that the ratios $\dfrac{AB}{AC}$ and $\dfrac{AD}{AE}$ are equal.

e.   Show that the ratios $\dfrac{AB}{BC}$ and $\dfrac{AD}{DE}$ are equal.

6.2 Unit Analysis 1: Length

In this section we will become more familiar with the units used to measure length. We will look at the U.S. system of measurement and the metric system of measurement.

Measuring the length of an object is done by assigning a number to its length. To let other people know what that number represents, we include with it a unit of measure. The most common units used to represent length in the U.S. system are inches, feet, yards, and miles. The basic unit of length is the foot. The other units are defined in terms of feet, as Table 1 shows.

Table 1

12 inches (in.) =	1 foot (ft)
1 yard (yd) =	3 feet
1 mile (mi) =	5,280 feet

As you can see from the table, the abbreviations for inches, feet, yards, and miles are in., ft, yd, and mi, respectively. What we haven't indicated, even though you may not have realized it, is what 1 foot represents. We have defined all our units associated with length in terms of feet, but we haven't said what a foot is.

There is a long history of the evolution of what is now called a foot. At different times in the past, a foot has represented different arbitrary lengths. Currently, a foot is defined to be exactly 0.3048 meter (the basic measure of length in the metric system), where a meter is 1,650,763.73 wavelengths of the orange–red line in the spectrum of krypton-86 in a vacuum (this doesn't mean much to me either). The reason a foot and a meter are defined this way is that we always want them to measure the same length. Because the wavelength of the orange–red line in the spectrum of krypton-86 will always remain the same, so will the length that a foot represents.

Now that we have said what we mean by 1 foot (even though we may not understand the technical definition), we can go on and look at some examples that involve converting from one kind of unit to another.

EXAMPLE 1   Convert 5 feet to inches.

SOLUTION   Because 1 foot = 12 inches, we can multiply 5 by 12 inches to get

$$5 \text{ feet} = 5 \times 12 \text{ inches}$$
$$= 60 \text{ inches} \qquad \blacksquare$$

This method of converting from feet to inches probably seems fairly simply. But as we go further in this chapter, the conversions from one kind of unit to another will become more complicated. For these more complicated problems, we need another way to show conversions so that we can be certain to end them with the correct unit of measure. For example, since 1 ft = 12 in., we can say that there are 12 in. per 1 ft or 1 ft per 12 in. That is:

$$\frac{12 \text{ in.}}{1 \text{ ft}} \longleftarrow \text{Per} \quad \text{ or } \quad \frac{1 \text{ ft}}{12 \text{ in.}} \longleftarrow \text{Per}$$

Practice Problems

1.   Convert 8 feet to inches.

Answer

1.   96 in.

We call the expressions $\frac{12 \text{ in.}}{1 \text{ ft}}$ and $\frac{1 \text{ ft}}{12 \text{ in.}}$ *conversion factors*. The fraction bar is read as "per." Both these conversion factors are really just the number 1. That is:

$$\frac{12 \text{ in.}}{1 \text{ ft}} = \frac{12 \text{ in.}}{12 \text{ in.}} = 1$$

We already know that multiplying a number by 1 leaves the number unchanged. So, to convert from one unit to the other, we can multiply by one of the conversion factors without changing value. Both the conversion factors above say the same thing about the units feet and inches. They both indicate that there are 12 inches in every foot. The one we choose to multiply by depends on what units we are starting with and what units we want to end up with. If we start with feet and we want to end up with inches, we multiply by the conversion factor

$$\frac{12 \text{ in.}}{1 \text{ ft}}$$

The units of feet will divide out and leave us with inches.

$$5 \text{ feet} = 5 \text{ ft} \times \frac{12 \text{ in.}}{1 \text{ ft}}$$
$$= 5 \times 12 \text{ in.}$$
$$= 60 \text{ in.}$$

The key to this method of conversion lies in setting the problem up so that the correct units divide out to simplify the expression. We are treating units such as feet in the same way we treated factors when reducing fractions. If a factor is common to the numerator and the denominator, we can divide it out and simplify the fraction. The same idea holds for units such as feet.

We can rewrite Table 1 so that it shows the conversion factors associated with units of length, as shown in Table 2.

Table 2

UNITS OF LENGTH IN THE U.S. SYSTEM

THE RELATIONSHIP BETWEEN	IS	TO CONVERT FROM ONE TO THE OTHER, MULTIPLY BY
feet and inches	12 in. = 1 ft	$\frac{12 \text{ in.}}{1 \text{ ft}}$ or $\frac{1 \text{ ft}}{12 \text{ in.}}$
feet and yards	1 yd = 3 ft	$\frac{3 \text{ ft}}{1 \text{ yd}}$ or $\frac{1 \text{ yd}}{3 \text{ ft}}$
feet and miles	1 mi = 5,280 ft	$\frac{5{,}280 \text{ ft}}{1 \text{ mi}}$ or $\frac{1 \text{ mi}}{5{,}280 \text{ ft}}$

EXAMPLE 2 The most common ceiling height in houses is 8 feet. How many yards is this?

8 ft

Note

We will use this method of converting from one kind of unit to another throughout the rest of this chapter. You should practice using it until you are comfortable with it and can use it correctly. However, it is not the only method of converting units. You may see shortcuts that will allow you to get results more quickly. Use shortcuts if you wish, so long as you can consistently get correct answers and are not using your shortcuts because you don't understand our method of conversion. Use the method of conversion as given here until you are good at it; then use shortcuts if you want to.

2. The roof of a two-story house is 26 feet above the ground. How many yards is this?

Answer

2. $8\frac{2}{3}$ yd, or 8.67 yd

SOLUTION To convert 8 feet to yards, we multiply by the conversion factor $\frac{1 \text{ yd}}{3 \text{ ft}}$ so that feet will divide out and we will be left with yards.

$$8 \text{ ft} = 8 \cancel{\text{ft}} \times \frac{1 \text{ yd}}{3 \cancel{\text{ft}}} \qquad \textbf{Multiply by correct conversion factor}$$

$$= \frac{8}{3} \text{ yd} \qquad \mathbf{8 \times \frac{1}{3} = \frac{8}{3}}$$

$$= 2\frac{2}{3} \text{ yd} \qquad \textbf{Or 2.67 yd to the nearest hundredth} \quad \blacksquare$$

EXAMPLE 3 A football field is 100 yards long. How many inches long is a football field?

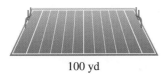

100 yd

SOLUTION In this example we must convert yards to feet and then feet to inches. (To make this example more interesting, we are pretending we don't know that there are 36 inches in a yard.) We choose the conversion factors that will allow all the units except inches to divide out.

$$100 \text{ yd} = 100 \cancel{\text{yd}} \times \frac{3 \cancel{\text{ft}}}{1 \cancel{\text{yd}}} \times \frac{12 \text{ in.}}{1 \cancel{\text{ft}}}$$

$$= 100 \times 3 \times 12 \text{ in.}$$

$$= 3{,}600 \text{ in.} \quad \blacksquare$$

Metric Units of Length

In the metric system the standard unit of length is a meter. A meter is a little longer than a yard (about 3.4 inches longer). The other units of length in the metric system are written in terms of a meter. The metric system uses prefixes to indicate what part of the basic unit of measure is being used. For example, in *milli*meter the prefix *milli* means "one thousandth" of a meter. Table 3 gives the meanings of the most common metric prefixes.

Table 3

THE MEANING OF METRIC PREFIXES

PREFIX	MEANING
milli	0.001
centi	0.01
deci	0.1
deka	10
hecto	100
kilo	1,000

We can use these prefixes to write the other units of length and conversion factors for the metric system, as given in Table 4.

3. How many inches are in 220 yards?

Answer

3. 7,920 in.

Table 4

METRIC UNITS OF LENGTH

THE RELATIONSHIP BETWEEN	IS	TO CONVERT FROM ONE TO THE OTHER, MULTIPLY BY
millimeters (mm) and meters (m)	1,000 mm = 1 m	$\dfrac{1{,}000 \text{ mm}}{1 \text{ m}}$ or $\dfrac{1 \text{ m}}{1{,}000 \text{ mm}}$
centimeters (cm) and meters	100 cm = 1 m	$\dfrac{100 \text{ cm}}{1 \text{ m}}$ or $\dfrac{1 \text{ m}}{100 \text{ cm}}$
decimeters (dm) and meters	10 dm = 1 m	$\dfrac{10 \text{ dm}}{1 \text{ m}}$ or $\dfrac{1 \text{ m}}{10 \text{ dm}}$
dekameters (dam) and meters	1 dam = 10 m	$\dfrac{10 \text{ m}}{1 \text{ dam}}$ or $\dfrac{1 \text{ dam}}{10 \text{ m}}$
hectometers (hm) and meters	1 hm = 100 m	$\dfrac{100 \text{ m}}{1 \text{ hm}}$ or $\dfrac{1 \text{ hm}}{100 \text{ m}}$
kilometers (km) and meters	1 km = 1,000 m	$\dfrac{1{,}000 \text{ m}}{1 \text{ km}}$ or $\dfrac{1 \text{ km}}{1{,}000 \text{ m}}$

We use the same method to convert between units in the metric system as we did with the U.S. system. We choose the conversion factor that will allow the units we start with to divide out, leaving the units we want to end up with.

4. Convert 67 centimeters to meters.

EXAMPLE 4   Convert 25 millimeters to meters.

SOLUTION   To convert from millimeters to meters, we multiply by the conversion factor $\dfrac{1 \text{ m}}{1{,}000 \text{ mm}}$:

$$25 \text{ mm} = 25 \text{ m\!m} \times \frac{1 \text{ m}}{1{,}000 \text{ m\!m}}$$
$$= \frac{25 \text{ m}}{1{,}000}$$
$$= 0.025 \text{ m} \qquad \blacksquare$$

5. Convert 78.4 mm to decimeters.

EXAMPLE 5   Convert 36.5 centimeters to decimeters.

SOLUTION   We convert centimeters to meters and then meters to decimeters:

$$36.5 \text{ cm} = 36.5 \text{ c\!m} \times \frac{1 \text{ m}}{100 \text{ c\!m}} \times \frac{10 \text{ dm}}{1 \text{ m}}$$
$$= \frac{36.5 \times 10}{100} \text{ dm}$$
$$= 3.65 \text{ dm} \qquad \blacksquare$$

The most common units of length in the metric system are millimeters, centimeters, meters, and kilometers. The other units of length we have listed in our table of metric lengths are not as widely used. The method we have used to convert from one unit of length to another in Examples 2–5 is called *unit analysis*. If you take a chemistry class, you will see it used many times. The same is true of many other science classes as well.

Answers
4. 0.67 m   **5.** 0.784 dm

We can summarize the procedure used in unit analysis with the following steps:

Steps Used in Unit Analysis

1. Identify the units you are starting with.
2. Identify the units you want to end with.
3. Find conversion factors that will bridge the starting units and the ending units.
4. Set up the multiplication problem so that all units except the units you want to end with will divide out.

EXAMPLE 6 A sheep farmer is making new lambing pens for the upcoming lambing season. Each pen is a rectangle 6 feet wide and 8 feet long. The fencing material he wants to use sells for $1.36 per foot. If he is planning to build five separate lambing pens (they are separate because he wants a walkway between them), how much will he have to spend for fencing material?

SOLUTION To find the amount of fencing material he needs for one pen, we find the perimeter of a pen.

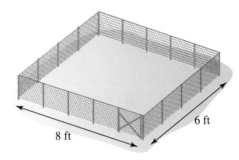

Perimeter = 6 + 6 + 8 + 8 = 28 feet

We set up the solution to the problem using unit analysis. Our starting unit is *pens* and our ending unit is *dollars*. Here are the conversion factors that will form a bridge between pens and dollars:

1 pen = 28 feet of fencing
1 foot of fencing = 1.36 dollars

Next we write the multiplication problem, using the conversion factors, that will allow all the units except dollars to divide out:

$$5 \text{ pens} = 5 \text{ pens} \times \frac{28 \text{ feet of fencing}}{1 \text{ pen}} \times \frac{1.36 \text{ dollars}}{1 \text{ foot of fencing}}$$
$$= 5 \times 28 \times 1.36 \text{ dollars}$$
$$= \$190.40$$

6. The farmer in Example 6 decides to build six pens instead of five and upgrades his fencing material so that it costs $1.72 per foot. How much does it cost him to build the six pens?

476

7. Assume that the mistake in the advertisement is that feet per second should read feet per minute. Is 1,100 feet per minute a reasonable speed for a chair lift?

EXAMPLE 7 In 1993, a ski resort in Vermont advertised their new high-speed chair lift as "the world's fastest chair lift, with a speed of 1,100 feet per second." Show why the speed cannot be correct.

SOLUTION   To solve this problem, we can convert feet per second into miles per hour, a unit of measure we are more familiar with on an intuitive level. Here are the conversion factors we will use:

$$1 \text{ mile} = 5{,}280 \text{ feet}$$
$$1 \text{ hour} = 60 \text{ minutes}$$
$$1 \text{ minute} = 60 \text{ seconds}$$

$$1100 \text{ ft/second} = \frac{1100 \text{ feet}}{1 \text{ second}} \times \frac{1 \text{ mile}}{5{,}280 \text{ feet}} \times \frac{60 \text{ seconds}}{1 \text{ minute}} \times \frac{60 \text{ minutes}}{1 \text{ hour}}$$
$$= \frac{1100 \times 60 \times 60 \text{ miles}}{5{,}280 \text{ hours}}$$
$$= 750 \text{ miles/hour} \qquad \blacksquare$$

Answer

7. 12.5 mph is a reasonable speed for a chair lift.

Problem Set 6.2

Make the following conversions in the U.S. system by multiplying by the appropriate conversion factors. Write your answers as whole numbers or mixed numbers.

1. 5 ft to inches

2. 9 ft to inches

3. 10 ft to inches

4. 20 ft to inches

5. 2 yd to feet

6. 8 yd to feet

7. 4.5 yd to inches

8. 9.5 yd to inches

9. 27 in. to feet

10. 36 in. to feet

11. 19 ft to yards

12. 100 ft to yards

13. 48 in. to yards

14. 56 in. to yards

Make the following conversions in the metric system by multiplying by the appropriate conversion factors. Write your answers as whole numbers or decimals.

15. 18 m to centimeters

16. 18 m to millimeters

17. 4.8 km to meters

18. 8.9 km to meters

19. 5 dm to centimeters

20. 12 dm to millimeters

21. 248 m to kilometers

22. 969 m to kilometers

23. 67 cm to millimeters

24. 67 mm to centimeters

25. 3,498 cm to meters

26. 4,388 dm to meters

27. 63.4 cm to decimeters

28. 89.5 cm to decimeters

Applying the Concepts

29. Softball If the distance between first and second base in softball is 60 feet, how many yards is it from first to second base?

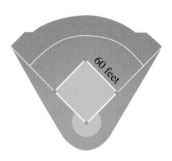

30. Tower Height A transmitting tower is 100 feet tall. How many inches is that?

31. High Jump If a person high jumps 6 feet 8 inches, how many inches is the jump?

32. Desk Width A desk is 48 inches wide. What is the width in yards?

33. Ceiling Height Suppose the ceiling of a home is 2.44 meters above the floor. Express the height of the ceiling in centimeters.

34. Notebook Width Standard-sized notebook paper is 21.6 centimeters wide. Express this width in millimeters.

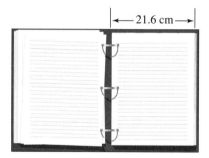

35. Dollar Width A dollar bill is about 6.5 centimeters wide. Express this width in millimeters.

36. Pencil Length Most new pencils are 19 centimeters long. Express this length in meters.

37. Surveying A unit of measure sometimes used in surveying is the *chain*. There are 80 chains in 1 mile. How many chains are in 37 miles?

38. Surveying Another unit of measure used in surveying is a *link*; 1 link is about 8 inches. About how many links are there in 5 feet?

39. Metric System A very small unit of measure in the metric system is the *micron* (abbreviated μm). There are 1,000 μm in 1 millimeter. How many microns are in 12 centimeters?

40. Metric System Another very small unit of measure in the metric system is the *angstrom* (abbreviated Å). There are 10,000,000 Å in 1 millimeter. How many angstroms are in 15 decimeters?

41. **Horse Racing** In horse racing, 1 *furlong* is 220 yards. How many feet are in 12 furlongs?

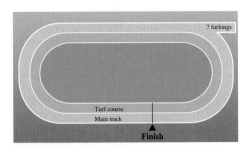

43. **Speed Limit** The maximum speed limit on part of Highway 101 in California is 55 miles/hour. Convert 55 miles/hour to feet/second. (Round to the nearest tenth.)

45. **Track and Field** A person who runs the 100-yard dash in 10.5 seconds has an average speed of 9.52 yards/second. Convert 9.52 yards/second to miles/hour. (Round to the nearest tenth.)

47. **Speed of a Bullet** The bullet from a rifle leaves the barrel traveling 1,500 feet/second. Convert 1,500 feet/second to miles/hour. (Round to the nearest whole number.)

49. **Farming** A farmer is fencing a pasture that is $\frac{1}{2}$ mile wide and 1 mile long. If the fencing material sells for $1.15 per foot, how much will it cost him to fence all four sides of the pasture?

51. **Farming** A 4-H Club group is raising lambs to show at the County Fair. Each lamb eats $\frac{1}{8}$ of a bale of alfalfa a day. If the alfalfa costs $10.50 per bale, how much will it cost to feed one lamb for 120 days?

42. **Sailing** A *fathom* is 6 feet. How many yards are in 19 fathoms?

44. **Speed Limit** The maximum speed limit on part of Highway 5 in California is 65 miles/hour. Convert 65 miles/hour to feet/second. (Round to the nearest tenth.)

46. **Track and Field** A person who runs a mile in 8 minutes has an average speed of 0.125 miles/minute. Convert 0.125 miles/minute to miles/hour. (Round to the nearest tenth.)

48. **Speed of a Bullet** A bullet from a machine gun on a B-17 Flying Fortress in World War II had a muzzle speed of 1,750 feet/second. Convert 1,750 feet/second to miles/hour. (Round to the nearest whole number.)

50. **Cost of Fencing** A family with a swimming pool puts up a chain-link fence around the pool. The fence forms a rectangle 12 yards wide and 24 yards long. If the chain-link fence sells for $2.50 per foot, how much will it cost to fence all four sides of the pool?

52. **Farming** A 4-H Club group is raising pigs to show at the County Fair. Each pig eats 2.4 pounds of grain a day. If the grain costs $5.25 per pound, how much will it cost to feed one pig for 60 days?

Calculator Problems

Set up the following conversions as you have been doing. Then perform the calculations on a calculator.

53. Change 751 miles to feet.

54. Change 639.87 centimeters to meters.

55. Change 4,982 yards to inches.

56. Change 379 millimeters to kilometers.

57. Mount Whitney is the highest point in California. It is 14,494 feet above sea level. Give its height in miles to the nearest tenth.

58. The tallest mountain in the United States is Mount McKinley in Alaska. It is 20,320 feet tall. Give its height in miles to the nearest tenth.

59. California has 3,427 miles of shoreline. How many feet is this?

60. The tip of the television tower at the top of the Empire State Building in New York City is 1,472 feet above the ground. Express this height in miles to the nearest hundredth.

Review Problems

The review problems in this chapter cover the work we have done previously with fractions and mixed numbers. We will begin with a review of multiplication and division of fractions and mixed numbers. Write your answers as whole numbers, proper fractions, or mixed numbers.

Find each product. (Multiply.)

61. $\dfrac{2}{3} \cdot \dfrac{1}{2}$

62. $\dfrac{7}{9} \cdot \dfrac{3}{14}$

63. $-8\left(-\dfrac{3}{4}\right)$

64. $-12 \cdot \dfrac{1}{3}$

65. $1\dfrac{1}{2} \cdot 2\dfrac{1}{3}$

66. $\dfrac{1}{6} \cdot 4\dfrac{2}{3}$

Find each quotient. (Divide.)

67. $\dfrac{3}{4} \div \dfrac{1}{8}$

68. $\dfrac{3}{5} \div \dfrac{6}{25}$

69. $-4 \div \left(-\dfrac{2}{3}\right)$

70. $1 \div \left(-\dfrac{1}{3}\right)$

71. $1\dfrac{3}{4} \div 2\dfrac{1}{2}$

72. $\dfrac{9}{8} \div 1\dfrac{7}{8}$

6.3 Unit Analysis II: Area and Volume

Figure 1 below gives a summary of the geometric objects we have worked with in previous chapters, along with the formulas for finding the area of each object.

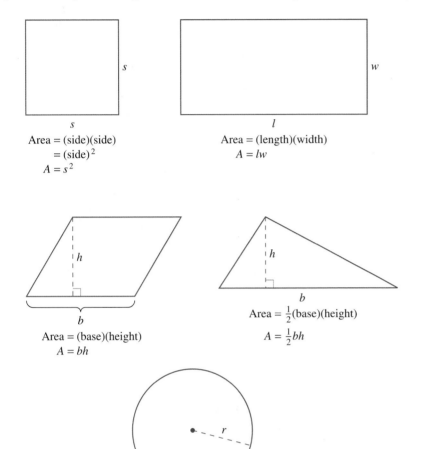

Area = (side)(side)
= (side)^2
$A = s^2$

Area = (length)(width)
$A = lw$

Area = (base)(height)
$A = bh$

Area = $\frac{1}{2}$(base)(height)
$A = \frac{1}{2}bh$

Area = π(radius)^2
$A = \pi r^2$

Figure 1   Areas of common geometric shapes

In Example 2 of Section 1.10, we found the number of square inches in one square foot. Here is the same problem again, but this time we work it using unit analysis.

EXAMPLE 1   Find the number of square inches in 1 square foot.

SOLUTION   We can think of 1 square foot as 1 ft^2 = 1 ft × ft. To convert from feet to inches, we use the conversion factor 1 foot = 12 inches. Because the unit foot appears twice in 1 ft^2, we multiply by our conversion factor twice.

$$1 \text{ ft}^2 = 1 \text{ ft} \times \text{ft} \times \frac{12 \text{ in.}}{1 \text{ ft}} \times \frac{12 \text{ in.}}{1 \text{ ft}} = 12 \times 12 \text{ in.} \times \text{in.} = 144 \text{ in}^2 \qquad \blacksquare$$

Now that we know that 1 ft^2 is the same as 144 in^2, we can use this fact as a conversion factor to convert between square feet and square inches. Depending on which units we are converting from, we would use either

$$\frac{144 \text{ in}^2}{1 \text{ ft}^2} \quad \text{or} \quad \frac{1 \text{ ft}^2}{144 \text{ in}^2}$$

2. If the poster in Example 2 is surrounded by a frame 6 inches wide, find the number of square feet of wall space covered by the framed poster.

EXAMPLE 2   A rectangular poster measures 36 inches by 24 inches. How many square feet of wall space will the poster cover?

SOLUTION   One way to work this problem is to find the number of square inches the poster covers, and then convert square inches to square feet.

$$\text{Area of poster} = \text{length} \times \text{width} = 36 \text{ in.} \times 24 \text{ in.} = 864 \text{ in.}^2$$

To finish the problem, we convert square inches to square feet:

$$864 \text{ in}^2 = 864 \text{ in}^2 \times \frac{1 \text{ ft}^2}{144 \text{ in}^2}$$
$$= \frac{864}{144} \text{ ft}^2$$
$$= 6 \text{ ft}^2$$

Table 1 gives the most common units of area in the U.S. system of measurement, along with the corresponding conversion factors.

Table 1

U.S. UNITS OF AREA

THE RELATIONSHIP BETWEEN	IS	TO CONVERT FROM ONE TO THE OTHER, MULTIPLY BY
square inches and square feet	$144 \text{ in}^2 = 1 \text{ ft}^2$	$\frac{144 \text{ in}^2}{1 \text{ ft}^2}$ or $\frac{1 \text{ ft}^2}{144 \text{ in}^2}$
square yards and square feet	$9 \text{ ft}^2 = 1 \text{ yd}^2$	$\frac{9 \text{ ft}^2}{1 \text{ yd}^2}$ or $\frac{1 \text{ yd}^2}{9 \text{ ft}^2}$
acres and square feet	$1 \text{ acre} = 43{,}560 \text{ ft}^2$	$\frac{43{,}560 \text{ ft}^2}{1 \text{ acre}}$ or $\frac{1 \text{ acre}}{43{,}560 \text{ ft}^2}$
acres and square miles	$640 \text{ acres} = 1 \text{ mi}^2$	$\frac{640 \text{ acres}}{1 \text{ mi}^2}$ or $\frac{1 \text{ mi}^2}{640 \text{ acres}}$

3. The same dressmaker orders a bolt of material that is 1.5 yards wide and 45 yards long. How many square feet of material were ordered?

EXAMPLE 3   A dressmaker orders a bolt of material that is 1.5 yards wide and 30 yards long. How many square feet of material were ordered?

SOLUTION   The area of the material in square yards is

$$A = 1.5 \times 30$$
$$= 45 \text{ yd}^2$$

Converting this to square feet, we have

$$45 \text{ yd}^2 = 45 \text{ yd}^2 \times \frac{9 \text{ ft}^2}{1 \text{ yd}^2}$$
$$= 405 \text{ ft}^2$$

Answers
2. 12 ft²   **3.** 607.5 ft²

EXAMPLE 4   A farmer has 75 acres of land. How many square feet of land does the farmer have?

SOLUTION   Changing acres to square feet, we have

$$75 \text{ acres} = 75 \text{ acres} \times \frac{43{,}560 \text{ ft}^2}{1 \text{ acre}}$$

$$= 75 \times 43{,}560 \text{ ft}^2$$

$$= 3{,}267{,}000 \text{ ft}^2$$

4. A farmer has 55 acres of land. How many square feet of land does the farmer have?

EXAMPLE 5   A new shopping center is to be constructed on 256 acres of land. How many square miles is this?

SOLUTION   Multiplying by the conversion factor that will allow acres to divide out, we have

$$256 \text{ acres} = 256 \text{ acres} \times \frac{1 \text{ mi}^2}{640 \text{ acres}}$$

$$= \frac{256}{640} \text{ mi}^2$$

$$= 0.4 \text{ mi}^2$$

5. A school is to be constructed on 960 acres of land. How many square miles is this?

Units of area in the metric system are considerably simpler than those in the U.S. system because metric units are given in terms of powers of 10. Table 2 lists the conversion factors that are most commonly used.

Table 2

METRIC UNITS OF AREA

THE RELATIONSHIP BETWEEN	IS	TO CONVERT FROM ONE TO THE OTHER, MULTIPLY BY
square millimeters and square centimeters	$1 \text{ cm}^2 = 100 \text{ mm}^2$	$\dfrac{100 \text{ mm}^2}{1 \text{ cm}^2}$ or $\dfrac{1 \text{ cm}^2}{100 \text{ mm}^2}$
square centimeters and square decimeters	$1 \text{ dm}^2 = 100 \text{ cm}^2$	$\dfrac{100 \text{ cm}^2}{1 \text{ dm}^2}$ or $\dfrac{1 \text{ dm}^2}{100 \text{ cm}^2}$
square decimeters and square meters	$1 \text{ m}^2 = 100 \text{ dm}^2$	$\dfrac{100 \text{ dm}^2}{1 \text{ m}^2}$ or $\dfrac{1 \text{ m}^2}{100 \text{ dm}^2}$
square meters and ares (a)	$1 \text{ a} = 100 \text{ m}^2$	$\dfrac{100 \text{ m}^2}{1 \text{ a}}$ or $\dfrac{1 \text{ a}}{100 \text{ m}^2}$
ares and hectares (ha)	$1 \text{ ha} = 100 \text{ a}$	$\dfrac{100 \text{ a}}{1 \text{ ha}}$ or $\dfrac{1 \text{ ha}}{100 \text{ a}}$

Answers
4. 2,395,800 ft²   **5.** 1.5 mi²

6. How many square centimeters are in 1 square meter?

EXAMPLE 6 How many square millimeters are in 1 square meter?

SOLUTION We start with 1 m² and end up with square millimeters:

$$1\ m^2 = 1\ \cancel{m^2} \times \frac{100\ \cancel{dm^2}}{1\ \cancel{m^2}} \times \frac{100\ \cancel{cm^2}}{1\ \cancel{dm^2}} \times \frac{100\ mm^2}{1\ \cancel{cm^2}}$$
$$= 100 \times 100 \times 100\ mm^2$$
$$= 1{,}000{,}000\ mm^2$$

Units of Measure for Volume

Table 3 lists the units of volume in the U.S. system and their conversion factors.

Table 3

UNITS OF VOLUME IN THE U.S. SYSTEM

THE RELATIONSHIP BETWEEN	IS	TO CONVERT FROM ONE TO THE OTHER, MULTIPLY BY
cubic inches (in³) and cubic feet (ft³)	1 ft³ = 1,728 in³	$\dfrac{1{,}728\ in^3}{1\ ft^3}$ or $\dfrac{1\ ft^3}{1{,}728\ in^3}$
cubic feet and cubic yards (yd³)	1 yd³ = 27 ft³	$\dfrac{27\ ft^3}{1\ yd^3}$ or $\dfrac{1\ yd^3}{27\ ft^3}$
fluid ounces (fl oz) and pints (pt)	1 pt = 16 fl oz	$\dfrac{16\ fl\ oz}{1\ pt}$ or $\dfrac{1\ pt}{16\ fl\ oz}$
pints and quarts (qt)	1 qt = 2 pt	$\dfrac{2\ pt}{1\ qt}$ or $\dfrac{1\ qt}{2\ pt}$
quarts and gallons (gal)	1 gal = 4 qt	$\dfrac{4\ qt}{1\ gal}$ or $\dfrac{1\ gal}{4\ qt}$

7. How many pints are in a 5-gallon pail?

EXAMPLE 7 What is the capacity (volume) in pints of a 1-gallon container of milk?

SOLUTION We change from gallons to quarts and then quarts to pints by multiplying by the appropriate conversion factors as given in Table 3.

$$1\ gal = 1\ \cancel{gal} \times \frac{4\ \cancel{qt}}{1\ \cancel{gal}} \times \frac{2\ pt}{1\ \cancel{qt}}$$
$$= 1 \times 4 \times 2\ pt$$
$$= 8\ pt$$

A 1-gallon container has the same capacity as 8 1-pint containers.

8. A dairy herd produces 2,000 quarts of milk each day. How many 10-gallon containers will this milk fill?

EXAMPLE 8 A dairy herd produces 1,800 quarts of milk each day. How many gallons is this equivalent to?

SOLUTION Converting 1,800 quarts to gallons, we have

$$1{,}800\ qt = 1{,}800\ \cancel{qt} \times \frac{1\ gal}{4\ \cancel{qt}}$$
$$= \frac{1{,}800}{4}\ gal$$
$$= 450\ gal$$

We see that 1,800 quarts is equivalent to 450 gallons.

Answers
6. 10,000 cm²   **7.** 40 pt
8. 50 containers

In the metric system the basic unit of measure for volume is the liter. A liter is the volume enclosed by a cube that is 10 cm on each edge, as shown in Figure 2. We can see that a liter is equivalent to 1,000 cm³.

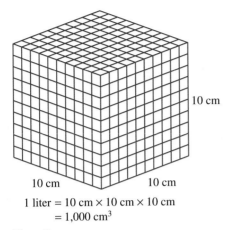

10 cm

10 cm   10 cm

1 liter = 10 cm × 10 cm × 10 cm
= 1,000 cm³

Figure 2

The other units of volume in the metric system use the same prefixes we encountered previously. The units with prefixes centi, deci, and deka are not as common as the others, so in Table 4 we include only liters, milliliters, hectoliters, and kiloliters.

Table 4

METRIC UNITS OF VOLUME

THE RELATIONSHIP BETWEEN	IS	TO CONVERT FROM ONE TO THE OTHER, MULTIPLY BY
milliliters (mL) and liters	1 liter (L) = 1,000 mL	$\dfrac{1{,}000 \text{ mL}}{1 \text{ liter}}$ or $\dfrac{1 \text{ liter}}{1{,}000 \text{ mL}}$
hectoliters (hL) and liters	100 liters = 1 hL	$\dfrac{100 \text{ liters}}{1 \text{ hL}}$ or $\dfrac{1 \text{ hL}}{100 \text{ liters}}$
kiloliters (kL) and liters	1,000 liters (L) = 1 kL	$\dfrac{1{,}000 \text{ liters}}{1 \text{ kL}}$ or $\dfrac{1 \text{ kL}}{1{,}000 \text{ liters}}$

Here is an example of conversion from one unit of volume to another in the metric system.

EXAMPLE 9 A sports car has a 2.2-liter engine. What is the displacement (volume) of the engine in milliliters?

SOLUTION   Using the appropriate conversion factor from Table 4, we have

$$2.2 \text{ liters} = 2.2 \text{ liters} \times \frac{1{,}000 \text{ mL}}{1 \text{ liter}}$$

$$= 2.2 \times 1{,}000 \text{ mL}$$

$$= 2{,}200 \text{ mL}$$

Note
As you can see from the table and the discussion above, a cubic centimeter (cm³) and a milliliter (mL) are equal. Both are one thousandth of a liter. It is also common in some fields (like medicine) to abbreviate the term cubic centimeter as cc. Although we will use the notation mL when discussing volume in the metric system, you should be aware that 1 mL = 1 cm³ = 1 cc.

9. A 3.5-liter engine will have a volume of how many milliliters?

Answer
9. 3,500 mL

Problem Set 6.3

Use the tables given in this section to make the following conversions. Be sure to show the conversion factor used in each case.

1. 3 ft^2 to square inches

2. 5 ft^2 to square inches

3. 288 in^2 to square feet

4. 720 in^2 to square feet

5. 30 acres to square feet

6. 92 acres to square feet

7. 2 mi^2 to acres

8. 7 mi^2 to acres

9. 1,920 acres to square miles

10. 3,200 acres to square miles

11. 12 yd^2 to square feet

12. 20 yd^2 to square feet

13. 17 cm^2 to square millimeters

14. 150 mm^2 to square centimeters

15. 2.8 m^2 to square centimeters

16. 10 dm^2 to square millimeters

17. 1,200 mm^2 to square meters

18. 19.79 cm^2 to square meters

19. 5 a to square meters

20. 12 a to square centimeters

21. 7 ha to ares

22. 3.6 ha to ares

23. 342 a to hectares

24. 986 a to hectares

Applying the Concepts

25. Sports The diagrams below show the dimensions of playing fields for the National Football League (NFL), the Canadian Football League, and Arena Football. Find the area of each field and then convert each area to acres. Round answers to the nearest hundredth.

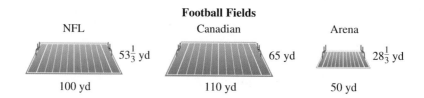

Football Fields

NFL — $53\frac{1}{3}$ yd, 100 yd   Canadian — 65 yd, 110 yd   Arena — $28\frac{1}{3}$ yd, 50 yd

26. Soccer The rules for soccer state that the playing field must be from 100 to 120 yards long and 55 to 75 yards wide. The 1999 Women's World Cup was played at the Rose Bowl on a playing field 116 yards long and 72 yards wide. The diagram below shows the smallest possible soccer field, the largest possible soccer field, and the soccer field at the Rose Bowl. Find the area of each one and then convert the area of each to acres. Round to the nearest tenth, if necessary.

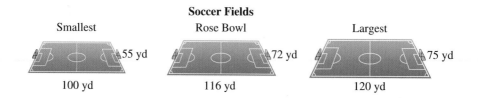

Soccer Fields

Smallest — 55 yd, 100 yd   Rose Bowl — 72 yd, 116 yd   Largest — 75 yd, 120 yd

27. Swimming Pool A public swimming pool measures 100 meters by 30 meters and is rectangular. What is the area of the pool in ares?

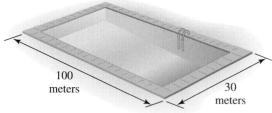

100 meters
30 meters

28. Construction A family decides to put tiles in the entryway of their home. The entryway has an area of 6 square meters. If each tile is 5 centimeters by 5 centimeters, how many tiles will it take to cover the entryway?

29. Landscaping A landscaper is putting in a brick patio. The area of the patio is 110 square meters. If the bricks measure 10 centimeters by 20 centimeters, how many bricks will it take to make the patio? Assume no space between bricks.

30. Sewing A dressmaker is using a pattern that requires 2 square yards of material. If the material is on a bolt that is 54 inches wide, how long a piece of material must be cut from the bolt to be sure there is enough material for the pattern?

Make the following conversions using the conversion factors given in Tables 3 and 4.

31. 5 yd³ to cubic feet

32. 3.8 yd³ to cubic feet

33. 3 pt to fluid ounces

34. 8 pt to fluid ounces

35. 2 gal to quarts

36. 12 gal to quarts

37. 2.5 gal to pints

38. 7 gal to pints

39. 15 qt to fluid ounces

40. 5.9 qt to fluid ounces

41. 64 pt to gallons

42. 256 pt to gallons

43. 12 pt to quarts

44. 18 pt to quarts

45. 243 ft³ to cubic yards

46. 864 ft³ to cubic yards

47. 5 L to milliliters

48. 9.6 L to milliliters

49. 127 mL to liters

50. 93.8 mL to liters

51. 4 kL to milliliters

52. 3 kL to milliliters

53. 14.92 kL to liters

54. 4.71 kL to liters

55. Filling Coffee Cups If a regular-size coffee cup holds about $\frac{1}{2}$ pint, about how many cups can be filled from a 1-gallon coffee maker?

56. Filling Glasses If a regular-size drinking glass holds about 0.25 liter of liquid, how many glasses can be filled from a 750-milliliter container?

57. Capacity of a Refrigerator A refrigerator has a capacity of 20 cubic feet. What is the capacity of the refrigerator in cubic inches?

58. Volume of a Tank The gasoline tank on a car holds 18 gallons of gas. What is the volume of the tank in quarts?

59. Filling Glasses How many 8-fluid-ounce glasses of water will it take to fill a 3-gallon aquarium?

60. Filling a Container How many 5-milliliter test tubes filled with water will it take to fill a 1-liter container?

Calculator Problems

Set up the following problems as you have been doing. Then use a calculator to perform the actual calculations. Round off to two decimal places where appropriate.

61. Geography Lake Superior is the largest of the Great Lakes. It covers 31,700 square miles of area. What is the area of Lake Superior in acres?

62. Geography The state of California consists of 156,360 square miles of land and 2,330 square miles of water. Write the total area (both land and water) in acres.

63. Geography Death Valley National Monument contains 2,067,795 acres of land. How many square miles is this?

64. Geography The Badlands National Monument in South Dakota was established in 1929. It covers 243,302 acres of land. What is the area in square miles?

65. Convert 93.4 qt to gallons.

66. Convert 7,362 fl oz to gallons.

67. How many cubic feet are contained in 796 cubic yards?

68. The engine of a car has a displacement of 440 cubic inches. What is the displacement in cubic feet?

69. Volume of Water The Grand Coulee Dam holds 10,585,000 cubic yards of water. What is the volume of water in cubic feet?

70. Volume of Water Hoover Dam was built in 1936 on the Colorado River in Nevada. It holds a volume of 4,400,000 cubic yards of water. What is this volume in cubic feet?

Review Problems

The following problems review addition and subtraction with fractions and mixed numbers.

Find each sum. (Add.)

71. $\dfrac{3}{8} + \dfrac{1}{4}$

72. $\dfrac{1}{2} + \dfrac{1}{4}$

73. $3\dfrac{1}{2} + 5\dfrac{1}{2}$

74. $6\dfrac{7}{8} + 1\dfrac{5}{8}$

Find each difference. (Subtract.)

75. $\dfrac{7}{15} - \dfrac{2}{15}$

76. $\dfrac{5}{8} - \dfrac{1}{4}$

77. $\dfrac{5}{36} - \dfrac{1}{48}$

78. $\dfrac{7}{39} - \dfrac{2}{65}$

79. $5\dfrac{1}{6} - 2\dfrac{1}{9}$

80. $8\dfrac{1}{9} - 3\dfrac{1}{6}$

6.4 Unit Analysis III: Weight

The most common units of weight in the U.S. system are ounces, pounds, and tons. The relationships among these units are given in Table 1.

Table 1

UNITS OF WEIGHT IN THE U.S. SYSTEM

THE RELATIONSHIP BETWEEN	IS	TO CONVERT FROM ONE TO THE OTHER, MULTIPLY BY
ounces (oz) and pounds (lb)	1 lb = 16 oz	$\dfrac{16 \text{ oz}}{1 \text{ lb}}$ or $\dfrac{1 \text{ lb}}{16 \text{ oz}}$
pounds and tons (T)	1 T = 2,000 lb	$\dfrac{2{,}000 \text{ lb}}{1 \text{ T}}$ or $\dfrac{1 \text{ T}}{2{,}000 \text{ lb}}$

EXAMPLE 1 Convert 12 pounds to ounces.

SOLUTION Using the conversion factor from the table, and applying the method we have been using, we have

$$12 \text{ lb} = 12 \text{ lb} \times \frac{16 \text{ oz}}{1 \text{ lb}}$$
$$= 12 \times 16 \text{ oz}$$
$$= 192 \text{ oz}$$

12 pounds is equivalent to 192 ounces. ■

EXAMPLE 2 If a certain cut of meat costs $2.49 per pound, how much will 3 lb 12 oz of this meat cost?

SOLUTION Because the cost is given in terms of pounds, we must convert 3 lb 12 oz to pounds. We can do this by first changing 12 oz to pounds:

$$12 \text{ oz} = 12 \text{ oz} \times \frac{1 \text{ lb}}{16 \text{ oz}}$$
$$= \frac{12}{16} \text{ lb}$$
$$= 0.75 \text{ lb}$$

Because 12 oz is equivalent to 0.75 lb, we have

$$3 \text{ lb } 12 \text{ oz} = 3.75 \text{ lb}$$

Multiplying by the price per pound, we have

$$\underset{\substack{\uparrow \\ \textbf{Price per} \\ \textbf{pound}}}{\$2.49} \times \underset{\substack{\uparrow \\ \textbf{Number} \\ \textbf{of pounds}}}{3.75 \text{ lb}} = \underset{\substack{\uparrow \\ \textbf{Total} \\ \textbf{cost}}}{\$9.34} \qquad \textbf{Rounded to the nearest cent}$$

It costs $9.34 for 3 lb 12 oz of meat that sells for $2.49 per pound. ■

In the metric system the basic unit of weight is a gram. We use the same prefixes we have already used to write the other units of weight in terms of grams. Table 2 lists the most common metric units of weight and their conversion factors.

Practice Problems

1. Convert 15 pounds to ounces.

2. A certain kind of cheese costs $2.09 per pound. How much will 2 lb 4 oz cost? Round to the nearest cent.

Answers
1. 240 oz   2. $4.70

Table 2

METRIC UNITS OF WEIGHT

THE RELATIONSHIP BETWEEN	IS	TO CONVERT FROM ONE TO THE OTHER, MULTIPLY BY
milligrams (mg) and grams (g)	1 g = 1,000 mg	$\dfrac{1,000 \text{ mg}}{1 \text{ g}}$ or $\dfrac{1 \text{ g}}{1,000 \text{ mg}}$
centigrams (cg) and grams	1 g = 100 cg	$\dfrac{100 \text{ cg}}{1 \text{ g}}$ or $\dfrac{1 \text{ g}}{100 \text{ cg}}$
kilograms (kg) and grams	1,000 g = 1 kg	$\dfrac{1,000 \text{ g}}{1 \text{ kg}}$ or $\dfrac{1 \text{ kg}}{1,000 \text{ g}}$
metric tons (t) and kilograms	1,000 kg = 1 t	$\dfrac{1,000 \text{ kg}}{1 \text{ t}}$ or $\dfrac{1 \text{ t}}{1,000 \text{ kg}}$

3. Convert 5 kilograms to milligrams.

EXAMPLE 3   Convert 3 kilograms to centigrams.

SOLUTION   We convert kilograms to grams and then grams to centigrams:

$$3 \text{ kg} = 3 \text{ kg} \times \frac{1,000 \text{ g}}{1 \text{ kg}} \times \frac{100 \text{ cg}}{1 \text{ g}}$$
$$= 3 \times 1,000 \times 100 \text{ cg}$$
$$= 300,000 \text{ cg}$$

■

4. A bottle of vitamin C contains 75 tablets. If each tablet contains 200 milligrams of vitamin C, what is the total number of grams of vitamin C in the bottle?

EXAMPLE 4   A bottle of vitamin C contains 50 tablets. Each tablet contains 250 milligrams of vitamin C. What is the total number of grams of vitamin C in the bottle?

SOLUTION   We begin by finding the total number of milligrams of vitamin C in the bottle. Since there are 50 tablets, and each contains 250 mg of vitamin C, we can multiply 50 by 250 to get the total number of milligrams of vitamin C:

$$\text{Milligrams of vitamin C} = 50 \times 250 \text{ mg}$$
$$= 12,500 \text{ mg}$$

Next we convert 12,500 mg to grams:

$$12,500 \text{ mg} = 12,500 \text{ mg} \times \frac{1 \text{ g}}{1,000 \text{ mg}}$$
$$= \frac{12,500}{1,000} \text{ g}$$
$$= 12.5 \text{ g}$$

The bottle contains 12.5 g of vitamin C.

■

Answers
3. 5,000,000 mg   **4.** 15 g

Problem Set 6.4

Use the conversion factors in Tables 1 and 2 to make the following conversions.

1. 8 lb to ounces

2. 5 lb to ounces

3. 2 T to pounds

4. 5 T to pounds

5. 192 oz to pounds

6. 176 oz to pounds

7. 1,800 lb to tons

8. 10,200 lb to tons

9. 1 T to ounces

10. 3 T to ounces

11. $3\frac{1}{2}$ lb to ounces

12. $5\frac{1}{4}$ lb to ounces

13. $6\frac{1}{2}$ T to pounds

14. $4\frac{1}{5}$ T to pounds

15. 2 kg to grams

16. 5 kg to grams

17. 4 cg to milligrams

18. 3 cg to milligrams

19. 2 kg to centigrams

20. 5 kg to centigrams

21. 5.08 g to centigrams

22. 7.14 g to centigrams

23. 450 cg to grams

24. 979 cg to grams

25. 478.95 mg to centigrams

26. 659.43 mg to centigrams

27. 1,578 mg to grams

28. 1,979 mg to grams

29. 42,000 cg to kilograms

30. 97,000 cg to kilograms

31. 9 lb 4 oz to pounds

32. 4 lb 8 oz to pounds

33. Cost of Meat If meat costs $2.03 per pound, how much will 4 lb 8 oz cost?

34. Cost of Meat If meat costs $2.09 per pound, how much will 9 lb 4 oz cost?

35. Milligrams of Vitamins A bottle contains 30 vitamin tablets. If each tablet contains 500 milligrams of vitamins, how many total grams of vitamins are in the bottle?

36. Cans of Soup A box contains six cans of soup. Each can weighs 305 grams. What is the total weight of the cans in kilograms?

Calculator Problems

Do any calculations required to work the following problems on a calculator.

37. Convert 3.98 lb to ounces.

38. Convert 59.46 oz to pounds. (Round to the nearest hundredth.)

39. Convert 12 lb 9 oz to pounds.

40. Convert 12 lb 7 oz to pounds.

41. A special cut of meat costs $4.95 per pound. How much will 15 lb 7 oz of this meat cost?

42. If a man weighs $157\frac{1}{4}$ lb, what is his weight in ounces?

Review Problems

The following problems review material from Section 5.5.

Write each decimal as a fraction in lowest terms.

43. 0.18   **44.** 0.04   **45.** 0.09   **46.** 0.045   **47.** 0.8

Write each fraction as a decimal.

48. $\frac{3}{4}$ **49.** $\frac{9}{10}$ **50.** $\frac{17}{20}$ **51.** $\frac{1}{8}$ **52.** $\frac{3}{5}$

6.5 Converting between the Two Systems

Because most of us have always used the U.S. system of measurement in our everyday lives, we are much more familiar with it on an intuitive level than we are with the metric system. We have an intuitive idea of how long feet and inches are, how much a pound weighs, and what a square yard of material looks like. The metric system is actually much easier to use than the U.S. system. The reason some of us have such a hard time with the metric system is that we don't have the feel for it that we do for the U.S. system. We have trouble visualizing how long a meter is or how much a gram weighs. The following list is intended to give you something to associate with each basic unit of measurement in the metric system:

1. A meter is just a little longer than a yard.

2. The length of the edge of a sugar cube is about 1 centimeter.

3. A liter is just a little larger than a quart.

4. A sugar cube has a volume of approximately 1 milliliter.

5. A paper clip weighs about 1 gram.

6. A 2-pound can of coffee weighs about 1 kilogram.

Table 1

ACTUAL CONVERSION FACTORS BETWEEN THE METRIC AND U.S. SYSTEMS OF MEASUREMENT

THE RELATIONSHIP BETWEEN	IS	TO CONVERT FROM ONE TO THE OTHER, MULTIPLY BY
Length		
inches and centimeters	2.54 cm = 1 in.	$\dfrac{2.54 \text{ cm}}{1 \text{ in.}}$ or $\dfrac{1 \text{ in.}}{2.54 \text{ cm}}$
feet and meters	1 m = 3.28 ft	$\dfrac{3.28 \text{ ft}}{1 \text{ m}}$ or $\dfrac{1 \text{ m}}{3.28 \text{ ft}}$
miles and kilometers	1.61 km = 1 mi	$\dfrac{1.61 \text{ km}}{1 \text{ mi}}$ or $\dfrac{1 \text{ mi}}{1.61 \text{ km}}$
Area		
square inches and square centimeters	$6.45 \text{ cm}^2 = 1 \text{ in}^2$	$\dfrac{6.45 \text{ cm}^2}{1 \text{ in}^2}$ or $\dfrac{1 \text{ in}^2}{6.45 \text{ cm}^2}$
square meters and square yards	$1.196 \text{ yd}^2 = 1 \text{ m}^2$	$\dfrac{1.196 \text{ yd}^2}{1 \text{ m}^2}$ or $\dfrac{1 \text{ m}^2}{1.196 \text{ yd}^2}$
acres and hectares	1 ha = 2.47 acres	$\dfrac{2.47 \text{ acres}}{1 \text{ ha}}$ or $\dfrac{1 \text{ ha}}{2.47 \text{ acres}}$
Volume		
cubic inches and milliliters	$16.39 \text{ mL} = 1 \text{ in}^3$	$\dfrac{16.39 \text{ mL}}{1 \text{ in}^3}$ or $\dfrac{1 \text{ in}^3}{16.39 \text{ mL}}$
liters and quarts	1.06 qt = 1 liter	$\dfrac{1.06 \text{ qt}}{1 \text{ liter}}$ or $\dfrac{1 \text{ liter}}{1.06 \text{ qt}}$
gallons and liters	3.79 liters = 1 gal	$\dfrac{3.79 \text{ liters}}{1 \text{ gal}}$ or $\dfrac{1 \text{ gal}}{3.79 \text{ liters}}$
Weight		
ounces and grams	28.3 g = 1 oz	$\dfrac{28.3 \text{ g}}{1 \text{ oz}}$ or $\dfrac{1 \text{ oz}}{28.3 \text{ g}}$
kilograms and pounds	2.20 lb = 1 kg	$\dfrac{2.20 \text{ lb}}{1 \text{ kg}}$ or $\dfrac{1 \text{ kg}}{2.20 \text{ lb}}$

There are many other conversion factors that we could have included in Table 1. We have listed only the most common ones. Almost all of them are approximations. That is, most of the conversion factors are decimals that have been rounded to the nearest hundredth. If we want more accuracy, we obtain a table that has more digits in the conversion factors.

Practice Problems
1. Convert 10 inches to centimeters.

EXAMPLE 1 Convert 8 inches to centimeters.

SOLUTION Choosing the appropriate conversion factor from Table 1, we have

$$8 \text{ in.} = 8 \text{ in.} \times \frac{2.54 \text{ cm}}{1 \text{ in.}}$$
$$= 8 \times 2.54 \text{ cm}$$
$$= 20.32 \text{ cm}$$

2. Convert 9 meters to feet.

EXAMPLE 2 Convert 80.5 kilometers to miles.

SOLUTION Using the conversion factor that takes us from kilometers to miles, we have

$$80.5 \text{ km} = 80.5 \text{ km} \times \frac{1 \text{ mi}}{1.61 \text{ km}}$$
$$= \frac{80.5}{1.61} \text{ mi}$$
$$= 50 \text{ mi}$$

So 50 miles is equivalent to 80.5 kilometers. If we travel at 50 miles per hour in a car, we are moving at the rate of 80.5 kilometers per hour.

3. Convert 15 gallons to liters.

EXAMPLE 3 Convert 3 liters to pints.

SOLUTION Because Table 1 doesn't list a conversion factor that will take us directly from liters to pints, we first convert liters to quarts, and then convert quarts to pints.

$$3 \text{ liters} = 3 \text{ liters} \times \frac{1.06 \text{ qt}}{1 \text{ liter}} \times \frac{2 \text{ pt}}{1 \text{ qt}}$$
$$= 3 \times 1.06 \times 2 \text{ pt}$$
$$= 6.36 \text{ pt}$$

4. The engine in a car has a 2.2-liter displacement. What is the displacement in cubic inches (to the nearest cubic inch)?

EXAMPLE 4 The engine in a car has a 2-liter displacement. What is the displacement in cubic inches?

SOLUTION We convert liters to milliliters and then milliliters to cubic inches:

$$2 \text{ liters} = 2 \text{ liters} \times \frac{1,000 \text{ mL}}{1 \text{ liter}} \times \frac{1 \text{ in}^3}{16.39 \text{ mL}}$$
$$= \frac{2 \times 1,000}{16.39} \text{ in}^3 \qquad \textbf{This calculation should be done on a calculator}$$
$$= 122 \text{ in}^3 \qquad \textbf{To the nearest cubic inch}$$

Answers
1. 25.4 cm   2. 29.52 ft
3. 56.85 liters 4. 134 in^3

EXAMPLE 5 If a person weighs 125 pounds, what is her weight in kilograms?

SOLUTION   Converting from pounds to kilograms, we have

$$125 \text{ lb} = 125 \text{ lb} \times \frac{1 \text{ kg}}{2.20 \text{ lb}}$$

$$= \frac{125}{2.20} \text{ kg}$$

$$= 56.8 \text{ kg} \quad \textbf{To the nearest tenth}$$

5. A person who weighs 165 pounds weighs how many kilograms?

Temperature

We end this section with a review of temperature in both systems of measurement.

In the U.S. system we measure temperature on the Fahrenheit scale. On this scale, water boils at 212 degrees and freezes at 32 degrees. When we write 32 degrees measured on the Fahrenheit scale, we use the notation

     32°F   (read, "32 degrees Fahrenheit")

In the metric system the scale we use to measure temperature is the Celsius scale (formerly called the centigrade scale). On this scale, water boils at 100 degrees and freezes at 0 degrees. When we write 100 degrees measured on the Celsius scale, we use the notation

     100°C   (read, "100 degrees Celsius")

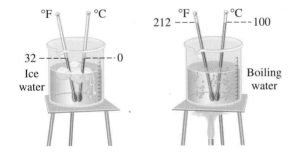

Table 2 is intended to give you a feel for the relationship between the two temperature scales. Table 3 gives the formulas, in both symbols and words, that are used to convert between the two scales.

Answer
5. 75 kg

Table 2

SITUATION	TEMPERATURE FAHRENHEIT	TEMPERATURE CELSIUS
Water freezes	32°F	0°C
Room temperature	68°F	20°C
Normal body temperature	98.6°F	37°C
Water boils	212°F	100°C
Bake cookies	365°F	185°C
Broil meat	554°F	290°C

Table 3

TO CONVERT FROM	FORMULA IN SYMBOLS	FORMULA IN WORDS
Fahrenheit to Celsius	$C = \dfrac{5(F - 32)}{9}$	Subtract 32, multiply by 5, and then divide by 9.
Celsius to Fahrenheit	$F = \dfrac{9}{5}C + 32$	Multiply by $\dfrac{9}{5}$, and then add 32.

The following examples show how we use the formulas given in Table 3.

6. Convert 40°C to degrees Fahrenheit.

EXAMPLE 6   Convert 120°C to degrees Fahrenheit.

SOLUTION   We use the formula

$$F = \frac{9}{5}C + 32$$

and replace C with 120:

$$\text{When} \quad C = 120$$
$$\text{the formula} \quad F = \frac{9}{5}C + 32$$
$$\text{becomes} \quad F = \frac{9}{5}(120) + 32$$
$$F = 216 + 32$$
$$F = 248$$

We see that 120°C is equivalent to 248°F; they both mean the same temperature.

7. A child is running a temperature of 101.6°F. What is her temperature, to the nearest tenth of a degree, on the Celsius scale?

EXAMPLE 7   A man with the flu has a temperature of 102°F. What is his temperature on the Celsius scale?

SOLUTION

$$\text{When} \quad F = 102$$
$$\text{the formula} \quad C = \frac{5(F - 32)}{9}$$
$$\text{becomes} \quad C = \frac{5(102 - 32)}{9}$$
$$C = \frac{5(70)}{9}$$
$$C = 38.9 \qquad \textbf{Rounded to the nearest tenth}$$

The man's temperature, rounded to the nearest tenth, is 38.9°C on the Celsius scale.

Answers

6. 104°F   **7.** 38.7°C

Problem Set 6.5

Use Tables 1 and 3 to make the following conversions.

1. 6 in. to centimeters

2. 1 ft to centimeters

3. 4 m to feet

4. 2 km to feet

5. 6 m to yards

6. 15 mi to kilometers

7. 20 mi to meters (to the nearest meter)

8. 600 m to yards

9. 5 m² to square yards

10. 2 in² to square centimeters

11. 10 ha to acres

12. 50 a to acres

13. 500 in³ to milliliters

14. 400 in³ to liters

15. 2 L to quarts

16. 15 L to quarts

17. 20 gal to liters

18. 15 gal to liters

19. 12 oz to grams

20. 1 lb to grams

21. 15 kg to pounds

22. 10 kg to ounces

23. 185°C to degrees Fahrenheit

24. 32°C to degrees Fahrenheit

25. 86°F to degrees Celsius

26. 122°F to degrees Celsius

Calculator Problems

Set up the following problems as we have set up the examples in this section. Then use a calculator for the calculations and round your answers to the nearest hundredth.

27. 10 cm to inches

28. 100 mi to kilometers

29. 25 ft to meters

30. 400 mL to cubic inches

31. 49 qt to liters

32. 65 L to gallons

33. 500 g to ounces

34. 100 lb to kilograms

35. Weight Give your weight in kilograms.

36. Height Give your height in meters and centimeters.

37. Sports The 100-yard dash is a popular race in track. How far is 100 yards in meters?

38. Engine Displacement A 351-cubic-inch engine has a displacement of how many liters?

39. Sewing 25 square yards of material is how many square meters?

40. Weight How many grams does a 5 lb 4 oz roast weigh?

41. Speed 55 miles per hour is equivalent to how many kilometers per hour?

42. Capacity A 1-quart container holds how many liters?

43. Sports A high jumper jumps 6 ft 8 in. How many meters is this?

44. Farming A farmer owns 57 acres of land. How many ares is that?

45. Body Temperature A person has a temperature of 101°F. What is the person's temperature, to the nearest tenth, on the Celsius scale?

46. Air Temperature If the temperature outside is 30°C, is it a better day for water skiing or for snow skiing?

Review Problems

The problems below review the relationship between fractions and decimals from Section 5.5.

Write each fraction or mixed number as an equivalent decimal number.

47. $\frac{3}{4}$

48. $\frac{2}{5}$

49. $5\frac{1}{2}$

50. $8\frac{1}{4}$

51. $\frac{3}{100}$

52. $\frac{2}{50}$

Write each decimal as an equivalent proper fraction or mixed number.

53. 0.34

54. 0.08

55. 2.4

56. 5.05

57. 1.75

58. 3.125

6.6 Proportions

In this section we will extend our work with ratios to include *proportions*. When two ratios are equal, such as 3 to 2 and 6 to 4, we say that they are *proportional*. We define a proportion as follows:

DEFINITION

A statement that two ratios are equal is called a **proportion.** If $\frac{a}{b}$ and $\frac{c}{d}$ are two equal ratios, then the statement

$$\frac{a}{b} = \frac{c}{d}$$

is called a proportion.

Each of the four numbers in a proportion is called a *term* of the proportion. We number the terms of a proportion as follows:

$$\begin{array}{l} \textbf{First term} \longrightarrow a \quad c \longleftarrow \textbf{Third term} \\ \textbf{Second term} \longrightarrow b \quad d \longleftarrow \textbf{Fourth term} \end{array}$$

The first and fourth terms of a proportion are called the *extremes,* and the second and third terms of a proportion are called the *means*.

$$\textbf{Means} \longrightarrow \frac{a}{b} = \frac{c}{d} \longleftarrow \textbf{Extremes}$$

EXAMPLE 1   In the proportion $\frac{3}{4} = \frac{6}{8}$, name the four terms, the means, and the extremes.

SOLUTION   The terms are numbered as follows:

First term $= 3$ Third term $= 6$
Second term $= 4$ Fourth term $= 8$

The means are 4 and 6; the extremes are 3 and 8.   ■

The only additional thing we need to know about proportions is the following property.

Fundamental Property of Proportions

In any proportion, the product of the extremes is equal to the product of the means. In symbols, it looks like this:

$$\text{If} \quad \frac{a}{b} = \frac{c}{d} \quad \text{then} \quad ad = bc$$

Practice Problems

1. In the proportion $\frac{2}{3} = \frac{6}{9}$, name the four terms, the means, and the extremes.

Answer

1. See solutions section.

2. Verify the fundamental property of proportions for the following proportions.

a. $\dfrac{5}{6} = \dfrac{15}{18}$

b. $\dfrac{13}{39} = \dfrac{1}{3}$

EXAMPLE 2   Verify the fundamental property of proportions for the following proportions.

a. $\dfrac{3}{4} = \dfrac{6}{8}$ **b.** $\dfrac{17}{34} = \dfrac{1}{2}$

SOLUTION   We verify the fundamental property by finding the product of the means and the product of the extremes in each case.

PROPORTION	PRODUCT OF THE MEANS	PRODUCT OF THE EXTREMES
a. $\dfrac{3}{4} = \dfrac{6}{8}$	$4 \cdot 6 = 24$	$3 \cdot 8 = 24$
b. $\dfrac{17}{34} = \dfrac{1}{2}$	$34 \cdot 1 = 34$	$17 \cdot 2 = 34$

For each proportion the product of the means is equal to the product of the extremes.   ■

We can use the fundamental property of proportions, along with a property we encountered in Section 4.3, to find a missing term in a proportion.

3. Find the missing term:

$\dfrac{3}{4} = \dfrac{9}{x}$

EXAMPLE 3   Find the missing term. (Solve for x.)

$$\dfrac{2}{3} = \dfrac{4}{x}$$

SOLUTION   Applying the fundamental property of proportions, we have

If $\dfrac{2}{3} = \dfrac{4}{x}$

then $2 \cdot x = 3 \cdot 4$ **The product of the extremes equals the product of the means**

$2x = 12$ **Multiply**

The result is an equation. We know from Section 4.3 that we can divide both sides of an equation by the same nonzero number without changing the solution to the equation. In this case we divide both sides by 2 to solve for x:

$2x = 12$

$\dfrac{2x}{2} = \dfrac{12}{2}$ **Divide both sides by 2**

$x = 6$ **Simplify each side**

The missing term is 6. We can check our work by using the fundamental property of proportions:

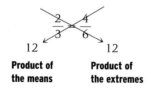

12   12

Product of the means   **Product of the extremes**

Because the product of the means and the product of the extremes are equal, our work is correct.   ■

Note

In some of these problems you will be able to see what the solution is just by looking the problem over. In those cases it is still best to show all the work involved in solving the proportion. It is good practice for the more difficult problems.

EXAMPLE 4 Solve for y: $\dfrac{5}{y} = \dfrac{10}{13}$

SOLUTION We apply the fundamental property and solve as we did in Example 3:

If $\quad \dfrac{5}{y} = \dfrac{10}{13}$

then $\quad 5 \cdot 13 = y \cdot 10 \quad$ **The product of the extremes equals the product of the means**

$65 = 10y \quad$ **Multiply 5 · 13**

$\dfrac{65}{10} = \dfrac{\cancel{10}y}{\cancel{10}} \quad$ **Divide both sides by 10**

$6.5 = y \quad$ **65 ÷ 10 = 6.5**

The missing term is 6.5. We could check our result by substituting 6.5 for y in the original proportion and then finding the product of the means and the product of the extremes. ∎

EXAMPLE 5 Find n if $\dfrac{n}{3} = \dfrac{0.4}{8}$.

SOLUTION We proceed as we did in the previous two examples:

If $\quad \dfrac{n}{3} = \dfrac{0.4}{8}$

then $\quad n \cdot 8 = 3(0.4) \quad$ **The product of the extremes equals the product of the means**

$8n = 1.2 \quad$ **3(0.4) = 1.2**

$\dfrac{8n}{8} = \dfrac{1.2}{8} \quad$ **Divide both sides by 8**

$n = 0.15 \quad$ **1.2 ÷ 8 = 0.15**

The missing term is 0.15. ∎

EXAMPLE 6 Solve for x: $\dfrac{\frac{2}{3}}{5} = \dfrac{x}{6}$

SOLUTION We begin by multiplying the means and multiplying the extremes:

If $\quad \dfrac{\frac{2}{3}}{5} = \dfrac{x}{6}$

then $\quad \dfrac{2}{3} \cdot 6 = 5 \cdot x \quad$ **The product of the extremes equals the product of the means**

$4 = 5 \cdot x \quad$ $\dfrac{2}{3} \cdot 6 = 4$

$\dfrac{4}{5} = \dfrac{\cancel{5} \cdot x}{\cancel{5}} \quad$ **Divide both sides by 5**

$\dfrac{4}{5} = x$

The missing term is $\dfrac{4}{5}$, or 0.8. ∎

The procedure for finding a missing term in a proportion is always the same. We first apply the fundamental property of proportions to find the product of the extremes and the product of the means. We then divide both sides of the resulting equation by the number being multiplied by the missing term. We then simplify the answer if possible.

4. Solve for y: $\dfrac{2}{y} = \dfrac{8}{19}$

5. Find n if $\dfrac{n}{6} = \dfrac{0.3}{15}$.

6. Solve for x: $\dfrac{\frac{3}{4}}{7} = \dfrac{x}{8}$

Answers

4. 4.75 **5.** 0.12 **6.** $\dfrac{6}{7}$

Applications of Proportions

Proportions can be used to solve a variety of word problems. The examples that follow show some of these word problems. In each case we will translate the word problem into a proportion and then solve the proportion using the method developed in this section.

7. A man drives his car 288 miles in 6 hours. If he continues at the same rate, how far will he travel in 10 hours?

EXAMPLE 7   A woman drives her car 270 miles in 6 hours. If she continues at the same rate, how far will she travel in 10 hours?

SOLUTION   We let x represent the distance traveled in 10 hours. Using x, we translate the problem into the following proportion:

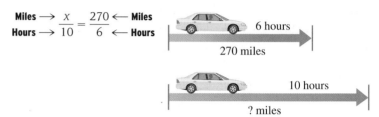

$$\text{Miles} \longrightarrow \frac{x}{10} = \frac{270}{6} \longleftarrow \text{Miles}$$
$$\text{Hours} \longrightarrow \qquad\qquad \longleftarrow \text{Hours}$$

Notice that the two ratios in the proportion compare the same quantities. That is, both ratios compare miles to hours. In words this proportion says:

x miles is to 10 hours as 270 miles is to 6 hours

$$\frac{x}{10} = \frac{270}{6}$$

Next, we solve the proportion.

$$x \cdot 6 = 10 \cdot 270$$
$$x \cdot 6 = 2{,}700$$
$$\frac{x \cdot \cancel{6}}{\cancel{6}} = \frac{2{,}700}{6}$$
$$x = 450 \text{ miles}$$

If the woman continues at the same rate, she will travel 450 miles in 10 hours.

    ■

8. A softball player gets 10 hits in the first 18 games of the season. If she continues at the same rate, how many hits will she get in 54 games?

EXAMPLE 8   A baseball player gets 8 hits in the first 18 games of the season. If he continues at the same rate, how many hits will he get in 45 games?

SOLUTION   We let x represent the number of hits he will get in 45 games. Then

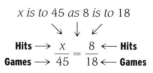

x is to 45 as 8 is to 18

$$\text{Hits} \longrightarrow \frac{x}{45} = \frac{8}{18} \longleftarrow \text{Hits}$$
$$\text{Games} \longrightarrow \qquad\qquad \longleftarrow \text{Games}$$

Notice again that the two ratios are comparing the same quantities, hits to games. We solve the proportion as follows:

$$18x = 360 \qquad \mathbf{45 \cdot 8 = 360}$$
$$\frac{\cancel{18}x}{\cancel{18}} = \frac{360}{18} \qquad \textbf{Divide both sides by 18}$$
$$x = 20 \qquad \mathbf{360 \div 18 = 20}$$

If he continues to hit at the rate of 8 hits in 18 games, he will get 20 hits in 45 games. ▪

EXAMPLE 9   A solution contains 4 milliliters of alcohol and 20 milliliters of water. If another solution is to have the same ratio of milliliters of alcohol to milliliters of water and must contain 25 milliliters of water, how much alcohol should it contain?

SOLUTION   We let x represent the number of milliliters of alcohol in the second solution. The problem translates to

x milliliters is to 25 milliliters as 4 milliliters is to 20 milliliters

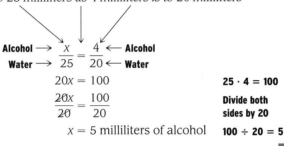

$$\begin{array}{c}\textbf{Alcohol} \longrightarrow \\ \textbf{Water} \longrightarrow\end{array} \frac{x}{25} = \frac{4}{20} \begin{array}{c}\longleftarrow \textbf{Alcohol} \\ \longleftarrow \textbf{Water}\end{array}$$
$$20x = 100 \qquad \mathbf{25 \cdot 4 = 100}$$
$$\frac{\cancel{20}x}{\cancel{20}} = \frac{100}{20} \qquad \begin{array}{l}\textbf{Divide both} \\ \textbf{sides by 20}\end{array}$$
$$x = 5 \text{ milliliters of alcohol} \qquad \mathbf{100 \div 20 = 5}$$

▪

EXAMPLE 10 The scale on a map indicates that 1 inch on the map corresponds to an actual distance of 85 miles. Two cities are 3.5 inches apart on the map. What is the actual distance between the two cities?

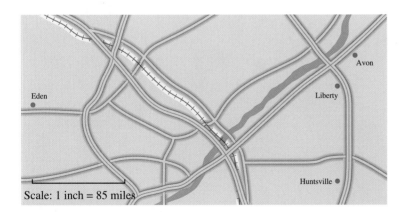

Scale: 1 inch = 85 miles

SOLUTION   We let x represent the actual distance between the two cities. The proportion is

$$\begin{array}{c}\textbf{Miles} \longrightarrow \\ \textbf{Inches} \longrightarrow\end{array} \frac{x}{3.5} = \frac{85}{1} \begin{array}{c}\longleftarrow \textbf{Miles} \\ \longleftarrow \textbf{Inches}\end{array}$$
$$x \cdot 1 = 3.5(85)$$
$$x = 297.5 \text{ miles}$$

▪

9. A solution contains 8 milliliters of alcohol and 20 milliliters of water. If another solution is to have the same ratio of milliliters of alcohol to water and must contain 35 milliliters of water, how much alcohol should it contain?

10. The scale on a map indicates that 1 inch on the map corresponds to an actual distance of 105 miles. Two cities are 4.75 inches apart on the map. What is the actual distance between the two cities?

Answers
9. 14 mL   **10.** 498.75 mi

11. A manufacturer knows that, during an average production run, out of every 120 parts produced by a certain machine, 8 will be defective. If the machine produces 300 parts, how many can be expected to be defective?

EXAMPLE 11 A manufacturer knows that, during an average production run, out of every 100 parts produced by a certain machine, 6 will be defective. If the machine produces 250 parts, how many can be expected to be defective?

SOLUTION   We let x represent the number of defective parts and solve the following proportion:

$$\frac{x}{250} = \frac{6}{100}$$

$$100x = 1{,}500$$

$$\frac{100x}{100} = \frac{1{,}500}{100}$$

$$x = 15$$

The manufacturer can expect to find 15 defective parts out of every 250 parts produced. ∎

Answer

11. 20 parts

Problem Set **6.6**

For each of the following proportions, name the means, name the extremes, and show that the product of the means is equal to the product of the extremes.

1. $\dfrac{1}{3} = \dfrac{5}{15}$

2. $\dfrac{6}{12} = \dfrac{1}{2}$

3. $\dfrac{10}{25} = \dfrac{2}{5}$

4. $\dfrac{5}{8} = \dfrac{10}{16}$

5. $\dfrac{\frac{1}{3}}{\frac{1}{2}} = \dfrac{4}{6}$

6. $\dfrac{2}{\frac{1}{4}} = \dfrac{4}{\frac{1}{2}}$

7. $\dfrac{0.5}{5} = \dfrac{1}{10}$

8. $\dfrac{0.3}{1.2} = \dfrac{1}{4}$

Find the missing term in each of the following proportions. Set up each problem like the examples in this section. Write your answers as fractions in lowest terms.

9. $\dfrac{2}{5} = \dfrac{4}{x}$

10. $\dfrac{3}{8} = \dfrac{9}{x}$

11. $\dfrac{1}{y} = \dfrac{5}{12}$

12. $\dfrac{2}{y} = \dfrac{6}{10}$

13. $\dfrac{x}{4} = \dfrac{3}{8}$

14. $\dfrac{x}{5} = \dfrac{7}{10}$

15. $\dfrac{5}{9} = \dfrac{x}{2}$

16. $\dfrac{3}{7} = \dfrac{x}{3}$

17. $\dfrac{3}{7} = \dfrac{3}{x}$

18. $\dfrac{2}{9} = \dfrac{2}{x}$

19. $\dfrac{25}{100} = \dfrac{x}{4}$

20. $\dfrac{30}{300} = \dfrac{x}{10}$

21. $\dfrac{\frac{1}{2}}{y} = \dfrac{\frac{1}{3}}{12}$

22. $\dfrac{\frac{2}{3}}{y} = \dfrac{\frac{1}{3}}{5}$

23. $\dfrac{n}{12} = \dfrac{\frac{1}{4}}{\frac{1}{2}}$

24. $\dfrac{n}{10} = \dfrac{\frac{3}{5}}{\frac{3}{8}}$

25. $\dfrac{10}{20} = \dfrac{20}{n}$

26. $\dfrac{8}{4} = \dfrac{4}{n}$

27. $\dfrac{x}{10} = \dfrac{10}{2}$

28. $\dfrac{x}{12} = \dfrac{12}{48}$

29. $\dfrac{1}{100} = \dfrac{y}{50}$

30. $\dfrac{3}{20} = \dfrac{y}{10}$

31. $\dfrac{0.4}{1.2} = \dfrac{1}{x}$

32. $\dfrac{5}{0.5} = \dfrac{20}{x}$

33. $\dfrac{0.3}{0.18} = \dfrac{n}{0.6}$

34. $\dfrac{0.01}{0.1} = \dfrac{n}{10}$

35. $\dfrac{0.5}{x} = \dfrac{1.4}{0.7}$

36. $\dfrac{0.3}{x} = \dfrac{2.4}{0.8}$

Applying the Concepts

Solve each of the following word problems by translating the statement into a proportion. Be sure to show the proportion used in each case.

37. Distance A woman drives her car 235 miles in 5 hours. At this rate how far will she travel in 7 hours?

38. Distance An airplane flies 1,260 miles in 3 hours. How far will it fly in 5 hours?

39. Basketball A basketball player scores 162 points in 9 games. At this rate how many points will he score in 20 games?

40. Football In the first 4 games of the season, a football team scores a total of 68 points. At this rate how many points will the team score in 11 games?

41. Mixture A solution contains 8 pints of antifreeze and 5 pints of water. How many pints of water must be added to 24 pints of antifreeze to get a solution with the same concentration?

42. Nutrition If 10 ounces of a certain breakfast cereal contains 3 ounces of sugar, how many ounces of sugar does 25 ounces of the same cereal contain?

43. Map Reading The scale on a map indicates that 1 inch corresponds to an actual distance of 95 miles. Two cities are 4.5 inches apart on the map. What is the actual distance between the two cities?

44. Map Reading A map is drawn so that every 2.5 inches on the map corresponds to an actual distance of 100 miles. If the actual distance between two cities is 350 miles, how far apart are they on the map?

45. Farming A farmer knows that of every 50 eggs his chickens lay, only 45 will be marketable. If his chickens lay 1,000 eggs in a week, how many of them will be marketable?

46. Manufacturing Of every 17 parts manufactured by a certain machine, only 1 will be defective. How many parts were manufactured by the machine if 8 defective parts were found?

Model Trains In the introduction to this chapter we indicated that the size of a model train relative to an actual train is referred to as its scale. Each scale is associated with a ratio as shown in the table below. For example, an HO model train has a ratio of 1 to 87, meaning it is $\frac{1}{87}$ as large as an actual train.

SCALE	RATIO
LGB	1 to 22.5
#1	1 to 32
O	1 to 43.5
S	1 to 64
HO	1 to 87
TT	1 to 120

47. Length of a Box Car How long is an actual box car that has an HO scale model 5 inches long? Give your answer in feet.

48. Length of a Flat Car How long is an actual flat car that has an LGB scale model 24 inches long? Give your answer in feet.

49. Travel Expenses A traveling salesman figures it costs 21¢ for every mile he drives his car. How much does it cost him a week to drive his car if he travels 570 miles a week?

50. Travel Expenses A family plans to drive their car during their annual vacation. The car can go 350 miles on a tank of gas, which is 18 gallons of gas. The vacation they have planned will cover 1,785 miles. How many gallons of gas will that take?

51. Nutrition A 6-ounce serving of grapefruit juice contains 159 grams of water. How many grams of water are in 10 ounces of grapefruit juice?

52. Nutrition If 100 grams of ice cream contains 13 grams of fat, how much fat is in 250 grams of ice cream?

Calculator Problems

Find the missing term in each proportion. Use a calculator to do the actual calculations. Write all answers in decimal form.

53. $\dfrac{168}{324} = \dfrac{56}{x}$

54. $\dfrac{280}{530} = \dfrac{112}{x}$

55. $\dfrac{429}{y} = \dfrac{858}{130}$

56. $\dfrac{573}{y} = \dfrac{2{,}292}{316}$

57. $\dfrac{n}{39} = \dfrac{533}{507}$

58. $\dfrac{n}{47} = \dfrac{1{,}003}{799}$

59. $\dfrac{756}{903} = \dfrac{x}{129}$

60. $\dfrac{321}{1{,}128} = \dfrac{x}{376}$

Use a proportion to set up each of the following problems. Then use a calculator to do the actual calculations.

61. Travel Expenses If a car travels 378.9 miles on 50 liters of gas, how many liters of gas will it take to go 692 miles if the car travels at the same rate? (Round to the nearest tenth.)

62. Nutrition If 125 grams of peas contains 26 grams of carbohydrates, how many grams of carbohydrates does 375 grams of peas contain?

63. Elections During a recent election, 47 of every 100 registered voters in a certain city voted. If there were 127,900 registered voters in that city, how many people voted?

64. Map Reading The scale on a map is drawn so that 4.5 inches corresponds to an actual distance of 250 miles. If two cities are 7.25 inches apart on the map, how many miles apart are they? (Round to the nearest tenth.)

Review Problems

The problems below are a review of some of the concepts we covered in Chapter 5.

Give the place value of the 5 in each number.

65. 250.14

66. 2.5014

Add or subtract as indicated.

67. $2.3 + 0.18 + 24.036$

68. $5 + 0.03 + 1.9$

69. $3.18 - 2.79$

70. $3.4 - 1.975$

Find the following products. (Multiply.)

71. 2.7×0.5

72. 0.7^2

73. 3.18×1.2

74. 0.3^4

Find the following quotients. (Divide.)

75. $2.8 \div 0.7$

76. $0.042 \div 0.21$

77. $24 \div 0.15$

78. $6.99 \div 2.33$

Chapter ⑥ Summary

Ratio [6.1]

The ratio of a to b is $\frac{a}{b}$. The ratio of two numbers is a way of comparing them using fraction notation.

1. The ratio of 6 to 8 is
$$\frac{6}{8}$$
which can be reduced to
$$\frac{3}{4}$$

Rates [6.1]

A ratio that compares two different quantities, like miles and hours, gallons and seconds, etc., is called a *rate*.

2. If a car travels 150 miles in 3 hours, then the ratio of miles to hours is considered a rate:
$$\frac{150 \text{ miles}}{3 \text{ hours}} = 50 \frac{\text{miles}}{\text{hour}}$$
$$= 50 \text{ miles per hour}$$

Unit Pricing [6.1]

The unit price of an item is the ratio of price to quantity.

3. If a 10-ounce package of frozen peas costs 69¢, then the price per ounce, or unit price, is
$$\frac{69 \text{ cents}}{10 \text{ ounces}} = 6.9 \frac{\text{cents}}{\text{ounce}}$$
$$= 6.9 \text{ cents per ounce}$$

Conversion Factors [6.2, 6.3, 6.4, 6.5]

To convert from one kind of unit to another, we choose an appropriate conversion factor from one of the tables given in this chapter. For example, if we want to convert 5 feet to inches, we look for conversion factors that give the relationship between feet and inches. There are two conversion factors for feet and inches:

$$\frac{12 \text{ in.}}{1 \text{ ft}} \quad \text{and} \quad \frac{1 \text{ ft}}{12 \text{ in.}}$$

4. Convert 5 feet to inches.
$$5 \text{ ft} = 5 \cancel{\text{ ft}} \times \frac{12 \text{ in.}}{1 \cancel{\text{ ft}}}$$
$$= 5 \times 12 \text{ in.}$$
$$= 60 \text{ in.}$$

Proportion [6.6]

A proportion is an equation that indicates that two ratios are equal.

The numbers in a proportion are called *terms* and are numbered as follows:

First term $\longrightarrow \dfrac{a}{b} = \dfrac{c}{d} \longleftarrow$ Third term
Second term $\qquad\qquad\qquad\quad \longleftarrow$ Fourth term

The first and fourth terms are called the *extremes.* The second and third terms are called the *means*.

$$\text{Means} \longrightarrow \frac{a}{b} = \frac{c}{d} \longleftarrow \text{Extremes}$$

5. The following is a proportion:
$$\frac{6}{8} = \frac{3}{4}$$

Fundamental Property of Proportions [6.6]

In any proportion the product of the extremes is equal to the product of the means. In symbols,

$$\text{If} \quad \frac{a}{b} = \frac{c}{d} \quad \text{then} \quad ad = bc$$

6. Find x: $\dfrac{2}{5} = \dfrac{8}{x}$

$2 \cdot x = 5 \cdot 8$

$2 \cdot x = 40$

$\dfrac{2 \cdot x}{2} = \dfrac{40}{2}$

$x = 20$

Finding a Missing Term in a Proportion [6.6]

To find the missing term in a proportion, we apply the fundamental property of proportions and solve the equation that results by dividing both sides by the number that is multiplied by the missing term. For instance, if we want to find the missing term in the proportion

$$\frac{2}{5} = \frac{8}{x}$$

we use the fundamental property of proportions to set the product of the extremes equal to the product of the means.

COMMON MISTAKE

1. A common mistake when working with ratios is to write the numbers in the wrong order when writing the ratio as a fraction. For example, the ratio 3 to 5 is equivalent to the fraction $\frac{3}{5}$. It cannot be written as $\frac{5}{3}$.

2. The most common mistake in converting from one kind of unit to another happens when we use the wrong conversion factor.

This is a mistake:

$$5 \text{ ft} = 5 \text{ ft} \times \frac{1 \text{ ft}}{12 \text{ in.}}$$

Conversion factor is used incorrectly. The units, ft, don't divide out, because they are both in the numerator. We should have multiplied by $\dfrac{12 \text{ in.}}{1 \text{ ft}}$.

Chapter **6** Review

Write each of the following ratios as a fraction in lowest terms. [6.1]

1. 9 to 30

2. 30 to 9

3. $\frac{3}{7}$ to $\frac{4}{7}$

4. $\frac{8}{5}$ to $\frac{8}{9}$

5. $2\frac{1}{3}$ to $1\frac{2}{3}$

6. 3 to $2\frac{3}{4}$

7. 0.6 to 1.2

8. 0.03 to 0.24

9. 2 feet to 2 yards [6.2]

10. 40 minutes to 1 hour [6.2]

The chart shows where each dollar spent on gasoline in the United States goes.

11. Ratio Find the ratio of money paid for taxes to money that goes to oil company profits. [6.1]

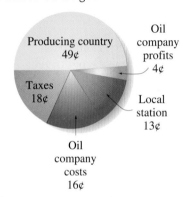

12. Ratio What is the ratio of the number of cents spent on oil company costs to the number of cents that goes to local stations? [6.1]

13. Ratio Give the ratio of oil company profits to oil company costs. [6.1]

14. Ratio Give the ratio of taxes to oil company costs and profits. [6.1]

15. Gas Mileage A car travels 285 miles on 15 gallons of gas. What is the rate of gas mileage in miles per gallon? [6.1]

16. Speed of Sound If it takes 2.5 seconds for sound to travel 2,750 feet, what is the speed of sound in feet per second? [6.1]

17. Unit Price If 6 ounces of frozen orange juice costs $1.38, what is the price per ounce? [6.1]

18. Unit Price A certain brand of frozen peas comes in two different-sized packages with the prices marked as follows:

16-ounce package $1.19
10-ounce package   99¢

Give the unit price for each package (to the nearest tenth of a cent) and indicate which is the better buy. [6.1]

Use the tables given in this chapter to make the following conversions. [6.2–6.5]

19. 12 ft to inches

20. 18 ft to yards

21. 49 cm to meters

22. 2 km to decimeters

23. 10 acres to square feet

24. 7,800 m² to ares

25. 4 ft² to square inches

26. 7 qt to pints

27. 24 qt to gallons

28. 5 L to milliliters

29. 8 lb to ounces

30. 2 lb 4 oz to ounces

31. 5 kg to grams

32. 5 t to kilograms

33. 4 in. to centimeters

34. 7 mi to kilometers

35. 7 L to quarts

36. 5 gal to liters

37. 5 oz to grams

38. 9 kg to pounds

39. 120°C to degrees Fahrenheit

40. 122°F to degrees Celsius

41. Construction A 12-square-meter patio is to be built using bricks that measure 10 centimeters by 20 centimeters. How many bricks will be needed to cover the patio? [6.3]

42. Capacity If a regular drinking glass holds 0.25 liter of liquid, how many glasses can be filled from a 6.5-liter container? [6.3]

43. Filling an Aquarium How many 8-fluid-ounce glasses of water will it take to fill a 5-gallon aquarium? [6.3]

44. Comparing Area On April 3, 2000, *USA Today* changed the size of its paper. Previous to this date, each page of the paper was $13\frac{1}{2}$ inches wide and $22\frac{1}{4}$ inches long, giving each page an area of $300\frac{3}{8}$ in². Convert this area to square feet. [6.3]

Find the missing term in each of the following proportions. [6.6]

45. $\dfrac{5}{7} = \dfrac{35}{x}$

46. $\dfrac{n}{18} = \dfrac{18}{54}$

47. $\dfrac{\frac{1}{2}}{10} = \dfrac{y}{2}$

48. $\dfrac{x}{1.8} = \dfrac{5}{1.5}$

49. Chemistry Suppose every 2,000 milliliters of a solution contains 24 milliliters of a certain drug. How many milliliters of solution are required to obtain 18 milliliters of the drug? [6.6]

50. Nutrition If $\frac{1}{2}$ cup of breakfast cereal contains 8 milligrams of calcium, how much calcium does $1\frac{1}{2}$ cups of the cereal contain? [6.6]

51. Weight Loss A man loses 8 pounds during the first 2 weeks of a diet. If he continues losing weight at the same rate, how long will it take him to lose 20 pounds? [6.6]

52. Men and Women If the ratio of men to women in a math class is 2 to 3, and there are 12 men in the class, how many women are in the class? [6.6]

Chapter **6** Test

Write each ratio as a fraction in lowest terms.

1. 24 to 18

2. $\frac{3}{4}$ to $\frac{5}{6}$

3. 5 to $3\frac{1}{3}$

4. 0.18 to 0.6

5. 36 seconds to 1 minute

A family of three budgeted the following amounts for some of their monthly bills:

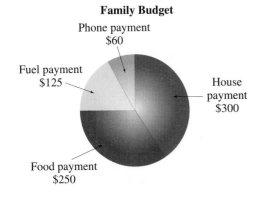

Family Budget

Phone payment
$60

Fuel payment
$125

House
payment
$300

Food payment
$250

6. Ratio Find the ratio of house payment to fuel payment.

7. Ratio Find the ratio of phone payment to food payment.

8. Gas Mileage A car travels 414 miles on 18 gallons of gas. What is the rate of gas mileage in miles per gallon?

9. Unit Price A certain brand of frozen orange juice comes in two different-sized cans with prices marked as shown. Give the unit price for each can, and indicate which is the better buy.

16-ounce can
$2.59

12-ounce can
$1.89

Use the tables in the chapter to make the following conversions.

10. 7 yd to feet

11. 750 m to kilometers

12. 3 acres to square feet

13. 432 in^2 to square feet

14. 10 L to milliliters

15. 5 mi to kilometers

16. 10 L to quarts

17. 80°F to degrees Celsius

18. Construction A 40-square-foot pantry floor is to be tiled using tiles that measure 8 inches by 8 inches. How many tiles will be needed to cover the pantry floor?

19. Filling an Aquarium How many 12-fluid-ounce glasses of water will it take to fill a 6-gallon aquarium?

20. Food Costs If coffee costs $5.60 per pound, how much will 3 lb 12 oz cost?

Find the missing term in each proportion.

21. $\dfrac{5}{6} = \dfrac{30}{x}$

22. $\dfrac{1.8}{6} = \dfrac{2.4}{x}$

23. Baseball A baseball player gets 9 hits in his first 21 games of the season. If he continues at the same rate, how many hits will he get in 56 games?

24. Map Reading The scale on a map indicates that 1 inch on the map corresponds to an actual distance of 60 miles. Two cities are $2\frac{1}{4}$ inches apart on the map. What is the actual distance between the two cities?

25. Cooking If the ratio of salt to sugar in a recipe is 3 to 5 and the recipe calls for 15 tablespoons of salt, how many tablespoons of sugar are needed?

26. Model Trains In the introduction to this chapter we indicated that the size of a model train relative to an actual train is referred to as its scale. Each scale is associated with a ratio as shown in the table below. For example, an HO model train has a ratio of 1 to 87, meaning it is $\frac{1}{87}$ as large as an actual train.

SCALE	RATIO
LGB	1 to 22.5
#1	1 to 32
O	1 to 43.5
S	1 to 64
HO	1 to 87
TT	1 to 120

If all six scales model the same box car, which one will have the largest box car?

How many times larger is the O scale than the HO scale?

Chapters 1-6 Cumulative Review

Simplify.

1. $(2x^2 - 6x - 9) + (3x^2 - 5x + 8)$

2. $x - \dfrac{5}{6}$

3. $613 - 297$

4. $\left(3\dfrac{1}{2} + \dfrac{1}{3}\right)\left(4\dfrac{1}{6} - \dfrac{2}{3}\right)$

5. $(4x + 9)^2$

6. $53(807)$

7. $\dfrac{5}{8} \div (-10)$

8. Round 37.6451 to the nearest hundredth.

9. Change $4\dfrac{7}{8}$ to an improper fraction.

10. Write the number 38,609 in words.

11. Identify the property or properties used in the following:
$5(x + 9) = 5(x) + 5(9)$

Simplify:

12. $6(3)^3 - 9(2)^2$

13. $\sqrt{\dfrac{36}{49}}$

14. $4\sqrt{12} - 6\sqrt{27}$

15. $5x^2 \cdot 8x^5$

16. $\dfrac{9 - 5}{-9 + 5}$

17. $-(-6)$

Write each ratio as a fraction in lowest terms.

18. 24 seconds to 1 minute

19. $\dfrac{2}{3}$ to $\dfrac{3}{4}$

20. Find the value of the expression $2y - 15$ when $y = 6$.

21. Translate into symbols: Four times x subtracted from twice the sum of x and nine.

Solve.

22. $\dfrac{7}{8} = \dfrac{49}{x}$

23. $7x + 0.21 = x - 0.69$

24. $\dfrac{5}{6}x = 25$

25. $5(2x - 7) = 8x - 5 - 4x$

Use the tables in Chapter 6 to make the following conversions.

26. 350 m to kilometers

27. 14 gal to liters

28. Write $\dfrac{14}{25}$ as a decimal.

29. Use the equation $2x - 7y = 10$ to find y when $x = -2$.

30. Reduce $\dfrac{99xy}{36x}$.

31. Temperature On Thursday, Arturo notices that the temperature reaches a high of 9° above 0 and a low of 8° below 0. What is the difference between the high and low temperatures for Thursday?

32. Geometry The length of a rectangle is 3 times the width. If the perimeter is 48 meters, find the length and width.

33. Ratio If the ratio of men to women in a self-defense class is 3 to 4, and there are 15 men in the class, how many women are in the class?

34. Surfboard Length A surfing company decides that a surfboard would be more efficient if its length were reduced by $3\frac{5}{8}$ inches. If the original length was 7 feet $\frac{3}{16}$ inches, what will be the new length of the board (in inches)?

35. Average Distance A bicyclist on a cross-country trip travels 72 miles the first day, 113 miles the second day, 108 miles the third day, and 95 miles the fourth day. What is her average distance traveled during the four days?

36. Area and Perimeter Find the area and perimeter of the triangle below.

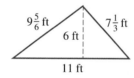

37. Cost of Chocolate If white chocolate sells for $4.32 per pound, how much will 2.5 pounds cost?

38. Number Line The distance between two numbers on the number line is 9. If one of the numbers is −4, what are the two possibilities for the other number?

39. Basketball Shots Erica makes a total of 8 two-point baskets in her first 18 games of the season. If she continues at the same rate, how many two-point baskets will she make in 45 games?

40. Graph $y = 3x + 2$.

Percent

INTRODUCTION

In preceding chapters we have used various methods to compare quantities. For example, in Chapter 6, we used the unit price of items to find the best buy. In this chapter we will study percent. When we use percent, we are comparing everything to 100, because percent means "per hundred."

The bar chart in Figure 1 shows the production costs for each of the first four *Star Wars* movies. As you can see, the largest increase in production costs occurred between *Return of the Jedi*, in 1983, and *The Phantom Menace*, in 1999. To see how this increase compares with the other increases, we look at the percent increase in production costs. The bar chart in Figure 2 shows the percent increase in these costs from each *Star Wars* movie to the next

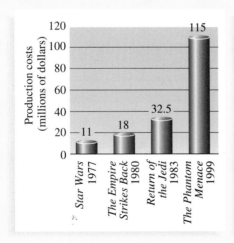

Figure 1

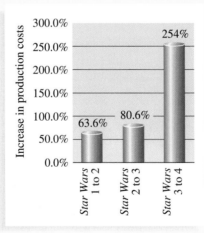

Figure 2

Because percent means "per hundred," comparing increases (and decreases) using percent gives us a way of standardizing our comparisons.

7.1 Percents, Decimals, and Fractions

Introduction . . .

The size of categories in the pie chart below are given as percents. The whole pie chart is represented by 100%. In general, 100% of something is the whole thing.

Factors producing more traffic today

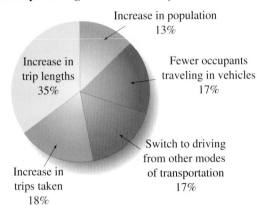

Increase in population
13%

Fewer occupants
traveling in vehicles
17%

Increase in
trip lengths
35%

Switch to driving
from other modes
of transportation
17%

Increase in
trips taken
18%

In this section we will look at the meaning of percent, and we will learn to change decimals to percents and percents to decimals.

The Meaning of Percent

Percent means "per hundred." Writing a number as a percent is a way of comparing the number with 100. For example, the number 42% (the % symbol is read "percent") is the same as 42 one-hundredths. That is:

$$42\% = \frac{42}{100}$$

Percents are really fractions (or ratios) with denominator 100.
Here are some examples that show the meaning of percent.

EXAMPLE 1 $50\% = \dfrac{50}{100}$ ■

EXAMPLE 2 $75\% = \dfrac{75}{100}$ ■

EXAMPLE 3 $25\% = \dfrac{25}{100}$ ■

EXAMPLE 4 $33\% = \dfrac{33}{100}$ ■

EXAMPLE 5 $6\% = \dfrac{6}{100}$ ■

EXAMPLE 6 $160\% = \dfrac{160}{100}$ ■

Changing Percents to Decimals

To change a percent to a decimal number, we simply use the meaning of percent.

7. Change 25.2% to a decimal.

EXAMPLE 7 Change 35.2% to a decimal.

SOLUTION We drop the % symbol and write 35.2 over 100.

$$35.2\% = \frac{35.2}{100}$$ **Use the meaning of % to convert to a fraction with denominator 100**

$$= 0.352$$ **Divide 35.2 by 100** ■

We see from Example 7 that 35.2% is the same as the decimal 0.352. The result is that the % symbol has been dropped and the decimal point has been moved two places to the *left*. Because % always means "per hundred," we will always end up moving the decimal point two places to the left when we change percents to decimals. Because of this, we can write the following rule.

> **RULE**
>
> To change a percent to a decimal, drop the % symbol and move the decimal point two places to the *left*.

Here are some examples to illustrate how to use this rule.

Change each percent to a decimal.

8. 40%

EXAMPLE 8 25% = 0.25 ■

9. 80%

EXAMPLE 9 75% = 0.75   **Notice that the results in Examples 8, 9, and 10 are consistent with the results in Examples 1, 2, and 3** ■

10. 15%

EXAMPLE 10 50% = 0.50 ■

11. 5.6%

EXAMPLE 11 6.8% = 0.068   **Notice here that we put a 0 in front of the 6 so we can move the decimal point two places to the left** ■

12. 4.86%

EXAMPLE 12 3.62% = 0.0362 ■

13. 0.6%

EXAMPLE 13 0.4% = 0.004   **This time we put two 0's in front of the 4 in order to be able to move the decimal point two places to the left** ■

14. 0.58%

EXAMPLE 14 0.63% = 0.0063 ■

Answers

7. 0.252   **8.** 0.40   **9.** 0.80
10. 0.15   **11.** 0.056   **12.** 0.0486
13. 0.006   **14.** 0.0058

Changing Decimals to Percents

Now we want to do the opposite of what we just did in Examples 7–14. We want to change decimals to percents. We know that 42% written as a decimal is 0.42, which means that in order to change 0.42 back to a percent, we must move the decimal point two places to the *right* and use the % symbol:

$0.42 = 42\%$ **Notice that we don't show the new decimal point if it is at the end of the number**

> ### RULE
>
> To change a decimal to a percent, we move the decimal point two places to the *right* and use the % symbol.

Examples 15–20 show how we use this rule.

Write each decimal as a percent.

EXAMPLE 15 $0.27 = 27\%$ ■

15. 0.35

EXAMPLE 16 $4.89 = 489\%$ ■

16. 5.77

EXAMPLE 17 $0.2 = 0.20 = 20\%$ **Notice here that we put a 0 after the 2 so we can move the decimal point two places to the right** ■

17. 0.4

EXAMPLE 18 $0.09 = 09\% = 9\%$ **Notice that we can drop the 0 at the left without changing the value of the number** ■

18. 0.03

EXAMPLE 19 $25 = 25.00 = 2,500\%$ **Here, we put two 0's after the 5 so we can move the decimal point two places to the right** ■

19. 45

EXAMPLE 20 $0.49\% = 49\%$ ■

20. 0.69

As you can see from the examples above, percent is just a way of comparing numbers to 100. To multiply decimals by 100, we move the decimal point two places to the right. To divide by 100, we move the decimal point two places to the left. Because of this, it is a fairly simple procedure to change percents to decimals and decimals to percents.

Answers
15. 35%   **16.** 577%   **17.** 40%
18. 3%   **19.** 4,500%   **20.** 69%

21. Change 82% to a fraction in lowest terms.

Who pays health care bills

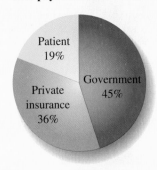

Patient
19%

Government
45%

Private
insurance
36%

22. Change 6.5% to a fraction in lowest terms.

23. Change $42\frac{1}{2}$% to a fraction in lowest terms.

Change each percent to a fraction in lowest terms.

24. 65%

Answers

21. $\frac{41}{50}$ **22.** $\frac{13}{200}$ **23.** $\frac{17}{40}$

24. $\frac{13}{20}$

Changing Percents to Fractions

To change a percent to a fraction, drop the % symbol and write the original number over 100.

EXAMPLE 21 The pie chart in the margin is from the Web site of *USA Today*. Change each percent to a fraction.

SOLUTION   In each case, we drop the percent symbol and write the number over 100. Then we reduce to lowest terms if possible.

$$19\% = \frac{19}{100} \qquad 45\% = \frac{45}{100} = \frac{9}{20} \qquad 36\% = \frac{36}{100} = \frac{9}{25}$$

$$\underset{\textbf{reduce}}{\uparrow} \qquad\qquad\qquad \underset{\textbf{reduce}}{\uparrow}$$

∎

EXAMPLE 22 Change 4.5% to a fraction in lowest terms.

SOLUTION   We begin by writing 4.5 over 100:

$$4.5\% = \frac{4.5}{100}$$

We now multiply the numerator and the denominator by 10 so the numerator will be a whole number:

$$\frac{4.5}{100} = \frac{4.5 \times \textbf{10}}{100 \times \textbf{10}} \qquad \textbf{Multiply the numerator and}$$
$$\textbf{the denominator by 10}$$
$$= \frac{45}{1{,}000}$$
$$= \frac{9}{200} \qquad \textbf{Reduce to lowest terms}$$

∎

EXAMPLE 23 Change $32\frac{1}{2}$% to a fraction in lowest terms.

SOLUTION   Writing $32\frac{1}{2}$% over 100 produces a complex fraction. We change $32\frac{1}{2}$ to an improper fraction and simplify:

$$32\frac{1}{2}\% = \frac{32\frac{1}{2}}{100}$$
$$= \frac{\frac{65}{2}}{100} \qquad \textbf{Change } 32\frac{1}{2} \textbf{ to the improper fraction } \frac{65}{2}$$
$$= \frac{65}{2} \times \frac{1}{100} \qquad \textbf{Dividing by 100 is the same as multiplying by } \frac{1}{100}$$
$$= \frac{\cancel{5} \cdot 13 \cdot 1}{2 \cdot \cancel{5} \cdot 20} \qquad \textbf{Multiplication}$$
$$= \frac{13}{40} \qquad \textbf{Reduce to lowest terms}$$

∎

EXAMPLE 24 $35\% = \frac{35}{100}$
$$= \frac{7}{20}$$

∎

EXAMPLE 25 $25.4\% = \dfrac{25.4}{100}$

$= \dfrac{254}{1,000}$ **Multiply the numerator and the denominator by 10**

$= \dfrac{127}{500}$ **Reduce to lowest terms**

25. 35.6%

Changing Fractions to Percents

To change a fraction to a percent, we can change the fraction to a decimal and then change the decimal to a percent.

EXAMPLE 26 Change $\frac{7}{10}$ to a percent.

SOLUTION   We can change $\frac{7}{10}$ to a decimal by dividing 7 by 10:

$$\begin{array}{r} 0.7 \\ 10\overline{)7.0} \\ \underline{7\,0} \\ 0 \end{array}$$

We then change the decimal 0.7 to a percent by moving the decimal point two places to the *right* and using the % symbol:

$$0.7 = 70\%$$

26. Change $\dfrac{9}{10}$ to a percent.

You may have noticed that we could have saved some time by simply writing $\frac{7}{10}$ as an equivalent fraction with denominator 100. That is:

$$\frac{7}{10} = \frac{7 \cdot \mathbf{10}}{10 \cdot \mathbf{10}} = \frac{70}{100} = 70\%$$

This is a good way to convert fractions like $\frac{7}{10}$ to percents. It works well for fractions with denominators of 2, 4, 5, 10, 25, and 50, because they are easy to change to fractions with denominators of 100.

EXAMPLE 27 Change $\frac{3}{8}$ to a percent.

SOLUTION   We write $\frac{3}{8}$ as a decimal by dividing 3 by 8. We then change the decimal to a percent by moving the decimal point two places to the right and using the % symbol.

27. Change $\dfrac{5}{8}$ to a percent.

$$\frac{3}{8} = 0.375 = 37.5\% \qquad \begin{array}{r} .375 \\ 8\overline{)3.000} \\ \underline{2\,4} \\ 60 \\ \underline{56} \\ 40 \\ \underline{40} \\ 0 \end{array}$$

Answers

25. $\dfrac{89}{250}$ **26.** 90%   **27.** 62.5%

28. Change $\frac{7}{12}$ to a percent.

EXAMPLE 28 Change $\frac{5}{12}$ to a percent.

SOLUTION   We begin by dividing 5 by 12:

$$
\begin{array}{r}
.4166 \\
12\overline{)5.000} \\
\underline{4\ 8} \\
20 \\
\underline{12} \\
80 \\
\underline{72} \\
80 \\
\underline{72}
\end{array}
$$

Because the 6's repeat indefinitely, we can use mixed number notation to write

$$
\frac{5}{12} = 0.41\overline{6} = 41\frac{2}{3}\%
$$

Or, rounding, we can write

$$
\frac{5}{12} = 41.7\% \quad \textbf{To the nearest tenth of a percent} \qquad \blacksquare
$$

Note
When rounding off, let's agree to round off to the nearest thousandth and then move the decimal point. Our answers in percent form will then be accurate to the nearest tenth of a percent, as in Example 28.

29. Change $3\frac{3}{4}$ to a percent.

EXAMPLE 29 Change $2\frac{1}{2}$ to a percent.

SOLUTION   We first change to a decimal and then to a percent:

$$
2\frac{1}{2} = 2.5
$$
$$
= 250\% \qquad \blacksquare
$$

   Table 1 lists some of the most commonly used fractions and decimals and their equivalent percents.

Table 1

FRACTION	DECIMAL	PERCENT
$\frac{1}{2}$	0.5	50%
$\frac{1}{4}$	0.25	25%
$\frac{3}{4}$	0.75	75%
$\frac{1}{3}$	$0.33\frac{1}{3}$	$33\frac{1}{3}\%$
$\frac{2}{3}$	$0.66\frac{2}{3}$	$66\frac{2}{3}\%$
$\frac{1}{5}$	0.2	20%
$\frac{2}{5}$	0.4	40%
$\frac{3}{5}$	0.6	60%
$\frac{4}{5}$	0.8	80%

Answers

28. $58\frac{1}{3}\% \approx 58.3\%$ **29.** 375%

Problem Set 7.1

Write each percent as a fraction with denominator 100.

1. 20% **2.** 40% **3.** 60% **4.** 80% **5.** 24% **6.** 48%

7. 65% **8.** 35%

Change each percent to a decimal.

9. 23% **10.** 34% **11.** 92% **12.** 87% **13.** 9% **14.** 7%

15. 3.4% **16.** 5.8% **17.** 6.34% **18.** 7.25% **19.** 0.9% **20.** 0.6%

Change each decimal to a percent.

21. 0.23 **22.** 0.34 **23.** 0.92 **24.** 0.87 **25.** 0.45 **26.** 0.54

27. 0.03 **28.** 0.04 **29.** 0.6 **30.** 0.9 **31.** 0.8 **32.** 0.5

33. 0.27 **34.** 0.62 **35.** 1.23 **36.** 2.34

Change each percent to a fraction in lowest terms.

37. 60%   **38.** 40%   **39.** 75%   **40.** 25%   **41.** 4%   **42.** 2%

43. 26.5%   **44.** 34.2%   **45.** 71.87%   **46.** 63.6%   **47.** 0.75%   **48.** 0.45%

49. $6\frac{1}{4}$%   **50.** $5\frac{1}{4}$%   **51.** $33\frac{1}{3}$%   **52.** $66\frac{2}{3}$%

Change each fraction or mixed number to a percent.

53. $\frac{1}{2}$ **54.** $\frac{1}{4}$ **55.** $\frac{3}{4}$ **56.** $\frac{2}{3}$ **57.** $\frac{1}{3}$ **58.** $\frac{1}{5}$

59. $\frac{4}{5}$ **60.** $\frac{1}{6}$ **61.** $\frac{7}{8}$ **62.** $\frac{1}{8}$ **63.** $\frac{7}{50}$ **64.** $\frac{9}{25}$

65. $3\frac{1}{4}$ **66.** $2\frac{1}{8}$ **67.** $1\frac{1}{2}$ **68.** $1\frac{3}{4}$

69. $\frac{21}{43}$ to the nearest tenth of a percent   **70.** $\frac{36}{49}$ to the nearest tenth of a percent

Applying the Concepts

71. Physiology The human body is between 50% and 75% water. Write each of these percents as a decimal.

72. Alcohol Consumption In the United States, 2.7% of those over 15 years of age drink more than 6.3 ounces of alcohol per day. In France, the same figure is 9%. Write each of these percents as a decimal.

73. Death Rate Cancer is responsible for 18.9% of all the deaths in the United States each year. Accidents account for only 8.1% of all deaths. Write each of these percents as a decimal.

74. Commuting The pie chart below shows how we, as a nation, commute to work. Change each percent to a fraction in lowest terms.

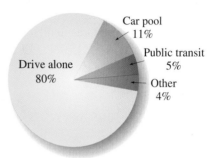

How we commute

Car pool 11%
Public transit 5%
Drive alone 80%
Other 4%

75. Nutrition Although nutritionally breakfast is the most important meal of the day, only $\frac{1}{5}$ of the people in the United States consistently eat breakfast. What percent of the population is this?

76. Children in School In Belgium, 96% of all children between 3 and 6 years of age go to school. In Sweden, the same figure is only 25%. In the United States, the figure is 60%. Write each of these percents as a fraction in lowest terms.

77. Student Enrollment The pie chart below is from Chapter 3. Change each fraction to a percent.

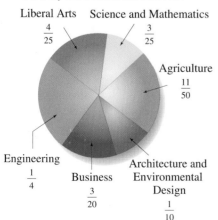

Cal Poly enrollment for Fall 1999

Liberal Arts $\frac{4}{25}$ Science and Mathematics $\frac{3}{25}$

Agriculture $\frac{11}{50}$

Engineering $\frac{1}{4}$

Business $\frac{3}{20}$

Architecture and Environmental Design $\frac{1}{10}$

78. An object weighed on the moon will be only $\frac{1}{6}$ as heavy as it is on the earth. Write this number as a percent.

Calculator Problems

Use a calculator to write each fraction as a decimal, and then change the decimal to a percent. Round all answers to the nearest tenth of a percent.

79. $\dfrac{29}{37}$

80. $\dfrac{18}{83}$

81. $\dfrac{6}{51}$

82. $\dfrac{8}{95}$

83. $\dfrac{236}{327}$

84. $\dfrac{568}{732}$

85. Women in the Military During World War II, $\frac{1}{12}$ of the Soviet armed forces were women, whereas today only $\frac{1}{450}$ of the Russian armed forces are women. Change both fractions to percents (to the nearest tenth of a percent).

86. Number of Teachers The ratio of the number of teachers to the number of students in secondary schools in Japan is 1 to 17. In the United States, the ratio is 1 to 19. Write each of these ratios as a fraction and then as a percent.

Review Problems

These review problems cover some material that will be needed in Section 7.2.

Multiply.

87. 0.25(300)

88. 0.06(2.95)

89. 0.53(16)

90. 0.0725(3,000)

Divide. Round the answers to the nearest thousandth, if necessary.

91. $\dfrac{25}{0.05}$

92. $\dfrac{7}{0.14}$

93. $\dfrac{45}{0.25}$

94. $\dfrac{62}{0.06}$

Solve for n.

95. $3 \cdot n = 12$

96. $16 = 9 \cdot n$

Extending the Concepts

97. Give a written explanation of the process you would use to change the fraction $\frac{2}{5}$ to a percent.

98. Give a written explanation of the process you would use to change 40% to a fraction in lowest terms.

99. 75% is how many times as large as 0.75%?

100. 60% is how many times as large as 0.30%?

7.2 Basic Percent Problems

Introduction . . .

The American Dietetic Association (ADA) recommends eating foods in which the number of calories from fat is less than 30% of the total number of calories. Foods that satisfy this requirement are considered healthy foods. Is the nutrition label shown below from a food that the ADA would consider healthy? This is the type of question we will be able to answer after we have worked through the examples in this section.

Nutrition Facts

Serving Size 1/2 cup (65g)
Servings Per Container: 8

Amount Per Serving

Calories 150	Calories from fat 90

	% Daily Value*
Total Fat 10g	16%
Saturated Fat 6g	32%
Cholesterol 35mg	12%
Sodium 30mg	1%
Total Carbohydrate 14g	5%
Dietary Fiber 0g	0%
Sugars 11g	
Protein 2g	

Vitamin A 6%	●	Vitamin C 0%
Calcium 6%	●	Iron 0%

*Percent Daily Values are based on a 2,000 calorie diet.

Figure 1   Nutrition label from vanilla ice cream

This section is concerned with three kinds of word problems that are associated with percents. Here is an example of each type:

Type A: What number is 15% of 63?
Type B: What percent of 42 is 21?
Type C: 25 is 40% of what number?

The method we use to solve all three types of problems involves translating the sentences into equations and then solving the equations. The following translations are used to write the sentences as equations:

ENGLISH	MATHEMATICS
is	=
of	· (multiply)
a number	n
what number	n
what percent	n

The word *is* always translates to an = sign, the word *of* almost always means multiply, and the number we are looking for can be represented with a letter, such as n or x.

EXAMPLE 1   What number is 15% of 63?

SOLUTION   We translate the sentence into an equation as follows:

What number is 15% *of* 63?
$$n = 0.15 \cdot 63$$

To do arithmetic with percents, we have to change to decimals.* That is why 15% is rewritten as 0.15. Solving the equation, we have

$$n = 0.15 \cdot 63$$
$$n = 9.45$$

> 15% of 63 is 9.45

EXAMPLE 2   What percent of 42 is 21?

SOLUTION   We translate the sentence as follows:

What percent of 42 *is* 21?
$$n \cdot 42 = 21$$

We solve for n by dividing both sides by 42.

$$\frac{n \cdot 42}{42} = \frac{21}{42}$$
$$n = \frac{21}{42}$$
$$n = 0.50$$

Because the original problem asked for a percent, we change 0.50 to a percent:

$$n = 50\%$$

> 21 is 50% of 42

EXAMPLE 3   25 is 40% of what number?

SOLUTION   Following the procedure from the first two examples, we have

25 *is* 40% *of what number?*
$$25 = 0.40 \cdot n$$

*It is sometimes convenient to change percents to fractions instead of decimals. You may want to try some of the examples in this section yourself, using fractions for percents where we have used decimals.

Again, we changed 40% to 0.40 so we can do the arithmetic involved in the problem. Dividing both sides of the equation by 0.40, we have

$$\frac{25}{0.40} = \frac{\cancel{0.40} \cdot n}{\cancel{0.40}}$$

$$\frac{25}{0.40} = n$$

$$62.5 = n$$

> 25 is 40% of 62.5

As you can see, all three types of percent problems are solved in a similar manner. We write *is* as =, *of* as ·, and *what number* as n. The resulting equation is then solved to obtain the answer to the original question.

EXAMPLE 4   What number is 43.5% of 25?

$$n = 0.435 \cdot 25$$

$$n = 10.9 \qquad \text{\textbf{Rounded to the nearest tenth}}$$

> 10.9 is 43.5% of 25

4. What number is 63.5% of 45? (Round to the nearest tenth.)

EXAMPLE 5   What percent of 78 is 31.9?

$$n \cdot 78 = 31.9$$

$$\frac{n \cdot \cancel{78}}{\cancel{78}} = \frac{31.9}{78}$$

$$n = \frac{31.9}{78}$$

$$n = 0.409 \qquad \text{\textbf{Rounded to the nearest thousandth}}$$

$$n = 40.9\%$$

> 31.9 is 40.9% of 78

5. What percent of 85 is 11.9?

6. 62 is 39% of what number? (Round to the nearest tenth.)

EXAMPLE 6   34 is 29% of what number?

$$34 = 0.29 \cdot n$$

$$\frac{34}{0.29} = \frac{0.29 \cdot n}{0.29}$$

$$\frac{34}{0.29} = n$$

$$117.2 = n \qquad \textbf{Rounded to the nearest tenth}$$

34 is 29% of 117.2

7. The nutrition label below is from a package of vanilla frozen yogurt. What percent of the total number of calories are fat calories? Round your answer to the nearest tenth of a percent.

EXAMPLE 7   As we mentioned in the introduction to this section, the American Dietetic Association recommends eating foods in which the number of calories from fat is less than 30% of the total number of calories. According to the nutrition label below, what percent of the total number of calories are fat calories?

Nutrition Facts

Serving Size 1/2 cup (98g)
Servings Per Container: 4

Amount Per Serving

Calories 160	Calories from fat 25

	% Daily Value*
Total Fat 2.5g	**4%**
Saturated Fat 1.5g	**7%**
Cholesterol 45mg	**15%**
Sodium 55mg	**2%**
Total Carbohydrate 26g	**9%**
Dietary Fiber 0g	**0%**
Sugars 19g	
Protein 8g	

Vitamin A 0%	●	Vitamin C 0%
Calcium 25%	●	Iron 0%

*Percent Daily Values are based on a 2,000 calorie diet.

Nutrition Facts

Serving Size 1/2 cup (65g)
Servings Per Container: 8

Amount Per Serving

Calories 150	Calories from fat 90

	% Daily Value*
Total Fat 10g	**16%**
Saturated Fat 6g	**32%**
Cholesterol 35mg	**12%**
Sodium 30mg	**1%**
Total Carbohydrate 14g	**5%**
Dietary Fiber 0g	**0%**
Sugars 11g	
Protein 2g	

Vitamin A 6%	●	Vitamin C 0%
Calcium 6%	●	Iron 0%

*Percent Daily Values are based on a 2,000 calorie diet.

Figure 2   Nutrition label from vanilla ice cream

SOLUTION   To solve this problem, we must write the question in the form of one of the three basic percent problems shown in Examples 1–6. Because there are 90 calories from fat and a total of 150 calories, we can write the question this way: 90 is what percent of 150?

Now that we have written the question in the form of one of the basic percent problems, we simply translate it into an equation. Then we solve the equation.

90 *is what percent of* 150?

$$90 = n \cdot 150$$

$$\frac{90}{150} = n$$

$$n = 0.60 = 60\%$$

The number of calories from fat in this package of ice cream is 60% of the total number of calories. Thus the ADA would not consider this to be a healthy food.

Answers

6. 159.0

7. 15.6% of the calories are from fat. (So far as fat content is concerned, the frozen yogurt is a healthier choice than the ice cream.)

Problem Set 7.2

Solve each of the following problems.

1. What number is 25% of 32?

2. What number is 10% of 80?

3. What number is 20% of 120?

4. What number is 15% of 75?

5. What number is 54% of 38?

6. What number is 72% of 200?

7. What number is 11% of 67?

8. What number is 2% of 49?

9. What percent of 24 is 12?

10. What percent of 80 is 20?

11. What percent of 50 is 5?

12. What percent of 20 is 4?

13. What percent of 36 is 9?

14. What percent of 70 is 14?

15. What percent of 8 is 6?

16. What percent of 16 is 9?

17. 32 is 50% of what number?

18. 16 is 20% of what number?

19. 10 is 20% of what number?

20. 11 is 25% of what number?

21. 37 is 4% of what number?

22. 90 is 80% of what number?

23. 8 is 2% of what number?

24. 6 is 3% of what number?

The following problems can be solved by the same method you used in Problems 1–24.

25. What is 6.4% of 87?

26. What is 10% of 102?

27. 25% of what number is 30?

28. 10% of what number is 22?

29. 28% of 49 is what number?

30. 97% of 28 is what number?

31. 27 is 120% of what number?

32. 24 is 150% of what number?

33. 65 is what percent of 130?

34. 26 is what percent of 78?

35. What is 0.4% of 235,671?

36. What is 0.8% of 721,423?

37. 4.89% of 2,000 is what number?

38. 3.75% of 4,000 is what number?

Applying the Concepts

Nutrition For each nutrition label in Problems 39–42, find what percent of the total number of calories comes from fat calories. Then indicate whether the label is from a food considered healthy by the American Dietetic Association. Round to the nearest tenth of a percent if necessary.

39. Spaghetti

Nutrition Facts

Serving Size 2 oz. (56g per 1/8 of pkg) dry
Servings Per Container: 8

Amount Per Serving

Calories 210	Calories from fat 10

	% Daily Value*
Total Fat 1g	**2%**
Saturated Fat 0g	**0%**
Polyunsaturated Fat 0.5g	
Monounsaturated Fat 0g	
Cholesterol 0mg	**0%**
Sodium 0mg	**0%**
Total Carbohydrate 42g	**14%**
Dietary Fiber 2g	**7%**
Sugars 3g	
Protein 7g	

Vitamin A 0%	●	Vitamin C 0%
Calcium 0%	●	Iron 10%
Thiamin 30%	●	Riboflavin 10%
Niacin 15%	●	

*Percent Daily Values are based on a 2,000 calorie diet

40. Canned Italian tomatoes

Nutrition Facts

Serving Size 1/2 cup (121g)
Servings Per Container: about 3 1/2

Amount Per Serving

Calories 25	Calories from fat 0

	% Daily Value*
Total Fat 0g	**0%**
Saturated Fat 0g	**0%**
Cholesterol 0mg	**0%**
Sodium 300mg	**12%**
Potassium 145mg	**4%**
Total Carbohydrate 4g	**2%**
Dietary Fiber 1g	**4%**
Sugars 4g	
Protein 1g	

Vitamin A 20%	●	Vitamin C 15%
Calcium 4%	●	Iron 15%

*Percent Daily Values are based on a 2,000 calorie diet.

41. Shredded romano cheese

Nutrition Facts

Serving Size 2 tsp (5g)
Servings Per Container: 34

Amount Per Serving

Calories 20	Calories from fat 10

	% Daily Value*
Total Fat 1.5g	**2%**
Saturated Fat 1g	**5%**
Cholesterol 5mg	**2%**
Sodium 70mg	**3%**
Total Carbohydrate 0g	**0%**
Fiber 0g	**0%**
Sugars 0g	
Protein 2g	

Vitamin A 0%	●	Vitamin C 0%
Calcium 4%	●	Iron 0%

*Percent Daily Values are based on a 2,000 calorie diet.

42. Tortilla chips

Nutrition Facts

Serving Size 1 oz (28g/About 12 chips)
Servings Per Container: about 2

Amount Per Serving

Calories 140	Calories from fat 60

	% Daily Value*
Total Fat 7g	**1%**
Saturated Fat 1g	**6%**
Cholesterol 0mg	**0%**
Sodium 170mg	**7%**
Total Carbohydrate 18g	**6%**
Dietary Fiber 1g	**4%**
Sugars less than 1g	
Protein 2g	

Vitamin A 0%	●	Vitamin C 0%
Calcium 4%	●	Iron 2%

*Percent Daily Values are based on a 2,000 calorie diet.

Calculator Problems

Set up each of the following problems using the method developed in this section. Then do all the calculations on a calculator.

43. What number is 62.5% of 398?

44. What number is 4.25% of 907?

45. What percent of 789 is 204? (Round to the nearest tenth of a percent.)

46. What percent of 123 is 119? (Round to the nearest tenth of a percent.)

47. 19 is 38.4% of what number? (Round to the nearest tenth.)

48. 249 is 16.9% of what number? (Round to the nearest tenth.)

Review Problems

The problems below are review from Section 3.2.

Factor each of the following into a product of prime factors:

49. 45

50. 12

51. 105

52. 75

53. 36

54. 72

55. 210

56. 150

Extending the Concepts: Writing Mathematics

57. Write a basic percent problem, the solution to which can be found by solving the equation $n = 0.25(350)$.

58. Write a basic percent problem, the solution to which can be found by solving the equation $n = 0.35(250)$.

59. Write a basic percent problem, the solution to which can be found by solving the equation $n \cdot 24 = 16$.

60. Write a basic percent problem, the solution to which can be found by solving the equation $n \cdot 16 = 24$.

61. Write a basic percent problem, the solution to which can be found by solving the equation $46 = 0.75 \cdot n$.

62. Write a basic percent problem, the solution to which can be found by solving the equation $75 = 0.46 \cdot n$.

7.3 General Applications of Percent

Introduction . . .

As you know from watching television and reading the newspaper, we encounter percent in many situations in everyday life. A 1995 newspaper article discussing the effects of a cholesterol-lowering drug stated that the drug in question "lowered levels of LDL cholesterol by an average of 35%." As we progress through this chapter, we will become more and more familiar with percent, and as a result, we will be better equipped to understand statements like the one above concerning cholesterol.

   In this section we continue our study of percent by doing more of the translations that were introduced at the end of Section 7.2. The better you are at working the problems in Section 7.2, the easier it will be for you to get started on the problems in this section.

EXAMPLE 1   On a 120-question test a student answered 96 correctly. What percent of the problems did the student work correctly?

SOLUTION   We have 96 correct answers out of a possible 120. The problem can be restated as

$$96 \text{ is what percent of } 120?$$
$$96 = n \cdot 120$$

$$\frac{96}{120} = \frac{n \cdot 120}{120} \qquad \textbf{Divide both sides by 120}$$

$$n = \frac{96}{120} \qquad \begin{array}{l}\textbf{Switch the left and right} \\ \textbf{sides of the equation}\end{array}$$

$$n = 0.80 \qquad \textbf{Divide 96 by 120}$$

$$= 80\% \qquad \textbf{Rewrite as a percent}$$

When we write a test score as a percent, we are comparing the original score to an equivalent score on a 100-question test. That is, 96 correct out of 120 is the same as 80 correct out of 100.   ■

EXAMPLE 2   How much HCl (hydrochloric acid) is in a 60-milliliter bottle that is marked 80% HCl?

SOLUTION   If the bottle is marked 80% HCl, that means 80% of the solution is HCl and the rest is water. Because the bottle contains 60 milliliters, we can restate the question as:

$$What \text{ is } 80\% \text{ of } 60?$$
$$n = 0.80 \cdot 60$$
$$n = 48$$

There are 48 milliliters of HCl in 60 milliliters of 80% HCl solution.   ■

Practice Problems

1. On a 150-question test a student answered 114 correctly. What percent of the problems did the student work correctly?

2. How much HCl is in a 40-milliliter bottle that is marked 75% HCl?

Answers
1. 76%   **2.** 30 milliliters

3. If 42% of the students in a certain college are female and there are 3,360 female students, what is the total number of students in the college?

EXAMPLE 3 If 48% of the students in a certain college are female and there are 2,400 female students, what is the total number of students in the college?

SOLUTION   We restate the problem as:

2,400 *is* 48% *of what number?*
$$\downarrow \quad \downarrow \quad \downarrow \quad \downarrow \checkmark$$
$$2{,}400 = 0.48 \cdot n$$
$$\frac{2{,}400}{0.48} = \frac{0.48 \cdot n}{0.48} \qquad \textbf{Divide both sides by 0.48}$$
$$n = \frac{2{,}400}{0.48} \qquad \textbf{Switch the left and right sides of the equation}$$
$$n = 5{,}000$$

There are 5,000 students. ■

4. Suppose in Example 4 that 35% of the students receive a grade of A. How many of the 300 students is that?

EXAMPLE 4 If 25% of the students in elementary algebra courses receive a grade of A, and there are 300 students enrolled in elementary algebra this year, how many students will receive A's?

SOLUTION   After reading the question a few times, we find that it is the same as this question:

What number is 25% of 300?
$$\searrow \quad \downarrow \quad \downarrow \quad \downarrow \quad \downarrow$$
$$n = 0.25 \cdot 300$$
$$n = 75$$

Thus, 75 students will receive A's in elementary algebra. ■

   Almost all application problems involving percents can be restated as one of the three basic percent problems we listed in Section 7.2. It takes some practice before the restating of application problems becomes automatic. You may have to review Section 7.2 and Examples 1–4 above several times before you can translate word problems into mathematical expressions yourself.

Answers
3. 8,000   **4.** 105

Problem Set 7.3

Solve each of the following problems by first restating it as one of the three basic percent problems of Section 7.2. In each case, be sure to show the equation.

1. **Test Scores** On a 120-question test a student answered 84 correctly. What percent of the problems did the student work correctly?

2. **Test Scores** An engineering student answered 81 questions correctly on a 90-question trigonometry test. What percent of the questions did she answer correctly? What percent were answered incorrectly?

3. **Basketball** A basketball player made 63 out of 75 free throws. What percent is this?

4. **Family Budget** A family spends $450 every month on food. If the family's income each month is $1,800, what percent of the family's income is spent on food?

5. **Chemistry** How much HCl (hydrochloric acid) is in a 60-milliliter bottle that is marked 75% HCl?

6. **Chemistry** How much acetic acid is in a 5-liter container of acetic acid and water that is marked 80% acetic acid? How much is water?

60 ml
HCl
75%

7. **Farming** A farmer owns 28 acres of land. Of the 28 acres, only 65% can be farmed. How many acres are available for farming? How many are not available for farming?

8. **Number of Students** Of the 420 students enrolled in a basic math class, only 30% are first-year students. How many are first-year students? How many are not?

9. Number of Students If 48% of the students in a certain college are female and there are 1,440 female students, what is the total number of students in the college?

10. Mixture Problem A solution of alcohol and water is 80% alcohol. The solution is found to contain 32 milliliters of alcohol. How many milliliters total (both alcohol and water) are in the solution?

11. Number of Graduates Suppose 60% of the graduating class in a certain high school goes on to college. If 240 students from this graduating class are going on to college, how many students are there in the graduating class?

12. Defective Parts In a shipment of airplane parts, 3% are known to be defective. If 15 parts are found to be defective, how many parts are in the shipment?

13. Number of Students There are 3,200 students at our school. If 52% of them are women, how many women students are there at our school?

14. Number of Students In a certain school, 75% of the students in first-year chemistry have had algebra. If there are 300 students in first-year chemistry, how many of them have had algebra?

15. Population In a city of 32,000 people, there are 10,000 people under 25 years of age. What percent of the population is under 25 years of age?

16. Number of Students If 45 people enrolled in a psychology course but only 35 completed it, what percent of the students completed the course? (Round to the nearest tenth of a percent.)

Calculator Problems

The following problems are similar to Problems 1–16. They should be set up in the same way. Then the actual calculations should be done on a calculator.

17. Number of People Of 7,892 people attending an outdoor concert in Los Angeles, 3,972 are over 18 years of age. What percent is this? (Round to the nearest whole-number percent.)

18. Manufacturing A car manufacturer estimates that 25% of the new cars sold in one city have defective engine mounts. If 2,136 new cars are sold in that city, how many will have defective engine mounts?

19. Mixture Problem Suppose a solution of alcohol and water contains 31.8 milliliters of alcohol. If the solution is 75.6% alcohol, how many milliliters of solution are there? (Round to the nearest whole number.)

20. House Payments A married couple makes $43,698 a year. If 23.4% of their income is used for house payments, how much did they spend on house payments for the year?

Review Problems

The problems below are review from Sections 3.3 and 3.7.

Multiply.

21. $\dfrac{1}{2} \cdot \dfrac{2}{5}$

22. $\dfrac{3}{4} \cdot \dfrac{1}{3}$

23. $\dfrac{3}{4} \cdot \dfrac{5}{9}$

24. $\dfrac{5}{6} \cdot \dfrac{12}{13}$

25. $2 \cdot \dfrac{3}{8}$

26. $3 \cdot \dfrac{5}{12}$

27. $1\dfrac{1}{4} \cdot \dfrac{8}{15}$

28. $2\dfrac{1}{3} \cdot \dfrac{9}{10}$

Extending the Concepts: Batting Averages

Batting averages in baseball are given as decimal numbers, rounded to the nearest thousandth. For example, on July 1, 1999, Derek Jeter had the highest batting average in the American League. At that time, he had 110 hits in 292 times at bat. His batting average was .377, which is found by dividing the number of hits by the number of times he was at bat and then rounding to the nearest thousandth.

$$\text{Batting average} = \frac{\text{number of hits}}{\text{number of times at bat}} = \frac{110}{292} \approx .377$$

Because we can write any decimal number as a percent, we can convert batting averages to percents and use our knowledge of percent to solve problems. Looking at Derek Jeter's batting average as a percent, we can say that he will get a hit 37.7% of the times he is at bat.

Each of the following problems can be solved by converting batting averages to percents and translating the problem into one of our three basic percent problems.

29. On July 1, 1999, Larry Walker had the highest batting average in the National League with 91 hits in 239 times at bat. What percent of the time Larry Walker is at bat can we expect him to get a hit?

30. Sammy Sosa had 87 hits in 296 times at bat. What percent of the time can we expect Sosa to get a hit?

31. On July 1, 1999, Mark McGwire was batting .259. If he had been at bat 255 times, how many hits did he have? (Remember his batting average has been rounded to the nearest thousandth.)

32. On July 1, 1999, Chipper Jones was batting .293. If he had been at bat 294 times, how many hits did he have? (Remember, his batting average has been rounded to the nearest thousandth.)

33. How many hits must Derek Jeter have in his next 50 times at bat to maintain a batting average of at least .377?

34. How many hits must Larry Walker have in his next 50 times at bat to maintain a batting average of at least .381?

7.4   Sales Tax and Commission

Introduction . . .

When I was preparing to produce the videotapes that accompany this book, I did some comparison shopping on the camera I was interested in. One store quoted me a total price of $3,200, which included the sales tax. At that time, the sales tax rate in California was 7.5%. In order to find the price of the camera before the sales tax was added on, I had to divide $3,200 by 1.075. After you are finished with this section, you will understand why this is the appropriate calculation in this situation.

Total price $3,200
"Price includes sales tax"

To solve the problems in this section, we will first restate them in terms of the problems we have already learned how to solve. Then we will solve the restated problems just as we solved the problems in Sections 7.2 and 7.3.

Sales Tax

EXAMPLE 1   Suppose the sales tax rate in Mississippi is 6% of the purchase price. If the price of a used refrigerator is $550, how much sales tax must be paid?

SOLUTION   Because the sales tax is 6% of the purchase price, and the purchase price is $550, the problem can be restated as:

What is 6% of $550?

We solve this problem, as we did in Section 7.2, by translating it into an equation:

$$\text{What is } 6\% \text{ of } \$550?$$
$$n = 0.06 \cdot 550$$
$$n = 33$$

The sales tax is $33. The total price of the refrigerator would be

Purchase price		Sales tax		Total price
$550	+	$33	=	$583

> Total price = Purchase price + Sales tax

Practice Problems

1. What is the sales tax on a new washing machine if the machine is purchased for $625 and the sales tax rate is 6%?

Note
In Example 1, the *sales tax rate* is 6%, and the *sales tax* is $33. In most everyday communications, people say "The sales tax is 6%," which is incorrect. The 6% is the tax *rate*, and the $33 is the actual tax.

Answer
1. $37.50

2. Suppose the sales tax rate is 3%. If the sales tax on a 10-speed bicycle is $4.35, what is the purchase price, and what is the total price of the bicycle?

EXAMPLE 2 Suppose the sales tax rate is 4%. If the sales tax on a 10-speed bicycle is $5.44, what is the purchase price, and what is the total price of the bicycle?

SOLUTION   We know that 4% of the purchase price is $5.44. We find the purchase price first by restating the problem as:

$$\$5.44 \text{ is } 4\% \text{ of what number?}$$
$$5.44 = 0.04 \cdot n$$

We solve the equation by dividing both sides by 0.04:

$$\frac{5.44}{0.04} = \frac{0.04 \cdot n}{0.04} \quad \textbf{Divide both sides by 0.04}$$

$$n = \frac{5.44}{0.04} \quad \textbf{Switch the left and right sides of the equation}$$

$$n = 136 \quad \textbf{Divide}$$

The purchase price is $136. The total price is the sum of the purchase price and the sales tax.

Purchase price	=	$136.00
Sales tax	=	5.44
Total price	=	$141.44

∎

3. Suppose the purchase price of two speakers is $197 and the sales tax is $11.82. What is the sales tax rate?

EXAMPLE 3 Suppose the purchase price of a stereo system is $396 and the sales tax is $19.80. What is the sales tax rate?

SOLUTION   We restate the problem as:

$$\$19.80 \text{ is what percent of } \$396?$$
$$19.80 = n \cdot 396$$

To solve this equation, we divide both sides by 396:

$$\frac{19.80}{396} = \frac{n \cdot 396}{396} \quad \textbf{Divide both sides by 396}$$

$$n = \frac{19.80}{396} \quad \textbf{Switch the left and right sides of the equation}$$

$$n = 0.05 \quad \textbf{Divide}$$

$$n = 5\% \quad \textbf{0.05 = 5\%}$$

The sales tax rate is 5%.   ∎

Commission

Many salespeople work on a *commission* basis. That is, their earnings are a percentage of the amount they sell. The *commission rate* is a percent, and the actual commission they receive is a dollar amount.

EXAMPLE 4 A real estate agent gets 6% of the price of each house she sells. If she sells a house for $89,500, how much money does she earn?

SOLUTION   The commission is 6% of the price of the house, which is $89,500. We restate the problem as:

What is 6% of $89,500?

$$n = 0.06 \cdot 89{,}500$$
$$n = 5{,}370$$

The commission is $5,370. (Actually, most real estate salespeople work for a real estate broker. In most cases, the broker gets half of the commission. If that were true in this case, the agent would get $2,685 and the broker would get $2,685.) ■

EXAMPLE 5 Suppose a car salesperson's commission rate is 12%. If the commission on one of the cars is $1,836, what is the purchase price of the car?

SOLUTION   12% of the sales price is $1,836. The problem can be restated as:

12% of what number is $1,836?

$$0.12 \cdot n = 1{,}836$$
$$\frac{0.12 \cdot n}{0.12} = \frac{1{,}836}{0.12} \quad \textbf{Divide both sides by 0.12}$$
$$n = 15{,}300$$

The car sells for $15,300. ■

EXAMPLE 6 If the commission on a $600 dining room set is $90, what is the commission rate?

SOLUTION   The commission rate is a percentage of the selling price. What we want to know is:

$90 is what percent of $600?

$$90 = n \cdot 600$$
$$\frac{90}{600} = \frac{n \cdot 600}{600} \quad \textbf{Divide both sides by 600}$$
$$n = \frac{90}{600} \quad \begin{array}{l}\textbf{Switch the left and right}\\\textbf{sides of the equation}\end{array}$$
$$n = 0.15 \quad \textbf{Divide}$$
$$n = 15\% \quad \textbf{Change to a percent}$$

The commission rate is 15%. ■

4. A real estate agent gets 3% of the price of each house she sells. If she sells a house for $115,000, how much money does she earn?

5. An appliance salesperson's commission rate is 10%. If the commission on one of the ovens is $115, what is the purchase price of the oven?

6. If the commission on a $750 sofa is $105, what is the commission rate?

Answers
4. $3,450   **5.** $1,150   **6.** 14%

Using the Blueprint for Problem Solving

In our next example we apply the blueprint for problem solving to solve a problem that is a little more complicated than the problems above.

EXAMPLE 7 The total price for one of the cameras I used to produce the videotapes that accompany this book was $3,200. The total price is the purchase price of the camera plus the sales tax. At that time, the sales tax rate in California was 7.5%. What was the purchase price of the camera?

Total price $3,200
"Price includes sales tax"

SOLUTION   We solve this problem using our six-step Blueprint for Problem Solving:

Step 1 *Read and list.*
Known items: The total bill is $3,200. The sales tax rate is 7.5%, which is 0.075 in decimal form.
Unknown item: The purchase price of the video equipment

Step 2 *Assign a variable and translate information.*
If we let x = the purchase price of the video equipment, then to calculate the sales tax we multiply the purchase price x by the sales tax rate:

$$\text{Sales tax} = (\text{Sales tax rate})(\text{Purchase price})$$
$$= 0.075x$$

Step 3 *Reread and write an equation.*

$$\text{Purchase price} + \text{Sales tax} = \text{Total price}$$
$$x \quad + \quad 0.075x \ = \quad 3{,}200$$

Step 4 *Solve the equation.*

$$x + 0.075x = 3{,}200$$
$$1.075x = 3{,}200$$
$$x = \frac{3{,}200}{1.075}$$
$$x = 2{,}976.74 \qquad \text{to the nearest hundredth}$$

Step 5 *Write the answer.*
The purchase price of the equipment was $2,976.74.

Step 6 *Reread and check.*
The purchase price of the equipment is $2,976.74. The tax on this is 0.075(2,976.74) = $223.26. Adding the purchase price and the sales tax, we have the total bill of $3,200.00. ∎

---

7. Find the amount of sales tax paid if the total cost for a video editing system is $6,864 with a sales tax rate of 7.25%.

Answer
7. $464.00

USING TECHNOLOGY

The Internet

The Internet has become a valuable tool that you will use many times during your college career. Below is the opening page from my Web site, which you can reach on the Internet at www.mckeague.com. If you go there, you can find links to some of the sites I used to research many of the problems you see in this book.

Problem Set 7.4

These problems should be solved by the method shown in this section. In each case show the equation needed to solve the problem. Write neatly, and show your work.

1. **Sales Tax** Suppose the sales tax rate in Mississippi is 7% of the purchase price. If a new food processor sells for $750, how much is the sales tax?

2. **Sales Tax** If the sales tax rate is 5% of the purchase price, how much sales tax is paid on a television that sells for $980?

3. **Sales Tax and Purchase Price** Suppose the sales tax rate in Michigan is 6%. How much is the sales tax on a $45 concert ticket? What is the total price?

4. **Sales Tax and Purchase Price** Suppose the sales tax rate in Hawaii is 4%. How much sales tax is charged on a new car if the purchase price is $16,400? What is the total price?

5. **Total Price** The sales tax rate is 4%. If the sales tax on a 10-speed bicycle is $6, what is the purchase price? What is the total price?

6. **Total Price** The sales tax on a new microwave oven is $30. If the sales tax rate is 5%, what is the purchase price? What is the total price?

7. **Tax Rate** Suppose the purchase price of a dining room set is $450. If the sales tax is $22.50, what is the sales tax rate?

8. **Tax Rate** If the purchase price of a case of California wine is $24 and the sales tax is $1.50, what is the sales tax rate?

Dining Room Set

Price	$450.00
Tax	22.50
You pay	$472.50

9. **Commission** A real estate agent has a commission rate of 3%. If a piece of property sells for $94,000, what is his commission?

10. **Commission** A tire salesperson has a 12% commission rate. If he sells a set of radial tires for $400, what is his commission?

11. **Commission and Purchase Price** Suppose a salesperson gets a commission rate of 12% on the lawnmowers she sells. If the commission on one of the mowers is $24, what is the purchase price of the lawnmower?

12. **Commission and Purchase Price** If an appliance salesperson gets 9% commission on all the appliances she sells, what is the price of a refrigerator if her commission is $67.50?

13. **Commission Rate** If the commission on an $800 washer is $112, what is the commission rate?

14. **Commission Rate** A salesperson makes a commission of $3,600 on a $90,000 house he sells. What is his commission rate?

Calculator Problems

The following problems are similar to Problems 1–14. Set them up in the same way, but use a calculator for the calculations.

15. **Sales Tax** The sales tax rate on a certain item is 5.5%. If the purchase price is $216.95, how much is the sales tax? (Round to the nearest cent.)

16. **Purchase Price** If the sales tax rate is 4.75% and the sales tax is $18.95, what is the purchase price? What is the total price? (Both answers should be rounded to the nearest cent.)

17. **Tax Rate** The purchase price for a new suit is $229.50. If the sales tax is $10.33, what is the sales tax rate? (Round to the nearest tenth of a percent.)

18. **Commission** If the commision rate for a mobile home salesperson is 11%, what is the commission on the sale of a $15,794 mobile home?

Men's Suit

Price $229.50
Tax   10.33
You pay $239.83

19. Selling Price Suppose the commission rate on the sale of used cars is 13%. If the commission on one of the cars is $519.35, what did the car sell for?

20. Commission Rate If the commission on the sale of $79.40 worth of clothes is $14.29, what is the commission rate? (Round to the nearest percent.)

21. Buying a Car A new car is advertised for sale for a total price of $17,481.75, which includes sales tax. If the sales tax rate is 7.25%, what is the purchase price of the car?

22. Buying a Textbook A student purchases a textbook on line for a total price of $84.53, which includes sales tax. If the sales tax rate is 7%, how much was the purchase price of the book?

23. Items Sold Every item in the Just a Dollar store is priced at $1.00. When Sylvia opens the store, there is $125.50 in the cash register. When she counts the money in the cash register at the end of the day, the total is $1058.60. If the sales tax rate is 8.5%, how many items were sold that day?

24. Sales Tax A woman owns a small, cash-only business in a state that requires her to charge a 6% sales tax on each item she sells. At the beginning of the day she has $250 in the cash register. At the end of the day she has $1,204 in the register. How much money should she send to the state government for the sales tax she collected?

Review Problems

The problems below review some basic concepts of division with fractions and mixed numbers from Sections 3.4 and 3.7.

Divide.

25. $\dfrac{1}{3} \div \dfrac{2}{3}$

26. $\dfrac{2}{3} \div \dfrac{1}{3}$

27. $2 \div \dfrac{3}{4}$

28. $3 \div \dfrac{1}{2}$

29. $\dfrac{3}{8} \div \dfrac{1}{4}$

30. $\dfrac{5}{9} \div \dfrac{2}{3}$

31. $2\dfrac{1}{4} \div \dfrac{1}{2}$

32. $1\dfrac{1}{4} \div 2\dfrac{1}{2}$

Extending the Concepts: Luxury Taxes

In 1990, Congress passed a law, which took effect on January 1, 1991, requiring an additional tax of 10% on a portion of the purchase price of certain luxury items. (In 1996 the law was amended so that this tax on luxury automobiles will expire in 2003.) For expensive cars, it was paid on the part of the purchase price that exceeded $30,000. For example, if you purchased a Jaguar XJ-S for $53,000, you would pay a luxury tax of 10% of $23,000, because the purchase price, $53,000, is $23,000 above $30,000.

33. If you purchased a Jaguar XJ-S for $53,000 on February 1, 1991, in California, where the sales tax rate was 6%, how much would you pay in luxury tax and how much would you pay in sales tax?

34. If you purchased a Mercedes 300E for $43,500 on January 20, 1991, in California, where the sales tax rate was 6%, how much more would you pay in sales tax than luxury tax?

35. How much would you have saved if you had purchased the Jaguar mentioned in Problem 33 on December 31, 1990?

36. How much would you have saved if you bought a car with a purchase price of $45,000 on December 31, 1990, instead of January 1, 1991?

37. Suppose you bought a car in 1991. How much did you save on a car with a sticker price of $31,500, if you persuaded the car dealer to reduce the price to $29,900?

38. Suppose you bought a car in 1991. One of the cars you were interested in had a sticker price of $35,500, while another had a sticker price of $28,500. If you expected to pay full price for either car, how much did you save if you bought the less expensive car?

7.5 Percent Increase or Decrease and Discount

The table and bar chart below show some statistics compiled by insurance companies regarding stopping distances for automobiles traveling at 20 miles per hour on ice.

	STOPPING DISTANCE	PERCENT DECREASE
Regular tires	150 ft	0
Snow tires	151 ft	–1%
Studded snow tires	120 ft	19%
Reinforced tire chains	75 ft	50%

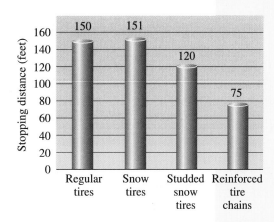

Source: Copyrighted table courtesy of *The Casualty Adjuster's Guide.*

Many times it is more effective to state increases or decreases in terms of percents, rather than the actual number, because with percent we are comparing everything to 100.

EXAMPLE 1 If a person earns $22,000 a year and gets a 5% increase in salary, what is the new salary?

SOLUTION We can find the dollar amount of the salary increase by finding 5% of $22,000:

$$0.05 \times 22,000 = 1,100$$

The increase in salary is $1,100. The new salary is the old salary plus the raise:

$22,000	**Old salary**
+ 1,100	**Raise (5% of $22,000)**
$23,100	**New salary**

EXAMPLE 2 In 1986, there were approximately 1,793,000 arrests for driving under the influence of alcohol or drugs (DUI) in the United States. By 1997, the number of arrests for DUI had decreased 17.6% from the 1986 number. How many people were arrested for DUI in 1997? Round the answer to the nearest thousand.

SOLUTION The decrease in the number of arrests is 17.6% of 1,793,000, or

$$0.176 \times 1,793,000 = 315,568$$

Practice Problems

1. A person earning $18,000 a year gets a 7% increase in salary. What is the new salary?

2. In 1986, there were approximately 271,000 drunk drivers under correctional supervision (prison, jail, or probation). By 1997, that number had increased 89%. How many drunk drivers were under correctional supervision in 1997? Round to the nearest thousand.

Answers

1. $19,260 **2.** 512,000

Subtracting this number from 1,793,000, we have the number of DUI arrests in 1997

1,793,000	**Number of arrests in 1986**
− 315,568	**Decrease of 17.6%**
1,477,432	**Number of arrests in 1997**

To the nearest thousand, there were approximately 1,477,000 arrests for DUI in 1997. ■

3. Shoes that usually sell for $35 are on sale for $28. What is the percent decrease in price?

EXAMPLE 3 Shoes that usually sell for $25 are on sale for $21. What is the percent decrease in price?

SOLUTION   We must first find the decrease in price. Subtracting the sale price from the original price, we have

$$\$25 - \$21 = \$4$$

The decrease is $4. To find the percent decrease (from the original price), we have

$$\$4 \text{ is what percent of } \$25?$$
$$4 = n \cdot 25$$
$$\frac{4}{25} = \frac{n \cdot 25}{25} \qquad \text{Divide both sides by 25}$$
$$n = \frac{4}{25} \qquad \begin{array}{l}\text{Switch the left and right}\\\text{sides of the equation}\end{array}$$
$$n = 0.16 \qquad \text{Divide}$$
$$n = 16\% \qquad \text{Change to a percent}$$

The shoes that sold for $25 have been reduced by 16% to $21. In a problem like this, $25 is the *original* (or *marked*) price, $21 is the *sale price*, $4 is the *discount*, and 16% is the *rate of discount*. ■

4. During a sale, a microwave oven that usually sells for $550 is marked "15% off." What is the discount? What is the sale price?

EXAMPLE 4 During a clearance sale, a suit that usually sells for $300 is marked "25% off." What is the discount? What is the sale price?

SOLUTION   To find the discount, we restate the problem as:

$$What \text{ is } 25\% \text{ of } 300?$$
$$n = 0.25 \cdot 300$$
$$n = 75$$

The discount is $75. The sale price is the original price less the discount:

$300	**Original price**
− 75	**Less the discount (25% of $300)**
$225	**Sale price**

■

EXAMPLE 5   A man buys a washing machine on sale. The machine usually sells for $450, but it is on sale at 12% off. If the sales tax rate is 5%, how much is the total bill for the washer?

SOLUTION   First we have to find the sale price of the washing machine, and we begin by finding the discount:

What is 12% *of* $450?
$$n = 0.12 \cdot 450$$
$$n = 54$$

The washing machine is marked down $54. The sale price is

$450	**Original price**
− 54	**Discount (12% of $450)**
$396	**Sale price**

Because the sales tax rate is 5%, we find the sales tax as follows:

What is 5% *of* 396?
$$n = 0.05 \cdot 396$$
$$n = 19.80$$

The sales tax is $19.80. The total price the man pays for the washing machine is

$396.00	**Sale price**
+ 19.80	**Sales tax**
$415.80	**Total price**

5. A woman buys a new coat on sale. The coat usually sells for $45, but it is on sale at 15% off. If the sales tax rate is 5%, how much is the total bill for the coat?

Answer
5. $40.16

Problem Set 7.5

Solve each of these problems using the method developed in this section.

1. **Salary Increase** If a person earns $23,000 a year and gets a 7% increase in salary, what is the new salary?

2. **Salary Increase** A computer programmer's yearly income of $57,000 is increased by 8%. What is the dollar amount of the increase, and what is her new salary?

3. **Tuition Increase** The yearly tuition at a college is presently $3,000. Next year it is expected to increase by 17%. What will the tuition at this school be next year?

4. **Price Increase** A supermarket increased the price of cheese that sold for $1.98 per pound by 3%. What is the new price for a pound of this cheese? (Round to the nearest cent.)

5. **Car Value** In one year a new car decreased in value by 20%. If it sold for $16,500 when it was new, what was it worth after one year?

6. **Calorie Content** A certain light beer has 20% fewer calories than the regular beer. If the regular beer has 120 calories per bottle, how many calories are in the same-sized bottle of the light beer?

7. **Salary Increase** A person earning $3,500 a month gets a raise of $350 per month. What is the percent increase in salary?

8. **Rate Increase** A student reader is making $6.50 per hour and gets a $0.70 raise. What is the percent increase? (Round to the nearest tenth of a percent.)

9. **Shoe Sale** Shoes that usually sell for $25 are on sale for $20. What is the percent decrease in price?

10. **Enrollment Decrease** The enrollment in a certain elementary school was 410 in 2000. In 2001, the enrollment in the same school was 328. Find the percent decrease in enrollment from 2000 to 2001.

11. **Internet Access Rates** The bar chart below was in a Los Angeles newspaper in 1999. It shows the projected growth of digital subscriber lines (DSL) and cable modems through 2003.

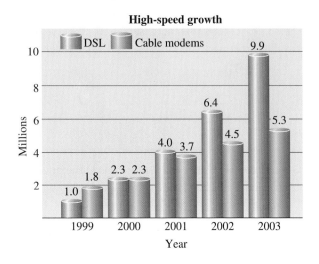

High-speed growth

a. What is the percent increase in the number of cable modems from 1999 to 2002?

b. What is the percent increase in the number of DSL from 1999 to 2002?

c. In 2002, how many more DSL will there be than cable modems?

12. **Dieting** At the beginning of a 2-month crash diet, Jo Ann weighed 140 pounds. At the end of 2 months, she weighed 119 pounds. What is the percent decrease in weight from the beginning of the diet to the end of the diet?

13. **Discount** During a clearance sale, a three-piece suit that usually sells for $300 is marked "15% off." What is the discount? What is the sale price?

14. **Sale Price** On opening day, a new music store offers a 12% discount on all electric guitars. If the regular price on a guitar is $550, what is the sale price?

15. **Total Price** A man buys a washing machine that is on sale. The washing machine usually sells for $450 but is on sale at 20% off. If the sales tax rate in his state is 6%, how much is the total bill for the washer?

16. **Total Price** A bedroom set that normally sells for $1,450 is on sale for 10% off. If the sales tax rate is 5%, what is the total price of the bedroom set if it is bought while on sale?

Calculator Problems

Set up the following problems the same way you set up Problems 1–16. Then use a calculator to do the calculations.

17. Salary Increase A teacher making $43,752 per year gets a 6.5% raise. What is the new salary?

18. Utility Increase A homeowner had a $15.90 electric bill in December. In January the bill was $17.81. Find the percent increase in the electric bill from December to January. (Round to the nearest whole number.)

19. Soccer The rules for soccer state that the playing field must be from 100 to 120 yards long and 55 to 75 yards wide. The 1999 Women's World Cup was played at the Rose Bowl on a playing field 116 yards long and 72 yards wide. The diagram below shows the smallest possible soccer field, the largest possible soccer field, and the soccer field at the Rose Bowl.

Soccer Fields

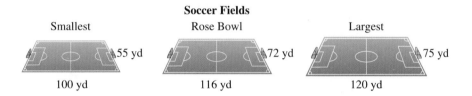

Smallest   55 yd   100 yd

Rose Bowl   72 yd   116 yd

Largest   75 yd   120 yd

a. Percent Increase A team plays on the smallest field, then plays in the Rose Bowl. What is the percent increase in the area of the playing field from the smallest field to the Rose Bowl? Round to the nearest tenth of a percent.

b. Percent Increase A team plays a soccer game in the Rose Bowl. The next game is on a field with the largest dimensions. What is the percent increase in the area of the playing field from the Rose Bowl to the largest field? Round to the nearest tenth of a percent.

20. Football The diagrams below show the dimensions of playing fields for the National Football League (NFL), the Canadian Football League (CFL), and Arena Football.

Football Fields

NFL $53\frac{1}{3}$ yd   100 yd

Canadian   65 yd   110 yd

Arena $28\frac{1}{3}$ yd   50 yd

a. Percent Increase In 1999 Kurt Warner made a successful transition from Arena Football to the NFL, winning the Most Valuable Player award. What was the percent increase in the area of the fields he played on in moving from Arena Football to the NFL? Round to the nearest percent.

b. Percent Decrease Doug Flutie played in the Canadian Football League before moving to the NFL. What was the percent decrease in the area of the fields he played on in moving from the CFL to the NFL? Round to the nearest tenth of a percent.

Review Problems

The problems below review some basic concepts of addition of fractions and mixed numbers from Sections 3.5 and 3.8.

Add each of the following and reduce all answers to lowest terms.

21. $\dfrac{1}{3} + \dfrac{2}{3}$

22. $\dfrac{3}{8} + \dfrac{1}{8}$

23. $\dfrac{1}{2} + \dfrac{1}{4}$

24. $\dfrac{1}{5} + \dfrac{3}{10}$

25. $\dfrac{3}{4} + \dfrac{2}{3}$

26. $\dfrac{3}{8} + \dfrac{1}{6}$

27. $2\dfrac{1}{2} + 3\dfrac{1}{2}$

28. $3\dfrac{1}{4} + 2\dfrac{1}{8}$

Extending the Concepts: Batting Averages

As you know from the Extending the Concepts problems in Problem Set 7.3, on July 1, 1999, Derek Jeter had 110 hits in 292 times at bat for a batting average of .377. On the same date, Larry Walker had accumulated 91 hits in 239 times at bat. Suppose that on July 2, Derek Jeter had 2 hits in 4 times at bat, and Larry Walker had only 1 hit in 3 times at bat.

29. What was Derek Jeter's batting average for the game on July 2?

30. What was Derek Jeter's overall batting average after the game on July 2?

31. By how much did Derek Jeter's batting average increase from July 1 to July 2?

32. What is the percent increase in Derek Jeter's batting average from July 1 to July 2?

33. What was Larry Walker's batting average for the game on July 2?

34. What was Larry Walker's overall batting average after the game on July 2?

35. By how much did Larry Walker's batting average decrease from July 1 to July 2?

36. What is the percent decrease in Larry Walker's batting average from July 1 to July 2?

7.6 Interest

Anyone who has borrowed money from a bank or other lending institution, or who has invested money in a savings account, is aware of *interest*. Interest is the amount of money paid for the use of money. If we put $500 in a savings account that pays 6% annually, the interest will be 6% of $500, or 0.06(500) = $30. The amount we invest ($500) is called the *principal,* the percent (6%) is the *interest rate,* and the money earned ($30) is the *interest.*

Practice Problems

EXAMPLE 1   A man invests $2,000 in a savings plan that pays 7% per year. How much money will be in the account at the end of 1 year?

1.   A man invests $3,000 in a savings plan that pays 8% per year. How much money will be in the account at the end of 1 year?

SOLUTION   We first find the interest by taking 7% of the principal, $2,000:

$$\text{Interest} = 0.07(\$2,000)$$
$$= \$140$$

The interest earned in 1 year is $140. The total amount of money in the account at the end of a year is the original amount plus the $140 interest:

$2,000	**Original investment (principal)**
+ 140	**Interest (7% of $2,000)**
$2,140	**Amount after 1 year**

The amount in the account after 1 year is $2,140.   ∎

EXAMPLE 2   A farmer borrows $8,000 from his local bank at 12%. How much does he pay back to the bank at the end of the year to pay off the loan?

2.   If a woman borrows $7,500 from her local bank at 12% interest, how much does she pay back to the bank if she pays off the loan in 1 year?

SOLUTION   The interest he pays on the $8,000 is

$$\text{Interest} = 0.12(\$8,000)$$
$$= \$960$$

At the end of the year, he must pay back the original amount he borrowed ($8,000) plus the interest at 12%:

$8,000	**Amount borrowed (principal)**
+ 960	**Interest at 12%**
$8,960	**Total amount to pay back**

The total amount that the farmer pays back is $8,960.   ∎

There are many situations in which interest on a loan is figured on other than a yearly basis. Many short-term loans are for only 30 or 60 days. In these cases we can use a formula to calculate the interest that has accumulated. This type of interest is called *simple interest*. The formula is

$$I = P \cdot R \cdot T$$

where

I = Interest
P = Principal
R = Interest rate (this is the percent)
T = Time (in years, 1 year = 360 days)

Answers
1. $3,240   2. $8,400

We could have used this formula to find the interest in Examples 1 and 2. In those two cases, T is 1. When the length of time is in days rather than years, it is common practice to use 360 days for 1 year, and we write T as a fraction. Examples 3 and 4 illustrate this procedure.

3. Another student takes out a loan like the one in Example 3. This loan is for $700 at 4%. How much interest does this student pay if the loan is paid back in 90 days?

EXAMPLE 3 A student takes out an emergency loan for tuition, books, and supplies. The loan is for $600 at an interest rate of 4%. How much interest does the student pay if the loan is paid back in 60 days?

SOLUTION   The principal P is $600, the rate R is 4% = 0.04, and the time T is $\frac{60}{360}$. Notice that T must be given in years, and 60 days = $\frac{60}{360}$ year. Applying the formula, we have

$$I = P \cdot R \cdot T$$

$$I = 600 \times 0.04 \times \frac{60}{360}$$

$$I = 600 \times 0.04 \times \frac{1}{6} \qquad \mathbf{\frac{60}{360}} = \mathbf{\frac{1}{6}}$$

$$I = 4 \qquad\qquad \mathbf{Multiplication}$$

The interest is $4. ∎

4. Suppose $1,200 is deposited in an account that pays 9.5% interest per year. If all the money is withdrawn after 120 days, how much money is withdrawn?

EXAMPLE 4 A woman deposits $900 in an account that pays 6% annually. If she withdraws all the money in the account after 90 days, how much does she withdraw?

SOLUTION   We have $P = \$900$, $R = 0.06$, and $T = 90$ days $= \frac{90}{360}$ year. Using these numbers in the formula, we have

$$I = P \cdot R \cdot T$$

$$I = 900 \times 0.06 \times \frac{90}{360}$$

$$I = 900 \times 0.06 \times \frac{1}{4} \qquad \mathbf{\frac{90}{360}} = \mathbf{\frac{1}{4}}$$

$$I = 13.5 \qquad\qquad \mathbf{Multiplication}$$

The interest earned in 90 days is $13.50. If the woman withdraws all the money in her account, she will withdraw

$900.00	**Original amount (principal)**
+ 13.50	**Interest for 90 days**
$913.50	**Total amount withdrawn**

The woman has withdrawn $913.50. ∎

A second common kind of interest is *compound interest*. Compound interest includes interest paid on interest. We can use what we know about simple interest to help us solve problems involving compound interest.

5. If $5,000 is put into an account that pays 6% annually, how much money is in the account at the end of 2 years?

EXAMPLE 5 A homemaker puts $3,000 into a savings account that pays 7% compounded annually. How much money is in the account at the end of 2 years?

SOLUTION   Because the account pays 7% annually, the simple interest at the end of 1 year is 7% of $3,000:

$$\text{Interest after 1 year} = 0.07(\$3,000)$$
$$= \$210$$

Answers
3. $7   **4.** $1,238   **5.** $5,618

Because the interest is paid annually, at the end of 1 year the total amount of money in the account is

$3,000   **Original amount**
+   210   **Interest for 1 year**
$3,210   **Total in account after 1 year**

The interest paid for the second year is 7% of this new total, or

Interest paid the second year = 0.07($3,210)
= $224.70

At the end of 2 years, the total in the account is

$3,210.00   **Amount at the beginning of year 2**
+   224.70   **Interest paid for year 2**
$3,434.70   **Account after 2 years**

At the end of 2 years, the account totals $3,434.70. The total interest earned during this 2-year period is $210 (first year) + $224.70 (second year) = $434.70.   ■

Note
If the interest earned in Example 5 were calculated using the formula for simple interest, $I = P \cdot R \cdot T$, the amount of money in the account at the end of two years would be $3,420.00.

You may have heard of savings and loan companies that offer interest rates that are compounded quarterly. If the interest rate is, say, 6% and it is compounded quarterly, then after every 90 days ($\frac{1}{4}$ of a year) the interest is added to the account. If it is compounded semiannually, then the interest is added to the account every 6 months. Most accounts have interest rates that are compounded daily, which means the simple interest is computed daily and added to the account.

EXAMPLE 6   If $10,000 is invested in a savings account that pays 6% compounded quarterly, how much is in the account at the end of a year?

SOLUTION   The interest for the first quarter ($\frac{1}{4}$ of a year) is calculated using the formula for simple interest:

$I = P \cdot R \cdot T$

$I = \$10,000 \times 0.06 \times \dfrac{1}{4}$ **First quarter**

$I = \$150$

At the end of the first quarter, this interest is added to the original principal. The new principal is $10,000 + $150 = $10,150. Again we apply the formula to calculate the interest for the second quarter:

$I = \$10,150 \times 0.06 \times \dfrac{1}{4}$ **Second quarter**

$I = \$152.25$

The principal at the end of the second quarter is $10,150 + $152.25 = $10,302.25. The interest earned during the third quarter is

$I = \$10,302.25 \times 0.06 \times \dfrac{1}{4}$ **Third quarter**

$I = \$154.53$ **To the nearest cent**

The new principal is $10,302.25 + $154.53 = $10,456.78. Interest for the fourth quarter is

$I = \$10,456.78 \times 0.06 \times \dfrac{1}{4}$ **Fourth quarter**

$I = \$156.85$ **To the nearest cent**

6. If $20,000 is invested in an account that pays 8% compounded quarterly, how much is in the account at the end of a year?

Answer
6. $21,648.64

The total amount of money in this account at the end of 1 year is

$$\$10,456.78 + \$156.85 = \$10,613.63$$ ∎

For comparison, let's compute the total amount of money that would be in the account in Example 6 at the end of 1 year using simple interest. The interest earned during the year would be

$$I = \$10,000 \times 0.06 \times 1$$
$$= \$600$$

The total amount in the account at the end of the year would be

$$\$10,000 + \$600 = \$10,600$$

which is $13.63 less than what we calculated using compound interest. Compounding interest means more money to the saver.

USING TECHNOLOGY

Compound Interest from a Formula

We can summarize the work above with a formula that allows us to calculate compound interest for any interest rate and any number of compounding periods. If we invest P dollars at an annual interest rate r, compounded n times a year, then the amount of money in the account after t years is given by the formula

$$A = P\left(1 + \frac{r}{n}\right)^{nt}$$

Using numbers from Example 6 to illustrate, we have

$$P = \text{Principal} = \$10,000$$
$$r = \text{annual interest rate} = 0.06$$
$$n = \text{number of compounding periods} = 4$$
$$\qquad \text{(interest is compounded quarterly)}$$
$$t = \text{number of years} = 1$$

Substituting these numbers into the formula above, we have

$$A = 10,000\left(1 + \frac{0.06}{4}\right)^{4 \cdot 1}$$
$$= 10,000(1 + 0.015)^4$$
$$= 10,000(1.015)^4$$

To simplify this last expression on a calculator, we have

Scientific calculator: 10,000 $\boxed{\times}$ 1.015 $\boxed{y^x}$ 4 $\boxed{=}$

Graphing calculator: 10,000 $\boxed{\times}$ 1.015 $\boxed{\wedge}$ 4 $\boxed{\text{ENTER}}$

In either case, the answer is $10,613.636, which rounds to $10,613.64.

Note

The reason that this answer is different than the result we obtained in Example 6 is that, in Example 6, we rounded each calculation as we did it. The calculator will keep all the digits in all of the intermediate calculations.

Problem Set 7.6

These problems are similar to the examples found in this section. They should be set up and solved in the same way. (Problems 1–12 involve simple interest.)

1. Savings Account A man invests $2,000 in a savings plan that pays 8% per year. How much money will be in the account at the end of 1 year?

2. Savings Account How much simple interest is earned on $5,000 if it is invested for 1 year at 5%?

3. Savings Account A savings account pays 7% per year. How much interest will $9,500 invested in such an account earn in a year?

4. Savings Account A local bank pays 5.5% annual interest on all savings accounts. If $600 is invested in this account, how much will be in the account at the end of a year?

5. Bank Loan A farmer borrows $8,000 from his local bank at 7%. How much does he pay back to the bank at the end of the year when he pays off the loan?

6. Bank Loan If $400 is borrowed at a rate of 12% for 1 year, how much is the interest?

7. Bank Loan A bank lends one of its customers $2,000 at 8% for 1 year. If the customer pays the loan back at the end of the year, how much does he pay the bank?

8. Bank Loan If a loan of $2,000 at 20% for 1 year is to be paid back in one payment at the end of the year, how much does he pay the bank?

9. Student Loan A student takes out an emergency loan for tuition, books, and supplies. The loan is for $600 with an interest rate of 5%. How much interest does the student pay if the loan is paid back in 60 days?

10. Short-Term Loan If a loan of $1,200 at 9% is paid off in 90 days, what is the interest?

11. Savings Account A woman deposits $800 in a savings account that pays 5%. If she withdraws all the money in the account after 120 days, how much does she withdraw?

12. Savings Account $1,800 is deposited in a savings account that pays 6%. If the money is withdrawn at the end of 30 days, how much interest is earned?

The problems that follow involve compound interest.

13. Compound Interest A woman puts $5,000 into a savings account that pays 6% compounded annually. How much money is in the account at the end of 2 years?

14. Compound Interest A savings account pays 5% compounded annually. If $10,000 is deposited in the account, how much is in the account at the end of 2 years?

15. Compound Interest If $8,000 is invested in a savings account that pays 5% compounded quarterly, how much is in the account at the end of a year?

16. Compound Interest Suppose $1,200 is invested in a savings account that pays 6% compounded semiannually. How much is in the account at the end of $1\frac{1}{2}$ years?

Calculator Problems

The following problems should be set up in the same way in which Problems 1–16 have been set up. Then the calculations should be done on a calculator.

17. Savings Account A woman invests $917.26 in a savings account that pays 6.25% annually. How much is in the account at the end of a year?

18. Business Loan The owner of a clothing store borrows $6,210 for 1 year at 11.5% interest. If he pays the loan back at the end of the year, how much does he pay back?

19. Compound Interest Suppose $10,000 is invested in each account below. In each case find the amount of money in the account at the end of 5 years.
  a. Annual interest rate = 6%, compounded quarterly

  b. Annual interest rate = 6%, compounded monthly

  c. Annual interest rate = 5%, compounded quarterly

  d. Annual interest rate = 5%, compounded monthly

20. Compound Interest Suppose $5,000 is invested in each account below. In each case find the amount of money in the account at the end of 10 years.
  a. Annual interest rate = 5%, compounded quarterly

  b. Annual interest rate = 6%, compounded quarterly

  c. Annual interest rate = 7%, compounded quarterly

  d. Annual interest rate = 8%, compounded quarterly

Review Problems

The problems below will allow you to review subtraction of fractions and mixed numbers from Sections 3.5 and 3.8.

21. $\dfrac{3}{4} - \dfrac{1}{4}$

22. $\dfrac{9}{10} - \dfrac{7}{10}$

23. $\dfrac{5}{8} - \dfrac{1}{4}$

24. $\dfrac{7}{10} - \dfrac{1}{5}$

25. $\dfrac{1}{3} - \dfrac{1}{4}$

26. $\dfrac{9}{12} - \dfrac{1}{5}$

27. $3\dfrac{1}{4} - 2$

28. $5\dfrac{1}{6} - 3\dfrac{1}{4}$

Extending the Concepts

The following problems are a random assortment of percent problems. Use any of the methods developed in this chapter to solve them.

29. Alcohol Content During the month of June 1991, G. Heilemann Brewing Company announced plans to market a new malt liquor, which had 50% more alcohol per can than regular beer. If regular beer is 3.8% alcohol, what is the percent alcohol of the new malt liquor? (In July 1991, the company canceled its plans for the new malt liquor, after the Federal Bureau of Alcohol, Tobacco and Firearms withdrew its approval for the beer's label because it was to be called *PowerMaster*. Federal law prohibits companies from marketing beer on the basis of strength.)

30. Budget Cutting In 1991, the Pentagon began cutting its budget by closing military bases. The newspaper *USA Today* reported that the Pentagon would save $1.7 billion annually by cutting 25% from its budget. If that is the case, what was the annual budget for the Pentagon before they began closing bases?

31. Movie Making The bar chart below is from the introduction to this chapter. It shows the production costs for each of the first four *Star Wars* movies. Find the percent increase in production costs from each *Star Wars* movie to the next, and then compare your results with the results given in the introduction to this chapter.

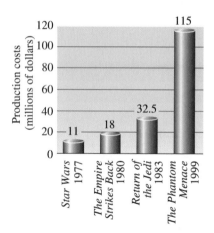

32. Movie Making The table below shows how much money each of the *Star Wars* movies brought in during the first weekend they were shown. Find the percent increase in opening weekend income from each *Star Wars* movie to the next. Round to the nearest percent.

OPENING WEEKEND INCOME	
Star Wars (1977)	$1,554,000
The Empire Strikes Back (1980)	$6,415,000
Return of the Jedi (1983)	$30,490,000
The Phantom Menace (1999)	$64,810,000

Chapter 7 Summary

Examples

The Meaning of Percent [7.1]

Percent means "per hundred." It is a way of comparing numbers to the number 100.

1. 42% means 42 per hundred or $\frac{42}{100}$

Changing Percents to Decimals [7.1]

To change a percent to a decimal, drop the percent symbol (%), and move the decimal point two places to the *left*.

2. 75% = 0.75

Changing Decimals to Percents [7.1]

To change a decimal to a percent, move the decimal point two places to the *right,* and use the % symbol.

3. 0.25 = 25%

Changing Percents to Fractions [7.1]

To change a percent to a fraction, drop the % symbol, and use a denominator of 100. Reduce the resulting fraction to lowest terms if necessary.

4. $6\% = \frac{6}{100} = \frac{3}{50}$

Changing Fractions to Percents [7.1]

To change a fraction to a percent, either write the fraction as a decimal and then change the decimal to a percent, or write the fraction as an equivalent fraction with denominator 100, drop the 100, and use the % symbol.

5. $\frac{3}{4} = 0.75 = 75\%$

or

$\frac{9}{10} = \frac{90}{100} = 90\%$

Basic Word Problems Involving Percents [7.2]

There are three basic types of word problems:

Type A: What number is 14% of 68?

Type B: What percent of 75 is 25?

Type C: 25 is 40% of what number?

To solve them, we write *is* as =, *of* as · (multiply), and *what number* or *what percent* as *n*. We then solve the resulting equation to find the answer to the original question.

6. Translating to equations, we have:

Type A: $n = 0.14(68)$

Type B: $75n = 25$

Type C: $25 = 0.40n$

Applications of Percent [7.3, 7.4, 7.5, 7.6]

There are many different kinds of application problems involving percent. They include problems on income tax, sales tax, commission, discount, percent increase and decrease, and interest. Generally, to solve these problems, we restate them as an equivalent problem of Type A, B, or C above. Problems involving simple interest can be solved using the formula

$$I = P \cdot R \cdot T$$

where I = interest, P = principal, R = interest rate, and T = time (in years). It is standard procedure with simple interest problems to use 360 days = 1 year.

COMMON MISTAKES

1. A common mistake is forgetting to change a percent to a decimal when working problems that involve percents in the calculations. We always change percents to decimals before doing any calculations.

2. Moving the decimal point in the wrong direction when converting percents to decimals or decimals to percents is another common mistake. Remember, *percent* means "per hundred." Rewriting a number expressed as a percent as a decimal will make the numerical part smaller.

    $25\% = 0.25$

Chapter **7** Review

Write each percent as a decimal. [7.1]

1. 35%   **2.** 17.8%   **3.** 5%   **4.** 0.2%

Write each decimal as a percent. [7.1]

5. 0.95   **6.** 0.8   **7.** 0.495   **8.** 1.65

Write each percent as a fraction or mixed number in lowest terms. [7.1]

9. 75%   **10.** 4%   **11.** 145%   **12.** 2.5%

Write each fraction or mixed number as a percent. [7.1]

13. $\dfrac{3}{10}$ **14.** $\dfrac{5}{8}$ **15.** $\dfrac{2}{3}$ **16.** $4\dfrac{3}{4}$

Solve the following problems.

17. What number is 60% of 28? [7.2]

18. What percent of 38 is 19? [7.2]

19. 24 is 30% of what number? [7.2]

20. Survey Suppose 45 out of 60 people surveyed believe a college education will increase a person's earning potential. What percent believe this? [7.3]

21. Commission A salesperson gets a 12% commission rate on all appliances she sells. If she sells $600 in appliances in 1 day, what is her commission? [7.4]

22. Discount A lawnmower that usually sells for $175 is marked down to $140. What is the discount? What is the discount rate? [7.5]

Power mower

Manufacturer's Suggested Price: $175

Sale Price: $140

23. Total Price A sewing machine that normally sells for $600 is on sale for 25% off. If the sales tax rate is 6%, what is the total price of the sewing machine if it is purchased during the sale? [7.4]

24. Home Mortgage If the interest rate on a home mortgage is 9%, then each month you pay 0.75% of the unpaid balance in interest. If the unpaid balance on one such loan is $60,000 at the beginning of a month, how much interest must be paid that month? [7.6]

25. Percent Increase At the beginning of the summer, the price for a gallon of regular gasoline is $1.25. By the end of summer, the price has increased 16%. What is the new price of a gallon of regular gasoline? Round to the nearest cent. [7.5]

26. Percent Decrease A gallon of regular gasoline is selling for $1.45 in September. If the price decreases 14% in October, what is the new price for a gallon of regular gasoline? Round to the nearest cent. [7.5]

Gas Prices

Regular	$1.25
Unleaded	$1.30
Super	$1.35
Diesel	$1.27

27. Medical Costs The table below shows the average yearly cost of visits to the doctor, as reported in *USA Today* in 1999. What is the percent increase in cost from 1990 to 2000? Round to the nearest tenth of a percent. [7.5]

28. Commission A real estate agent gets a commission of 6% on all houses he sells. If his total sales for December are $420,000, how much money does he make? [7.4]

MEDICAL COSTS

YEAR	AVERAGE ANNUAL COST
1990	$583
1995	$739
2000	$906
2005	$1,172

29. Simple Interest If $1,800 is invested at 7% simple interest for 120 days, how much interest is earned? [7.6]

30. Compound Interest How much interest will be earned on a savings account that pays 8% compounded semi-annually if $1,000 is invested for 2 years? [7.6]

Chapter **7** Test

Write each percent as a decimal.

1. 18%   **2.** 4%   **3.** 0.5%

Write each decimal as a percent.

4. 0.45   **5.** 0.7   **6.** 1.35

Write each percent as a fraction or a mixed number in lowest terms.

7. 65%   **8.** 146%   **9.** 3.5%

Write each number as a percent.

10. $\dfrac{7}{20}$ **11.** $\dfrac{3}{8}$ **12.** $1\dfrac{3}{4}$

13. What number is 75% of 60?   **14.** What percent of 40 is 18?

15. 16 is 20% of what number?

16. Driver's Test On a 25-question driver's test, a student answered 23 questions correctly. What percent of the questions did the student answer correctly?

17. Commission A salesperson gets an 8% commission rate on all computers she sells. If she sells $12,000 in computers in 1 day, what is her commission?

18. Discount A washing machine that usually sells for $250 is marked down to $210. What is the discount? What is the discount rate?

19. Total Price A tennis racket that normally sells for $280 is on sale for 25% off. If the sales tax rate is 5%, what is the total price of the tennis racket if it is purchased during the sale?

Manufacturer's
Suggested
Price: $250

Sale Price:
$210

Manufacturer's
Suggested
Price: $280

Sale:
Now 25% off!

20. Simple Interest If $5000 is invested at 8% simple interest for 3 months, how much interest is earned?

21. Compound Interest How much interest will be earned on a savings account that pays 10% compounded annually, if $12,000 is invested for 2 years?

22. Comparing Area On April 3, 2000, *USA Today* changed the size of its paper. Previous to this date, each page of the paper was $13\frac{1}{2}$ inches wide and $22\frac{1}{4}$ inches long, giving each page an area of 300.375 in². The new paper size is $1\frac{1}{4}$ inches narrower and $\frac{1}{2}$ inch longer, giving each page in the new paper an area of 278.6875 in². Find the percent decrease in the area of each page. Round to the nearest tenth of a percent.

Chapters 1-7 Cumulative Review

Simplify.

1. $5{,}309 + 687$

2. $8 + \dfrac{5}{x}$

3. $11.09 - 6.531$

4. $4\dfrac{1}{8} - 1\dfrac{3}{4}$

Multiply.

5. $(3x - 5)(6x + 7)$

6. $5a^3(2a^2 - 7)$

Divide.

7. $314\overline{)13{,}188}$

8. $\dfrac{6}{32} \div \dfrac{9}{48}$

9. Round the number 435,906 to the nearest ten thousand.

10. Write 0.48 as a fraction in lowest terms.

11. Change $\dfrac{76}{12}$ to a mixed number in lowest terms.

12. Use the equation $3x + 4y = 24$ to find x when $y = -3$.

13. Write the decimal 0.8 as a percent.

Use the table given in Chapter 6 to make the following conversion.

14. 7 kilograms to pounds

15. Write 124% as a fraction or mixed number in lowest terms.

16. What percent of 60 is 21?

Simplify.

17. $\sqrt{32xy^2}$

18. $8x + 9 - 9x - 14$

19. $-|-7|$

20. $\dfrac{-3(-8) + 4(-2)}{11 - 9}$

21. $19 - 5(7 - 4)$

22. $5\sqrt{49} + 3\sqrt{81}$

Solve.

23. $\dfrac{3}{8}y = 21$

24. $-3(2x - 1) = 3(x + 5)$

25. $\dfrac{3.6}{4} = \dfrac{4.5}{x}$

26. Write the following ratio as a fraction in lowest terms: 0.04 to 0.32

27. Subtract $2x - 9$ from $6x - 8$.

28. Identify the property or properties used in the following: $(6 + 8) + 2 = 6 + (8 + 2)$.

29. Surface Area Find the surface area of a rectangular solid with length 7 inches, width 3 inches, and height 2 inches.

30. Age Ben is 8 years older than Ryan. In 6 years the sum of their ages will be 38. How old are they now?

31. Gas Mileage A truck travels 432 miles on 27 gallons of gas. What is the rate of gas mileage in miles per gallon?

32. Discount A surfboard that usually sells for $400 is marked down to $320. What is the discount? What is the discount rate?

33. Geometry Find the length of the hypotenuse of a right triangle with sides of 5 and 12 meters.

34. Cost of Coffee If coffee costs $6.40 per pound, how much will 2 lb 4 oz cost?

35. Interest If $1,400 is invested at 6% simple interest for 90 days, how much interest is earned?

36. Wildflower Seeds C. J. works in a nursery, and one of his tasks is filling packets of wildflower seeds. If each packet is to contain $\frac{1}{4}$ pound of seeds, how many packets can be filled from 16 pounds of seeds?

37. Checking Account Balance Rosa has a balance of $469 in her checking account when she writes a check for $376 for her car payment. Then she writes another check for $138 for textbooks. Write a subtraction problem that gives the new balance in her checking account. What is the new balance?

38. Commission A car stereo salesperson receives a commission of 8% on all units he sells. If his total sales for March are $9,800, how much money in commission will he make?

39. Volume How many 8-fluid-ounce glasses of water will it take to fill a 15-gallon aquarium?

40. Internet Access Speed The table below gives the speed of the most common modems used for Internet access. The abbreviation bps stands for bits per second. Use the template to construct a bar chart of the information in the table.

MODEM SPEEDS	
MODEM TYPE	SPEED (bps)
28K	28,000
56K	56,000
ISDN	128,000
Cable	512,000
DSL	786,000

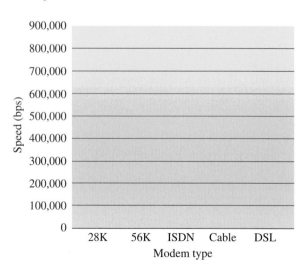

One Hundred Addition Facts

The following 100 problems should be done mentally. You should be able to find these sums quickly and accurately. Do all 100 problems, and then check your answers. Make a list of each problem you missed, and then go over the list as many times as it takes to memorize the correct answers. Once this has been done, go back and work all 100 problems again. Repeat this process until you get all 100 problems correct.

Add.

1. 0 +4	**2.** 1 +9	**3.** 2 +7	**4.** 0 +3	**5.** 4 +1	**6.** 2 +5	**7.** 5 +6	**8.** 2 +6
9. 9 +8	**10.** 4 +9	**11.** 3 +7	**12.** 9 +4	**13.** 4 +5	**14.** 1 +4	**15.** 0 +1	**16.** 6 +4
17. 5 +3	**18.** 9 +5	**19.** 5 +8	**20.** 0 +7	**21.** 5 +1	**22.** 1 +0	**23.** 6 +5	**24.** 5 +2
25. 5 +7	**26.** 3 +6	**27.** 7 +0	**28.** 4 +6	**29.** 1 +5	**30.** 3 +1	**31.** 1 +7	**32.** 7 +6
33. 3 +3	**34.** 8 +0	**35.** 1 +8	**36.** 5 +4	**37.** 2 +8	**38.** 0 +5	**39.** 5 +9	**40.** 9 +9
41. 0 +0	**42.** 6 +9	**43.** 7 +4	**44.** 8 +2	**45.** 7 +3	**46.** 8 +5	**47.** 8 +4	**48.** 6 +7
49. 8 +1	**50.** 1 +3	**51.** 7 +5	**52.** 0 +1	**53.** 8 +3	**54.** 6 +6	**55.** 9 +6	**56.** 8 +9
57. 3 +0	**58.** 6 +2	**59.** 2 +1	**60.** 6 +0	**61.** 4 +8	**62.** 6 +1	**63.** 2 +0	**64.** 7 +9
65. 2 +9	**66.** 9 +3	**67.** 3 +8	**68.** 7 +2	**69.** 8 +8	**70.** 4 +4	**71.** 8 +7	**72.** 2 +4
73. 3 +2	**74.** 4 +7	**75.** 9 +7	**76.** 1 +2	**77.** 6 +3	**78.** 2 +2	**79.** 9 +2	**80.** 9 +1
81. 3 +5	**82.** 1 +1	**83.** 5 +5	**84.** 7 +7	**85.** 0 +8	**86.** 7 +8	**87.** 2 +3	**88.** 3 +9
89. 6 +8	**90.** 4 +0	**91.** 0 +6	**92.** 4 +3	**93.** 7 +1	**94.** 0 +9	**95.** 9 +0	**96.** 8 +6
97. 3 +4	**98.** 5 +0	**99.** 1 +6	**100.** 4 +2				

One Hundred Multiplication Facts

The following 100 problems should be done mentally. You should be able to find these products quickly and accurately. Do all 100 problems, and then check your answers. Make a list of each problem you missed, and then go over the list as many times as it takes to memorize the correct answers. Once this has been done, go back and work all 100 problems again. Repeat this process until you get all 100 problems correct.

Multiply.

1. 3×4
2. 5×4
3. 0×5
4. 0×6
5. 2×5
6. 1×7
7. 4×4
8. 7×2

9. 7×1
10. 5×9
11. 2×6
12. 5×3
13. 3×5
14. 4×5
15. 8×1
16. 3×7

17. 5×8
18. 0×3
19. 6×0
20. 0×4
21. 4×6
22. 8×2
23. 5×7
24. 0×9

25. 2×7
26. 6×1
27. 1×8
28. 4×7
29. 3×8
30. 6×8
31. 5×4
32. 7×3

33. 0×8
34. 4×8
35. 2×9
36. 5×6
37. 1×9
38. 2×8
39. 0×7
40. 6×2

41. 4×9
42. 8×0
43. 3×0
44. 9×6
45. 3×4
46. 8×3
47. 7×9
48. 1×9

49. 1×3
50. 9×9
51. 2×4
52. 8×7
53. 5×2
54. 7×8
55. 1×4
56. 0×2

57. 9×8
58. 2×3
59. 9×5
60. 7×7
61. 6×5
62. 0×1
63. 9×3
64. 8×9

65. 9×7
66. 6×4
67. 3×1
68. 1×5
69. 9×6
70. 1×2
71. 4×1
72. 2×2

73. 6×3
74. 5×1
75. 9×2
76. 6×9
77. 7×0
78. 4×0
79. 7×6
80. 5×0

81. 0×0
82. 3×9
83. 1×1
84. 8×6
85. 9×0
86. 7×4
87. 8×4
88. 2×0

89. 9×1
90. 8×5
91. 4×2
92. 8×3
93. 7×5
94. 2×1
95. 6×6
96. 3×2

97. 4×3
98. 1×6
99. 6×7
100. 1×0

Negative Exponents

Up to this point, all the exponents we have worked with have been positive numbers. We want to extend the type of numbers we can use for exponents to include negative numbers as well. The definition that follows allows us to work with exponents that are negative numbers.

> **DEFINITION**
>
> **Negative Exponents** If r is a positive integer, then $a^{-r} = \dfrac{1}{a^r}$

Note
Because division by 0 is undefined, our definition includes the restriction $a \neq 0$.

The definition indicates that negative exponents give us reciprocals, as the following examples illustrate.

EXAMPLE 1 Simplify: 2^{-3}

SOLUTION   The first step is to rewrite the expression with a positive exponent, by using our definition. After that, we simplify.

$$2^{-3} = \frac{1}{2^3} \quad \textbf{Definition of negative exponents}$$

$$= \frac{1}{8} \quad \textbf{The cube of 2 is 8}$$

EXAMPLE 2 Simplify: 5^{-2}

SOLUTION   First we write the expression with a positive exponent, then we simplify.

$$5^{-2} = \frac{1}{5^2} \quad \textbf{Definition of negative exponents}$$

$$= \frac{1}{25} \quad \textbf{The square of 5 is 25}$$

As you can see from our first two examples, when we apply the definition for negative exponents to expressions containing negative exponents, we end up with reciprocals.

Practice Problems
1. Simplify: 2^{-4}

2. Simplify: 4^{-2}

Answers
1. $\dfrac{1}{16}$ 2. $\dfrac{1}{16}$

3. Simplify: $(-2)^{-3}$

EXAMPLE 3 Simplify: $(-3)^{-2}$

SOLUTION   The fact that the base in this problem is negative does not change the procedure we use to simplify:

$$(-3)^{-2} = \frac{1}{(-3)^2} \quad \textbf{Definition of negative exponents}$$

$$= \frac{1}{9} \qquad \textbf{The square of } -3 \textbf{ is 9}$$

To check your understanding of negative exponents, look over the two lines below.

$$2^1 = 2 \qquad 2^2 = 4 \qquad 2^3 = 8 \qquad 2^4 = 16$$
$$2^{-1} = \frac{1}{2} \qquad 2^{-2} = \frac{1}{4} \qquad 2^{-3} = \frac{1}{8} \qquad 2^{-4} = \frac{1}{16}$$

The properties of exponents we have developed so far hold for negative exponents as well as positive exponents. For example, Property 1, our multiplication property for exponents, is still written as

$$a^r \cdot a^s = a^{r+s}$$

but now r and s can be negative numbers also.

4. Simplify: $3^4 \cdot 3^{-7}$

EXAMPLE 4 Simplify: $2^5 \cdot 2^{-7}$

SOLUTION   This is multiplication with the same base, so we add exponents.

$$2^5 \cdot 2^{-7} = 2^{5+(-7)} \quad \textbf{Property 1 for exponents}$$

$$= 2^{-2} \qquad \textbf{Addition}$$

$$= \frac{1}{2^2} \qquad \textbf{Definition of negative exponents}$$

$$= \frac{1}{4} \qquad \textbf{The square of 2 is 4}$$

When we simplify expressions containing negative exponents, let's agree that the final expression contain only positive exponents.

5. Simplify: $x^6 \cdot x^{-10}$

EXAMPLE 5 Simplify: $x^9 \cdot x^{-12}$

SOLUTION   Again, because we have the product of two expressions with the same base, we use Property 1 for exponents to add exponents.

$$x^9 \cdot x^{-12} = x^{9+(-12)} \quad \textbf{Property 1 for exponents}$$

$$= x^{-3} \qquad \textbf{Add exponents}$$

$$= \frac{1}{x^3} \qquad \textbf{Definition of negative exponents}$$

Division with Exponents

To develop our next property of exponents, we use the definition for positive exponents. Consider the expression $\frac{x^6}{x^4}$. We can simplify by expanding the numerator and denominator and then reducing to lowest terms by dividing out common factors.

Answers

3. $-\frac{1}{8}$ **4.** $\frac{1}{27}$ **5.** $\frac{1}{x^4}$

$$\frac{x^6}{x^4} = \frac{x \cdot x \cdot x \cdot x \cdot x \cdot x}{x \cdot x \cdot x \cdot x}$$ **Expand numerator and denominator**

$$= \frac{x \cdot x \cdot x \cdot x \cdot x \cdot x}{x \cdot x \cdot x \cdot x}$$ **Divide out common factors**

$$= x \cdot x$$

$$= x^2$$ **Write answer with exponent 2**

Note that the exponent in the answer is the difference of the exponents in the original problem. More specifically, if we subtract the exponent in the denominator from the exponent in the numerator, we obtain the exponent in the answer. This discussion leads us to another property of exponents.

Division Property for Expressions Containing Exponents

If a is any number other than zero, and r and s are integers, then

$$\frac{a^r}{a^s} = a^{r-s}$$

In words: To divide two numbers with the same base, subtract the exponent in the denominator from the exponent in the numerator, and use the common base as the base in the answer.

EXAMPLE 6 Simplify: $\dfrac{2^5}{2^8}$

SOLUTION Using our new property, we subtract the exponent in the denominator from the exponent in the numerator. The result is an expression containing a negative exponent.

$$\frac{2^5}{2^8} = 2^{5-8}$$ **Division property for exponents**

$$= 2^{-3}$$ **Subtraction**

$$= \frac{1}{2^3}$$ **Definition of negative exponents**

$$= \frac{1}{8}$$ **The cube of 2 is 8**

To further justify our new property of exponents, we can rework the problem shown in Example 6, without using the division property for exponents, to see that we obtain the same answer.

$$\frac{2^5}{2^8} = \frac{2 \cdot 2 \cdot 2 \cdot 2 \cdot 2}{2 \cdot 2 \cdot 2 \cdot 2 \cdot 2 \cdot 2 \cdot 2 \cdot 2}$$ **Expand numerator and denominator**

$$= \frac{2 \cdot 2 \cdot 2 \cdot 2 \cdot 2}{2 \cdot 2 \cdot 2 \cdot 2 \cdot 2 \cdot 2 \cdot 2 \cdot 2}$$ **Divide out common factors**

$$= \frac{1}{2 \cdot 2 \cdot 2}$$

$$= \frac{1}{8}$$ **Multiply**

As you can see, the answer matches the answer obtained by using the division property for exponents.

6. Simplify: $\dfrac{2^6}{2^8}$

Answer

6. $\dfrac{1}{4}$

7. Divide: $10^{-5} \div 10^{3}$

EXAMPLE 7 Divide: $10^{-3} \div 10^{5}$

SOLUTION   We begin by writing the division problem in fractional form. Then we apply our division property.

$$10^{-3} \div 10^{5} = \frac{10^{-3}}{10^{5}} \qquad \textbf{Write problem in fractional form}$$

$$= 10^{-3-5} \qquad \textbf{Division property for exponents}$$

$$= 10^{-8} \qquad \textbf{Subtraction}$$

$$= \frac{1}{10^{8}} \qquad \textbf{Definition of negative exponents}$$

We can leave the answer in exponential form, as it is, or we can expand the denominator to obtain

$$= \frac{1}{100,000,000} \qquad \blacksquare$$

8. Simplify: $\dfrac{10^{-6}}{10^{-8}}$

EXAMPLE 8 Simplify: $\dfrac{10^{-8}}{10^{-6}}$

SOLUTION   Again, we are dividing expressions that have the same base. To find the exponent on the answer, we subtract the exponent in the denominator from the exponent in the numerator.

$$\frac{10^{-8}}{10^{-6}} = 10^{-8-(-6)} \qquad \textbf{Division property for exponents}$$

$$= 10^{-2} \qquad \textbf{Subtraction}$$

$$= \frac{1}{10^{2}} \qquad \textbf{Definition of negative exponents}$$

$$= \frac{1}{100} \qquad \textbf{Answer in expanded form} \qquad \blacksquare$$

9. Simplify: $4x^{-3} \cdot 7x$

EXAMPLE 9 Simplify: $3x^{-4} \cdot 5x$

SOLUTION   To begin to simplify this expression, we regroup using the commutative and associative properties. That way, the numbers 3 and 5 are grouped together, as are the powers of x. Note also how we write x as x^{1}, so we can see the exponent.

$$3x^{-4} \cdot 5x = 3x^{-4} \cdot 5x^{1} \qquad \textbf{Write \textit{x} as \textit{x}}^{1}$$

$$= (3 \cdot 5)\,(x^{-4} \cdot x^{1}) \qquad \textbf{Commutative and associative properties}$$

$$= 15x^{-4+1} \qquad \textbf{Multiply 3 and 5, then add exponents}$$

$$= 15x^{-3} \qquad \textbf{The sum of } -4 \textbf{ and 1 is } -3$$

$$= \frac{15}{x^{3}} \qquad \textbf{Definition of negative exponents} \qquad \blacksquare$$

10. Simplify: $\dfrac{x^{-5} \cdot x^{2}}{x^{-8}}$

EXAMPLE 10 Simplify: $\dfrac{x^{-4} \cdot x^{7}}{x^{-2}}$

SOLUTION   We simplify the numerator first by adding exponents.

$$\frac{x^{-4} \cdot x^{7}}{x^{-2}} = \frac{x^{-4+7}}{x^{-2}} \qquad \textbf{Multiplication property for exponents}$$

$$= \frac{x^{3}}{x^{-2}} \qquad \textbf{The sum of } -4 \textbf{ and 7 is 3}$$

$$= x^{3-(-2)} \qquad \textbf{Division property for exponents}$$

$$= x^{5} \qquad \textbf{Subtracting } -2 \textbf{ is equivalent to adding } +2 \qquad \blacksquare$$

Answers

7. $\dfrac{1}{10^{8}}$ **8.** 100   **9.** $\dfrac{28}{x^{2}}$

10. x^{5}

Problem Set **A.3**

Write each expression with positive exponents, then simplify.

1. 2^{-5}

2. 3^{-4}

3. 10^{-2}

4. 10^{-3}

5. x^{-3}

6. x^{-4}

7. $(-4)^{-2}$

8. $(-2)^{-4}$

9. $(-5)^{-3}$

10. $(-4)^{-3}$

11. $2^6 \cdot 2^{-8}$

12. $3^5 \cdot 3^{-8}$

13. $10^{-2} \cdot 10^5$

14. $10^{-3} \cdot 10^6$

15. $x^{-4} \cdot x^{-3}$

16. $x^{-3} \cdot x^{-7}$

17. $3^8 \cdot 3^{-6}$

18. $2^9 \cdot 2^{-6}$

19. $2 \cdot 2^{-4}$

20. $3^{-5} \cdot 3$

21. $10^{-3} \cdot 10^8 \cdot 10^{-2}$

22. $10^{-3} \cdot 10^9 \cdot 10^{-4}$

23. $x^{-5} \cdot x^{-4} \cdot x^{-3}$

24. $x^{-7} \cdot x^{-9} \cdot x^{-6}$

25. $2^{-3} \cdot 2 \cdot 2^3$

26. $3 \cdot 3^5 \cdot 3^{-5}$

27. $\dfrac{2^9}{2^7}$

28. $\dfrac{3^8}{3^4}$

29. $\dfrac{2^7}{2^9}$

30. $\dfrac{3^5}{3^7}$

31. $\dfrac{x^4}{x^3}$

32. $\dfrac{x^7}{x^5}$

33. $\dfrac{x^3}{x^4}$

34. $\dfrac{x^5}{x^7}$

35. $10^{-2} \div 10^5$

36. $10^{-4} \div 10^2$

37. $10^{-5} \div 10^2$

38. $10^{-3} \div 10$

39. $\dfrac{10^{-2}}{10^{-5}}$

40. $\dfrac{10^{-4}}{10^{-7}}$

41. $x^{12} \div x^{-12}$

42. $x^6 \div x^{-6}$

43. $2^{12} \div 2^{10}$

44. $3^{11} \div 3^9$

45. $\dfrac{10}{10^{-3}}$

46. $\dfrac{10^2}{10^{-5}}$

47. $3x^{-2} \cdot 5x^7$

48. $5x^{-3} \cdot 8x^6$

49. $2x^{-2} \cdot 3x^{-3}$

50. $4x^{-6} \cdot 5x^{-4}$

51. $(4x^{-4})(-3x^{-3})$

52. $(3x^2)(-4x^{-1})$

53. $7x \cdot 3x^{-4} \cdot 2x^5$

54. $6x \cdot 2x^{-3} \cdot 3x^4$

55. $\dfrac{x^{-3} \cdot x^7}{x^{-1}}$

56. $\dfrac{x^{-4} \cdot x^6}{x^{-2}}$

57. $\dfrac{x^{-2} \cdot x^{-8}}{x^{-12}}$

58. $\dfrac{x^{-5} \cdot x^{-7}}{x^{-15}}$

Scientific Notation

There are many disciplines that deal with very large numbers and others that deal with very small numbers. For example, in astronomy, distances commonly are given in light-years. A light-year is the distance that light will travel in one year. It is approximately

5,880,000,000,000 miles

It can be difficult to perform calculations with numbers in this form because of the number of zeros present. Scientific notation provides a way of writing very large, or very small, numbers in a more manageable form.

> **DEFINITION**
>
> A number is in scientific notation when it is written as the product of a number between 1 and 10 and an integer power of 10. A number written in scientific notation has the form
>
> $$n \times 10^r$$
>
> where $1 \le n < 10$ and $r =$ an integer.

Practice Problems

1. Write 27,500 in scientific notation.

EXAMPLE 1 The speed of light is 186,000 miles per second. Write 186,000 in scientific notation.

SOLUTION To write this number in scientific notation, we rewrite it as the product of a number between 1 and 10 and a power of 10. To do so, we move the decimal point 5 places to the left so that it appears between the 1 and the 8, giving us 1.86. Then we multiply this number by 10^5. The number that results has the same value as our original number, but is written in scientific notation. Here is our result:

$$186,000 = 1.86 \times 10^5$$

Both numbers have exactly the same value. The number on the left is written in *standard form,* while the number on the right is written in scientific notation. ∎

EXAMPLE 2 If your pulse rate is 60 beats per minute, then your heart will beat 8.64×10^4 times each day. Write 8.64×10^4 in standard form.

2. Write 7.89×10^5 in standard form.

SOLUTION Because 10^4 is 10,000, we can think of this as simply a multiplication problem. That is,

$$8.64 \times 10^4 = 8.64 \times 10,000 = 86,400$$

Answers

1. 2.75×10^4 **2.** 789,000

Looking over our result, we can think of the exponent 4 as indicating the number of places we need to move the decimal point in order to write our number in standard form. Because our exponent is positive 4, we move the decimal point from its original position, between the 8 and the 6, four places to the right. If we need to add any zeros on the right we do so. The result is the standard form of our number, 86,400. ∎

Next, we turn our attention to writing small numbers in scientific notation. To do so, we use the negative exponents developed in the previous appendix. For example, the number 0.00075, when written in scientific notation, is equivalent to 7.5×10^{-4}. Here's why:

$$7.5 \times 10^{-4} = 7.5 \times \frac{1}{10^4} = 7.5 \times \frac{1}{10,000} = \frac{7.5}{10,000} = 0.00075$$

The table below lists some other numbers both in scientific notation and in standard form.

3. Fill in the missing numbers in the table below:

STANDARD FORM		SCIENTIFIC NOTATION
a.	24,500	=	
b.		=	5.6×10^5
c.	0.000789	=	
d.		=	4.8×10^{-3}

EXAMPLE 3 Each pair of numbers in the table below is equal.

STANDARD FORM		SCIENTIFIC NOTATION
376,000	=	3.76×10^5
49,500	=	4.95×10^4
3,200	=	3.2×10^3
591	=	5.91×10^2
46	=	4.6×10^1
8	=	8×10^0
0.47	=	4.7×10^{-1}
0.093	=	9.3×10^{-2}
0.00688	=	6.88×10^{-3}
0.0002	=	2×10^{-4}
0.000098	=	9.8×10^{-5}

∎

As we read across the table, for each pair of numbers, notice how the decimal point in the number on the right is placed so that the number containing the decimal point is always a number between 1 and 10. Correspondingly, the exponent on 10 keeps track of how many places the decimal point was moved in converting from standard form to scientific notation. In general, when the exponent is positive, we are working with a large number. On the other hand, when the exponent is negative, we are working with a small number. (By small number, we mean a number that is less than 1, but larger than 0.)

We end this section with a diagram that shows two numbers, one large and one small, that are converted to scientific notation.

$$376,000 = 3.76 \times 10^5$$

Moved 5 places

Keeps track of the 5 places we moved the decimal point

Decimal point originally here

$$0.00688 = 6.88 \times 10^{-3}$$

Moved 3 places

Keeps track of the 3 places we moved the decimal point

Answers

3. a. 2.45×10^4

b. 560,000

c. 7.89×10^{-4}

d. 0.0048

Problem Set A.4

Write each number in scientific notation.

1. 425,000　　　**2.** 635,000　　　**3.** 6,780,000　　　**4.** 5,490,000

5. 540　　　**6.** 930　　　**7.** 11,000　　　**8.** 29,000

9. 89,000,000　　　**10.** 37,000,000

Write each number in standard form.

11. 3.84×10^4　　　**12.** 3.84×10^7　　　**13.** 5.71×10^7　　　**14.** 5.71×10^5

15. 6×10^2　　　**16.** 6×10^1　　　**17.** 3.3×10^3　　　**18.** 3.3×10^2

19. 8.913×10^7　　　**20.** 8.913×10^5

Write each number in scientific notation.

21. 0.00035　　　**22.** 0.0000035　　　**23.** 0.0007　　　**24.** 0.007

25. 0.012　　　**26.** 0.12　　　**27.** 0.06035　　　**28.** 0.0006035

29. 0.1276　　　**30.** 0.001276

Write each number in standard form.

31. 8.3×10^{-4}

32. 8.3×10^{-7}

33. 6.25×10^{-2}

34. 6.25×10^{-4}

35. 1.1×10^{-5}

36. 1.1×10^{-3}

37. 3.125×10^{-1}

38. 3.125×10^{-2}

39. 5×10^{-3}

40. 5×10^{-5}

Fill in the missing numbers in the table that follows.

STANDARD FORM		SCIENTIFIC NOTATION
41. 4,200,000	=	
42.	=	3.3×10^{6}
43.	=	2.75×10^{8}
44. 526,000,000	=	
45. 0.00496	=	
46.	=	3.58×10^{-3}
47.	=	5.1×10^{-4}
48. 0.000082	=	
49. 47,892,000	=	
50.	=	5.7832×10^{7}

51. Age If you are 25 years old, you have been alive for over 788,000,000 seconds. Write this number in scientific notation.

52. Photography Some cameras used in scientific studies have the ability to take a picture every 0.000000167 of a second. Write this number in scientific notation.

53. Lifetime Earnings If you had a job where you earned $15.00 per hour, by working full-time for the next 35 years, you would have made over 1.092×10^{6} dollars. Write this number in standard form.

54. Savings If you put $1,000 into a savings account every year from the time you are 25 years old until you are 55 years old, you will have over 1.8×10^{5} dollars in the account when you reach 55 years of age (assuming 10% annual interest). Write $$1.8 \times 10^{5}$ in standard form.

55. Super Bowl Advertising The table below shows the cost of a 30-second television ad for three of the Super Bowls. Complete the table by writing each number in scientific notation.

GAME	YEAR	AD COST	COST IN SCIENTIFIC NOTATION
Super Bowl I	1967	$42,000	
Super Bowl XXII	1985	$525,000	
Super Bowl XXXIII	1999	$1,600,000	

More about Scientific Notation

In this appendix, we extend our work with scientific notation to include multiplication and division with numbers written in scientific notation. To work the problems in this appendix, we use the material presented in the previous two appendixes, along with the commutative and associative properties of multiplication and the rule for multiplication with fractions. Here is our first example.

EXAMPLE 1 Multiply: $(3.5 \times 10^8)(2.2 \times 10^{-5})$

SOLUTION   First we apply the commutative and associative properties to rearrange the numbers, so that the decimal numbers are grouped together and the powers of 10 are also.

$$(3.5 \times 10^8)(2.2 \times 10^{-5}) = (3.5)(2.2) \times (10^8)(10^{-5})$$

Next, we multiply the decimal numbers together and then the powers of ten. To multiply the powers of ten, we add exponents.

$$= 7.7 \times 10^{8+(-5)}$$
$$= 7.7 \times 10^3$$ ■

EXAMPLE 2 Find the product of 130,000,000 and 0.000005. Write your answer in scientific notation.

SOLUTION   We begin by writing both numbers in scientific notation. Then we proceed as we did in Example 1: We group the numbers between 1 and 10 separately from the powers of 10.

$$(130,000,000)(0.000005) = (1.3 \times 10^8)(5 \times 10^{-6})$$
$$= (1.3)(5) \times (10^8)(10^{-6})$$
$$= 6.5 \times 10^2$$ ■

Our next examples involve division with numbers in scientific notation.

EXAMPLE 3 Divide: $\dfrac{8 \times 10^3}{4 \times 10^{-6}}$

SOLUTION   To separate the numbers between 1 and 10 from the powers of 10, we "undo" the multiplication and write the problem as the product of two fractions. Doing so looks like this:

$$\frac{8 \times 10^3}{4 \times 10^{-6}} = \frac{8}{4} \times \frac{10^3}{10^{-6}} \qquad \textbf{Write as two separate fractions}$$

Practice Problems
1. Multiply: $(2.5 \times 10^6)(1.4 \times 10^2)$

2. Find the product of 2,200,000 and 0.00015.

3. Divide: $\dfrac{6 \times 10^5}{2 \times 10^{-4}}$

Answers
1. 3.5×10^8 2. 3.3×10^2
3. 3×10^9

Next, we divide 8 by 4 to obtain 2. Then we divide 10^3 by 10^{-6} by subtracting exponents.

$$= 2 \times 10^{3-(-6)} \quad \textbf{Divide}$$
$$= 2 \times 10^9 \qquad \qquad \blacksquare$$

4. Divide: $\dfrac{0.0038}{19,000,000}$

EXAMPLE 4 Divide: $\dfrac{0.00045}{1,500,000}$

SOLUTION To begin, write each number in scientific notation.

$$\frac{0.00045}{1,500,000} = \frac{4.5 \times 10^{-4}}{1.5 \times 10^6} \quad \textbf{Write numbers in scientific notation}$$

Next, as in the previous example, we write the problem as two separate fractions in order to group the numbers between 1 and 10 together, as well as the powers of 10.

$$= \frac{4.5}{1.5} \times \frac{10^{-4}}{10^6} \quad \textbf{Write as two separate fractions}$$
$$= 3 \times 10^{-4-6} \quad \textbf{Divide}$$
$$= 3 \times 10^{-10} \quad \textbf{-4 - 6 = -4 + (-6) = -10} \qquad \blacksquare$$

5. Simplify: $\dfrac{(6.8 \times 10^{-4})(3.9 \times 10^2)}{7.8 \times 10^{-6}}$

EXAMPLE 5 Simplify: $\dfrac{(6.8 \times 10^5)(3.9 \times 10^{-7})}{7.8 \times 10^{-4}}$

SOLUTION We group the numbers between 1 and 10 separately from the powers of 10:

$$\frac{(6.8 \times 10^5)(3.9 \times 10^{-7})}{7.8 \times 10^{-4}} = \frac{(6.8)(3.9)}{7.8} \times \frac{(10^5)(10^{-7})}{10^{-4}}$$
$$= 3.4 \times 10^{5+(-7)-(-4)}$$
$$= 3.4 \times 10^2 \qquad \blacksquare$$

6. Simplify: $\dfrac{(0.000035)(45,000)}{0.000075}$

EXAMPLE 6 Simplify: $\dfrac{(35,000)(0.0045)}{7,500,000}$

SOLUTION We write each number in scientific notation, then we proceed as we have in the examples above.

$$\frac{(35,000)(0.0045)}{7,500,000} = \frac{(3.5 \times 10^4)(4.5 \times 10^{-3})}{7.5 \times 10^6}$$
$$= \frac{(3.5)(4.5)}{7.5} \times \frac{(10^4)(10^{-3})}{10^6}$$
$$= 2.1 \times 10^{4+(-3)-6}$$
$$= 2.1 \times 10^{-5} \qquad \blacksquare$$

Answers
4. 2×10^{-10} **5.** 3.4×10^4
6. 2.1×10^4

Problem Set A.5

Find each product. Write all answers in scientific notation.

1. $(2 \times 10^4)(3 \times 10^6)$

2. $(3 \times 10^3)(1 \times 10^5)$

3. $(2.5 \times 10^7)(6 \times 10^3)$

4. $(3.8 \times 10^6)(5 \times 10^3)$

5. $(7.2 \times 10^3)(9.5 \times 10^{-6})$

6. $(8.5 \times 10^5)(4.2 \times 10^{-9})$

7. $(36,000)(450,000)$

8. $(25,000)(620,000)$

9. $(4,200)(0.00009)$

10. $(0.0000065)(86,000)$

Find each quotient. Write all answers in scientific notation.

11. $\dfrac{3.6 \times 10^5}{1.8 \times 10^2}$

12. $\dfrac{9.3 \times 10^{15}}{3.0 \times 10^5}$

13. $\dfrac{8.4 \times 10^{-6}}{2.1 \times 10^3}$

14. $\dfrac{6.0 \times 10^{-10}}{1.5 \times 10^3}$

15. $\dfrac{3.5 \times 10^5}{7.0 \times 10^{-10}}$

16. $\dfrac{1.6 \times 10^7}{8.0 \times 10^{-14}}$

17. $\dfrac{540,000}{9,000}$

18. $\dfrac{750,000,000}{250,000}$

19. $\dfrac{0.00092}{46,000}$

20. $\dfrac{0.00000047}{235,000}$

Simplify each expression, and write all answers in scientific notation.

21. $\dfrac{(3 \times 10^7)(8 \times 10^4)}{6 \times 10^5}$

22. $\dfrac{(4 \times 10^9)(6 \times 10^5)}{8 \times 10^3}$

23. $\dfrac{(2 \times 10^{-3})(6 \times 10^{-5})}{3 \times 10^{-4}}$

24. $\dfrac{(4 \times 10^{-5})(9 \times 10^{-10})}{6 \times 10^{-6}}$

25. $\dfrac{(3.5 \times 10^{-4})(4.2 \times 10^5)}{7 \times 10^3}$

26. $\dfrac{(2.4 \times 10^{-6})(3.6 \times 10^3)}{9 \times 10^5}$

27. $\dfrac{(0.00087)(40,000)}{1,160,000}$

28. $\dfrac{(0.0045)(24,000)}{270,000}$

29. $\dfrac{(525)(0.0000032)}{0.0025}$

30. $\dfrac{(465)(0.000004)}{0.0093}$

Solutions to Selected Practice Problems

Solutions to all practice problems that require more than one step are shown here. Before you look back here to see where you have made a mistake, you should try the problem you are working on twice. If you do not get the correct answer the second time you work the problem, then the solution here should show you where you went wrong.

Chapter 1

Section 1.2

4. $(x + 5) + 9 = x + (5 + 9)$
$ = x + 14$

5. $6 + (8 + y) = (6 + 8) + y$
$ = 14 + y$

6. $(1 + x) + 4 = (x + 1) + 4$
$ = x + (1 + 4)$
$ = x + 5$

7. $8 + (a + 4) = 8 + (4 + a)$
$ = (8 + 4) + a$
$ = 12 + a$

8. a. $n = 8$, since $8 + 9 = 17$
b. $n = 8$, since $8 + 2 = 10$
c. $n = 1$, since $8 + 1 = 9$
d. $n = 6$, since $16 = 6 + 10$

9. $(x + 5) + 4 = 10$
$x + (5 + 4) = 10$
$x + 9 = 10$
$x = 1$

10. $(a + 4) + 2 = 7 + 9$
$a + (4 + 2) = 16$
$a + 6 = 16$
$a = 10$

Section 1.3

1. $\quad 63 = 6$ tens $+ 3$ ones
$\underline{+25 = 2 \text{ tens} + 5 \text{ ones}}$
$ 8$ tens $+ 8$ ones
Answer: 88

2. $\quad 342 = 3$ hundreds $+ 4$ tens $+ 2$ ones
$\underline{+605 = 6 \text{ hundreds} + 0 \text{ tens} + 5 \text{ ones}}$
$ 9$ hundreds $+ 4$ tens $+ 7$ ones
Answer: 947

3. $\quad 375 = 3$ hundreds $+ 7$ tens $+ 5$ ones
$\quad 121 = 1$ hundred $ + 2$ tens $+ 1$ one
$\underline{+473 = 4 \text{ hundreds} + 7 \text{ tens} + 3 \text{ ones}}$
$ 8$ hundreds $+ 16$ tens $+ 9$ ones $= 9$ hundreds $+ 6$ tens $+ 9$ ones
Answer: 969

4. $\overset{1\ 1\ \ 1\ 1}{57,904}$
$7,193$
$\underline{655}$
$65,752$

5. a. $7 + 7 + 7 + 7 = 28$ ft **b.** $88 + 88 + 33 + 33 = 242$ in. **c.** $44 + 66 + 77 = 187$ yd

Section 1.4

5. a. Food $ 5,296
$\phantom{\text{a. }}$ Car $\quad\underline{\ 4,847}$
$\phantom{\text{a. }}$ Total $10,143 = $10,140$ to the nearest ten dollars

b. Savings $2,149
$\phantom{\text{b. }}$ Taxes $\quad\underline{6,137}$
$\phantom{\text{b. }}$ Total $\quad $8,286 = $8,300$ to the nearest hundred dollars

c. House $\quad$10,200
$\phantom{\text{c. }}$ Taxes $\qquad$6,137
$\phantom{\text{c. }}$ Misc. $\qquad$6,142
$\phantom{\text{c. }}$ Car $\qquad$4,847
$\phantom{\text{c. }}$ Savings $\quad\underline{2,149}$
$\phantom{\text{c. }}$ Total $\qquad $29,475 = $29,000$ to the nearest thousand dollars

6. We round each of the four numbers in the sum to the nearest thousand, and then we add the rounded numbers.

$5,287 \quad$ rounds to $\qquad 5,000$
$2,561 \quad$ rounds to $\qquad 3,000$
$888 \quad$ rounds to $\qquad 1,000$
$\underline{+4,898 \quad \text{rounds to} \quad + \ 5,000}$
$14,000$

We estimate the answer to this problem to be approximately 14,000. The actual answer, found by adding the original, unrounded numbers, is 13,634.

7.

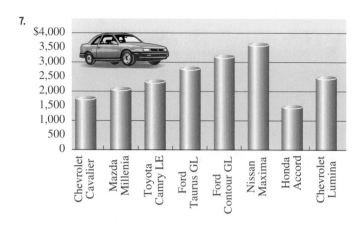

Section 1.5

1. $684 = 6$ hundreds $+ 8$ tens $+ 4$ ones

$-431 = \underline{4 \text{ hundreds} \quad 3 \text{ tens} \quad 1 \text{ one}}$

$\qquad 2$ hundreds $+ 5$ tens $+ 3$ ones

Answer: 253

2.
```
  6,857
−   405
  6,452
```

3. $63 = 6$ tens $+ 3$ ones $= 5$ tens $+ 13$ ones

$-47 = \underline{4 \text{ tens} \quad 7 \text{ ones}} = \underline{4 \text{ tens} \quad 7 \text{ ones}}$

$\qquad\qquad\qquad\qquad\qquad 1$ ten $+ 6$ ones

Answer: 16

4.
```
   5 15
   6̶8̶6
 −283
  373
```

5.
```
  $127
 −  52
  $ 75
```

Section 1.6

5. $4(7x) = (4 \cdot 7)x$
$\qquad\quad = 28x$

6. $7(9y) = (7 \cdot 9)y$
$\qquad\quad = 63y$

7. a. $n = 7$, since $5 \cdot 7 = 35$ **b.** $n = 9$, since $8 \cdot 9 = 72$ **c.** $n = 7$, since $49 = 7 \cdot 7$ **d.** $n = 3$, since $27 = 9 \cdot 3$

8. $5 \cdot 100 = 100 + 100 + 100 + 100 + 100$
$\qquad\qquad = 500$

9. $4 \cdot 7,000 = 7,000 + 7,000 + 7,000 + 7,000$
$\qquad\qquad\quad = 28,000$

Section 1.7

1. $3(5 + 7) = 3(5) + 3(7)$
$\qquad\qquad = 15 + 21$
$\qquad\qquad = 36$

2. $5(6 + 8) = 5(6) + 5(8)$
$\qquad\qquad = 30 + 40$
$\qquad\qquad = 70$

3. $4(x + 2) = 4 \cdot x + 4 \cdot 2$
$\qquad\qquad = 4x + 8$

4. $7(a + 9) = 7 \cdot a + 7 \cdot 9$
$\qquad\qquad = 7a + 63$

5. $4(7x + 2) = 4(7x) + 4 \cdot 2$
$\qquad\qquad\quad = (4 \cdot 7)x + 4 \cdot 2$
$\qquad\qquad\quad = 28x + 8$

6. $7(9y + 6) = 7(9y) + 7 \cdot 6$
$\qquad\qquad\quad = (7 \cdot 9)y + 7 \cdot 6$
$\qquad\qquad\quad = 63y + 42$

7. *Long Method*   *Shortcut Method*
```
  50 + 7          5
               57
×     8        ×  8
   56          456
 +400
  456
```

8. *Long Method*   *Shortcut Method*
```
 500 + 70 + 2      4 1
                 572
×          6      ×  6
         12      3,432
        420
      +3,000
       3,432
```

9.
```
      45
     ×62
      90 ← 2(45) = 90
   +2,700 ← 60(45) = 2,700
    2,790
```

10.
```
     356
    ×641
     356 ← 1(356) = 356
  14,240 ← 40(356) = 14,240
 213,600 ← 600(356) = 213,600
 228,196
```

11.
```
     365
    ×550
  18,250
 182,500
 200,750 mg
```

12. $36(\$12) = \$432$ Total weekly earnings
$\$432 - \$109 = \$323$ Take-home pay

13. Fat: $3(10) = 30$ grams of fat; sodium: $3(160) = 480$ milligrams of sodium

14. Bowling for 3 hours burns $3(265) = 795$ calories. Eating two bags of chips means you are consuming $2(3)(160) = 960$ calories. No; bowling won't burn all the calories.

Section 1.8

1.
$$
\begin{array}{r}
72 \\
4\overline{)288} \\
\underline{28} \\
08 \\
\underline{8} \\
0
\end{array}
$$

2.
$$
\begin{array}{r}
283 \\
24\overline{)6{,}792} \\
\underline{48} \\
1\,99 \\
\underline{1\,92} \\
72 \\
\underline{72} \\
0
\end{array}
$$

3. 69 R 20, or $69\frac{20}{27}$
$$
\begin{array}{r}
27\overline{)1{,}883} \\
\underline{1\,62} \\
263 \\
\underline{243} \\
20
\end{array}
$$

4. 156   The family spent $156 per day
$$
\begin{array}{r}
12\overline{)1{,}872} \\
\underline{1\,2} \\
67 \\
\underline{60} \\
72 \\
\underline{72} \\
0
\end{array}
$$

Section 1.9

1. Base 5, exponent 2; 5 to the second power, or 5 squared   **2.** Base 2, exponent 3; 2 to the third power, or 2 cubed

3. Base 1, exponent 4; 1 to the fourth power   **4.** $5^2 = 5 \cdot 5 = 25$ **5.** $9^2 = 9 \cdot 9 = 81$ **6.** $2^3 = 2 \cdot 2 \cdot 2 = 8$

7. $1^4 = 1 \cdot 1 \cdot 1 \cdot 1 = 1$ **8.** $2^5 = 2 \cdot 2 \cdot 2 \cdot 2 \cdot 2 = 32$ **9.** $7^1 = 7$ **10.** $4^1 = 4$ **11.** $9^0 = 1$ **12.** $1^0 = 1$

13. $5 \cdot 7 - 3 \cdot 6 = 35 - 18$
$ = 17$

14. $7 + 3(6 + 4) = 7 + 3(10)$
$ = 7 + 30$
$ = 37$

15. $6 \cdot 3^2 + 64 \div 2^4 - 2 = 6 \cdot 9 + 64 \div 16 - 2$
$ = 54 + 4 - 2$
$ = 58 - 2$
$ = 56$

16. $5 + 3[24 - 5(6 - 2)] = 5 + 3[24 - 5(4)]$
$ = 5 + 3[24 - 20]$
$ = 5 + 3[4]$
$ = 5 + 12$
$ = 17$

17.
$$
\begin{array}{r}
187 \\
273 \\
150 \\
173 \\
\underline{227} \\
1010
\end{array}
$$
$\dfrac{1010}{5} = 202$ miles

18. First we place them in order from smallest to largest

150   173   187   227   273

Because there are 5 numbers, the one in the middle, 187, is the median.

19. The numbers are already in order from smallest to largest. Because there is an even number of numbers, we find the mean of the middle two:

$$\frac{42{,}635 + 44{,}475}{2} = \frac{87{,}110}{2} = 43{,}555$$

The median is $43,555.

20. The most frequently occurring score is 74. It occurs three times. The mode is 74.

Section 1.10

1. $A = 10 \cdot 2 + 4 \cdot 4$
$ = 20 + 16$
$ = 36 \text{ ft}^2$

2. $A = (1 \text{ yd}) \cdot (1 \text{ yd})$
$ = (3 \text{ ft}) \cdot (3 \text{ ft})$
$ = (3 \cdot 3) \text{ ft}^2$
$ = 9 \text{ ft}^2$

3. Area of small square $= 4 \cdot 4 = 16 \text{ ft}^2$
Area of large square $= 12 \cdot 12 = 144 \text{ ft}^2$
Since $9 \cdot 16 = 144$, the area of the larger square is 9 times the area of the small square.

4. $V = 15 \cdot 12 \cdot 8 = 1{,}440 \text{ ft}^3$

5. **a.**   Surface area $= 2(15 \cdot 8) + 2(8 \cdot 12) + (15 \cdot 12) = 612 \text{ ft}^2$
   b.   2 gallons will cover it, with some paint left over

Chapter 2

Section 2.2

1. $2 + (-5) = -3$ **2.** $-2 + 5 = 3$ **3.** $-2 + (-5) = -7$ **4.** $2 + 6 = 8$ **5.** $2 + (-6) = -4$ **6.** $-2 + 6 = 4$

7. $-2 + (-6) = -8$ **8.** $15 + 12 = 27$
$ 15 + (-12) = 3$
$ -15 + 12 = -3$
$ -15 + (-12) = -27$

9. $12 + (-3) + (-7) + 5 = 9 + (-7) + 5$
$ = 2 + 5$
$ = 7$

10. $[-2 + (-12)] + [7 + (-5)] = [-14] + [2]$
$ = -12$

Section 2.3

1. $7 - 3 = 7 + (-3)$ **2.** $-7 - 3 = -7 + (-3)$ **3.** $-8 - 6 = -8 + (-6)$ **4.** $10 - (-6) = 10 + 6$ **5.** $-10 - (-15) = -10 + 15$
$ = 4$ $ = -10$ $ = -14$ $ = 16$ $ = 5$

6. a. $8 - 5 = 8 + (-5)$ **b.** $-8 - 5 = -8 + (-5)$ **c.** $8 - (-5) = 8 + 5$ **d.** $-8 - (-5) = -8 + 5$ **e.** $12 - 10 = 12 + (-10)$
$ = 3$ $ = -13$ $ = 13$ $ = -3$ $ = 2$

f. $-12 - 10 = -12 + (-10)$ **g.** $12 - (-10) = 12 + 10$ **h.** $-12 - (-10) = -12 + 10$ **7.** $-4 + 6 - 7 = -4 + 6 + (-7)$
$ = -22$ $ = 22$ $ = -2$ $ = 2 + (-7)$
$ = -5$

8. $15 - (-5) - 8 = 15 + 5 + (-8)$ **9.** $-8 - 2 = -8 + (-2)$ **10.** $7 - (-5) = 7 + 5$ **11.** $7 - (-3) = 7 + 3$
$ = 20 + (-8)$ $ = -10$ $ = 12$ $ = 10$
$ = 12$

12. $-8 - (-6) = -8 + 6$ **13.** $42 - (-42) = 42 + 42 = 84°F$
$ = -2$

Section 2.4

1. $2(-6) = (-6) + (-6)$ **2.** $-2(6) = 6(-2) = (-2) + (-2) + (-2) + (-2) + (-2) + (-2)$ **3.** $-2(-6) = 12$
$ = -12$ $ = -12$

10. $-5(2)(-4) = -10(-4)$ **11.** $(-5)^2 = (-5)(-5)$ **12.** $-2[5 + (-8)] = -2[-3]$ **13.** $-3 + 4(-7 + 3) = -3 + 4(-4)$
$ = 40$ $ = 25$ $ = 6$ $ = -3 + (-16)$
$ = -19$

14. $-3(5) + 4(-4) = -15 + (-16)$ **15.** $-2(3 - 5) - 7(-2 - 4) = -2(-2) - 7(-6)$ **16.** $(-6 - 1)(4 - 9) = (-7)(-5)$
$ = -31$ $ = 4 - (-42)$ $ = 35$
$ = 4 + 42$
$ = 46$

Section 2.5

6. $\dfrac{8(-5)}{-4} = \dfrac{-40}{-4}$ **7.** $\dfrac{-20 + 6(-2)}{7 - 11} = \dfrac{-20 + (-12)}{-4}$ **8.** $-3(4^2) + 10 \div (-5) = -3(16) + 10 \div (-5)$ **9.** $-80 \div 2 \div 10 = -40 \div 10$
$\phantom{\dfrac{8(-5)}{-4}} = 10$ $\phantom{\dfrac{-20 + 6(-2)}{7 - 11}} = \dfrac{-32}{-4}$ $ = -48 + (-2)$ $ = -4$
$\phantom{\dfrac{-20 + 6(-2)}{7 - 11}} = 8$ $ = -50$

Section 2.6

1. $5(7a) = (5 \cdot 7)a$ **2.** $-3(9x) = (-3 \cdot 9)x$ **3.** $5(-8y) = [5(-8)]y$ **4.** $6 + (9 + x) = (6 + 9) + x$
$ = 35a$ $ = -27x$ $ = -40y$ $ = 15 + x$

5. $(3x + 7) + 4 = 3x + (7 + 4)$ **6.** $6(x + 4) = 6(x) + 6(4)$ **7.** $7(a - 5) = 7(a) - 7(5)$ **8.** $6(4x + 5) = 6(4x) + 6(5)$
$ = 3x + 11$ $ = 6x + 24$ $ = 7a - 35$ $ = (6 \cdot 4)x + 6(5)$
$ = 24x + 30$

9. $3(8a - 4) = 3(8a) - 3(4)$ **10.** $8(3x + 4y) = 8(3x) + 8(4y)$ **11.** $(2a)^3 = (2a)(2a)(2a)$ **12.** $(7xy)^2 = (7xy)(7xy)$
$ = 24a - 12$ $ = 24x + 32y$ $ = (2 \cdot 2 \cdot 2)(a \cdot a \cdot a)$ $ = (7 \cdot 7)(x \cdot x)(y \cdot y)$
$ = 8a^3$ $ = 49x^2y^2$

13. $(3x)^2(7xy)^2 = (3x)(3x)(7xy)(7xy)$ **14.** $(5x)^3(2x)^2 = (5x)(5x)(5x)(2x)(2x)$ **15.** $A = s^2 = 12^2 = 144 \text{ ft}^2$
$ = (3 \cdot 3 \cdot 7 \cdot 7)(x \cdot x \cdot x \cdot x)(y \cdot y)$ $ = (5 \cdot 5 \cdot 5 \cdot 2 \cdot 2)(x \cdot x \cdot x \cdot x \cdot x)$ $ P = 4s = 4(12) = 48 \text{ ft}$
$ = 441x^4y^2$ $ = 500x^5$

16. $A = lw = 100(53) = 5,300 \text{ yd}^2$
$P = 2l + 2w = 2(100) + 2(53) = 200 + 106 = 306 \text{ yd}$

Section 2.7

1. $x^3 \cdot x^5 = (x \cdot x \cdot x)(x \cdot x \cdot x \cdot x \cdot x)$ **2.** $4x^2 \cdot 6x^5 = (4 \cdot 6)(x^2 \cdot x^5)$ **3.** $2x^5 \cdot 3x^4 \cdot 4x^3 = (2 \cdot 3 \cdot 4)(x^5 \cdot x^4 \cdot x^3)$
$ = x \cdot x \cdot x \cdot x \cdot x \cdot x \cdot x \cdot x$ $ = (4 \cdot 6)(x^{2+5})$ $ = (2 \cdot 3 \cdot 4)(x^{5+4+3})$
$ = x^8$ $ = 24x^7$ $ = 24x^{12}$

4. $(10xy^2)(9x^3y^5) = (10 \cdot 9)(x \cdot x^3)(y^2 \cdot y^5)$ **5.** $(2^2)^3 = 2^{2 \cdot 3}$ **6.** $(x^3)^4 \cdot (x^5)^2 = x^{3 \cdot 4} \cdot x^{5 \cdot 2}$ **7.** $(5xy)^3 = 5^3 \cdot x^3 \cdot y^3$
$ = (10 \cdot 9)(x^{1+3})(y^{2+5})$ $ = 2^6$ $ = x^{12} \cdot x^{10}$ $ = 125x^3y^3$
$ = 90x^4y^7$ $ = 64$ $ = x^{12+10}$
$ = x^{22}$

8. $(4x^5y^2)^3 = 4^3(x^5)^3(y^2)^3$
$\qquad = 64x^{15}y^6$

9. $(2x^3y^4)^3(3xy^5)^2 = 2^3(x^3)^3(y^4)^3 \cdot 3^2x^2(y^5)^2$
$\qquad = 8x^9y^{12} \cdot 9x^2y^{10}$
$\qquad = (8 \cdot 9)(x^9x^2)(y^{12}y^{10})$
$\qquad = 72x^{11}y^{22}$

Section 2.8

1. $(5x^2 - 3x + 2) + (2x^2 + 10x - 9) = (5x^2 + 2x^2) + (-3x + 10x) + (2 - 9)$
$\qquad\qquad\qquad\qquad\qquad\qquad = (5 + 2)x^2 + (-3 + 10)x + (2 - 9)$
$\qquad\qquad\qquad\qquad\qquad\qquad = 7x^2 + 7x - 7$

$\qquad\qquad 5x^2 - 3x + 2$
$\qquad\qquad \underline{2x^2 + 10x - 9}$
$\qquad\qquad 7x^2 + 7x - 7$

2. $(3y^2 + 9y - 5) + (6y^2 - 4) = (3y^2 + 6y^2) + 9y + (-5 - 4)$
$\qquad\qquad\qquad\qquad\qquad\quad = (3 + 6)y^2 + 9y + (-5 - 4)$
$\qquad\qquad\qquad\qquad\qquad\quad = 9y^2 + 9y - 9$

$\qquad\qquad 3y^2 + 9y - 5$
$\qquad\qquad \underline{6y^2 \qquad - 4}$
$\qquad\qquad 9y^2 + 9y - 9$

3. $(8x^3 + 4x^2 + 3x + 2) + (4x^2 + 5x + 6) = 8x^3 + (4x^2 + 4x^2) + (3x + 5x) + (2 + 6)$
$\qquad\qquad\qquad\qquad\qquad\qquad\qquad\qquad = 8x^3 + (4 + 4)x^2 + (3 + 5)x + (2 + 6)$
$\qquad\qquad\qquad\qquad\qquad\qquad\qquad\qquad = 8x^3 + 8x^2 + 8x + 8$

$\qquad\qquad 8x^3 + 4x^2 + 3x + 2$
$\qquad\qquad \underline{\qquad\quad 4x^2 + 5x + 6}$
$\qquad\qquad 8x^3 + 8x^2 + 8x + 8$

4. $(5x^2 - 2x + 7) - (4x^2 + 8x - 4) = 5x^2 - 2x + 7 - 4x^2 - 8x + 4$
$\qquad\qquad\qquad\qquad\qquad\qquad = (5x^2 - 4x^2) + (-2x - 8x) + (7 + 4)$
$\qquad\qquad\qquad\qquad\qquad\qquad = (5 - 4)x^2 + (-2 - 8)x + (7 + 4)$
$\qquad\qquad\qquad\qquad\qquad\qquad = 1x^2 - 10x + 11$
$\qquad\qquad\qquad\qquad\qquad\qquad = x^2 - 10x + 11$

5. $(3y^3 - 2y^2 + 7y - 6) - (8y^3 - 6y^2 + 4y - 8) = 3y^3 - 2y^2 + 7y - 6 - 8y^3 + 6y^2 - 4y + 8$
$\qquad\qquad\qquad\qquad\qquad\qquad\qquad\qquad\qquad = (3y^3 - 8y^3) + (-2y^2 + 6y^2) + (7y - 4y) + (-6 + 8)$
$\qquad\qquad\qquad\qquad\qquad\qquad\qquad\qquad\qquad = (3 - 8)y^3 + (-2 + 6)y^2 + (7 - 4)y + (-6 + 8)$
$\qquad\qquad\qquad\qquad\qquad\qquad\qquad\qquad\qquad = -5y^3 + 4y^2 + 3y + 2$

6. $(-2x^2 + 5x - 1) - (6x^2 - 2x + 5) = -2x^2 + 5x - 1 - 6x^2 + 2x - 5$
$\qquad\qquad\qquad\qquad\qquad\qquad = (-2x^2 - 6x^2) + (5x + 2x) + (-1 - 5)$
$\qquad\qquad\qquad\qquad\qquad\qquad = -8x^2 + 7x - 6$

7. When $x = -3$
the polynomial $5x^2 - 3x + 8$
becomes $5(-3)^2 - 3(-3) + 8 = 5(9) + 9 + 8$
$\qquad\qquad\qquad\qquad\qquad\qquad = 45 + 9 + 8$
$\qquad\qquad\qquad\qquad\qquad\qquad = 62$

8. When $n = 1, 2n + 1 = 2 \cdot 1 + 1 = 3$
When $n = 2, 2n + 1 = 2 \cdot 2 + 1 = 5$
When $n = 3, 2n + 1 = 2 \cdot 3 + 1 = 7$
When $n = 4, 2n + 1 = 2 \cdot 4 + 1 = 9$

9. a. When $n = 1, 3n = 3 \cdot 1 = 3$
When $n = 2, 3n = 3 \cdot 2 = 6$
When $n = 3, 3n = 3 \cdot 3 = 9$
When $n = 4, 3n = 3 \cdot 4 = 12$

b. When $n = 1, n^3 = 1^3 = 1$
When $n = 2, n^3 = 2^3 = 8$
When $n = 3, n^3 = 3^3 = 27$
When $n = 4, n^3 = 4^3 = 64$

Section 2.9

1. $x^3(x^5 + x^7) = x^3 \cdot x^5 + x^3 \cdot x^7$
$\qquad\qquad\quad = x^8 + x^{10}$

2. $(x^5 + x^7)x^3 = x^5 \cdot x^3 + x^7 \cdot x^3$
$\qquad\qquad\quad = x^8 + x^{10}$

3. $5x^2(6x^3 - 4) = 5x^2 \cdot 6x^3 - 5x^2 \cdot 4$
$\qquad\qquad\quad = (5 \cdot 6)(x^2 \cdot x^3) - (5 \cdot 4)x^2$
$\qquad\qquad\quad = 30x^5 - 20x^2$

4. $5a^3b^5(2a^2 + 7b^2) = 5a^3b^5 \cdot 2a^2 + 5a^3b^5 \cdot 7b^2$
$\qquad\qquad\qquad\quad = (5 \cdot 2)(a^3 \cdot a^2)b^5 + (5 \cdot 7)(a^3)(b^5 \cdot b^2)$
$\qquad\qquad\qquad\quad = 10a^5b^5 + 35a^3b^7$

5. $(x + 2)(x + 6) = (x + 2)x + (x + 2)6$
$\qquad\qquad\qquad = x \cdot x + 2 \cdot x + x \cdot 6 + 2 \cdot 6$
$\qquad\qquad\qquad = x^2 + 2x + 6x + 12$
$\qquad\qquad\qquad = x^2 + 8x + 12$

6. $(x - 2)(x + 6) = (x - 2)x + (x - 2)6$
$\qquad\qquad\qquad = x \cdot x - 2 \cdot x + x \cdot 6 - 2 \cdot 6$
$\qquad\qquad\qquad = x^2 - 2x + 6x - 12$
$\qquad\qquad\qquad = x^2 + 4x - 12$

7. $(3x - 2)(5x + 4) = (3x - 2) \cdot 5x + (3x - 2) \cdot 4$
$\qquad\qquad\qquad\quad = 3x \cdot 5x - 2 \cdot 5x + 3x \cdot 4 - 2 \cdot 4$
$\qquad\qquad\qquad\quad = 15x^2 - 10x + 12x - 8$
$\qquad\qquad\qquad\quad = 15x^2 + 2x - 8$

8. $(x + 3)^2 = (x + 3)(x + 3)$
$\qquad\qquad = (x + 3) \cdot x + (x + 3) \cdot 3$
$\qquad\qquad = x \cdot x + 3 \cdot x + x \cdot 3 + 3 \cdot 3$
$\qquad\qquad = x^2 + 3x + 3x + 9$
$\qquad\qquad = x^2 + 6x + 9$

9. $(3x - 5)^2 = (3x - 5)(3x - 5)$
$\qquad\qquad = (3x - 5)[3x + (-5)]$
$\qquad\qquad = (3x - 5) \cdot 3x + (3x - 5)(-5)$
$\qquad\qquad = 3x \cdot 3x - 5 \cdot 3x + 3x(-5) - 5(-5)$
$\qquad\qquad = 9x^2 - 15x - 15x + 25$
$\qquad\qquad = 9x^2 - 30x + 25$

10.

	x	4
x	x^2	$4x$
2	$2x$	8

$x^2 + 4x + 2x + 8 = x^2 + 6x + 8$

11.

	$3x$	7
$2x$	$6x^2$	$14x$
5	$15x$	35

$6x^2 + 14x + 15x + 35 = 6x^2 + 29x + 35$

Chapter 3

Section 3.1

6. $\dfrac{2}{3} = \dfrac{2 \cdot 4}{3 \cdot 4} = \dfrac{8}{12}$ **7.** $\dfrac{2}{3} = \dfrac{2 \cdot 4x}{3 \cdot 4x} = \dfrac{8x}{12x}$ **8.** $\dfrac{15}{20} = \dfrac{15 \div 5}{20 \div 5} = \dfrac{3}{4}$

Section 3.2

1. 37 and 59 are prime numbers; 39 is divisible by 3 and 13; 51 is divisible by 3 and 17   **2.** $90 = 9 \cdot 10$

$= 3 \cdot 3 \cdot 2 \cdot 5$
$= 2 \cdot 3^2 \cdot 5$

4. $\dfrac{12}{18} = \dfrac{12 \div 6}{18 \div 6} = \dfrac{2}{3}$ **5.** $\dfrac{15}{20} = \dfrac{3 \cdot \cancel{5}}{2 \cdot 2 \cdot \cancel{5}} = \dfrac{3}{4}$ **6.** $\dfrac{30}{35} = \dfrac{2 \cdot 3 \cdot \cancel{5}}{\cancel{5} \cdot 7} = \dfrac{6}{7}$ **7.** $\dfrac{8}{72} = \dfrac{2 \cdot 2 \cdot 2 \cdot 1}{2 \cdot 2 \cdot 2 \cdot 3 \cdot 3} = \dfrac{1}{9}$ **8.** $\dfrac{5}{50} = \dfrac{\cancel{5} \cdot 1}{\cancel{5} \cdot 2 \cdot 5} = \dfrac{1}{10}$

9. $\dfrac{120}{25} = \dfrac{2 \cdot 2 \cdot 2 \cdot 3 \cdot \cancel{5}}{5 \cdot \cancel{5}} = \dfrac{24}{5}$ **10.** $\dfrac{54x}{90xy} = \dfrac{2 \cdot 3 \cdot 3 \cdot 3 \cdot x}{2 \cdot 3 \cdot 3 \cdot 5 \cdot x \cdot y} = \dfrac{3}{5y}$ **11.** $\dfrac{306a^2}{228a} = \dfrac{2 \cdot 3 \cdot 3 \cdot 17 \cdot \cancel{a} \cdot a}{2 \cdot 2 \cdot 3 \cdot 19 \cdot \cancel{a}} = \dfrac{51a}{38}$

Section 3.3

1. $\dfrac{2}{3} \cdot \dfrac{5}{9} = \dfrac{10}{27}$ **2.** $-\dfrac{2}{5} \cdot 7 = -\dfrac{2}{5} \cdot \dfrac{7}{1}$ **3.** $\dfrac{1}{3}\left(\dfrac{4}{5} \cdot \dfrac{1}{3}\right) = \dfrac{1}{3}\left(\dfrac{4}{15}\right)$ **4.** $\dfrac{1}{4}(4y) = \left(\dfrac{1}{4} \cdot 4\right)y$

$\qquad\qquad\qquad = -\dfrac{14}{5} \qquad\qquad\qquad = \dfrac{4}{45} \qquad\qquad\qquad = 1 \cdot y$

$\qquad\qquad\qquad\qquad\qquad\qquad\qquad\qquad\qquad\qquad\qquad\qquad = y$

5. $\dfrac{12}{25} \cdot \dfrac{5}{6} = \dfrac{12 \cdot 5}{25 \cdot 6}$ **6.** $\dfrac{8}{3} \cdot \dfrac{9}{24} = \dfrac{8 \cdot 9}{3 \cdot 24}$ **7.** $\dfrac{yz^2}{x} \cdot \dfrac{x^3}{yz} = \dfrac{\cancel{y} \cdot z \cdot z \cdot x \cdot x \cdot x}{x \cdot \cancel{y} \cdot z}$ **8.** $\dfrac{3}{4} \cdot \dfrac{8}{3} \cdot \dfrac{1}{6} = \dfrac{3 \cdot 8 \cdot 1}{4 \cdot 3 \cdot 6}$

$\quad = \dfrac{(2 \cdot 2 \cdot 3) \cdot \cancel{5}}{(\cancel{5} \cdot 5) \cdot (2 \cdot 3)}$ $\quad = \dfrac{(2 \cdot 2 \cdot 2) \cdot (3 \cdot 3)}{3 \cdot (2 \cdot 2 \cdot 2 \cdot 3)}$ $\quad = \dfrac{z \cdot x \cdot x}{1}$ $\quad = \dfrac{3 \cdot (2 \cdot 2 \cdot 2) \cdot 1}{(2 \cdot 2) \cdot 3 \cdot (2 \cdot 3)}$

$\quad = \dfrac{2}{5}$ $\quad = \dfrac{1}{1}$ $\quad = x^2z$ $\quad = \dfrac{1}{3}$

$\qquad\qquad\qquad = 1$

9. $\left(\dfrac{2}{3}\right)^2 = \dfrac{2}{3} \cdot \dfrac{2}{3}$ **10.** $\left(\dfrac{3}{4}\right)^2 \cdot \dfrac{1}{2} = \dfrac{3}{4} \cdot \dfrac{3}{4} \cdot \dfrac{1}{2}$ **11.** $\dfrac{2}{3} \cdot \dfrac{1}{2} = \dfrac{2 \cdot 1}{3 \cdot 2}$ **12.** $\dfrac{2}{3}(-12) = \dfrac{2}{3}\left(-\dfrac{12}{1}\right)$ **13.** $A = \dfrac{1}{2}(7)(10)$

$\quad = \dfrac{4}{9}$ $\quad = \dfrac{9}{32}$ $\quad = \dfrac{1}{3}$ $\quad = -\dfrac{2 \cdot 2 \cdot 2 \cdot 3}{3 \cdot 1}$ $\quad = 35 \text{ in}^2$

$\qquad\qquad\qquad\qquad\qquad\qquad\qquad\qquad\qquad\qquad\qquad\qquad = -\dfrac{8}{1}$

$\qquad\qquad\qquad\qquad\qquad\qquad\qquad\qquad\qquad\qquad\qquad\qquad = -8$

14.

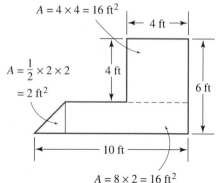

$A = 4 \times 4 = 16 \text{ ft}^2$

$A = \dfrac{1}{2} \times 2 \times 2$

$= 2 \text{ ft}^2$

$A = 8 \times 2 = 16 \text{ ft}^2$

Total area $= 2 + 16 + 16 = 34 \text{ ft}^2$

Section 3.4

1. $\dfrac{1}{3} \div \dfrac{1}{6} = \dfrac{1}{3} \cdot \dfrac{6}{1}$

$= \dfrac{6}{3}$

$= 2$

2. $\dfrac{5}{9} \div \dfrac{10}{3} = \dfrac{5}{9} \cdot \dfrac{3}{10}$

$= \dfrac{\cancel{5} \cdot \cancel{3}}{\cancel{3} \cdot 3 \cdot 2 \cdot \cancel{5}}$

$= \dfrac{1}{6}$

3. $-\dfrac{3}{4} \div 3 = -\dfrac{3}{4} \cdot \dfrac{1}{3}$

$= -\dfrac{1}{4}$

4. $-4 \div \left(-\dfrac{1}{5}\right) = -4(-5)$

$= 20$

5. $\dfrac{5}{32} \div \dfrac{10}{42} = \dfrac{5}{32} \cdot \dfrac{42}{10}$

$= \dfrac{\cancel{5} \cdot (\cancel{2} \cdot 3 \cdot 7)}{2 \cdot 2 \cdot 2 \cdot 2 \cdot 2 \cdot \cancel{2} \cdot \cancel{5}}$

$= \dfrac{21}{32}$

6. $\dfrac{12}{25} \div 6 = \dfrac{12}{25} \cdot \dfrac{1}{6}$

$= \dfrac{2 \cdot 2 \cdot 3 \cdot 1}{5 \cdot 5 \cdot 2 \cdot 3}$

$= \dfrac{2}{25}$

7. $-12 \div \left(-\dfrac{4}{3}\right) = -12\left(-\dfrac{3}{4}\right)$

$= 9$

8. $\dfrac{x^3}{y} \div \dfrac{x^2}{y^2} = \dfrac{x^3}{y} \cdot \dfrac{y^2}{x^2}$

$= \dfrac{x \cdot x \cdot x \cdot y \cdot \cancel{y}}{\cancel{y} \cdot x \cdot x}$

$= \dfrac{x \cdot y}{1}$

$= xy$

9. $\dfrac{5}{4} \div \dfrac{1}{8} + 8 = \dfrac{5}{4} \cdot \dfrac{8}{1} + 8$

$= 10 + 8$

$= 18$

10. $18 \div \left(\dfrac{3}{5}\right)^2 + 48 \div \left(\dfrac{2}{5}\right)^2 = 18 \div \dfrac{9}{25} + 48 \div \dfrac{4}{25}$

$= 18 \cdot \dfrac{25}{9} + 48 \cdot \dfrac{25}{4}$

$= 50 + 300$

$= 350$

11. $12 \div \dfrac{3}{4} = 12 \cdot \dfrac{4}{3}$

$= 4 \cdot 4$

$= 16$ blankets

Section 3.5

1. $\dfrac{3}{10} + \dfrac{1}{10} = \dfrac{3+1}{10}$

$= \dfrac{4}{10}$

$= \dfrac{2}{5}$

2. $\dfrac{a-5}{12} + \dfrac{3}{12} = \dfrac{a-5+3}{12}$

$= \dfrac{a-2}{12}$

3. $\dfrac{8}{x} - \dfrac{5}{x} = \dfrac{8-5}{x}$

$= \dfrac{3}{x}$

4. $\dfrac{5}{9} - \dfrac{8}{9} + \dfrac{5}{9} = \dfrac{5-8+5}{9}$

$= \dfrac{2}{9}$

5. $\left.\begin{array}{l} 18 = 2 \cdot 3 \cdot 3 \\ 14 = 2 \cdot 7 \end{array}\right\}$ $\text{LCD} = 2 \cdot 3 \cdot 3 \cdot 7$

$= 126$

6. $\dfrac{5}{18} + \dfrac{3}{14} = \dfrac{5 \cdot 7}{18 \cdot 7} + \dfrac{3 \cdot 9}{14 \cdot 9}$

$= \dfrac{35}{126} + \dfrac{27}{126}$

$= \dfrac{62}{126}$

$= \dfrac{31}{63}$

7. $\left.\begin{array}{l} 9 = 3 \cdot 3 \\ 15 = 3 \cdot 5 \end{array}\right\}$ $\text{LCD} = 3 \cdot 3 \cdot 5$

$= 45$

8. $\dfrac{2}{9} + \dfrac{4}{15} = \dfrac{2 \cdot 5}{9 \cdot 5} + \dfrac{4 \cdot 3}{15 \cdot 3}$

$= \dfrac{10}{45} + \dfrac{12}{45}$

$= \dfrac{22}{45}$

9. $\text{LCD} = 100;$ $\dfrac{8}{25} - \dfrac{3}{20} = \dfrac{8 \cdot 4}{25 \cdot 4} - \dfrac{3 \cdot 5}{20 \cdot 5}$

$= \dfrac{32}{100} - \dfrac{15}{100}$

$= \dfrac{17}{100}$

10. $\text{LCD} = 20;$ $\dfrac{x}{4} - \dfrac{1}{5} = \dfrac{x \cdot 5}{4 \cdot 5} - \dfrac{1 \cdot 4}{5 \cdot 4}$

$= \dfrac{5x}{20} - \dfrac{4}{20}$

$= \dfrac{5x - 4}{20}$

11. $\text{LCD} = 36;$ $\dfrac{1}{9} + \dfrac{1}{4} + \dfrac{1}{6} = \dfrac{1 \cdot 4}{9 \cdot 4} + \dfrac{1 \cdot 9}{4 \cdot 9} + \dfrac{1 \cdot 6}{6 \cdot 6}$

$= \dfrac{4}{36} + \dfrac{9}{36} + \dfrac{6}{36}$

$= \dfrac{19}{36}$

12. $2 - \dfrac{3}{4} = \dfrac{2}{1} - \dfrac{3}{4}$

$= \dfrac{2 \cdot 4}{1 \cdot 4} - \dfrac{3}{4}$

$= \dfrac{8}{4} - \dfrac{3}{4}$

$= \dfrac{5}{4}$

13. $\dfrac{5}{x} + \dfrac{2}{3} = \dfrac{5 \cdot 3}{x \cdot 3} + \dfrac{2 \cdot x}{3 \cdot x}$

$= \dfrac{15}{3x} + \dfrac{2x}{3x}$

$= \dfrac{15 + 2x}{3x}$

Section 3.6

1. $5\dfrac{2}{3} = 5 + \dfrac{2}{3}$

$= \dfrac{5}{1} + \dfrac{2}{3}$

$= \dfrac{5 \cdot 3}{1 \cdot 3} + \dfrac{2}{3}$

$= \dfrac{15}{3} + \dfrac{2}{3}$

$= \dfrac{17}{3}$

2. $3\dfrac{1}{6} = 3 + \dfrac{1}{6}$

$= \dfrac{3}{1} + \dfrac{1}{6}$

$= \dfrac{3 \cdot 6}{1 \cdot 6} + \dfrac{1}{6}$

$= \dfrac{18}{6} + \dfrac{1}{6}$

$= \dfrac{19}{6}$

3. $5\dfrac{2}{3} = \dfrac{(3 \cdot 5) + 2}{3}$

$= \dfrac{17}{3}$

4. $6\dfrac{4}{9} = \dfrac{(9 \cdot 6) + 4}{9}$

$= \dfrac{58}{9}$

5. $\begin{array}{r} 3 \\ 3\overline{)11} \\ \underline{9} \\ 2 \end{array}$ so $\dfrac{11}{3} = 3\dfrac{2}{3}$

6.
$$5\overline{)14}$$
$$\underline{10}$$
$$4$$
so $\dfrac{14}{5} = 2\dfrac{4}{5}$

7.
$$26\overline{)207}$$
$$\underline{182}$$
$$25$$
so $\dfrac{207}{26} = 7\dfrac{25}{26}$

8. $3 + \dfrac{2}{x} = \dfrac{3}{1} + \dfrac{2}{x}$

$= \dfrac{3 \cdot x}{1 \cdot x} + \dfrac{2}{x}$

$= \dfrac{3x}{x} + \dfrac{2}{x}$

$= \dfrac{3x + 2}{x}$

9. $x - \dfrac{2}{3} = \dfrac{x}{1} - \dfrac{2}{3}$

$= \dfrac{x \cdot 3}{1 \cdot 3} - \dfrac{2}{3}$

$= \dfrac{3x}{3} - \dfrac{2}{3}$

$= \dfrac{3x - 2}{3}$

Section 3.7

1. $2\dfrac{3}{4} \cdot 4\dfrac{1}{3} = \dfrac{11}{4} \cdot \dfrac{13}{3}$

$= \dfrac{143}{12}$

$= 11\dfrac{11}{12}$

2. $2 \cdot 3\dfrac{5}{8} = \dfrac{2}{1} \cdot \dfrac{29}{8}$

$= \dfrac{58}{8}$

$= 7\dfrac{2}{8}$

$= 7\dfrac{1}{4}$

3. $1\dfrac{3}{5} \div 3\dfrac{2}{5} = \dfrac{8}{5} \div \dfrac{17}{5}$

$= \dfrac{8}{5} \cdot \dfrac{5}{17}$

$= \dfrac{8}{17}$

4. $4\dfrac{5}{8} \div 2 = \dfrac{37}{8} \div \dfrac{2}{1}$

$= \dfrac{37}{8} \cdot \dfrac{1}{2}$

$= \dfrac{37}{16}$

$= 2\dfrac{5}{16}$

Section 3.8

1. $3\dfrac{2}{3} + 2\dfrac{1}{4} = 3 + \dfrac{2}{3} + 2 + \dfrac{1}{4}$

$= (3 + 2) + \left(\dfrac{2}{3} + \dfrac{1}{4}\right)$

$= 5 + \left(\dfrac{2 \cdot 4}{3 \cdot 4} + \dfrac{1 \cdot 3}{4 \cdot 3}\right)$

$= 5 + \left(\dfrac{8}{12} + \dfrac{3}{12}\right)$

$= 5 + \dfrac{11}{12} = 5\dfrac{11}{12}$

2. $5\dfrac{3}{4} = 5\dfrac{3 \cdot 5}{4 \cdot 5} = 5\dfrac{15}{20}$

$+6\dfrac{4}{5} = 6\dfrac{4 \cdot 4}{5 \cdot 4} = 6\dfrac{16}{20}$

$11\dfrac{31}{20} = 11 + 1\dfrac{11}{20} = 12\dfrac{11}{20}$

3. $6\dfrac{3}{4} = 6\dfrac{3 \cdot 2}{4 \cdot 2} = 6\dfrac{6}{8}$

$+2\dfrac{7}{8} = 2\dfrac{7}{8} = 2\dfrac{7}{8}$

$8\dfrac{13}{8} = 9\dfrac{5}{8}$

4. $2\dfrac{1}{3} = 2\dfrac{1 \cdot 4}{3 \cdot 4} = 2\dfrac{4}{12}$

$1\dfrac{1}{4} = 1\dfrac{1 \cdot 3}{4 \cdot 3} = 1\dfrac{3}{12}$

$+3\dfrac{11}{12} = 3\dfrac{11}{12} = 3\dfrac{11}{12}$

$6\dfrac{18}{12} = 7\dfrac{6}{12} = 7\dfrac{1}{2}$

5. $4\dfrac{7}{8}$

$-1\dfrac{5}{8}$

$3\dfrac{2}{8} = 3\dfrac{1}{4}$

6. $12\dfrac{7}{10} = 12\dfrac{7}{10} = 12\dfrac{7}{10}$

$-7\dfrac{2}{5} = -7\dfrac{2 \cdot 2}{5 \cdot 2} = -7\dfrac{4}{10}$

$5\dfrac{3}{10}$

7. $10 = 9\dfrac{7}{7}$

$-5\dfrac{4}{7} = -5\dfrac{4}{7}$

$4\dfrac{3}{7}$

8. $6\dfrac{1}{3} = \left(5 + \dfrac{3}{3}\right) + \dfrac{1}{3} = 5\dfrac{4}{3}$

$-2\dfrac{2}{3} = \qquad -2\dfrac{2}{3} = -2\dfrac{2}{3}$

$3\dfrac{2}{3}$

9. $6\dfrac{3}{4} = 6\dfrac{3 \cdot 3}{4 \cdot 3} = 6\dfrac{9}{12} = 5\dfrac{21}{12}$

$-2\dfrac{5}{6} = -2\dfrac{5 \cdot 2}{6 \cdot 2} = -2\dfrac{10}{12} = -2\dfrac{10}{12}$

$3\dfrac{11}{12}$

10. $17\dfrac{1}{8} = 17\dfrac{1 \cdot 5}{8 \cdot 5} = 17\dfrac{5}{40} = 16\dfrac{45}{40}$

$-12\dfrac{4}{5} = -12\dfrac{4 \cdot 8}{5 \cdot 8} = -12\dfrac{32}{40} = -12\dfrac{32}{40}$

$4\dfrac{13}{40}$

11. $10{,}000\left(2\dfrac{1}{8} - 1\dfrac{15}{16}\right) = 10{,}000\left(\dfrac{3}{16}\right) = \dfrac{30{,}000}{16} = \$1{,}875$

Section 3.9

1. $4 + \left(1\dfrac{1}{2}\right)\left(2\dfrac{3}{4}\right) = 4 + \left(\dfrac{3}{2}\right)\left(\dfrac{11}{4}\right)$

$= 4 + \dfrac{33}{8}$

$= \dfrac{32}{8} + \dfrac{33}{8}$

$= \dfrac{65}{8}$

$= 8\dfrac{1}{8}$

2. $\left(\dfrac{2}{3} + \dfrac{1}{6}\right)\left(2\dfrac{5}{6} + 1\dfrac{1}{3}\right) = \left(\dfrac{5}{6}\right)\left(4\dfrac{1}{6}\right)$

$= \dfrac{5}{6}\left(\dfrac{25}{6}\right)$

$= \dfrac{125}{36}$

$= 3\dfrac{17}{36}$

3. $\dfrac{3}{7} + \dfrac{1}{3}\left(1\dfrac{1}{2} + 4\dfrac{1}{2}\right)^2 = \dfrac{3}{7} + \dfrac{1}{3}(6)^2$

$= \dfrac{3}{7} + \dfrac{1}{3}(36)$

$= \dfrac{3}{7} + 12$

$= 12\dfrac{3}{7}$

4. $\dfrac{\frac{2}{3}}{\frac{5}{9}} = \dfrac{2}{3} \div \dfrac{5}{9}$

$= \dfrac{2}{3} \cdot \dfrac{9}{5}$

$= \dfrac{18}{15}$

$= \dfrac{6}{5} = 1\dfrac{1}{5}$

5. $\dfrac{\frac{1}{2} + \frac{3}{4}}{\frac{2}{3} - \frac{1}{4}} = \dfrac{12\left(\frac{1}{2} + \frac{3}{4}\right)}{12\left(\frac{2}{3} - \frac{1}{4}\right)}$

$= \dfrac{12 \cdot \frac{1}{2} + 12 \cdot \frac{3}{4}}{12 \cdot \frac{2}{3} - 12 \cdot \frac{1}{4}}$

$= \dfrac{6 + 9}{8 - 3}$

$= \dfrac{15}{5} = 3$

6. $\dfrac{4 + \frac{2}{3}}{3 - \frac{1}{4}} = \dfrac{12\left(4 + \frac{2}{3}\right)}{12\left(3 - \frac{1}{4}\right)}$

$= \dfrac{12 \cdot 4 + 12 \cdot \frac{2}{3}}{12 \cdot 3 - 12 \cdot \frac{1}{4}}$

$= \dfrac{48 + 8}{36 - 3}$

$= \dfrac{56}{33} = 1\dfrac{23}{33}$

7. $\dfrac{12\frac{1}{3}}{6\frac{2}{3}} = 12\dfrac{1}{3} \div 6\dfrac{2}{3}$

$= \dfrac{37}{3} \div \dfrac{20}{3}$

$= \dfrac{37}{3} \cdot \dfrac{3}{20}$

$= \dfrac{37}{20}$

$= 1\dfrac{17}{20}$

Chapter 4

Section 4.1

1. $6(x + 4) = 6(x) + 6(4)$
$= 6x + 24$

2. $-3(2x + 4) = -3(2x) + (-3)(4)$
$= -6x + (-12)$
$= -6x - 12$

3. $6x - 2 + 3x + 8 = 6x + 3x + (-2) + 8$
$= 9x + 6$

4. $2(4x + 3) + 7 = 2(4x) + 2(3) + 7$
$= 8x + 6 + 7$
$= 8x + 13$

5. $3(2x + 1) + 5(4x - 3) = 3(2x) + 3(1) + 5(4x) - 5(3)$
$= 6x + 3 + 20x - 15$
$= 26x - 12$

6. a. $x = 90° - 45° = 45°$
b. $x = 180° - 60° = 120°$

Section 4.2

1.
When	$x = 3$
the equation	$5x - 4 = 11$
becomes	$5(3) - 4 = 11$
or	$15 - 4 = 11$
	$11 = 11$

2.
When	$a = -3$
the equation	$6a - 3 = 2a + 4$
becomes	$6(-3) - 3 = 2(-3) + 4$
	$-18 - 3 = -6 + 4$
	$-21 = -2$

This is a false statement, so $a = -3$ is not a solution.

3.
$x + 5 = -2$
$x + 5 + (-5) = -2 + (-5)$
$x + 0 = -7$
$x = -7$

4.
$a - 2 = 7$
$a - 2 + 2 = 7 + 2$
$a + 0 = 9$
$a = 9$

5.
$y + 6 - 2 = 8 - 9$
$y + 4 = -1$
$y + 4 + (-4) = -1 + (-4)$
$y + 0 = -5$
$y = -5$

6. $5x - 3 - 4x = 4 - 7$
$x - 3 = -3$
$x - 3 + 3 = -3 + 3$
$x + 0 = 0$
$x = 0$

7.
$-5 - 7 = x + 2$
$-12 = x + 2$
$-12 + (-2) = x + 2 + (-2)$
$-14 = x + 0$
$-14 = x$

8.
$a - \dfrac{2}{3} = \dfrac{5}{6}$
$a - \dfrac{2}{3} + \dfrac{2}{3} = \dfrac{5}{6} + \dfrac{2}{3}$
$a = \dfrac{9}{6} = \dfrac{3}{2}$

Section 4.3

1. $\dfrac{1}{3}x = 5$
$3 \cdot \dfrac{1}{3}x = 3 \cdot 5$
$x = 15$

2. $\dfrac{1}{5}a + 3 = 7$
$\dfrac{1}{5}a + 3 + (-3) = 7 + (-3)$
$\dfrac{1}{5}a = 4$
$5 \cdot \dfrac{1}{5}a = 5 \cdot 4$
$a = 20$

3. $\dfrac{3}{5}y = 6$
$\dfrac{5}{3} \cdot \dfrac{3}{5}y = \dfrac{5}{3} \cdot 6$
$y = 10$

4. $-\dfrac{3}{4}x = \dfrac{6}{5}$
$-\dfrac{4}{3}\left(-\dfrac{3}{4}x\right) = -\dfrac{4}{3} \cdot \dfrac{6}{5}$
$x = -\dfrac{8}{5}$

5. $6x = -42$
$\dfrac{6x}{6} = \dfrac{-42}{6}$
$x = -7$

6.
$$-5x + 6 = -14$$
$$-5x + 6 + (-6) = -14 + (-6)$$
$$-5x = -20$$
$$\frac{-5x}{-5} = \frac{-20}{-5}$$
$$x = 4$$

7.
$$3x - 7x + 5 = 3 - 18$$
$$-4x + 5 = -15$$
$$-4x + 5 + (-5) = -15 + (-5)$$
$$-4x = -20$$
$$\frac{-4x}{-4} = \frac{-20}{-4}$$
$$x = 5$$

8.
$$-5 + 4 = 2x - 11 + 3x$$
$$-1 = 5x - 11$$
$$-1 + 11 = 5x - 11 + 11$$
$$10 = 5x$$
$$\frac{10}{5} = \frac{5x}{5}$$
$$2 = x$$

Section 4.4

1.
$$4(x + 3) = -8$$
$$4x + 12 = -8$$
$$4x + 12 + (-12) = -8 + (-12)$$
$$4x = -20$$
$$\frac{4x}{4} = \frac{-20}{4}$$
$$x = -5$$

2.
$$6a + 7 = 4a - 3$$
$$6a + (-4a) + 7 = 4a + (-4a) - 3$$
$$2a + 7 = -3$$
$$2a + 7 + (-7) = -3 + (-7)$$
$$2a = -10$$
$$\frac{2a}{2} = \frac{-10}{2}$$
$$a = -5$$

3.
$$5(x - 2) + 3 = -12$$
$$5x - 10 + 3 = -12$$
$$5x - 7 = -12$$
$$5x - 7 + 7 = -12 + 7$$
$$5x = -5$$
$$\frac{5x}{5} = \frac{-5}{5}$$
$$x = -1$$

4.
$$3(4x - 5) + 6 = 3x + 9$$
$$12x - 15 + 6 = 3x + 9$$
$$12x - 9 = 3x + 9$$
$$12x + (-3x) - 9 = 3x + (-3x) + 9$$
$$9x - 9 = 9$$
$$9x - 9 + 9 = 9 + 9$$
$$9x = 18$$
$$\frac{9x}{9} = \frac{18}{9}$$
$$x = 2$$

5.
$$\frac{x}{3} + \frac{x}{6} = 9$$
$$6\left(\frac{x}{3} + \frac{x}{6}\right) = 6(9)$$
$$6\left(\frac{x}{3}\right) + 6\left(\frac{x}{6}\right) = 6(9)$$
$$2x + x = 54$$
$$3x = 54$$
$$x = 18$$

6.
$$3x + \frac{1}{4} = \frac{5}{8}$$
$$8\left(3x + \frac{1}{4}\right) = 8\left(\frac{5}{8}\right)$$
$$8(3x) + 8\left(\frac{1}{4}\right) = 8\left(\frac{5}{8}\right)$$
$$24x + 2 = 5$$
$$24x = 3$$
$$x = \frac{1}{8}$$

7.
$$\frac{4}{x} + 3 = \frac{11}{5}$$
$$5x\left(\frac{4}{x} + 3\right) = 5x\left(\frac{11}{5}\right)$$
$$5x\left(\frac{4}{x}\right) + 5x(3) = 5x\left(\frac{11}{5}\right)$$
$$20 + 15x = 11x$$
$$20 = -4x$$
$$-5 = x$$

Section 4.5

1. Step 1 *Read and list.*

Known items: The numbers 3 and 10

Unknown item: The number in question

Step 2 *Assign a variable and translate the information.*

Let x = the number asked for in the problem.

Then "The sum of a number and 3" translates to $x + 3$.

Step 3 *Reread and write an equation.*

The sum of x and 3 is 10.

$$x + 3 \qquad = 10$$

Step 4 *Solve the equation.*

$$x + 3 = 10$$
$$x = 7$$

Step 5 *Write your answer.*

The number is 7.

Step 6 *Reread and check.*

The sum of **7** and 3 is 10.

2. Step 1 *Read and list.*

Known items: The numbers 4 and 34, twice a number, and three times a number

Unknown item: The number in question

Step 2 *Assign a variable and translate the information.*

Let x = the number asked for in the problem.

Then "The sum of twice a number and three times the number" translates to $2x + 3x$.

Step 3 *Reread and write an equation.*

4 added to   the sum of twice a number   is   34
and three times the number

$$4 + \qquad 2x + 3x \qquad = 34$$

Step 4 *Solve the equation.*

$$4 + 2x + 3x = 34$$
$$5x + 4 = 34$$
$$5x = 30$$
$$x = 6$$

Step 5 *Write your answer.*

The number is 6.

Step 6 *Reread and check.*

Twice **6** is 12 and three times **6** is 18. Their sum is $12 + 18 = 30$. Four added to this is 34. Therefore, 4 added to the sum of twice **6** and three times **6** is 34.

3. **Step 1** *Read and list.*

   Known items: Length is twice width;
   perimeter 42 cm

   Unknown items: The length and the width

   Step 2 *Assign a variable and translate the information.*

   Let x = the width. Since the length is twice the width, the length must be $2x$. Here is a picture.

   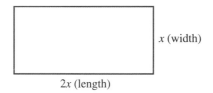

   x (width)

   $2x$ (length)

   Step 3 *Reread and write an equation.*

   The perimeter is the sum of the sides, and is also given as 42; therefore,

   $x + x + 2x + 2x = 42$

   Step 4 *Solve the equation.*

   $x + x + 2x + 2x = 42$

   $6x = 42$

   $x = 7$

   Step 5 *Write your answer.*

   The width is 7 centimeters and the length is $2(7) = 14$ centimeters.

   Step 6 *Reread and check.*

   The length, 14, is twice the width, 7. The perimeter is $7 + 7 + 14 + 14 = 42$ centimeters.

4. **Step 1** *Read and list.*

   Known items: Three angles are in a triangle. One is 3 times the smallest. The largest is 5 times the smallest.

   Unknown item: The three angles

   Step 2 *Assign a variable and translate the information.*

   Let x = the smallest angle. The other two angles are $3x$ and $5x$.

   Step 3 *Reread and write an equation.*

   The three angles must add up to 180°, so

   $x + 3x + 5x = 180°$

   Step 4 *Solve the equation.*

   $x + 3x + 5x = 180°$

   $9x = 180°$ Add similar terms on left side

   $x = 20°$ Divide each side by 9

   Step 5 *Write the answer.*

   The three angles are, 20°, $3(20°) = 60°$, and $5(20°) = 100°$.

   Step 6 *Reread and check.*

   The sum of the three angles is $20° + 60° + 100° = 180°$. One angle is 3 times the smallest, while the largest is 5 times the smallest.

5. **Step 1** *Read and list.*

   Known items: Joyce is 21 years older than Travis. Six years from now their ages will add to 49.

   Unknown items: Their ages now

   Step 2 *Assign a variable and translate the information.*

   Let x = Travis's age now; since Joyce is 21 years older than that, she is presently $x + 21$ years old.

   Step 3 *Reread and write an equation.*

	NOW	IN 6 YEARS
Joyce	$x + 21$	$x + 27$
Travis	x	$x + 6$

   $x + 27 + x + 6 = 49$

   Step 4 *Solve the equation.*

   $x + 27 + x + 6 = 49$

   $2x + 33 = 49$

   $2x = 16$

   $x = 8$

   Travis is now 8 years old,
   and Joyce is $8 + 21 = 29$ years old.

   Step 5 *Write your answer.*

   Travis is 8 years old and Joyce is 29 years old.

   Step 6 *Reread and check.*

   Joyce is 21 years older than Travis. In six years, Joyce will be 35 years old and Travis will be 14 years old. At that time, the sum of their ages will be $35 + 14 = 49$.

Section 4.6

1. When $P = 80$ and $w = 6$

the formula $P = 2l + 2w$

becomes $80 = 2l + 2(6)$

$80 = 2l + 12$

$68 = 2l$

$34 = l$

The length is 34 feet.

2. When $F = 77$

the formula $C = \dfrac{5}{9}(F - 32)$

becomes $C = \dfrac{5}{9}(77 - 32)$

$= \dfrac{5}{9}(45)$

$= \dfrac{5}{9} \cdot \dfrac{45}{1}$

$= \dfrac{225}{9}$

$= 25$ degrees Celsius

3. When $x = 0$

the formula $y = 2x + 6$

becomes $y = 2 \cdot 0 + 6$

$= 0 + 6$

$= 6$

4. When $x = -3$

the formula $2x + 3y = 4$

becomes $2(-3) + 3y = 4$

$-6 + 3y = 4$

$3y = 10$

$y = \dfrac{10}{3}$

5. With $x = 35°$ we use the formulas for finding the complement and the supplement of an angle:

The complement of $35°$ is $90° - 35° = 55°$

The supplement of $35°$ is $180° - 35° = 145°$

Section 4.7

1. When $x = 0$

the equation $3x + 5y = 15$

becomes $3 \cdot 0 + 5y = 15$

$5y = 15$

$y = 3$

which gives $(0, 3)$ as one solution

When $y = 0$

the equation $3x + 5y = 15$

becomes $3x + 5 \cdot 0 = 15$

$3x = 15$

$x = 5$

which means $(5, 0)$ is a second solution

When $x = -5$

the equation $3x + 5y = 15$

becomes $3(-5) + 5y = 15$

$-15 + 5y = 15$

$5y = 30$

$y = 6$

which gives $(-5, 6)$ as a third solution

2. When $x = 2$, we have

$5 \cdot 2 + 2y = 20$

$10 + 2y = 20$

$2y = 10$

$y = 5$

When $x = 0$, we have

$5 \cdot 0 + 2y = 20$

$2y = 20$

$y = 10$

When $y = 5$, we have

$5x + 2 \cdot 5 = 20$

$5x + 10 = 20$

$5x = 10$

$x = 2$

When $y = 0$, we have

$5x + 2 \cdot 0 = 20$

$5x = 20$

$x = 4$

3. When $x = 0$, we have

$y = \dfrac{1}{2} \cdot 0 + 1$

$y = 1$

When $x = 4$, we have

$y = \dfrac{1}{2} \cdot 4 + 1$

$y = 2 + 1$

$y = 3$

When $y = 7$, we have

$7 = \dfrac{1}{2}x + 1$

$6 = \dfrac{1}{2}x$

$12 = x$

When $y = -3$, we have

$-3 = \dfrac{1}{2}x + 1$

$-4 = \dfrac{1}{2}x$

$-8 = x$

4. $(1, 5)$ is not a solution. When we substitute 1 for x and 5 for y into $y = 5x - 6$, we get a false statement.

$5 = 5 \cdot 1 - 6$

$5 = 5 - 6$

$5 = -1$ a false statement

$(2, 4)$ is a solution. Substituting 2 for x and 4 for y in the equation $y = 5x - 6$ yields a true statement.

$4 = 5 \cdot 2 - 6$

$4 = 10 - 6$

$4 = 4$ a true statement

Section 4.8

1.

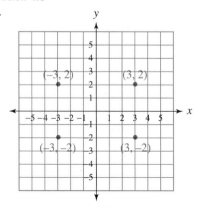

2.

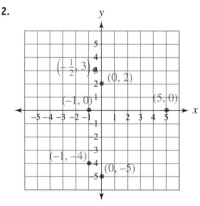

Section 4.9

1.

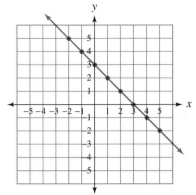

All the points shown on the graph have coordinates that add to 3.

2. When $x = 1, y = -2(1) + 1$
$$y = -2 + 1$$
$$y = -1$$
$(1, -1)$ is one solution
When $x = 0, y = -2(0) + 1$
$$y = 0 + 1$$
$$y = 1$$
$(0, 1)$ is a second solution
When $x = -1, y = -2(-1) + 1$
$$y = 2 + 1$$
$$y = 3$$
$(-1, 3)$ is our third solution

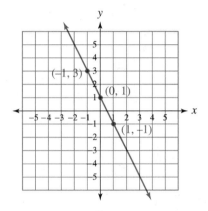

3. Let $x = -1: y = 2(-1) - 3$
$$y = -2 - 3$$
$$y = -5$$
$(-1, -5)$ is one solution
Let $x = 0: y = 2(0) - 3$
$$y = 0 - 3$$
$$y = -3$$
$(0, -3)$ is another solution
Let $x = 2: y = 2(2) - 3$
$$y = 4 - 3$$
$$y = 1$$
$(2, 1)$ is a third solution

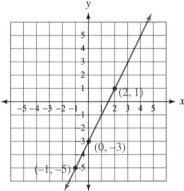

4. Let $x = 0: 3 \cdot 0 - 2y = 6$
$$0 - 2y = 6$$
$$-2y = 6$$
$$y = -3$$
$(0, -3)$ is one solution
Let $y = 0: 3x - 2 \cdot 0 = 6$
$$3x - 0 = 6$$
$$3x = 6$$
$$x = 2$$
$(2, 0)$ is a second solution
The third point is up to you to find.
Substituting -2, 4, or 6 for x will make your work easier. Do you know why?

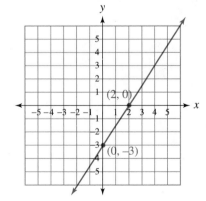

5. Any ordered pair with an x-coordinate of -3 is a
solution to the equation $x = -3$. Here are a few of them:
$(-3, -4)$, $(-3, -2)$, $(-3, 0)$, $(-3, 2)$, and $(-3, 4)$.

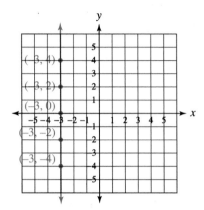

Chapter 5

Section 5.1

1. $700 + 80 + 5 + \dfrac{4}{10} + \dfrac{6}{100} + \dfrac{2}{1,000}$ **2. a.** Six hundredths   **b.** Seven tenths   **c.** Eight thousandths

3. a. Five and six hundredths   **b.** Four and seven tenths   **c.** Three and eight thousandths
4. Five and ninety-eight hundredths   **5.** Three hundred five and four hundred six thousandths

Section 5.2

1.
$$38.45 = 38\frac{45}{100} = 38\frac{450}{1,000}$$
$$+456.073 = 456\frac{73}{1,000} = 456\frac{73}{1,000}$$
$$494\frac{523}{1,000} = 494.523$$

2.   78.674
      -23.431
        55.243

3.   16.000
        0.033
        4.600
      $+ \ 0.080$
       20.713

4.   6.7000
      -2.0563
        4.6437

5.   7.000   10.567
      $+ \ 3.567$ and $- \ 5.890$
       10.567   4.677

8. $10.00   1 quarter + 2 dimes + 4 pennies = 0.25 + 0.20 + 0.04 = 0.49, which is too much change.
   $- \ \ 9.56$ One of the dimes should be a nickel. Tell the clerk that you have been given too much change.
   $ \ \ .44$

9.

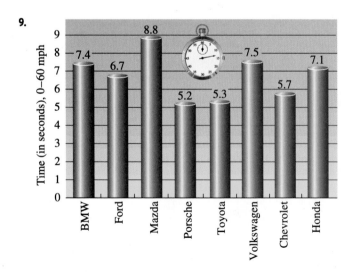

Section 5.3

1. $0.4 \times 0.6 = \dfrac{4}{10} \times \dfrac{6}{10}$

$= \dfrac{24}{100}$

$= 0.24$

2. $0.5 \times 0.007 = \dfrac{5}{10} \times \dfrac{7}{1,000}$

$= \dfrac{35}{10,000}$

$= 0.0035$

3. $3.5 \times 0.04 = 3\dfrac{5}{10} \times \dfrac{4}{100}$

$= \dfrac{35}{10} \times \dfrac{4}{100}$

$= \dfrac{140}{1,000}$

$= \dfrac{14}{100}$

$= 0.14$

4. $3 + 2 = 5$ digits to the right

5.
```
    4.03
  ×5.22
    806
   8060
 201500
 21.0366
```

8. **a.** $80 \times 6 = 480$
b. $40 \times 180 = 7,200$
c. $8^2 = 64$

9. $0.03(5.5 + 0.02) = 0.03(5.52)$
$= 0.1656$

10. $5.7 + 14(2.4)^2 = 5.7 + 14(5.76)$
$= 5.7 + 80.64$
$= 86.34$

11. $0.43 + 19(0.32) = 0.43 + 6.08$
$= \$6.51$

12. $6.32(36) + 9.48(14) = 227.52 + 132.72$
$= \$360.24$

13. $C = 3.14(3)$
$= 9.42$ cm

14. $C = 2(3.14)(5)$
$= 31.4$ ft

15. Radius $= \dfrac{1}{2}(20) = 10$ ft

$A = \pi r^2 = (3.14)(10)^2$
$= 314$ ft^2

16. Radius $= 2(0.125) = 0.250$

$V = \pi r^2 h$
$= (3.14)(0.250)^2(6)$
$= 1.178$ in^3

Section 5.4

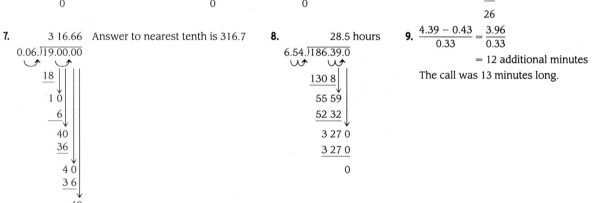

7. Answer to nearest tenth is 316.7

8. 28.5 hours

9. $\dfrac{4.39 - 0.43}{0.33} = \dfrac{3.96}{0.33}$

$= 12$ additional minutes

The call was 13 minutes long.

10.

CLASS	UNITS	GRADE	VALUE	GRADE POINTS
Algebra	5	B	3	$5 \times 3 = 15$
Chemistry	4	B	3	$4 \times 3 = 12$
English	3	A	4	$3 \times 4 = 12$
History	3	B	3	$3 \times 3 = 9$
Total Units:	15			Total Grade Points: 48

$\text{GPA} = \dfrac{48}{15} = 3.20$

Section 5.5

1. $\begin{array}{r} 0.6 \\ 5\overline{)3.0} \\ \underline{3\ 0} \\ 0 \end{array}$ so $\dfrac{3}{5} = 0.6$

2. $\begin{array}{r} 0.9166 \\ 12\overline{)11.0000} \\ \underline{10\ 8} \\ 20 \\ \underline{12} \\ 80 \\ \underline{72} \\ 80 \\ \underline{72} \\ 8 \end{array}$ so $\dfrac{11}{12} = 0.917$ to the nearest thousandth

3. $\begin{array}{r} 0.4545 \\ 11\overline{)5.0000} \\ \underline{4\ 4} \\ 60 \\ \underline{55} \\ 50 \\ \underline{44} \\ 60 \\ \underline{55} \\ 5 \end{array}$ so $\dfrac{5}{11} = 0.\overline{45}$

4. $0.48 = \dfrac{48}{100} = \dfrac{12}{25}$

5. $0.025 = \dfrac{25}{1,000} = \dfrac{1}{40}$

6. $12.8 = 12\dfrac{8}{10} = 12\dfrac{4}{5}$

7. $\dfrac{14}{25}(2.43 + 0.27) = 0.56(2.43 + 0.27)$
$= 0.56(2.70)$
$= 1.512$

8. $\dfrac{1}{4} + 0.25\left(\dfrac{3}{5}\right) = \dfrac{1}{4} + \dfrac{1}{4}\left(\dfrac{3}{5}\right)$
$= \dfrac{1}{4} + \dfrac{3}{20}$
$= \dfrac{5}{20} + \dfrac{3}{20}$
$= \dfrac{8}{20}$
$= \dfrac{2}{5}$ or 0.4

9. $\left(\dfrac{1}{3}\right)^3(5.4) + \left(\dfrac{1}{5}\right)^2(2.5) = \dfrac{1}{27}(5.4) + \dfrac{1}{25}(2.5)$
$= 0.2 + 0.1$
$= 0.3$

10. $35.50 - \dfrac{1}{4}(35.50) = \dfrac{3}{4}(35.50) = 26.625$
$= \$26.63$ to the nearest cent

11. $V = \pi r^2 h + \dfrac{1}{2} \cdot \dfrac{4}{3}\pi r^3$

$= (3.14)(10)^2(10) + \dfrac{1}{2} \cdot \dfrac{4}{3}(3.14)(10)^3$

$= 3,140 + \dfrac{2}{3}(3,140)$

$= 3,140 + 2,093.3$

$= 5,233.3 \text{ in}^3$

Section 5.6

1. $x - 3.4 = 6.7$
$x - 3.4 + \mathbf{3.4} = 6.7 + \mathbf{3.4}$
$x + 0 = 10.1$
$x = 10.1$

2. $4y = 3.48$
$\dfrac{4y}{\mathbf{4}} = \dfrac{3.48}{\mathbf{4}}$
$y = 0.87$

3. $\dfrac{1}{5}x - 2.4 = 8.3$
$\dfrac{1}{5}x - 2.4 + \mathbf{2.4} = 8.3 + \mathbf{2.4}$
$\dfrac{1}{5}x = 10.7$
$\mathbf{5}\left(\dfrac{1}{5}x\right) = \mathbf{5}(10.7)$
$x = 53.5$

4. $7a - 0.18 = 2a + 0.77$
$7a + (\mathbf{-2a}) - 0.18 = 2a + (\mathbf{-2a}) + 0.77$
$5a - 0.18 = 0.77$
$5a - 0.18 + \mathbf{0.18} = 0.77 + \mathbf{0.18}$
$5a = 0.95$
$\dfrac{5a}{5} = \dfrac{0.95}{5}$
$a = 0.19$

5. Let $x =$ the number of miles driven
$2(11) + 0.16x = 40.88$
$22 + 0.16x = 40.88$
$22 + (\mathbf{-22}) + 0.16x = 40.88 + (\mathbf{-22})$
$0.16x = 18.88$
$\dfrac{0.16x}{\mathbf{0.16}} = \dfrac{18.88}{\mathbf{0.16}}$
$x = 118$ miles

6. Let $x =$ the number of quarters. Then the number of dimes is $x + 7$.

	QUARTERS	DIMES
Number of	x	$x + 7$
Value of	$0.25x$	$0.10(x + 7)$

$0.25x + 0.10(x + 7) = 1.75$
$0.25x + 0.10x + 0.70 = 1.75$
$0.35x + 0.70 = 1.75$
$0.35x = 1.05$
$x = 3$

She has 3 quarters and 10 dimes.

Section 5.7

1. $4\sqrt{25} = 4 \cdot 5$
$= 20$

2. $\sqrt{36} + \sqrt{4} = 6 + 2$
$= 8$

3. $\sqrt{\dfrac{36}{100}} = \dfrac{6}{10}$
$= \dfrac{3}{5}$

4. $14\sqrt{36} = 14 \cdot 6$
$= 84$

5. $\sqrt{81} - \sqrt{25} = 9 - 5$
$= 4$

6. $\sqrt{\dfrac{64}{121}} = \dfrac{8}{11}$

7. $5\sqrt{14} \approx 5(3.7416574)$
$= 18.708287$
$= 18.7083$ to the nearest ten thousandth

8. $\sqrt{405} + \sqrt{147} \approx 20.124612 + 12.124356$
$= 32.248968$
$= 32.25$ to the nearest hundredth

9. $\sqrt{\dfrac{7}{12}} \approx \sqrt{0.5833333}$
≈ 0.7637626
$= 0.764$ to the nearest thousandth

10. a. $c = \sqrt{16^2 + 12^2}$
$= \sqrt{256 + 144}$
$= \sqrt{400}$
$c = 20$ cm

b. $c = \sqrt{5^2 + 5^2}$
$= \sqrt{25 + 25}$
$= \sqrt{50}$
$c = 7.07$ ft (to the nearest hundredth)

11. $c = \sqrt{12^2 + 5^2} = \sqrt{144 + 25}$
$= \sqrt{169}$
$= 13$ ft

Section 5.8

1. $\sqrt{63} = \sqrt{3 \cdot 3 \cdot 7}$
$= \sqrt{3 \cdot 3} \cdot \sqrt{7}$
$= 3\sqrt{7}$

2. $\sqrt{45x^2} = \sqrt{3 \cdot 3 \cdot 5 \cdot x \cdot x}$
$= 3 \cdot x \cdot \sqrt{5}$
$= 3x\sqrt{5}$

3. $\sqrt{300} = \sqrt{10 \cdot 10 \cdot 3}$
$= 10\sqrt{3}$

4. $\sqrt{50x^3} = \sqrt{5 \cdot 5 \cdot 2 \cdot x \cdot x \cdot x}$
$= 5 \cdot x \cdot \sqrt{2x}$
$= 5x\sqrt{2x}$

Section 5.9

1. $5\sqrt{3} - 2\sqrt{3} = (5 - 2)\sqrt{3} = 3\sqrt{3}$

2. $7\sqrt{y} + 3\sqrt{y} = (7 + 3)\sqrt{y} = 10\sqrt{y}$

3. $8\sqrt{5} - 2\sqrt{5} + 9\sqrt{5} = (8 - 2 + 9)\sqrt{5} = 15\sqrt{5}$

4. $\sqrt{27} + \sqrt{75} - \sqrt{12} = \sqrt{3 \cdot 3 \cdot 3} + \sqrt{5 \cdot 5 \cdot 3} - \sqrt{2 \cdot 2 \cdot 3}$
$= 3\sqrt{3} + 5\sqrt{3} - 2\sqrt{3}$
$= (3 + 5 - 2)\sqrt{3}$
$= 6\sqrt{3}$

5. $5\sqrt{45} - 3\sqrt{20} = 5\sqrt{3 \cdot 3 \cdot 5} - 3\sqrt{2 \cdot 2 \cdot 5}$
$= 5 \cdot 3\sqrt{5} - 3 \cdot 2\sqrt{5}$
$= 15\sqrt{5} - 6\sqrt{5}$
$= (15 - 6)\sqrt{5}$
$= 9\sqrt{5}$

6. $7\sqrt{18x^3} - 3\sqrt{50x^3} = 7\sqrt{3 \cdot 3 \cdot 2 \cdot x \cdot x \cdot x} - 3\sqrt{5 \cdot 5 \cdot 2 \cdot x \cdot x \cdot x}$
$= 7 \cdot 3 \cdot x\sqrt{2x} - 3 \cdot 5 \cdot x\sqrt{2x}$
$= 21x\sqrt{2x} - 15x\sqrt{2x}$
$= 6x\sqrt{2x}$

Chapter 6

Section 6.1

1. $\dfrac{32}{48} = \dfrac{2}{3}$

2. $\dfrac{\frac{3}{5}}{\frac{9}{10}} = \dfrac{3}{5} \cdot \dfrac{10}{9} = \dfrac{2}{3}$

3. $\dfrac{0.06}{0.12} = \dfrac{0.06 \times 100}{0.12 \times 100} = \dfrac{6}{12} = \dfrac{1}{2}$

4. $\dfrac{12}{16} = \dfrac{3}{4}$

5. Alcohol to water: $\dfrac{4}{12} = \dfrac{1}{3}$; water to alcohol: $\dfrac{12}{4} = \dfrac{3}{1}$; water to total solution: $\dfrac{12}{16} = \dfrac{3}{4}$

6. $\dfrac{107 \text{ miles}}{2 \text{ hours}} = 53.5$ miles/hour

7. $\dfrac{192 \text{ miles}}{6 \text{ gallons}} = 32$ miles/gallon

8. $\dfrac{48¢}{5.5 \text{ ounces}} = 8.7¢$/ounce; $\dfrac{75¢}{11.5 \text{ ounces}} = 6.5¢$/ounce; $\dfrac{219¢}{46 \text{ ounces}} = 4.8¢$/ounce
(Answers are rounded to the nearest tenth.)

Section 6.2

1. 8 ft $= 8 \times 12$ in.
$= 96$ in.

2. 26 ft $= 26 \text{ ft} \times \dfrac{1 \text{ yd}}{3 \text{ ft}}$
$= \dfrac{26}{3}$ yd
$= 8\dfrac{2}{3}$ yd, or 8.67 yd

3. 220 yd $= 220 \text{ yd} \times \dfrac{3 \text{ ft}}{1 \text{ yd}} \times \dfrac{12 \text{ in.}}{1 \text{ ft}}$
$= 220 \times 3 \times 12$ in.
$= 7{,}920$ in.

4. 67 cm $= 67 \text{ cm} \times \dfrac{1 \text{ m}}{100 \text{ cm}}$
$= \dfrac{67 \text{ m}}{100}$
$= 0.67$ m

5. 78.4 mm $= 78.4 \text{ mm} \times \dfrac{1 \text{ m}}{1{,}000 \text{ mm}} \times \dfrac{10 \text{ dm}}{1 \text{ m}}$
$= \dfrac{78.4 \times 10}{1{,}000}$ dm
$= 0.784$ dm

6. 6 pens $= 6 \text{ pens} \times \dfrac{28 \text{ feet of fencing}}{1 \text{ pen}} \times \dfrac{1.72 \text{ dollars}}{1 \text{ foot of fencing}}$
$= 6 \times 28 \times 1.72$ dollars
$= \$288.96$

7. $1{,}100$ feet per minute $= \dfrac{1{,}100 \text{ feet}}{1 \text{ minute}} \cdot \dfrac{1 \text{ mile}}{5{,}280 \text{ feet}} \cdot \dfrac{60 \text{ minutes}}{1 \text{ hour}}$

$= \dfrac{1{,}100 \cdot 60 \text{ miles}}{5{,}280 \text{ hours}}$

$= 12.5$ miles per hour, which is a reasonable speed for a chair lift.

Section 6.3

1. $1 \text{ yd}^2 = 1 \text{ yd} \times \text{yd} \times \dfrac{3 \text{ ft}}{1 \text{ yd}} \times \dfrac{3 \text{ ft}}{1 \text{ yd}} = 3 \times 3 \text{ ft} \times \text{ft} = 9 \text{ ft}^2$

2. Length $= 36$ in. $+ 12$ in. $= 48$ in.; Width $= 24$ in. $+ 12$ in. $= 36$ in.;
Area $= 48$ in. $\times 36$ in. $= 1{,}728 \text{ in}^2$

Area in square feet $= 1{,}728 \text{ in}^2 \times \dfrac{1 \text{ ft}^2}{144 \text{ in}^2} = \dfrac{1{,}728}{144} \text{ ft}^2 = 12 \text{ ft}^2$

3. $A = 1.5 \times 45$
$= 67.5 \text{ yd}^2$

$67.5 \text{ yd}^2 = 67.5 \text{ yd}^2 \times \dfrac{9 \text{ ft}^2}{1 \text{ yd}^2}$
$= 607.5 \text{ ft}^2$

4. 55 acres $= 55 \text{ acres} \times \dfrac{43{,}560 \text{ ft}^2}{1 \text{ acre}}$
$= 55 \times 43{,}560 \text{ ft}^2$
$= 2{,}395{,}800 \text{ ft}^2$

5. 960 acres $= 960 \text{ acres} \times \dfrac{1 \text{ mi}^2}{640 \text{ acres}}$
$= \dfrac{960}{640} \text{ mi}^2$
$= 1.5 \text{ mi}^2$

6. $1 \text{ m}^2 = 1 \text{ m}^2 \times \dfrac{100 \text{ dm}^2}{1 \text{ m}^2} \times \dfrac{100 \text{ cm}^2}{1 \text{ dm}^2}$
$= 10{,}000 \text{ cm}^2$

7. $5 \text{ gal} = 5 \text{ gal} \times \dfrac{4 \text{ qt}}{1 \text{ gal}} \times \dfrac{2 \text{ pt}}{1 \text{ qt}}$
$= 5 \times 4 \times 2 \text{ pt}$
$= 40 \text{ pt}$

8. $2{,}000 \text{ qt} = 2{,}000 \text{ qt} \times \dfrac{1 \text{ gal}}{4 \text{ qt}}$
$= \dfrac{2{,}000}{4} \text{ gal}$
$= 500 \text{ gal}$
The number of 10-gal containers in
500 gal is $\dfrac{500}{10} = 50$ containers.

9. 3.5 liters $= 3.5 \text{ liters} \times \dfrac{1{,}000 \text{ mL}}{1 \text{ liter}}$
$= 3.5 \times 1{,}000 \text{ mL}$
$= 3{,}500 \text{ mL}$

Section 6.4

1. $15 \text{ lb} = 15 \text{ lb} \times \dfrac{16 \text{ oz}}{1 \text{ lb}}$
$= 15 \times 16 \text{ oz}$
$= 240 \text{ oz}$

2. $2 \text{ lb } 4 \text{ oz} = 2 \text{ lb} + 4 \text{ oz} \times \dfrac{1 \text{ lb}}{16 \text{ oz}} = 2 \text{ lb} + \dfrac{4}{16} \text{ lb}$
$= 2 \text{ lb} + 0.25 \text{ lb}$
$= 2.25 \text{ lb}$
Total price for 2.25 lb is $\$2.09 \times 2.25 = \$4.70$ (rounded to nearest cent)

3. $5 \text{ kg} = 5 \text{ kg} \times \dfrac{1{,}000 \text{ g}}{1 \text{ kg}} \times \dfrac{1{,}000 \text{ mg}}{1 \text{ g}}$
$= 5 \times 1{,}000 \times 1{,}000 \text{ mg}$
$= 5{,}000{,}000 \text{ mg}$

4. Total number of milligrams in bottle $= 75 \times 200 = 15{,}000 \text{ mg}$

$15{,}000 \text{ mg} = 15{,}000 \text{ mg} \times \dfrac{1 \text{ g}}{1{,}000 \text{ mg}}$
$= \dfrac{15{,}000}{1{,}000} \text{ g}$
$= 15 \text{ g}$

Section 6.5

1. 10 in. $= 10 \text{ in.} \times \dfrac{2.54 \text{ cm}}{1 \text{ in.}}$
$= 10 \times 2.54 \text{ cm}$
$= 25.4 \text{ cm}$

2. $9 \text{ m} = 9 \text{ m} \times \dfrac{3.28 \text{ ft}}{1 \text{ m}}$
$= 9 \times 3.28 \text{ ft}$
$= 29.52 \text{ ft}$

3. $15 \text{ gal} = 15 \text{ gal} \times \dfrac{3.79 \text{ liters}}{1 \text{ gal}}$
$= 15 \times 3.79 \text{ liters}$
$= 56.85 \text{ liters}$

4. 2.2 liters $= 2.2 \text{ liters} \times \dfrac{1{,}000 \text{ mL}}{1 \text{ liter}} \times \dfrac{1 \text{ in}^3}{16.39 \text{ mL}}$

$= \dfrac{2.2 \times 1{,}000}{16.39} \text{ in}^3$

$= 134 \text{ in}^3$ (rounded to the nearest cubic inch)

5. $165 \text{ lb} = 165 \text{ lb} \times \dfrac{1 \text{ kg}}{2.20 \text{ lb}}$
$= \dfrac{165}{2.20} \text{ kg}$
$= 75 \text{ kg}$

6. $F = \dfrac{9}{5}(40) + 32$
$= 72 + 32$
$= 104°F$

7. $C = \dfrac{5(101.6 - 32)}{9}$

$= 38.7°C$ (rounded to the nearest tenth)

Section 6.6

1. First term = 2, second term = 3, third term = 6, fourth term = 9; means: 3 and 6; extremes: 2 and 9

2. a. $5 \cdot 18 = 90$ **b.** $13 \cdot 3 = 39$ **3.** $3 \cdot x = 4 \cdot 9$ **4.** $2 \cdot 19 = 8 \cdot y$ **5.** $15 \cdot n = 6(0.3)$ **6.** $\dfrac{3}{4} \cdot 8 = 7 \cdot x$

$6 \cdot 15 = 90$ $39 \cdot 1 = 39$ $3 \cdot x = 36$ $38 = 8 \cdot y$ $15 \cdot n = 1.8$ $6 = 7 \cdot x$

$$\frac{3 \cdot x}{3} = \frac{36}{3} \qquad \frac{38}{8} = \frac{8 \cdot y}{8} \qquad \frac{15 \cdot n}{15} = \frac{1.8}{15} \qquad \frac{6}{7} = \frac{7 \cdot x}{7}$$

$$x = 12 \qquad\qquad 4.75 = y \qquad\qquad n = 0.12 \qquad\qquad \frac{6}{7} = x$$

7. $\dfrac{x}{10} = \dfrac{288}{6}$ **8.** $\dfrac{x}{54} = \dfrac{10}{18}$ **9.** $\dfrac{x}{35} = \dfrac{8}{20}$ **10.** $\dfrac{x}{4.75} = \dfrac{105}{1}$ **11.** $\dfrac{x}{300} = \dfrac{8}{120}$

$x \cdot 6 = 10 \cdot 288$ $x \cdot 18 = 54 \cdot 10$ $x \cdot 20 = 35 \cdot 8$ $x \cdot 1 = 4.75(105)$ $x \cdot 120 = 300 \cdot 8$

$x \cdot 6 = 2{,}880$ $x \cdot 18 = 540$ $x \cdot 20 = 280$ $x = 498.75$ miles $x \cdot 120 = 2{,}400$

$$\frac{x \cdot 6}{6} = \frac{2{,}880}{6} \qquad \frac{x \cdot 18}{18} = \frac{540}{18} \qquad \frac{x \cdot 20}{20} = \frac{280}{20} \qquad\qquad \frac{x \cdot 120}{120} = \frac{2{,}400}{120}$$

$x = 480$ miles $x = 30$ hits $x = 14$ milliliters of alcohol $x = 20$ parts

Chapter 7

Section 7.1

21. $\dfrac{82}{100} = \dfrac{41}{50}$ **22.** $\dfrac{6.5}{100} = \dfrac{65}{1{,}000} = \dfrac{13}{200}$ **23.** $\dfrac{42\frac{1}{2}}{100} = \dfrac{\frac{85}{2}}{100}$ **24.** $\dfrac{65}{100} = \dfrac{13}{20}$ **25.** $\dfrac{35.6}{100} = \dfrac{356}{1{,}000} = \dfrac{89}{250}$ **26.** $\dfrac{9}{10} = 0.9 = 90\%$

$$= \frac{85}{2} \cdot \frac{1}{100}$$

$$= \frac{85}{200}$$

$$= \frac{17}{40}$$

27. $\dfrac{5}{8} = 0.625 = 62.5\%$ **28.** $\dfrac{7}{12} = 0.58\overline{3} = 58\dfrac{1}{3}\%$ or $\approx 58.3\%$ **29.** $3\dfrac{3}{4} = 3.75 = 375\%$

Section 7.2

1. $n = 0.25(74)$ **2.** $n \cdot 84 = 21$ **3.** $35 = 0.40 \cdot n$ **4.** $n = 0.635(45)$ **5.** $n \cdot 85 = 11.9$ **6.** $62 = 0.39 \cdot n$

$= 18.5$ $\dfrac{n \cdot 84}{84} = \dfrac{21}{84}$ $\dfrac{35}{0.40} = \dfrac{0.40 \cdot n}{0.40}$ $n \approx 28.6$ $\dfrac{n \cdot 85}{85} = \dfrac{11.9}{85}$ $\dfrac{62}{0.39} = \dfrac{0.39 \cdot n}{0.39}$

$n = 0.25$ $87.5 = n$ $n = 0.14$ $159.0 \approx n$

$n = 25\%$ $n = 14\%$

7. 25 is what percent of 160?

$25 = n \cdot 160$

$\dfrac{25}{160} = n$

$n = 0.156 = 15.6\%$ to the nearest tenth of a percent

Section 7.3

1. 114 is what percent of 150?   **2.** What is 75% of 40?   **3.** 3,360 is 42% of what number?   **4.** What is 35% of 300?

$114 = n \cdot 150$ $n = 0.75(40)$ $3{,}360 = 0.42 \cdot n$ $n = 0.35(300)$

$\dfrac{114}{150} = \dfrac{n \cdot 150}{150}$ $n = 30$ milliliters HCl $\dfrac{3{,}360}{0.42} = \dfrac{0.42 \cdot n}{0.42}$ $= 105$ students

$n = 0.76$ $n = 8{,}000$ students

$n = 76\%$

Section 7.4

1. What is 6% of $625?   **2.** $4.35 is 3% of what number?   **3.** $11.82 is what percent of $197?   **4.** What is 3% of $115,000?

$n = 0.06(625)$ $4.35 = 0.03 \cdot n$ $11.82 = n \cdot 197$ $n = 0.03(115{,}000)$

$n = \$37.50$ $n = \$145$ Purchase price $n = 0.06 = 6\%$ $n = \$3{,}450$

Total price $= \$145 + \$4.35$

$= \$149.35$

5. 10% of what number is $115?
$0.10 \cdot n = 115$
$n = \$1,150$

6. $105 is what percent of $750?
$105 = n \cdot 750$
$n = 0.14 = 14\%$

7. Let x = the retail price of the video equipment.
$x + 0.0725x = 6,864$
$1.0725x = 6,864$
$x = 6,400$
The tax on this is $0.0725(6,400) = \$464$.

Section 7.5

1. $0.07(18,000) = 1,260$
$\$18,000$ Old salary
$+\quad 1,260$ Raise
$\overline{\$19,260}$ New salary

2. $0.89(271,000) = 241,190$
$271,000$ Drunk drivers in 1986
$+241,190$ Increase
$\overline{512,190} = 512,000$ to the nearest thousand

3. $\$35 - \$28 = \$7$ Decrease
$7 is what percent of $35?
$7 = n \cdot 35$
$n = 0.20 = 20\%$ Decrease

4. What is 15% of $550?
$n = 0.15(550)$
$n = \$82.50$ Discount
$\$550.00$ Original price
$-\quad 82.50$ Less discount
$\overline{\$467.50}$ Sale price

5. What is 15% of $45?
$n = 0.15(45)$
$n = \$6.75$ Discount
$\$45.00$ Original price
$-\quad 6.75$ Less discount
$\overline{\$38.25}$ Sale price

What is 5% of $38.25?
$n = 0.05(38.25)$
$n = \$1.91$ to the nearest cent
$\$38.25$ Sale price
$+\quad 1.91$ Sales tax
$\overline{\$40.16}$ Total price

Section 7.6

1. Interest $= 0.08(\$3,000)$
$= \$240$
$\$3,000$ Principal
$+\quad 240$ Interest
$\overline{\$3,240}$ Amount after 1 year

2. Interest $= 0.12(\$7,500)$
$= \$900$
$\$7,500$ Principal
$+\quad 900$ Interest
$\overline{\$8,400}$ Total amount to pay back

3. $I = P \cdot R \cdot T$
$I = 700 \times 0.04 \times \dfrac{90}{360}$
$I = 700 \times 0.04 \times \dfrac{1}{4}$
$I = \$7$ Interest

4. $I = P \cdot R \cdot T$
$I = 1,200 \times 0.095 \times \dfrac{120}{360}$
$I = 1,200 \times 0.095 \times \dfrac{1}{3}$
$I = \$38$ Interest
$\$1,200$ Principal
$+\quad 38$ Interest
$\overline{\$1,238}$ Total amount withdrawn

5. Interest after 1 year is
$0.06(\$5,000) = \$300$
Total in account after 1 year is
$\$5,000$ Principal
$+\quad 300$ Interest
$\overline{\$5,300}$

Interest paid the second year is
$0.06(\$5,300) = \$318$
Total in account after 2 years is
$\$5,300$ Principal
$+\quad 318$ Interest
$\overline{\$5,618}$

6. Interest at the end of first quarter
$I = \$20,000 \times 0.08 \times \dfrac{1}{4} = \$400$
Total in account at end of first quarter
$\$20,000 + \$400 = \$20,400$
Interest for the second quarter
$I = \$20,400 \times 0.08 \times \dfrac{1}{4} = \$408$
Total in account at end of second quarter
$\$20,400 + \$408 = \$20,808$

Interest for the third quarter
$I = \$20,808 \times 0.08 \times \dfrac{1}{4} = \$416.16$
Total in account at the end of third quarter
$\$20,808 + \$416.16 = \$21,224.16$
Interest for the fourth quarter
$I = \$21,224.16 \times 0.08 \times \dfrac{1}{4} = \$424.48$ to the nearest cent
Total in account at end of 1 year
$\$21,224.16$
$+\quad 424.48$
$\overline{\$21,648.64}$

Answers to Odd-Numbered Problems

Chapter 1

Problem Set 1.1

1. 8 ones, 7 tens   **3.** 5 ones, 4 tens   **5.** 8 ones, 4 tens, 3 hundreds   **7.** 8 ones, 0 tens, 6 hundreds
9. 8 ones, 7 tens, 3 hundreds, 2 thousands   **11.** 9 ones, 6 tens, 5 hundreds, 3 thousands, 7 ten thousands, 2 hundred thousands
13. Ten thousands   **15.** Hundred millions   **17.** Ones   **19.** Hundred thousands   **21.** 600 + 50 + 8   **23.** 60 + 8
25. 4,000 + 500 + 80 + 7   **27.** 30,000 + 2,000 + 600 + 70 + 4   **29.** 3,000,000 + 400,000 + 60,000 + 2,000 + 500 + 70 + 7
31. 400 + 7   **33.** 30,000 + 60 + 8   **35.** 3,000,000 + 4,000 + 8   **37.** Twenty-nine   **39.** Forty   **41.** Five hundred seventy-three
43. Seven hundred seven   **45.** Seven hundred seventy   **47.** Twenty-three thousand, five hundred forty   **49.** Three thousand, four
51. Three thousand, forty   **53.** One hundred four million, sixty-five thousand, seven hundred eighty
55. Five billion, three million, forty thousand, eight   **57.** Two million, five hundred forty-six thousand, seven hundred thirty-one
59. 325   **61.** 5,432   **63.** 86,762   **65.** 2,000,200   **67.** 2,002,200   **69.** Thousands   **71.** 100,000 + 6,000 + 700 + 20 + 1
73. One million, three hundred eighty-four thousand, five hundred thirty

	POPULATION	
COUNTRY	**DIGITS**	**WORDS**
75. United States	275,000,000	Two hundred seventy-five million
77. Japan	127,000,000	One hundred twenty-seven million

83. 3 inches   **85.** 5 inches   **87.** 6   **89.** 3   **91.** 12

	POPULATION	
CITY	**DIGITS**	**WORDS**
79. Tokyo	30,000,000	Thirty million
81. Paris	8,800,000	Eight million, eight hundred thousand

Problem Set 1.2

1. 12   **3.** 14   **5.** 11   **7.** 13   **9.** 15   **11.** 16   **13.** 13   **15.** 18   **17.** 14   **19.** 12   **21.** 9 + 5   **23.** 8 + 3   **25.** 4 + 6
27. 5 + n   **29.** 1 + x   **31.** 1 + (2 + 3)   **33.** 2 + (1 + 6)   **35.** (1 + 9) + 1   **37.** 4 + (n + 1)   **39.** a + (b + c)   **41.** x + 11
43. a + 16   **45.** 5 + y   **47.** 16 + t   **49.** n = 5   **51.** n = 4   **53.** n = 5   **55.** n = 8   **57.** n = 9   **59.** n = 8
61.

FIRST NUMBER	SECOND NUMBER	THEIR SUM
a	b	a + b
1	4	5
2	3	5
3	2	5
4	1	5

63.

FIRST NUMBER	SECOND NUMBER	THEIR SUM
a	b	a + b
1	7	8
3	5	8
5	3	8
7	1	8

65. b   **67.** c   **69.** b   **71.** a   **73.** c

75. b   **77.** x = 5   **79.** a = 7   **81.** y = 10   **83.** x = 0   **85.** a = 3   **87.** t = 4   **89.** The sum of 4 and 9
91. The sum of 8 and 1   **93.** The sum of 2 and 3 is 5.   **95.** The sum of a and b is c.   **97.** 5 + 2   **99.** 5 + a   **101.** 6 + 8 = 14
103. 2 + 8   **105.** x + 4   **107.** t + 6
109. The commutative property allows us to change the order of numbers in a sum without changing the answer.
111. Subtraction is a commutative operation if changing the order of the numbers in a difference does not change the result. What do you think? Is 5 − 3 the same as 3 − 5?

Problem Set 1.3

1. 15   **3.** 14   **5.** 24   **7.** 15   **9.** 20   **11.** 68   **13.** 98   **15.** 7,297   **17.** 6,487   **19.** 96   **21.** 7,449   **23.** 65   **25.** 102
27. 875   **29.** 829   **31.** 10,391   **33.** 16,204   **35.** 155,554   **37.** 111,110   **39.** 17,391   **41.** 14,892   **43.** 180   **45.** 2,220
47. 18,285   **49.**

FIRST NUMBER	SECOND NUMBER	THEIR SUM
a	b	a + b
61	38	99
63	36	99
65	34	99
67	32	99

51.

FIRST NUMBER	SECOND NUMBER	THEIR SUM
a	b	a + b
9	16	25
36	64	100
81	144	225
144	256	400

53. 34 gallons   **55.** $349

57. $1,148   **59. a.** 34 **b.** 34 **c.** 34 **d.** 34   **61.** 12 in.   **63.** 16 ft   **65.** 26 yd   **67.** 18 in.   **69.** 76 in.   **71.** 168 ft

73. 11,619,201   **75.** 2,593,614   **77.** 51,341,722 **79.** 71,385,863   **81.** $w = 1$ and $l = 5$; $w = 2$ and $l = 4$; $w = 3$ and $l = 3$

83. Yes, when $w = l = 5$ ft   **85.** 5   **87.** 10

89. 14; An arithmetic sequence, starting with 5, in which each term comes from adding 3 to the term before it.

91. 85; The sequence starts with 10. Each number after that is 25 more than the number before it.

Problem Set 1.4

1. 40   **3.** 50   **5.** 50   **7.** 80   **9.** 460   **11.** 470   **13.** 56,780   **15.** 4,500   **17.** 500   **19.** 800   **21.** 900   **23.** 1,100

25. 5,000   **27.** 39,600   **29.** 5,000   **31.** 10,000   **33.** 1,000   **35.** 658,000   **37.** 510,000   **39.** 3,789,000

49. $1,400,000   **51.** 15,500,000   **53.** $4,680,000,000

| | ROUNDED TO THE NEAREST | | |
ORIGINAL NUMBER	TEN	HUNDRED	THOUSAND
41. 7,821	7,820	7,800	8,000
43. 5,999	6,000	6,000	6,000
45. 10,985	10,990	11,000	11,000
47. 99,999	100,000	100,000	100,000

55. $15,200   **57.** $31,000   **59.** 1,200   **61.** 1,900   **63.** 58,000   **65.** 160 miles per hour

67. Answers will vary, but 75 miles per hour is a good estimate.

69.

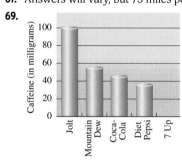

71.

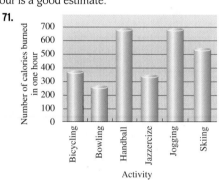

Problem Set 1.5

1. 32   **3.** 22   **5.** 10   **7.** 111   **9.** 312   **11.** 403   **13.** 1,111   **15.** 4,544   **17.** 15   **19.** 33   **21.** 5   **23.** 33   **25.** 95

27. 152   **29.** 274   **31.** 488   **33.** 538   **35.** 163   **37.** 1,610   **39.** 2,143

41.

FIRST NUMBER a	SECOND NUMBER b	THE DIFFERENCE OF a AND b a − b
25	15	10
24	16	8
23	17	6
22	18	4

43.

FIRST NUMBER a	SECOND NUMBER b	THE DIFFERENCE OF a AND b a − b
400	256	144
400	144	256
225	144	81
225	81	144

45. $255   **47.** 33 feet

49. 168 students   **51.** $574   **53.** $350   **55. a.**

YEAR	SALES ($ millions)
1996	386
1997	518
1998	573
1999	819

b. $301,000,000   **57.** 700   **59.** 3,000

61. 20,000   **63.** 50 miles per hour   **65.** Answers will vary, but 50 miles per hour is a good estimate.   **67.** 1,772,098

69. 1,062,000   **71.** 184,000

Problem Set 1.6

1. The product of 6 and 7   **3.** The product of 4 and 9   **5.** The product of 2 and n **7.** The product of 9 and 7 is 63.

9. The product of 2 and 9 is 18.   **11.** $8 \cdot 3$ **13.** $6 \cdot 6$ **15.** $7 \cdot n$ **17.** $6 \cdot 7 = 42$ **19.** $0 \cdot 6 = 0$ **21.** Products: $9 \cdot 7$ and 63

23. Products: 4(4) and 16   **25.** Factors: 2, 3, and 4   **27.** Factors: 2, 2, and 3   **29.** 9(5)   **31.** $7 \cdot 6$ **33.** 0(1)   **35.** $n \cdot 3$

37. $(2 \cdot 7) \cdot 6$ **39.** $(3 \times 9) \times 1$ **41.** $(2 \cdot 6) \cdot 5$ **43.** $(3 \cdot 4) \cdot n$ **45.** d   **47.** a   **49.** c   **51.** c   **53.** c   **55.** $42x$ **57.** $27x$

59. $28a$ **61.** $18y$ **63.** $n = 3$ **65.** $n = 9$ **67.** $n = 0$ **69.** $n = 8$ **71.** $n = 3$ **73.** $n = 7$ **75.** 60   **77.** 150   **79.** 900

81. 300   **83.** 600   **85.** 3,000   **87.** 5,000   **89.** 21,000   **91.** 81,000   **93.** 2,400,000   **95.** 300   **97.** 30,000   **99.** 720

101. 720,000   **103.** 400   **105.** $270   **107.** 225   **109.** 45   **111.** 380   **113.** $1,800   **115.** $312   **117.** 1 and 12; 2 and 6; 3 and 4

119. The product of 8 and 5 is 27 more than the sum of 8 and 5.

121. The commutative property of multiplication tells us that we can change the order of two numbers being multiplied together without changing the result.

123. A factor of a number is a whole number that can be multiplied by another whole number to produce the original number. For example, 5 is a factor of 15, while 6 is not a factor of 15.

Problem Set 1.7

1. $7(2) + 7(3) = 35$ **3.** $9(4) + 9(7) = 99$ **5.** $3x + 3$ **7.** $4a + 36$ **9.** $15x + 20$ **11.** $48y + 12$ **13.** 100   **15.** 228   **17.** 36

19. 1,440   **21.** 950   **23.** 1,725   **25.** 121   **27.** 1,552   **29.** 4,200   **31.** 66,248   **33.** 279,200   **35.** 12,321   **37.** 106,400

39. 198,592   **41.** 612,928   **43.** 333,180   **45.** 18,053,805   **47.** 263,646,976   **49.** 81,647,808

51.

FIRST NUMBER	SECOND NUMBER	THEIR PRODUCT
a	b	ab
11	11	121
11	22	242
22	22	484
22	44	968

53.

FIRST NUMBER	SECOND NUMBER	THEIR PRODUCT
a	b	ab
25	15	375
25	30	750
50	15	750
50	30	1,500

55. $3(3 + 4) = 3 \cdot 3 + 3 \cdot 4$

57. 5,060 miles   **59.** 42 animals   **61.** $11,400   **63.** $250   **65.** 2,081 calories   **67.** 280 calories   **69.** Yes   **71.** 8,000

73. 1,500,000   **75.** 1,400,000   **77.** 16,978,130   **79.** 43,605,972   **81.** 462,672 calories

83. 40; the sequence starts with 5. Each number after that comes from the number before it by multiplying by 2 each time.

85. 54; the sequence starts with 2. Each number after that is 3 times as large as the number before it.   **87.** Arithmetic

89. Geometric   **91.** Neither

Problem Set 1.8

1. $6 \div 3$ **3.** $45 \div 9$ **5.** $r \div s$ **7.** $20 \div 4 = 5$ **9.** $2 \cdot 3 = 6$ **11.** $9 \cdot 4 = 36$ **13.** $6 \cdot 8 = 48$ **15.** $7 \cdot 4 = 28$ **17.** 5   **19.** 8

21. Undefined   **23.** 45   **25.** 23   **27.** 1,530   **29.** 1,350   **31.** 18,000   **33.** 16,680   **35.** a   **37.** b   **39.** 1   **41.** 2   **43.** 4

45. 6   **47.** 45   **49.** 49   **51.** 432   **53.** 1,438   **55.**

FIRST NUMBER	SECOND NUMBER	THE QUOTIENT OF a AND b
a	b	$\dfrac{a}{b}$
100	25	4
100	26	3 R 22
100	27	3 R 19
100	28	3 R 16

57. 61 R 4   **59.** 90 R 1

61. 13 R 7   **63.** 234 R 6   **65.** 452 R 4   **67.** 35 R 35   **69.** $1,850   **71.** 16¢   **73.** 8 glasses   **75.** 6 glasses with 2 oz left over

77. 3 bottles   **79.** $12   **81.** 665 mg   **83.** 437   **85.** 3,247   **87.** 869   **89.** 5,684   **91.** 169 gal

Problem Set 1.9

1. Base 4; Exponent 5   **3.** Base 3; Exponent 6   **5.** Base 8; Exponent 2   **7.** Base 9; Exponent 1   **9.** Base 4; Exponent 0

11. 36   **13.** 8   **15.** 1   **17.** 1   **19.** 81   **21.** 10   **23.** 12   **25.** 1   **27.** 43   **29.** 16   **31.** 84   **33.** 416   **35.** 66   **37.** 21

39. 7   **41.** 124   **43.** 11   **45.** 91   **47.** $8(4 + 2) = 48$ **49.** $2(10 + 3) = 26$ **51.** $3(3 + 4) + 4 = 25$ **53.** $(20 \div 2) - 9 = 1$

55. $(8 \cdot 5) + (5 \cdot 4) = 60$ **57.** 255 calories   **59.** 465 calories   **61.** 30 calories   **63.** Big Mac has twice the calories.   **65.** 3

67. 6   **69.** 11   **71.** 50   **73.** 18   **75.** $16,000   **77.** 83   **79.** 5,932   **81.** 775,029   **83.** 63,259   **85.** 4   **87.** 16

91. 13, 21, 34

Problem Set 1.10

1. 25 in²   **3.** 84 m²   **5.** 60 ft²   **7.** 192 cm²   **9.** 2,200 ft²   **11.** 945 cm²   **13.** 64 cm³   **15.** 420 ft³   **17.** 162 in³   **19.** 96 cm²

21. 148 ft²   **23.** 7 ft   **25.** 9 ft   **27.** 124 tiles   **29.** 288 ft²

31. The area increases from 25 ft² to 49 ft², which is an increase of 24 ft².

33. a.

b.

PERIMETERS OF SQUARES		AREAS OF SQUARES	
LENGTH OF EACH SIDE (IN CENTIMETERS)	PERIMETER (IN CENTIMETERS)	LENGTH OF EACH SIDE (IN CENTIMETERS)	AREA (IN SQUARE CENTIMETERS)
1	4	1	1
2	8	2	4
3	12	3	9
4	16	4	16

c.

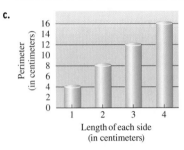 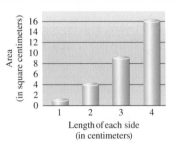

35. 12,321 ft² **37.** Volume = 1,331 ft³; surface area = 726 ft² **39.** 1,356,531 ft³

Chapter 1 Review

1. one thousand, three hundred seventy-six **3.** 5,245,652 **5.** 900,000 + 10,000 + 4,000 + 900 + 9 **7.** d **9.** c **11.** b
13. g **15.** 749 **17.** 8,272 **19.** 314 **21.** 3,149 **23.** 584 **25.** 3,717 **27.** 173 **29.** 428 **31.** 3,781,090
33. 3,800,000 **35.** 79 **37.** 222 **39.** 79 **41.** 3(4 + 6) = 30 **43.** 2(17 − 5) = 24 **45.** $488
47. Smallest: 310 yd; Rose Bowl: 376 yd; Largest: 390 yd **49.** $2,032 **51.** $1,938 **53.** $183 **55.** Perimeter = 40 ft; area = 84 ft²
57. 1,470 calories **59.** 250 more calories **61.** No **63.** Answers will vary.

Chapter 1 Test

1. Twenty thousand, three hundred forty-seven **2.** 2,045,006 **3.** 100,000 + 20,000 + 3,000 + 400 + 7 **4.** f **5.** c **6.** a
7. e **8.** 876 **9.** 16,383 **10.** 524 **11.** 3,085 **12.** 1,674 **13.** 22,258 **14.** 85 **15.** 21 **16.** 520,000 **17.** 11 **18.** 4
19. a. $264,300 **b.** $142,000 **c.** $125,000
  d. The median is probably the most representative measure of the "center" of these home prices. The mean is too heavily influenced by the two expensive homes.
20. 2(11 + 7) = 36 **21.** (20 ÷ 5) + 9 = 13 **22.** **23.** $235

URBAN AREA	AVERAGE HOURS IN GRIDLOCK PER YEAR
Los Angeles	82
Washington	76
Seattle-Everett	69
Atlanta	68
Boston	66

24. Perimeter = 14 feet; area = 12 ft² **25.** Volume = 96 in³; surface area = 136 in²

Chapter 2

Problem Set 2.1

1. 4 is less than 7. **3.** 5 is greater than −2. **5.** −10 is less than −3. **7.** 0 is greater than −4. **9.** 30 > −30 **11.** −10 < 0
13. −3 > −15 **15.** −3 **17.** 2 **19.** −75 **21.** 0 **23.** 5 **25.** −100 **27.** < **29.** > **31.** < **33.** < **35.** > **37.** <
39. < **41.** < **43.** > **45.** < **47.** 2 **49.** 100 **51.** 8 **53.** 231 **55.** 6 **57.** 200 **59.** 8 **61.** 231 **63.** 2 **65.** 8
67. −2 **69.** −8 **71.** 0 **73.** Positive **75.** −100 **77.** −20 **79.** −360 **81.** −61°F, −51F° **83.** −5°F, −15°F **85.** −7°F

87. 10°F and 25-mph wind   **89.**

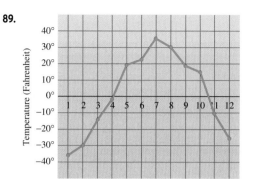

95. 0   **97.** 0   **99.** 5 + 3   **101.** (7 + 2) + 6

103. $x + 4$ **105.** $y + 5$ **107.** 4,313   **109.** 131,221   **111.** -3 **115.** The opposite of the absolute value of -3. It simplifies to -3.

Problem Set 2.2

1. 5   **3.** 1   **5.** -2 **7.** -6 **9.** 4   **11.** 4   **13.** -9 **15.** 15   **17.** -3 **19.** -11 **21.** -7 **23.** -3 **25.** -16 **27.** -8
29. -127 **31.** 49   **33.** 34   **35.**   **37.**   **39.** -4

35.

FIRST NUMBER	SECOND NUMBER	THEIR SUM
a	b	$a + b$
5	-3	2
5	-4	1
5	-5	0
5	-6	-1
5	-7	-2

37.

FIRST NUMBER	SECOND NUMBER	THEIR SUM
x	y	$x + y$
-5	-3	-8
-5	-4	-9
-5	-5	-10
-5	-6	-11
-5	-7	-12

41. 10   **43.** -445 **45.** 107   **47.** -1 **49.** -20 **51.** -17 **53.** -50 **55.** -7 **57.** 3   **59.** 50   **61.** -73 **63.** -11
65. 17   **67.** -21 **69.** -5 **71.** -4 **73.** 7   **75.** 10   **77.** a   **79.** b   **81.** d   **83.** c   **85.** \$10   **87.** \$74 + (−\$141) = −\$67
89. 3 + (−5) = −2   **91.** -7 and 13   **93.** $10 - x$ **95.** $y - 17$ **97.** 474   **99.** 8   **101.** 14   **103.** 730

Problem Set 2.3

1. 2   **3.** 2   **5.** -8 **7.** -5 **9.** 7   **11.** 12   **13.** 3   **15.** -7 **17.** -3 **19.** -13 **21.** -50 **23.** -100 **25.** 399
27. -21 **29.**   **31.**   **33.** -9

29.

FIRST NUMBER	SECOND NUMBER	THE DIFFERENCE OF x AND y
x	y	$x - y$
8	6	2
8	7	1
8	8	0
8	9	-1
8	10	-2

31.

FIRST NUMBER	SECOND NUMBER	THE DIFFERENCE OF x AND y
x	y	$x - y$
8	-6	14
8	-7	15
8	-8	16
8	-9	17
8	-10	18

35. -14 **37.** -1 **39.** 5   **41.** 11   **43.** -4 **45.** 8   **47.** 6   **49.** b   **51.** a   **53.** -100 **55.** b   **57.** a   **59.** 44°
61. $-\$22$ **63.** $-11 - (-22) = 11°F$ **65.** $3 - (-24) = 27°F$ **67.** $60 - (-26) = 86°F$ **69.** $-14 - (-26) = 12°F$ **71.** $3 \cdot 5$
73. $7x$ **75.** 5(3)   **77.** $(5 \cdot 7) \cdot 8$ **79.** $2(3) + 2(4) = 6 + 8 = 14$ **81.** 2,352   **83.** $3 - 5 \neq 5 - 3$
85. You have a balance of \$10 in your checkbook, and your write a check for \$12. Your new balance is −\$2.   **87.** $-5, -10$ **89.** 2, 6

Problem Set 2.4

1. -56 **3.** -60 **5.** 56   **7.** 81   **9.** -24 **11.** 30   **13.** -50 **15.** -24 **17.** 24   **19.** -6 **21.** 16   **23.** -125 **25.** 16
27.   **29.**   **31.**   **33.** -4

27.

NUMBER	SQUARE
x	x^2
-3	9
-2	4
-1	1
0	0
1	1
2	4
3	9

29.

FIRST NUMBER	SECOND NUMBER	THEIR PRODUCT
x	y	xy
6	2	12
6	1	6
6	0	0
6	-1	-6
6	-2	-12

31.

FIRST NUMBER	SECOND NUMBER	THEIR PRODUCT
a	b	ab
-5	3	-15
-5	2	-10
-5	1	-5
-5	0	0
-5	-1	5
-5	-2	10
-5	-3	15

35. 50   **37.** 1   **39.** -35 **41.** -22 **43.** -30 **45.** -25 **47.** 9   **49.** -13 **51.** 19   **53.** 6   **55.** -6 **57.** -4 **59.** -17

61. A gain of +$200   **63.** −16 degrees   **65.**

NATIONAL LEAGUE					
PITCHER, TEAM	**W**	**L**	**S**	**BS**	**PTS**
Trevor Hoffman, S.D.	3	1	39	1	119
Robb Nen, S.F.	7	3	31	3	95
Rod Beck, Chi.	1	2	36	6	94
Jeff Shaw, L.A.	2	5	36	6	90
Ugueth Urbina, Mon.	4	2	26	4	74
Billy Wagner, Hou.	3	3	23	2	65
Kerry Ligtenberg, Atl.	3	2	20	2	58
Mark Leiter, Phi.	6	3	22	8	56
Gregg Olson, Ari.	1	3	21	3	53
Bob Wickman, Mil.	5	6	20	4	50

67.

	VALUE	NUMBER	PRODUCT
Eagle	−2	0	0
Birdie	−1	7	−7
Par	0	7	0
Bogie	+1	3	+3
Double Bogie	+2	1	+2
		Total:	−2

If par for the course is 72, then his score was $72 − 2 = 70$.   **69.** a

71. d   **73.** a   **75.** b   **77.** $12 ÷ 6$ or $\dfrac{12}{6}$ **79.** $6 ÷ 3 = 2$ **81.** $5 · 2 = 10$ **83.** 89   **85.** a, c   **89.** −54, 162   **91.** 54, −162

93. 44   **95.** 19   **97.** −17

Problem Set 2.5

1. −3   **3.** −5   **5.** 3   **7.** 2   **9.** −4   **11.** −2   **13.** 0   **15.** 5   **17.**

FIRST NUMBER	SECOND NUMBER	THE QUOTIENT OF a AND b
a	b	$\dfrac{a}{b}$
100	−5	−20
100	−10	−10
100	−25	−4
100	−50	−2

19.

FIRST NUMBER	SECOND NUMBER	THE QUOTIENT OF a AND b
a	b	$\dfrac{a}{b}$
−100	−5	20
−100	5	−20
100	−5	−20
100	5	20

21. 1   **23.** −6   **25.** −2   **27.** −1   **29.** −1   **31.** 2   **33.** −3   **35.** −7

37. 30   **39.** 4   **41.** −5   **43.** −20   **45.** −5   **47.** −5   **49.** 35   **51.** 6   **53.** −1   **55.** c   **57.** a   **59.** d   **61.** 0

63. **a.** 0 **b.** 2 **c.** 2   **65.**   **67.** **a.** 2°F **b.** 5°F   **69.** $x + 3$ **71.** $(5 + 7) + a$

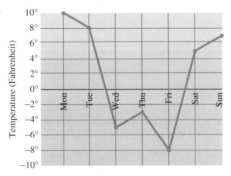

73. $(3 · 4)y$ **75.** $5(3) + 5(7)$ **77.** a **79.** x **81.** $2 ÷ 6 ≠ 6 ÷ 2$ **85.** −4   **87.** 4   **89.** 4   **91.** −1   **93.** −1

Problem Set 2.6

1. $20a$ **3.** $48a$ **5.** $−18x$ **7.** $−27x$ **9.** $−10y$ **11.** $−60y$ **13.** $5 + x$ **15.** $13 + x$ **17.** $10 + y$ **19.** $8 + y$ **21.** $5x + 6$
23. $6y + 7$ **25.** $12a + 21$ **27.** $7x + 28$ **29.** $7x + 35$ **31.** $6a − 42$ **33.** $2x − 2y$ **35.** $20 + 4x$ **37.** $6x + 15$ **39.** $18a + 6$
41. $12x − 6y$ **43.** $35 − 20y$ **45.** $8x$ **47.** $4a$ **49.** $4x$ **51.** $5y$ **53.** $−10a$ **55.** $−5x$ **57.** $36x^2$ **59.** $16x^2$ **61.** $27a^3$
63. $8a^3b^3$ **65.** $81x^2y^2$ **67.** $25x^2y^2z^2$ **69.** $1{,}296x^4y^2$ **71.** $576x^6$ **73.** $200x^5$ **75.** $7{,}200a^7$ **77.** $A = 36$ sq ft, $P = 24$ ft

79. $A = 81$ sq in., $P = 36$ in.   **81.** $A = 200$ sq in., $P = 60$ in.   **83.** $A = 300$ sq ft, $P = 74$ ft   **85.** 20°C   **87.** 5°C   **89.** −10°C
91. 4   **93.** −12   **95.** 12   **97.** −32   **99.** 32   **101.** −2

Problem Set 2.7

1. x^8 **3.** a^8 **5.** y^{15} **7.** x^5 **9.** x^9 **11.** a^{12} **13.** y^6 **15.** $14x^9$ **17.** $24y^8$ **19.** $21x^2$ **21.** $192x^{15}$ **23.** $24a^3$
25. $125y^{12}$ **27.** $6x^5y^8$ **29.** $56a^4b^5$ **31.** $24a^6b^6$ **33.** $28a^{12}b^{12}$ **35.** x^{10} **37.** a^{21} **39.** $125y^{12}$ **41.** $81x^8$ **43.** $8a^3b^3$
45. $625x^8y^{12}$ **47.** x^{14} **49.** a^{29} **51.** $72x^{12}y^{22}$ **53.** $175a^{12}b^9$ **55.** $a^{18}b^{34}$ **57.** 52 ft   **59.** 84 ft   **61.** 9 in.

Problem Set 2.8

1. $6x^2 + 6x + 8$ **3.** $5a^2 − 9a + 8$ **5.** $6x^3 + 4x^2 − 3x + 2$ **7.** $13t^2 − 2$ **9.** $33x^3 − 10x^2 − 4x − 4$ **11.** $2x^2 − 2x + 1$
13. $4y^3 − 8y^2 + 3y − 12$ **15.** $−4x − 4$ **17.** $x^2 − 30x + 64$ **19.** $7y^2 + 3y + 25$ **21.** 9   **23.** 0   **25.** 1   **27.** 2, 5, 8, 11
29. 4, 9, 16, 25   **31.** 10, 13, 18, 25   **33.**

n	1	2	3	4
$4n$	4	8	12	16

35.

n	1	2	3	4
$n − 3$	−2	−1	0	1

37. 169 ft^2 **39.** 210 in^2

41. 480 yd^2

Problem Set 2.9

1. $x^6 + x^5$ **3.** $x^6 + x^5$ **5.** $6x^6 − 14x^4$ **7.** $12x^6y^3 + 8x^2y^6$ **9.** $6x^6 − 14x^4$ **11.** $12x^2y^2 − 27xy$ **13.** $12x^2y^2 − 27xy$
15. $6x^5y^7 + 4x^6y^6 + 10x^4y^5$ **17.** $x^2 + 5x + 6$ **19.** $x^2 + x − 6$ **21.** $6x^2 + 19x + 10$ **23.** $3x^2 + 6x − 24$ **25.** $16x^2 − 9$
27. $12x^2 − x − 1$ **29.** $6x^2 − 23x + 20$ **31.** $x^2 + 6x + 8$ **33.** $6x^2 + 13x + 6$ **35.** $21x^2 + 34x + 8$ **37.** $x^2 + 6x + 9$
39. $x^2 + 12x + 36$ **41.** $9x^2 + 12x + 4$ **43.** $4x^2 + 12x + 9$ **45.** $49x^2 + 84x + 36$ **47.** $16x^2 + 8x + 1$ **49.** $x^2 − 10x + 25$
51. $4x^2 − 16x + 16$ **53.**

x	$(x + 3)^2$	$x^2 + 9$	$x^2 + 6x + 9$
1	16	10	16
2	25	13	25
3	36	18	36
4	49	25	49

55. $x^2 + 6x + 9$ **57.** $x^2 + 10x + 25$ **59.** $4x^2 + 12x + 9$

61. Volume $= 175$ in^3; surface area $= 190$ in^2 **63.** Volume $= 2,340$ ft^3; surface area $= 1,062$ ft^2

Chapter 2 Review

1. −17   **3.** >   **5.** >   **7.** 4   **9.** −2   **11.** −971   **13.** −17   **15.** −20   **17.** −3   **19.** 2   **21.** −8   **23.** 2   **25.** −42
27. −7   **29.** −11   **31.** −129   **33.** −2   **35.** 39   **37.** −19   **39.** True   **41.** False
43.

45. −1 and −15   **47.** 8 or 2   **49.** $24x$ **51.** $−30y$ **53.** $2x − 10$

55. $6a + 15b$ **57.** $125x^3$ **59.** x^8 **61.** $81x^4y^4$ **63.** $125x^{18}y^6$ **65.** $12x^2 − 3x − 3$ **67.** $3x − 12$ **69.** $x^5 + x^7$
71. $x^2 + 9x + 14$ **73.** $x^2 + 10x + 25$

Chapter 2 Test

1. −14   **2.** 5   **3.** >   **4.** >   **5.** 7   **6.** −2   **7.** −9   **8.** −6   **9.** −21   **10.** −36   **11.** 42   **12.** 54   **13.** −5   **14.** 5
15. 9   **16.** −8   **17.** −11   **18.** −15   **19.** 7   **20.** −4   **21.** −61   **22.** −7   **23.** 24   **24.** −5

25.

26. $-\$35$ **27.** $25°$ **28.** $30x^7$ **29.** $8a^5 - 12a^2$ **30.** $6x^2 + 2x - 20$

31. $x^2 + 6x + 9$ **32.** $5x^2 - 3x - 5$ **33.** $3x - 8$

Chapters 1-2 Cumulative Review

1. 7,714 **3.** 2,269 **5.** 45,084 **7.** $42a^5 - 54a^2$ **9.** $6x^2 + x - 40$ **11.** $64b^6$ **13.** 68 R 3 or 68 3/15 **15.** 4 **17.** 0

19. -11 **21.** -27 **23.** -15 **25.** 12 **27.** $9x^2 + 12x + 4$ **29.** $<$ **31.** $2(19 - 7) = 24$ **33.**

SPEED (MI/HR)	DISTANCE (FT)
20	22
30	**49**
40	**88**
50	137
60	198
70	269
80	**352**

35. $\$7$/hr **37.** 23 degrees **39.** $\$1,750$

Chapter 3

Problem Set 3.1

1. 1 **3.** 2 **5.** x **7.** a **9.** 5 **11.** 1 **13.** 12 **15.**

NUMERATOR a	DENOMINATOR b	FRACTION $\dfrac{a}{b}$
3	5	$\dfrac{3}{5}$
1	7	$\dfrac{1}{7}$
x	y	$\dfrac{x}{y}$
$x + 1$	x	$\dfrac{x + 1}{x}$

17. $\dfrac{3}{4}, \dfrac{1}{2}, \dfrac{9}{10}$

19. True **21.** False **23.**

HOW OFTEN WORKERS SEND NON-WORK-RELATED E-MAIL FROM THE OFFICE	FRACTION OF RESPONDENTS SAYING YES
never	$\dfrac{4}{25}$
1 to 5 times a day	$\dfrac{47}{100}$
5 to 10 times a day	$\dfrac{8}{25}$
more than 10 times a day	$\dfrac{1}{20}$

25. $\dfrac{3}{4}$ **27.** $\dfrac{43}{47}$ **29.** $\dfrac{4}{3}$ **31.** $\dfrac{13}{17}$ **33.** $\dfrac{4}{6}$ **35.** $\dfrac{5}{6}$

37. $\dfrac{8}{12}$ **39.** $\dfrac{8}{12}$ **41.**

YEAR	FRACTION	EQUIVALENT FRACTION
1985	$\dfrac{3}{20}$	$\dfrac{15}{100}$
1990	$\dfrac{4}{25}$	$\dfrac{16}{100}$
1995	$\dfrac{11}{50}$	$\dfrac{22}{100}$
2000	$\dfrac{8}{25}$	$\dfrac{32}{100}$

43. $\dfrac{2x}{12x}$ **45.** $\dfrac{9a}{24a}$ **47.** $\dfrac{20a}{24a}$

49.   Answers will vary

51. 3   **53.** 2   **55.** 37   **57.** $\frac{4}{5}$ **59.** $\frac{2}{15}$ **61.** $\frac{29}{43}$ **63.** $\frac{1,121}{1,791}$ **65.** $\frac{238}{527}$

67. a. $\frac{1}{2}$ **b.** $\frac{1}{2}$ **c.** $\frac{1}{4}$ **d.** $\frac{1}{4}$ **69.–77.**

79. d   **81.** a   **83.** Yes   **85.** $\frac{735}{2,205}$ **87.** 96

89. 1,896   **91.** 23   **93.** 32   **95.** 16   **97.** 18   **99.** False

Problem Set 3.2

1. Prime   **3.** Composite; 3, 5, and 7 are factors   **5.** Composite; 3 is a factor   **7.** Prime   **9.** $2^2 \cdot 3$ **11.** 3^4 **13.** $5 \cdot 43$

15. $3 \cdot 5$ **17.** $\frac{1}{2}$ **19.** $\frac{2}{3}$ **21.** $\frac{4}{5}$ **23.** $\frac{9}{5}$ **25.** $\frac{7}{11}$ **27.** $\frac{3x}{5}$ **29.** $\frac{1}{7}$ **31.** $\frac{7}{9}$ **33.** $\frac{7x}{5}$ **35.** $\frac{a}{5}$ **37.** $\frac{11}{7}$ **39.** $\frac{5z}{3}$

41. $\frac{8x}{9y}$ **43.** $\frac{42}{55}$ **45.** $\frac{17ac}{19b}$ **47.** $\frac{14}{33}$ **49. a.** $\frac{2}{17}$ **b.** $\frac{3}{26}$ **c.** $\frac{1}{9}$ **d.** $\frac{3}{28}$ **e.** $\frac{2}{19}$ **51. a.** $\frac{1}{45}$ **b.** $\frac{1}{30}$ **c.** $\frac{1}{18}$ **d.** $\frac{1}{15}$ **e.** $\frac{1}{10}$

53. a. $\frac{1}{3}$ **b.** $\frac{5}{6}$ **c.** $\frac{1}{5}$ **55.** $\frac{9}{16}$ **57.** $\frac{1}{4}$ **59.** $\frac{1}{6}$ **61.** $\frac{2}{3}$ **63–65.**

$$\frac{1}{2}=\frac{2}{4}=\frac{4}{8}=\frac{8}{16} \qquad \frac{3}{2}=\frac{6}{4}=\frac{12}{8}=\frac{24}{16}$$

67. b   **69.** c   **71.** $\frac{1}{3}$

$$\frac{1}{4}=\frac{2}{8}=\frac{4}{16} \qquad \frac{5}{4}=\frac{10}{8}=\frac{20}{16}$$

73. $\frac{1}{8}$ **75.** $\frac{3}{8}$ **77.** 410   **79.** 819   **81.** 7   **83.** 2   **85.** 72   **87.** 45   **89.** $5x + 20$ **91.** $12x + 15$

93. The answer must include the fact that the numerator and the denominator must be factored completely and then any factors common to them can be divided out.

95. Six of them: 3, 6, 9, 12, 15, and 18

97. No, because the square of a prime number will always be divisible by the original prime number.

Problem Set 3.3

1. $\frac{8}{15}$ **3.** $\frac{7}{8}$ **5.** -1 **7.** $\frac{27}{4}$ **9.** 1   **11.** $\frac{1}{24}$ **13.** $\frac{24}{125}$ **15.** 1

17.

FIRST NUMBER	SECOND NUMBER	THEIR PRODUCT
x	y	xy
$\frac{1}{2}$	$\frac{2}{3}$	$\frac{1}{3}$
$\frac{2}{3}$	$\frac{3}{4}$	$\frac{1}{2}$
$\frac{3}{4}$	$\frac{4}{5}$	$\frac{3}{5}$
$\frac{5}{a}$	$\frac{a}{6}$	$\frac{5}{6}$

19.

FIRST NUMBER	SECOND NUMBER	THEIR PRODUCT
x	y	xy
$\frac{1}{2}$	30	15
$\frac{1}{5}$	30	6
$\frac{1}{6}$	30	5
$\frac{1}{15}$	30	2

21. $\frac{1}{6}$ **23.** 9   **25.** 1   **27.** 8

29. $\frac{1}{15}$ **31.** $\frac{ac^2}{b}$ **33.** 9   **35.** 4   **37.** x **39.** y **41.** a **43.** $\frac{1}{4}$ **45.** $-\frac{8}{27}$ **47.** $\frac{1}{2}$ **49.** $\frac{9}{100}$ **51.** 3   **53.** 24   **55.** 4

57. 9   **59.** $\frac{3}{10}$; numerator should be 3, not 4.   **61. a.**

NUMBER	SQUARE
x	x^2
1	1
2	4
3	9
4	16
5	25
6	36
7	49

b. Either *larger* or *greater* will work.   **63.** 133 in^2

65. $\frac{4}{9}$ ft^2 **67.** 3 yd^2 **69.** 138 in^2 **71.** 3,476 students   **73.** $\frac{3}{8}$ cup

75. a.

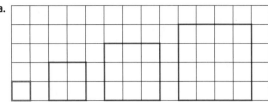

b.

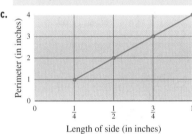

Table 1

PERIMETERS OF SQUARES

LENGTH OF EACH SIDE (IN INCHES)	PERIMETER (IN INCHES)
$\frac{1}{4}$	1
$\frac{2}{4} = \frac{1}{2}$	2
$\frac{3}{4}$	3
$\frac{4}{4} = 1$	4

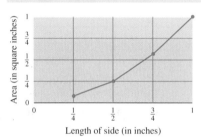

Table 2

AREAS OF SQUARES

LENGTH OF EACH SIDE (IN INCHES)	AREA (IN SQUARE INCHES)
$\frac{1}{4}$	$\frac{1}{16}$
$\frac{1}{2}$	$\frac{1}{4}$
$\frac{3}{4}$	$\frac{9}{16}$
1	1

c.

77. 2   **79.** 3   **81.** 0   **83.** $\frac{54,653}{59,136}$ **85.** $\frac{77,004}{5,141}$ **87.** 3   **89.** 2   **91.** 5   **93.** 3   **95.** $\frac{1}{27}$ **97.** $\frac{8}{27}$

Problem Set 3.4

1. $\frac{15}{4}$ **3.** $-\frac{4}{3}$ **5.** -9 **7.** 200   **9.** $-\frac{3}{8}$ **11.** 1   **13.** $\frac{49}{64}$ **15.** $-\frac{3}{4}$ **17.** $\frac{15}{16}$ **19.** $\frac{1}{6}$ **21.** 6   **23.** $\frac{x}{y}$ **25.** ab **27.** $4a$

29. 40   **31.** $\frac{7}{10}$ **33.** 13   **35.** 12   **37.** 186   **39.** 646   **41.** $\frac{3}{5}$ **43.** 40   **45.** $3 \div \frac{1}{5} = 3 \cdot \frac{5}{1} = 3 \cdot 5$ **47.** 14 blankets

49. 48 bags   **51.** 6   **53.** $\frac{14}{32} = \frac{7}{16}$ **55.** $\frac{7}{16}(4,064) = 7(254) = 1,778$ students   **57.** 28 cartons   **59.** 2   **61.** 2   **63.** 0

65. $\frac{3,479}{3,589}$ **67.** $\frac{493}{1,245}$ **69.** $\frac{21,113}{19,341}$ **71.** $\frac{3}{6}$ **73.** $\frac{9}{6}$ **75.** $\frac{4}{12}$ **77.** $\frac{8}{12}$ **79.** $\frac{10x}{12x}$ **81.** $\frac{8x}{12x}$

Problem Set 3.5

1. $\frac{2}{3}$ **3.** $-\frac{1}{4}$ **5.** $\frac{1}{2}$ **7.** $\frac{x-1}{3}$ **9.** $\frac{3}{2}$ **11.** $\frac{x+6}{2}$ **13.** $-\frac{3}{5}$ **15.** $\frac{10}{a}$

17.

FIRST NUMBER a	SECOND NUMBER b	THE SUM OF a AND b $a+b$
$\frac{1}{2}$	$\frac{1}{3}$	$\frac{5}{6}$
$\frac{1}{3}$	$\frac{1}{4}$	$\frac{7}{12}$
$\frac{1}{4}$	$\frac{1}{5}$	$\frac{9}{20}$
$\frac{1}{5}$	$\frac{1}{6}$	$\frac{11}{30}$

19.

FIRST NUMBER a	SECOND NUMBER b	THE SUM OF a AND b $a+b$
$\frac{1}{12}$	$\frac{1}{2}$	$\frac{7}{12}$
$\frac{1}{12}$	$\frac{1}{3}$	$\frac{5}{12}$
$\frac{1}{12}$	$\frac{1}{4}$	$\frac{4}{12} = \frac{1}{3}$
$\frac{1}{12}$	$\frac{1}{6}$	$\frac{3}{12} = \frac{1}{4}$

21. $\frac{7}{9}$ **23.** $\frac{7}{3}$ **25.** $\frac{1}{4}$ **27.** $\frac{7}{6}$

29. $\frac{5x+4}{20}$ **31.** $\frac{10+3x}{5x}$ **33.** $\frac{19}{24}$ **35.** $\frac{13}{60}$ **37.** $\frac{10a+1}{100}$ **39.** $\frac{3x+28}{7x}$ **41.** $\frac{29}{35}$ **43.** $\frac{949}{1,260}$ **45.** $\frac{13}{420}$ **47.** $\frac{41}{24}$

49. $\frac{7y+24}{6y}$ **51.** $\frac{5}{4}$ **53.** $\frac{160}{63}$ **55.** $\frac{5}{8}$ **57.** $\frac{9}{2}$ pints   **59.** $\frac{5}{8}(2,120) = 5(265) = \$1,325$ **61.** $\frac{2}{5}$

63.

GRADE	NUMBER OF STUDENTS	FRACTION OF STUDENTS
A	5	$\frac{1}{8}$
B	8	$\frac{1}{5}$
C	20	$\frac{1}{2}$
below C	7	$\frac{7}{40}$
Total	40	1

65. 10 lots   **67.** $\frac{3}{2}$ in.   **69.** $\frac{9}{5}$ ft   **71.** $\frac{66}{497}$ **73.** $\frac{610}{437}$ **75.** $\frac{9}{10}$

77. -8 **79.** 3   **81.** 2   **83.** xy^2 **85.** $\frac{2}{7}$ **87.** $\frac{15}{22}$ **89.** $\frac{7}{3}$ **91.** 3

Problem Set 3.6

1. $\frac{14}{3}$ **3.** $\frac{21}{4}$ **5.** $\frac{13}{8}$ **7.** $\frac{47}{3}$ **9.** $\frac{104}{21}$ **11.** $\frac{427}{33}$ **13.** $1\frac{1}{8}$ **15.** $4\frac{3}{4}$ **17.** $4\frac{5}{6}$ **19.** $3\frac{1}{4}$ **21.** $4\frac{1}{27}$ **23.** $28\frac{8}{15}$

25. $\frac{8x+3}{x}$ **27.** $\frac{2y-5}{y}$ **29.** $\frac{6x+5}{6}$ **31.** $\frac{3a-1}{3}$ **33.** $\frac{17}{8}$ **35.** \$6   **37.** $\frac{71}{12}$ **39.**

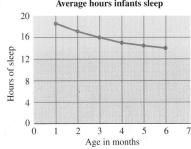

Average hours infants sleep

41. $\frac{1,359}{10}$ **43.** $\frac{824}{23}$ **45.** $\frac{2,512}{17}$ **47.** $\frac{826}{139}$ **49.** $\frac{1,018}{51}$ **51.** $\frac{9}{40}$ **53.** $\frac{3}{8}$ **55.** $-\frac{32}{35}$ **57.** $\frac{3}{2}=1\frac{1}{2}$ **59.** $\frac{4}{5}$ **61.** $\frac{ab}{c}$

Problem Set 3.7

1. $5\frac{1}{10}$ **3.** $13\frac{2}{3}$ **5.** $6\frac{93}{100}$ **7.** $5\frac{5}{6}$ **9.** $9\frac{3}{4}$ **11.** $3\frac{1}{5}$ **13.** $12\frac{1}{2}$ **15.** $9\frac{9}{20}$ **17.** $\frac{32}{45}$ **19.** $1\frac{2}{3}$ **21.** 4   **23.** $4\frac{3}{10}$

25. $\frac{1}{10}$ **27.** $3\frac{1}{5}$ **29.** $2\frac{1}{8}$ **31.** $7\frac{1}{2}$ **33.** $\frac{11}{13}$ **35.** $5\frac{1}{2}$ cups   **37.** $\frac{1}{3}\cdot 2\frac{1}{2}=\frac{5}{6}$ cup   **39.** $1\frac{1}{3}$ **41.** $1,087\frac{1}{5}$ cents

43. $164\frac{3}{4}$ mi   **45.** $4\frac{1}{2}$ yd   **47.** $2\frac{1}{4}$ yd^2 **49.** $9\frac{3}{8}$ in^2

51. Can 1 contains $157\frac{1}{2}$ calories, whereas Can 2 contains $87\frac{1}{2}$ calories. Can 1 contains 70 more calories than Can 2.

53. Can 1 contains 1,960 milligrams of sodium, whereas Can 2 contains 1,050 milligrams of sodium. Can 1 contains 910 more milligrams of sodium than Can 2.

55. $\frac{3,638}{9}$ **57.** $\frac{2,516}{437}$ **59.** $\frac{4,117}{629}$ **61.** 2   **63.** $\frac{25}{168}$ **65.** $\frac{10}{9}=1\frac{1}{9}$ **67.** $2\frac{1}{4}$ **69.** $3\frac{1}{16}$ **71.** $1\frac{1}{2}$ ft   **73.** $1\frac{1}{2}$ ft

Problem Set 3.8

1. $5\frac{4}{5}$ **3.** $12\frac{2}{5}$ **5.** $3\frac{4}{9}$ **7.** 12   **9.** $1\frac{3}{8}$ **11.** $14\frac{1}{6}$ **13.** $4\frac{1}{12}$ **15.** $2\frac{1}{12}$ **17.** $26\frac{7}{12}$ **19.** 12   **21.** $2\frac{1}{2}$ **23.** $8\frac{6}{7}$

25. $3\frac{3}{8}$ **27.** $10\frac{4}{15}$ **29.** $2\frac{1}{15}$ **31.** 9   **33.** $18\frac{1}{10}$ **35.** 14   **37.** 17   **39.** $24\frac{1}{24}$ **41.** $27\frac{6}{7}$ **43.** $37\frac{3}{20}$ **45.** $6\frac{1}{4}$ **47.** $9\frac{7}{10}$

49. $5\frac{1}{2}$ **51.** $\frac{2}{3}$ **53.** $1\frac{11}{12}$ **55.** $3\frac{11}{12}$ **57.** $5\frac{19}{20}$ **59.** $5\frac{1}{2}$ **61.** $\frac{13}{24}$ **63.** $3\frac{1}{2}$ **65.** $5\frac{29}{40}$ **67.** $12\frac{1}{4}$ in.   **69.** $\frac{3}{4}$ in.

71. $31\frac{1}{6}$ in.   **73.** $5\frac{7}{8}$ ¢   **75.** NFL: $P=306\frac{2}{3}$ yd; Canadian: $P=350$ yd; Arena: $P=156\frac{2}{3}$ yd   **77. a.** $2\frac{1}{2}$ **b.** \$250   **79.** \$300

81. $7\frac{1}{2}$ yd   **83.** $15\frac{3}{4}$ in.   **85.** $1,154\frac{790}{2,173}$ **87.** $279\frac{3,691}{5,777}$ **89.** 17   **91.** 14   **93.** 104   **95.** 96   **97.** 40

Problem Set 3.9

1. 7   **3.** 7   **5.** 2   **7.** 35   **9.** $\frac{7}{8}$ **11.** $8\frac{1}{3}$ **13.** $\frac{11}{36}$ **15.** $3\frac{2}{3}$ **17.** $6\frac{3}{8}$ **19.** $4\frac{5}{12}$ **21.** $\frac{8}{9}$ **23.** $\frac{1}{2}$ **25.** $1\frac{1}{10}$ **27.** 5

29. $\frac{3}{5}$ **31.** $\frac{7}{11}$ **33.** 5   **35.** $\frac{17}{28}$ **37.** $1\frac{7}{16}$ **39.** $\frac{13}{22}$ **41.** $\frac{5}{22}$ **43.** $\frac{15}{16}$ **45.** $1\frac{5}{17}$ **47.** $\frac{3}{29}$ **49.** $1\frac{34}{67}$ **51.** $\frac{346}{441}$ **53.** $5\frac{2}{5}$

55. 8   **57.** $115\frac{2}{3}$ yd   **59.** \$690   **61.** $\$\frac{9}{80}$ **63.** $-\$462\frac{1}{2}$ **65.** $\frac{2}{3}$ **67.** $-\frac{1}{6}$ **69.** $-\frac{1}{7}$ **71.** $9\frac{7}{9}$ **73.** $\frac{5x+12}{15}$ **75.** $\frac{4x-2}{x}$

77. $\frac{21+2a}{3a}$

Chapter 3 Review

1. $\frac{3}{4}$ **3.** $\frac{11a^2}{7}$ **5.** x **7.** $\frac{8}{21}$ **9.** $\frac{2}{3}$ **11.** a^2b **13.** $\frac{1}{2}$ **15.** $-\frac{7}{2}$ **17.** $\frac{1}{36}$ **19.** $\frac{29}{8}$ **21.** $\frac{4x-3}{4}$ **23.** $\frac{8}{13}$ **25.** 12

27. $17\frac{11}{12}$ **29.** $11\frac{2}{3}$ **31.** 5   **33.** $\frac{1}{2}$ **35.** 20 items   **37.** 9   **39.** $1\frac{7}{8}$ cups   **41.** $10\frac{1}{2}$ tablespoons

43. Area $= 25\frac{1}{5}$ ft²; perimeter $= 28$ ft

Chapter 3 Test

1.

$\frac{1}{8}$ $\frac{3}{8}$ $\frac{5}{8}$ $\frac{7}{8}$

2. a. $\frac{2}{3}$ **b.** $\frac{13y}{5}$ **3.** $18x$ **4.** $\frac{8}{35}$ **5.** $\frac{8}{27}$ **6.** $-\frac{1}{10}$ **7.** $\frac{2}{5}$ **8.** $\frac{3}{x}$ **9.** $-\frac{23}{5}$

10. $\frac{15+2x}{5x}$ **11.** $\frac{47}{36}$ **12.** $\frac{37}{7}$ **13.** $8\frac{3}{5}$ **14.** $\frac{5x+4}{x}$ **15.** $\frac{9}{2}$ **16.** $9\frac{17}{24}$ **17.** $3\frac{2}{3}$ **18.** $16\frac{3}{4}$ **19.** $9\frac{11}{12}$ **20.** $\frac{1}{2}$

21. 40 grapefruit   **22.** $27\frac{1}{2}$ in.   **23.** $9\frac{1}{3}$ cups   **24.** $3\frac{2}{15}$ ft   **25.** Area $= 23\frac{1}{3}$ ft²; perimeter $= 23\frac{1}{3}$ ft

Chapters 1–3 Cumulative Review

1. 17   **3.** 1844   **5.** $-6x$ **7.** $3t^3 - 4t^2 - 4t + 3$ **9.** 137,280   **11.** 7   **13.** z^{18} **15.** $y^2 - 36$ **17.** $y^2 - 6y + 9$ **19.** $\frac{1}{7}$

21. 1   **23.** $\frac{81}{125}$ **25.** $14\frac{7}{12}$ **27.** $\frac{55}{36}$ **29.** $2\frac{19}{36}$ **31.** 192 in²   **33.** -6 **35.** $\frac{2}{7xy^3}$ **37.** $-\frac{26}{45}$ **39.** 70 pictures

Chapter 4

Problem Set 4.1

1. $10x$ **3.** $4a$ **5.** y **7.** $-8x$ **9.** $3a$ **11.** $-5x$ **13.** $6x + 11$ **15.** $2x + 2$ **17.** $-a + 12$ **19.** $4y - 4$ **21.** $-2x + 4$
23. $8x - 6$ **25.** $5a + 9$ **27.** $-x + 3$ **29.** $17y + 3$ **31.** $a - 3$ **33.** $6x + 16$ **35.** $10x - 11$ **37.** $19y + 32$ **39.** $30y - 18$
41. $6x + 14$ **43.** $12x - 1$ **45.** $27a + 5$ **47.** 14   **49.** 27   **51.** -19 **53.** 14   **55.** 7   **57.** 1   **59.** -9 **61.** 18   **63.** 12
65. -10 **67.** 28   **69.** 40   **71.** 26   **73.** 4   **75.** Complement 65°, supplement 155°, acute   **77. a.** 32°F **b.** 22°F **c.** -18°F
79. a. \$27 **b.** \$47   **81.** -9 **83.** 6   **85.** a   **87.** c   **89.** c

Problem Set 4.2

1. Yes   **3.** Yes   **5.** No   **7.** Yes   **9.** No   **11.** 6   **13.** 11   **15.** -15 **17.** 1   **19.** -3 **21.** -1 **23.** $\frac{7}{5}$ **25.** $\frac{11}{12}$ **27.** $\frac{23}{26}$
29. -4 **31.** -3 **33.** 2   **35.** -6 **37.** -6 **39.** -1 **41.** 1   **43.** 3   **45.** 67°   **47. a.** 225 **b.** \$11,125
49. Equation: $x + 12 = 30$; Solution: 18   **51.** Equation: $8 - 5 = x + 7$; Solution: -4 **53.** $20x$ **55.** $-6y$ **57.** $-12a$ **59.** x
61. y **63.** a

Problem Set 4.3

1. 8   **3.** -6 **5.** -6 **7.** 6   **9.** 16   **11.** 16   **13.** $-\frac{3}{2}$ **15.** -7 **17.** -8 **19.** 6   **21.** 2   **23.** 3   **25.** 4   **27.** -24

29. 12   **31.** -8 **33.** 15   **35.** 3   **37.** -1 **39.** 1   **41.** 3   **43.** 1   **45.** -1 **47.** $-\frac{1}{3}$ **49.** -14 **51.** 4 three-pointers

53. 300 shares   **55.** $2x + 5 = 19; x = 7$ **57.** $5x - 6 = -9; x = -\dfrac{3}{5}$ **59.** $3x + 12$ **61.** $6a - 16$ **63.** $-15x + 3$ **65.** $3y - 9$

67. $16x - 6$

Problem Set 4.4

1. 3   **3.** 2   **5.** -3 **7.** -4 **9.** 1   **11.** 0   **13.** 2   **15.** -2 **17.** 3   **19.** -1 **21.** 7   **23.** -3 **25.** 1   **27.** -2 **29.** 4

31. 10   **33.** -5 **35.** $-\dfrac{1}{12}$ **37.** -3 **39.** 20   **41.** -1 **43.** 5   **45.** 4   **47.** 13   **49.** 12   **51.** $2(x + 6)$ **53.** $x - 4$

55. $2x + 5$

Problem Set 4.5

1. $x + 3$ **3.** $2x + 1$ **5.** $5x - 6$ **7.** $3(x + 1)$ **9.** $5(3x + 4)$ **11.** The number is 2.   **13.** The number is -2.
15. The number is 3.   **17.** The number is 5.   **19.** The number is -2.   **21.** The length is 10 m and the width is 5 m.
23. The length of one side is 8 cm.   **25.** The measures of the angles are 45°, 45°, and 90°.   **27.** The angles are 30°, 60°, and 90°.
29. Patrick is 33 years old, and Pat is 53 years old.   **31.** Sue is 35 years old, and Dale is 39 years old.   **33.** $x = 8, y = 6, z = 9$

35. 39 hours   **37.** $54   **39.** Yes   **41.** $\dfrac{17}{12}$ or $1\dfrac{5}{12}$ **43.** $\dfrac{1}{2}$ **45.** $2\dfrac{4}{5}$ **47.** $\dfrac{8}{5}$ or $1\dfrac{3}{5}$ **49.** $5\dfrac{1}{2}$ **51.** $\dfrac{65}{9}$ or $7\dfrac{2}{9}$

Problem Set 4.6

1. 704 ft^2 **3.** $\dfrac{9}{8} \text{ in}^2$ **5.** $240   **7.** $285   **9.** 12 ft   **11.** 8 ft   **13.** $140   **15.** 58 in.   **17.** $3\dfrac{1}{4} = \dfrac{13}{4}$ ft   **19.** $C = 100°C$; yes

21. $C = 20°C$; yes   **23.** 0°C   **25.** 5°F   **27.** 44 in.   **29.** 22 in.   **31.** 154 ft^2 **33.** $\dfrac{11}{14}$ ft^2 **35.** 360 in^3 **37.** 1 yd^3 **39.** $y = 7$

41. $y = -3$ **43.** $y = -2$ **45.** $x = 3$ **47.** $x = 5$ **49.** $x = 8$ **51.** $y = 1$ **53.** $y = 3$ **55.** $y = \dfrac{5}{2}$ **57.** $y = 0$ **59.** $y = \dfrac{13}{3}$

61. $y = 4$ **63.** $x = 0$ **65.** $x = \dfrac{13}{4}$ **67.** $x = 3$ **69.** Complement is 45°, supplement is 135°

71. Complement is 59°, supplement is 149°   **73.**

NATIONAL LEAGUE					
PITCHER, TEAM	W	L	S	BS	PTS
Trevor Hoffman, S.D.	3	1	39	1	119
Robb Nen, S.F.	7	3	31	3	95
Rod Beck, Chi.	1	2	36	6	94
Jeff Shaw, L.A.	2	5	36	6	90
Ugueth Urbina, Mon.	4	2	26	4	74
Billy Wagner, Hou.	3	3	23	2	65
Kerry Ligtenberg, Atl.	3	2	20	2	58

75. $1\dfrac{4}{5}$ tons

77. $\dfrac{3}{4}$ **79.** 3   **81.** 2   **83.** 6

Problem Set 4.7

1. $(0, 4), (3, 1), (-2, 6)$ **3.** $(0, 3), (2, 2), (18, -6)$ **5.** $(0, 4), (3, 0), (-3, 8)$ **7.** $(1, 1), (0, -3), (5, 17)$ **9.** $(0, 3), (2, 7), (-2, -1)$

11. $(2, 14), \left(\dfrac{6}{7}, 6\right), (0, 0)$ **13.** $(0, 0), (-2, 4), (2, -4)$ **15.** $(0, 0), (2, 1), (4, 2)$ **17.** $(-2, 3), (0, 2), (2, 1)$

19.

x	y
2	3
3	2
1	4
0	5

21.

x	y
0	0
-1	4
-2	8
1	-4

23.

x	y
-2	-6
2	0
4	3
0	-3

25.

x	y
-1	-7
1	5
-2	-13
0	-1

27. $(0, -2)$ **29.** $(0, 3)$ **31.** $(0, 0), (5, -5), (-3, 3)$

33. $(1, 5)$ **35.**

α	θ
0°	90°
30°	60°
45°	45°
60°	30°
75°	15°
90°	0°

37. b   **39.** Area = 6 ft^2; perimeter = 12 ft   **41.** Area = 12 in^2; perimeter = 18 in.

Problem Set 4.8

1.–17.

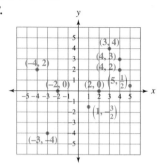

19.–25.

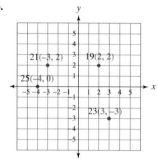

27. Yes **29.** No **31.** Yes **33.** No

35. (0, 50), (10, 110), and (30, 230); there are others

37.

Digital Camera Sales

39. xy^2 **41.** $\dfrac{1}{xy}$ **43.** $\dfrac{4x + 15}{20}$

45. $\dfrac{3x - 1}{x}$

Problem Set 4.9

1.

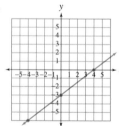

3.

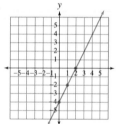

5.

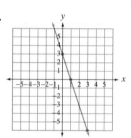

7.

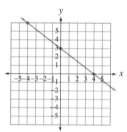

9.

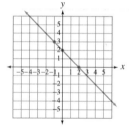

11.

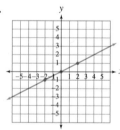

13.

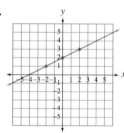

15.

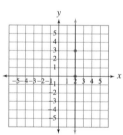

17.

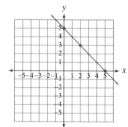

19.

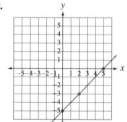

21.

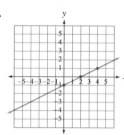

23.

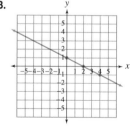

25. 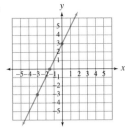 **27.** 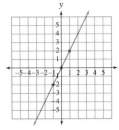 **29.** 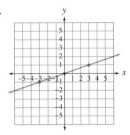 **31.**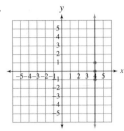

33. 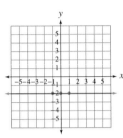 **35.** b **37.** 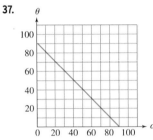 **39.**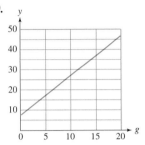

41. $\dfrac{33}{50}$ **43.** $\dfrac{56}{225}$ **45.** $\dfrac{1}{100}$

Chapter 4 Review

1. $17x$ **3.** $11a - 3$ **5.** $5x + 4$ **7.** $10a - 4$ **9.** 42 **11.** 1 **13.** Yes **15.** 9 **17.** 3 **19.** 9 **21.** 5 **23.** -1 **25.** $-\dfrac{1}{8}$

27. The number is -2. **29.** The number is -4. **31.** The length is 14 m, and the width is 7 m. **33.** $y = 6$ **35.** $y = 3$

37. $x = 0$ **39.** $x = 2$ **41.–47.** 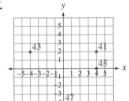 **49.** 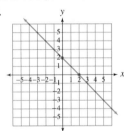 **51.**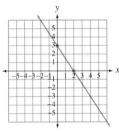

53. 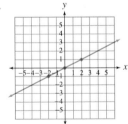 **55.** 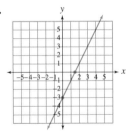 **57.**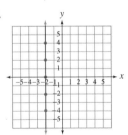

Chapter 4 Test

1. $6x - 5$ **2.** $3b - 4$ **3.** -3 **4.** 9 **5.** Yes **6.** 4 **7.** 27 **8.** -4 **9.** 1 **10.** $\dfrac{2}{3}$ **11.** 1 **12.** The number is -8.

13. The number is 3. **14.** The length is 9 cm, and the width is 5 cm. **15.** Susan is 11 years old, and Karen is 6 years old. **16.** 8

17. 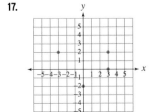 **18.** 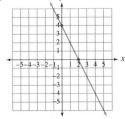 **19.** 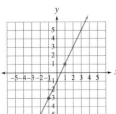 **20.**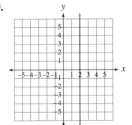

Chapters 1-4 Cumulative Review

1. 16,759   **3.** 12   **5.** $490\frac{1}{32}$ **7.** 21   **9.** -39 **11.** 13   **13.** $108x^{10}y^5$ **15.** $35x^2 + 50x + 15$ **17.** 153   **19.** $6\frac{2}{5}$

21. $-3z + 8$ **23.** $14\frac{4}{5}$ **25.** $a = -2$ **27.** 72 in., 207 in²   **29.** 47   **31.** 9,261 cm³   **33.** 42   **35.** (10, 4), (8, 2), (4, −2)

37. 220 in.   **39.** 10,368 in³

Chapter 5

Problem Set 5.1

1. Three tenths.   **3.** Fifteen thousandths   **5.** Three and four tenths   **7.** Fifty-two and seven tenths   **9.** $405\frac{36}{100}$ **11.** $9\frac{9}{1,000}$
13. $1\frac{234}{1,000}$ **15.** $\frac{305}{100,000}$ **17.** Tens   **19.** Tenths   **21.** Hundred thousandths   **23.** Ones   **25.** Hundreds   **27.** 0.55   **29.** 6.9
31. 11.11   **33.** 100.02   **35.** 3,000.003
37.-45.

	NUMBER	ROUNDED TO THE NEAREST			
		WHOLE NUMBER	TENTH	HUNDREDTH	THOUSANDTH
37.	47.5479	48	47.5	47.55	47.548
39.	0.8175	1	0.8	0.82	0.818
41.	0.1562	0	0.2	0.16	0.156
43.	2,789.3241	2,789	2,789.3	2,789.32	2,789.324
45.	99.9999	100	100.0	100.00	100.000

47. Three and eleven hundredths; two and five tenths   **49.** 186,282.40   **51.** Fifteen hundredths

53.

PRICE OF 1 GALLON OF REGULAR GASOLINE	
DATE	PRICE (DOLLARS)
2/7/00	1.383
2/14/00	1.401
2/21/00	1.444
2/28/00	1.518

55. 0.002 0.005 0.02 0.025 0.05 0.052   **57.** 7.451 7.54   **59.** $\frac{1}{4}$ **61.** $\frac{7}{20}$ **63.** $\frac{1}{8}$

65. $\frac{5}{8}$ **67.** $\frac{7}{8}$ **69.** 9.99   **71.** 10.05   **73.** 0.05   **75.** 0.01   **77.** $6\frac{31}{100}$ **79.** $6\frac{23}{50}$ **81.** $18\frac{123}{1,000}$ **83.** $\frac{2x + 3}{x}$ **85.** $\frac{4x - 3}{4}$

87. 2.85, 2.86, 2.87, 2.88, 2.89, 2.90, 2.91, 2.92, 2.93, 2.94   **89.** 2.5, 2.6, 2.7, 2.8, 2.9, 3.0, 3.1, 3.2, 3.3, 3.4

Problem Set 5.2

1. 6.19   **3.** 1.13   **5.** 1.49   **7.** 9.042   **9.** −1.979   **11.** 11.7843   **13.** 24.343   **15.** 24.111   **17.** 258.5414   **19.** 666.66
21. 11.11   **23.** −3.57   **25.** 4.22   **27.** −120.41   **29.** 44.933   **31.** −8.327   **33.** 530.865   **35.** 27.89   **37.** 35.64
39. 411.438   **41.** 6   **43.** 1   **45.** 3.1   **47.** 5.9   **49.** 3.272   **51.**   **53.** $116.82

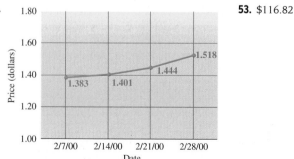

55. $0.29 per hour   **57.** $1,571.10   **59.** $5.43   **61.** $154.12   **63.** $4.83

65.

5.6 seconds   **67.** b   **69.** 219.0842   **71.** 69.34537   **73.** 1,035.7943   **75.** $\frac{3}{100}$

77. $\frac{51}{10,000}$ **79.** $1\frac{1}{2}$ **81.** 1,400   **83.** 2 in.   **85.** $3.25; three $1 bills and a quarter   **87.** 3.25

Problem Set 5.3

1. 0.28   **3.** 0.028   **5.** 0.0027   **7.** −0.78   **9.** 0.792   **11.** 0.0156   **13.** 24.29821   **15.** −0.03   **17.** 187.85   **19.** 0.002
21. −27.96   **23.** 0.43   **25.** 49,940   **27.** 9,876,540   **29.** 1.89   **31.** −0.0025   **33.** −4.9894   **35.** 7.3485   **37.** 4.4
39. 2.074   **41.** 3.58   **43.** 187.4   **45.** 116.64   **47.** 20.75   **49.** 0.126   **51.** Moves it two places to the right   **53.** $6.93
55. 1,319¢ or $13.19   **57.** $415.38   **59.** $7.10   **61.** $44.40   **63.** $293.04   **65.** $C = 25.12$ in.; $A = 50.24$ in^2
67. $C = 37.68$ in.; $A = 113.04$ in^2 **69.** $C = 24,492$ mi   **71.** 100.48 ft^3 **73.** 50.24 ft^3 **75.** 105,957.65   **77.** 202.60296
79. 0.421875   **81.** 1.2321   **83.** 1,879   **85.** 1,516 R 4   **87.** $\frac{3}{4}$ **89.** $\frac{1}{5}$ **91.** 90   **93.** x^2y^3 **95.** b^2c^3 **97.** $4x^2y^2$
99. 0.4; arithmetic; start 0.1; add 0.1   **101.** 0.0001; geometric; start 0.1; multiply by 0.1   **103.** 0.04, 0.008, 0.0016; decreasing
105. 1.1, 1.21, 1.331; increasing

Problem Set 5.4

1. 19.7   **3.** 6.2   **5.** 5.2   **7.** 11.04   **9.** −4.8   **11.** 9.7   **13.** 2.63   **15.** 4.24   **17.** 2.55   **19.** −1.35   **21.** 6.5   **23.** 9.9
25. 0.05   **27.** 89   **29.** 2.2   **31.** 1.35   **33.** 16.97   **35.** 0.25   **37.** 2.71   **39.** 11.69   **41.** False   **43.** True   **45.** 5 six-packs
49. $5.65/hr   **51.** 22.4 mi   **53.** 5 hr

47.

PLAYER	NUMBER OF TOURNAMENTS	TOTAL EARNINGS	AVERAGE EARNINGS PER TOURNAMENT
Karrie Web	25	$1,591,959	$63,678
Juli Inkster	24	$1,337,256	$55,719
Si Re Pak	27	$959,926	$35,553
Annika Sorenstam	22	$863,816	$39,264
Lorie Kane	30	$757,844	$25,261

55. 7 min   **57.** 2.73   **59.** 0.13   **61.** 0.77778   **63.** 307.20607   **65.** 0.70945   **67.** $\frac{3}{4}$ **69.** $\frac{2}{3y}$ **71.** $\frac{3xy}{4}$ **73.** $\frac{6}{10}$ **75.** $\frac{60}{100}$

77. $\frac{12x}{15x}$ **79.** $\frac{60}{15x}$ **81.** $\frac{18}{15x}$

Problem Set 5.5

1. 0.125   **3.** 0.625   **5.**

Fraction	$\frac{1}{4}$	$\frac{2}{4}$	$\frac{3}{4}$	$\frac{4}{4}$
Decimal	0.25	0.5	0.75	1

7.

Fraction	$\frac{1}{6}$	$\frac{2}{6}$	$\frac{3}{6}$	$\frac{4}{6}$	$\frac{5}{6}$	$\frac{6}{6}$
Decimal	$0.1\overline{6}$	$0.\overline{3}$	0.5	$0.\overline{6}$	$0.8\overline{3}$	1

9. 0.48   **11.** 0.4375

13. 0.92   **15.** 0.27   **17.** 0.09   **19.** 0.28   **21.**

Decimal	0.125	0.250	0.375	0.500	0.625	0.750	0.875
Fraction	$\frac{1}{8}$	$\frac{1}{4}$	$\frac{3}{8}$	$\frac{1}{2}$	$\frac{5}{8}$	$\frac{3}{4}$	$\frac{7}{8}$

23. $\frac{3}{20}$ **25.** $\frac{2}{25}$ **27.** $\frac{3}{8}$

29. $5\frac{3}{5}$ **31.** $5\frac{3}{50}$ **33.** $17\frac{13}{50}$ **35.** $1\frac{11}{50}$ **37.** 2.4   **39.** 1.7625   **41.** 3.02   **43.** 0.3   **45.** 0.072   **47.** 0.8   **49.** 1   **51.** 0.2

53. 0.25   **55.** $8.42   **57.** 4 glasses   **59.** $38.66   **61.**

	CHANGE IN STOCK PRICE	
DATE	GAIN/LOSS ($)	AS A DECIMAL ($) (TO THE NEAREST HUNDREDTH)
Monday, March 6, 2000	$\frac{3}{4}$	0.75
Tuesday, March 7, 2000	$-\frac{9}{16}$	−0.56
Wednesday, March 8, 2000	$\frac{3}{32}$	0.09
Thursday, March 9, 2000	$\frac{7}{32}$	0.22
Friday, March 10, 2000	$\frac{1}{16}$	0.06

63. 104.625 Cal   **65.** 33.49 mi^3 **67.** 248.35 in^3 **69.** 22.28 in^2 **71.** 226.1 ft^3 **73.** $0.\overline{7}$ **75.** $0.\overline{123}$ **77.** $10.38   **79.** Yes

81. −9   **83.** −49   **85.** −6   **87.** −3   **89.** 2   **91.** 1   **93.** 10.125, $11\frac{3}{8}$ **95.** $\frac{1}{10,000}$, 0.00001

Problem Set 5.6

1. −1.5   **3.** 0.77   **5.** 0.15   **7.** −0.35   **9.** −0.8   **11.** 2.05   **13.** −0.064   **15.** 12.03   **17.** 0.44   **19.** −0.175   **21.** 2
23. 3,000   **25.** The car was driven 87 miles.   **27.** The car was driven 147 miles.   **29.** Mary has 8 nickels and 18 dimes.
31. You have 16 dimes and 32 quarters.   **33.** The call was 16 minutes long.   **35.** Katie has 7 nickels, 10 dimes, and 12 quarters.
37. 125   **39.** 9   **41.** $\frac{1}{81}$ **43.** $\frac{25}{36}$ **45.** 0.25   **47.** 1.44   **49.** x^{13} **51.** $45a^7$ **53.** $225x^{10}y^6$

Problem Set 5.7

1. 8   **3.** 9   **5.** 6   **7.** 5   **9.** 15   **11.** 48   **13.** 45   **15.** 48   **17.** 15   **19.** 1   **21.** 78   **23.** 9   **25.** $\frac{4}{7}$ **27.** $\frac{3}{4}$
29. False   **31.** True   **33.** 10 in.   **35.** 13 ft   **37.** 6.40 in.   **39.** 17.49 m   **41.** 30 ft   **43.** 25 ft   **45.** 3.4641   **47.** 11.1803
49. 18.0000   **51.** 3.46   **53.** 11.18   **55.** 0.58   **57.** 0.58   **59.** 12.124   **61.** 9.327   **63.** 12.124   **65.** 12.124

67.

HEIGHT h(feet)	DISTANCE d(miles)
10	4
50	9
90	12
130	14
170	16
190	17

69. $\frac{2}{5}$ **71.** $9\frac{8}{25}$ **73.** 8   **75.** $2\frac{4}{5}$ **77.** Side = 6 in.; area = 24 in^2 **79.** 5 ft

Problem Set 5.8

1. $2\sqrt{3}$ **3.** $2\sqrt{5}$ **5.** $6\sqrt{2}$ **7.** $7\sqrt{2}$ **9.** $2\sqrt{7}$ **11.** $10\sqrt{2}$ **13.** $2x\sqrt{3}$ **15.** $5x\sqrt{2}$ **17.** $5x\sqrt{3x}$ **19.** $5x\sqrt{2x}$
21. $4xy\sqrt{2y}$ **23.** $9x^2\sqrt{3}$ **25.** $6xy^2\sqrt{2}$ **27.** $2xy\sqrt{3xy}$ **29.** $3\sqrt{2}$ in.   **31.** $4\sqrt{2}$ in.   **33.** $x\sqrt{2}$ in.   **35.** 1.25 sec
37. See Chapter Introduction   **39.** 8.4852814   **41.** 4.8989795   **43.** 11.18034   **45.** 55.901699

47.

x	$\sqrt{x}$	$2\sqrt{x}$	$\sqrt{4x}$
1	1	2	2
2	1.414	2.828	2.828
3	1.732	3.464	3.464
4	2	4	4

49.

x	$\sqrt{x}$	$3\sqrt{x}$	$\sqrt{9x}$
1	1	3	3
2	1.414	4.243	4.243
3	1.732	5.196	5.196
4	2	6	6

51. −3   **53.** −5   **55.** −2   **57.** 2   **59.** 0   **61.** 5   **63.** 10

Problem Set 5.9

1. $10\sqrt{3}$ **3.** $4\sqrt{5}$ **5.** $7\sqrt{x}$ **7.** $9\sqrt{7}$ **9.** $6\sqrt{y}$ **11.** $7\sqrt{2}$ **13.** $8\sqrt{3}$ **15.** $-2\sqrt{3}$ **17.** $8\sqrt{10}$ **19.** $x\sqrt{2}$ **21.** $17x\sqrt{5x}$
23. $7\sqrt{2}$ ft   **25.** $\sqrt{2} + \sqrt{3} \approx 3.1463$; $\sqrt{5} \approx 2.236$ **27.**

x	$\sqrt{x^2 + 9}$	$x + 3$
1	3.162	4
2	3.606	5
3	4.243	6
4	5	7
5	5.831	8
6	6.708	9

29.

x	$\sqrt{x + 3}$	$\sqrt{x} + \sqrt{3}$
1	2	2.732
2	2.236	3.146
3	2.450	3.464
4	2.646	3.732
5	2.828	3.968
6	3	4.182

31. 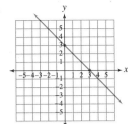   **33.** 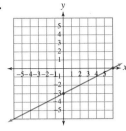   **35.**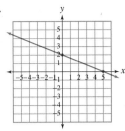

Chapter 5 Review

1. Thousandths   **3.** 37.0042   **5.** 98.77   **7.** 7.779   **9.** −21.5736   **11.** 1.23   **13.** $\frac{141}{200}$ **15.** 2.42   **17.** 79   **19.** −5.9

21. 6.4   **23.** 15   **25.** 47   **27.** $267.70   **29.** 4.875   **31. a.** 2.80 **b.** ≈0.27   **33.** 5.4 ft   **35.** $P = 12.4$ ft; $A = 5$ ft^2 **37.** $2\sqrt{3}$

39. $4\sqrt{2}$ **41.** $2x\sqrt{5}$ **43.** $3xy\sqrt{2y}$ **45.** $10\sqrt{3}$ **47.** $5\sqrt{3}$ **49.** $5\sqrt{6}$ **51.** $-9\sqrt{3}$

Chapter 5 Test

1. Five and fifty-three thousandths   **2.** Thousandths   **3.** 17.0406   **4.** 46.75   **5.** 8.18   **6.** 6.056   **7.** 35.568   **8.** 8.72

9. 0.92   **10.** $\frac{14}{25}$ **11.** 14.664   **12.** 4.69   **13.** 17.129   **14.** 0.26   **15.** 0.19   **16.** 36   **17.** $\frac{5}{9}$ **18.** $11.53   **19.** $19.04

20. $6.55   **21.** 5 in.   **22.** $6\sqrt{2}$ **23.** $4x\sqrt{3y}$ **24.** $8\sqrt{7}$ **25.** $-8\sqrt{3}$

Chapters 1–5 Cumulative Review

1. $\frac{6x + 5}{x}$ **3.** 16,072   **5.** $15a^5 - 10a^3$ **7.** 55.728   **9.** $\frac{1}{2}$ **11.** $15\frac{3}{4}$ **13.** No   **15.** −9   **17.** $3(13 + 4) = 51$ **19.** $x = -3$

21. 7   **23.** −0.34   **25.** $14\frac{7}{8}$ **27.** 28   **29.** 1   **31.** $1\frac{1}{8}$ **33.** 84   **35.** Carrie is 5 years old, and Martin is 12 years old.

37. 105 in^3 **39.** 32 students

Chapter 6

Problem Set 6.1

1. $\frac{4}{3}$ **3.** $\frac{16}{3}$ **5.** $\frac{2}{5}$ **7.** $\frac{1}{2}$ **9.** $\frac{3}{1}$ **11.** $\frac{7}{6}$ **13.** $\frac{7}{5}$ **15.** $\frac{5}{7}$ **17.** $\frac{8}{5}$ **19.** $\frac{1}{3}$ **21.** $\frac{1}{10}$ **23.** $\frac{3}{25}$ **25.** $\frac{1}{3}$

27. 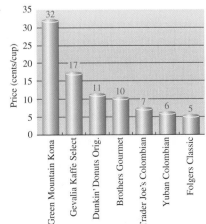   **29. a.** $\frac{5}{8}$ **b.** $\frac{1}{4}$ **c.** $\frac{3}{8}$ **d.** $\frac{5}{3}$ **31.** 55 mi/hr   **33.** 84 km/hr   **35.** 0.2 gal/sec

37. 14 L/min   **39.** 19 mi/gal   **41.** $4\frac{1}{3}$ mi/L   **43.** 16¢ per ounce   **45.** 4.95¢ per ounce

47. a. $\frac{2,314}{2,408} = 0.96$ **b.** $\frac{2,408}{2,314} = 1.04$ **c.** $\frac{2,314}{4,722} = 0.49$ **d.** $\frac{4,722}{2,408} = 1.96$ **49.** 54.03 mi/hr **51.** 9.3 mi/gal **53.** $\frac{8}{3}$, or $2\frac{2}{3}$

55. $\frac{1}{40}$ **57.** $\frac{73}{20}$, or $3\frac{13}{20}$ **59.** $\frac{62}{25}$, or $2\frac{12}{25}$ **61.** $\frac{4}{3}$, or $1\frac{1}{3}$ **63. a.** 10 **b.** 15 **c.** 5 **d.** Both equal $\frac{5}{4}$ **e.** Both equal $\frac{5}{3}$

Problem Set 6.2

1. 60 in. **3.** 120 in. **5.** 6 ft **7.** 162 in. **9.** $2\frac{1}{4}$ ft **11.** $6\frac{1}{3}$ yd **13.** $1\frac{1}{3}$ yd **15.** 1,800 cm **17.** 4,800 m **19.** 50 cm

21. 0.248 km **23.** 670 mm **25.** 34.98 m **27.** 6.34 dm **29.** 20 yd **31.** 80 in. **33.** 244 cm **35.** 65 mm

37. 2,960 chains **39.** 120,000 μm **41.** 7,920 ft **43.** 80.7 ft/sec **45.** 19.5 mi/hr **47.** 1,023 mi/hr **49.** $18,216

51. $157.50 **53.** 3,965,280 ft **55.** 179,352 in. **57.** 2.7 mi **59.** 18,094,560 ft **61.** $\frac{1}{3}$ **63.** 6 **65.** $3\frac{1}{2}$ **67.** 6 **69.** 6

71. $\frac{7}{10}$

Problem Set 6.3

1. 432 in^2 **3.** 2 ft^2 **5.** 1,306,800 ft^2 **7.** 1,280 acres **9.** 3 mi^2 **11.** 108 ft^2 **13.** 1,700 mm^2 **15.** 28,000 cm^2

17. 0.0012 m^2 **19.** 500 m^2 **21.** 700 a **23.** 3.42 ha

25. NFL: $A = 5,333\frac{1}{3}$ sq yd = 1.10 acres; Canadian: $A = 7,150$ sq yd = 1.48 acres; Arena: $A = 1,416\frac{2}{3}$ sq yd = 0.29 acres **27.** 30 a

29. 5,500 bricks **31.** 135 ft^3 **33.** 48 fl oz **35.** 8 qt **37.** 20 pt **39.** 480 fl oz **41.** 8 gal **43.** 6 qt **45.** 9 yd^3

47. 5,000 mL **49.** 0.127 L **51.** 4,000,000 mL **53.** 14,920 L **55.** 16 cups **57.** 34,560 in^3 **59.** 48 glasses

61. 20,288,000 acres **63.** 3,230.93 mi^2 **65.** 23.35 gal **67.** 21,492 ft^3 **69.** 285,795,000 ft^3 **71.** $\frac{5}{8}$ **73.** 9 **75.** $\frac{1}{3}$

77. $\frac{17}{144}$ **79.** $3\frac{1}{18}$

Problem Set 6.4

1. 128 oz **3.** 4,000 lb **5.** 12 lb **7.** 0.9 T **9.** 32,000 oz **11.** 56 oz **13.** 13,000 lb **15.** 2,000 g **17.** 40 mg

19. 200,000 cg **21.** 508 cg **23.** 4.5 g **25.** 47.895 cg **27.** 1.578 g **29.** 0.42 kg **31.** $9\frac{1}{4}$ lb, or 9.25 lb **33.** $9.14

35. 15 g **37.** 63.68 oz **39.** 12.5625 lb **41.** $76.42 **43.** $\frac{9}{50}$ **45.** $\frac{9}{100}$ **47.** $\frac{4}{5}$ **49.** 0.9 **51.** 0.125

Problem Set 6.5

1. 15.24 cm **3.** 13.12 ft **5.** 6.56 yd **7.** 32,200 m **9.** 5.98 yd^2 **11.** 24.7 acres **13.** 8,195 mL **15.** 2.12 qt **17.** 75.8 L

19. 339.6 g **21.** 33 lb **23.** 365°F **25.** 30°C **27.** 3.94 in. **29.** 7.62 m **31.** 46.23 L **33.** 17.67 oz **37.** 91.46 m

39. 20.90 m^2 **41.** 88.55 km/hr **43.** 2.03 m **45.** 38.3°C **47.** 0.75 **49.** 5.5 **51.** 0.03 **53.** $\frac{17}{50}$ **55.** $2\frac{2}{5}$ **57.** $1\frac{3}{4}$

Problem Set 6.6

1. Means: 3, 5; extremes: 1, 15; products: 15 **3.** Means: 25, 2; extremes: 10, 5; products: 50

5. Means: $\frac{1}{2}$, 4; extremes: $\frac{1}{3}$, 6; products: 2 **7.** Means: 5, 1; extremes: 0.5, 10; products: 5 **9.** 10 **11.** $\frac{12}{5}$ **13.** $\frac{3}{2}$ **15.** $\frac{10}{9}$

17. 7 **19.** 1 **21.** 18 **23.** 6 **25.** 40 **27.** 50 **29.** $\frac{1}{2}$ **31.** 3 **33.** 1 **35.** $\frac{1}{4}$ **37.** 329 mi **39.** 360 points **41.** 15 pt

43. 427.5 mi **45.** 900 eggs **47.** 36.25 feet **49.** $119.70 **51.** 265 g **53.** 108 **55.** 65 **57.** 41 **59.** 108 **61.** 91.3 liters

63. 60,113 people **65.** Tens **67.** 26.516 **69.** 0.39 **71.** 1.35 **73.** 3.816 **75.** 4 **77.** 160

Chapter 6 Review

1. $\frac{3}{10}$ **3.** $\frac{3}{4}$ **5.** $\frac{7}{5}$ **7.** $\frac{1}{2}$ **9.** $\frac{1}{3}$ **11.** $\frac{9}{2}$ **13.** $\frac{1}{4}$ **15.** 19 mi/gal **17.** 23¢ per oz **19.** 144 in. **21.** 0.49 m

23. 435,600 ft^2 **25.** 576 in^2 **27.** 6 gal **29.** 128 oz **31.** 5,000 g **33.** 10.16 cm **35.** 7.42 qt **37.** 141.5 g **39.** 248°F

41. 600 bricks **43.** 80 glasses **45.** 49 **47.** $\frac{1}{10}$ **49.** 1,500 mL **51.** 5 weeks

Chapter 6 Test

1. $\dfrac{4}{3}$ **2.** $\dfrac{9}{10}$ **3.** $\dfrac{3}{2}$ **4.** $\dfrac{3}{10}$ **5.** $\dfrac{3}{5}$ **6.** $\dfrac{12}{5}$ **7.** $\dfrac{6}{25}$ **8.** 23 mpg

9. 16-ounce can: 16.2¢/ounce; 12-ounce can: 15.8¢/ounce; 12-ounce can is the better buy   **10.** 21 ft   **11.** 0.75 km

12. 130,680 ft²   **13.** 3 ft²   **14.** 10,000 mL   **15.** 8.05 km   **16.** 10.6 qt   **17.** 26.7°C   **18.** 90 tiles   **19.** 64 glasses   **20.** $21.00

21. 36   **22.** 8   **23.** 24 hits   **24.** 135 miles   **25.** 25 tbsp   **26.** LGB; Twice as large

Chapters 1–6 Cumulative Review

1. $5x^2 - 11x - 1$ **3.** 316   **5.** $16x^2 + 72x + 81$ **7.** $-\dfrac{1}{16}$ **9.** $\dfrac{39}{8}$ **11.** Distributive property of multiplication   **13.** $\dfrac{6}{7}$ **15.** $40x^7$

17. 6   **19.** $\dfrac{8}{9}$ **21.** $2(x + 9) - 4x$ **23.** -0.15 **25.** 5   **27.** 53.06 L   **29.** -2 **31.** 17°   **33.** 20 women   **35.** 97 mi.

37. $10.80   **39.** 20 baskets

Chapter 7

Problem Set 7.1

1. $\dfrac{20}{100}$ **3.** $\dfrac{60}{100}$ **5.** $\dfrac{24}{100}$ **7.** $\dfrac{65}{100}$ **9.** 0.23   **11.** 0.92   **13.** 0.09   **15.** 0.034   **17.** 0.0634   **19.** 0.009   **21.** 23%   **23.** 92%

25. 45%   **27.** 3%   **29.** 60%   **31.** 80%   **33.** 27%   **35.** 123%   **37.** $\dfrac{3}{5}$ **39.** $\dfrac{3}{4}$ **41.** $\dfrac{1}{25}$ **43.** $\dfrac{53}{200}$ **45.** $\dfrac{7,187}{10,000}$ **47.** $\dfrac{3}{400}$

49. $\dfrac{1}{16}$ **51.** $\dfrac{1}{3}$ **53.** 50%   **55.** 75%   **57.** $33\dfrac{1}{3}$%   **59.** 80%   **61.** 87.5%   **63.** 14%   **65.** 325%   **67.** 150%   **69.** 48.8%

71. 0.50; 0.75   **73.** 0.189; 0.081   **75.** 20%   **77.**   **79.** 78.4%   **81.** 11.8%   **83.** 72.2%

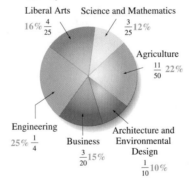

Cal Poly enrollment for Fall 1999

Liberal Arts $16\%\ \dfrac{4}{25}$ Science and Mathematics $\dfrac{3}{25}\ 12\%$

Agriculture $\dfrac{11}{50}\ 22\%$

Engineering $25\%\ \dfrac{1}{4}$ Business $\dfrac{3}{20}\ 15\%$ Architecture and Environmental Design $\dfrac{1}{10}\ 10\%$

85. 8.3%; 0.2%   **87.** 75   **89.** 8.48   **91.** 500   **93.** 180   **95.** 4

97. Divide 2 by 5 to obtain 0.40. Then move the decimal point two places to the right, and attach the % symbol. The result is 40%.

99. 75% is 100 times as large as 0.75%.

Problem Set 7.2

1. 8   **3.** 24   **5.** 20.52   **7.** 7.37   **9.** 50%   **11.** 10%   **13.** 25%   **15.** 75%   **17.** 64   **19.** 50   **21.** 925   **23.** 400

25. 5.568   **27.** 120   **29.** 13.72   **31.** 22.5   **33.** 50%   **35.** 942.684   **37.** 97.8   **39.** 4.8% calories from fat; healthy

41. 50% calories from fat; not healthy   **43.** 248.75   **45.** 25.9%   **47.** 49.5   **49.** $3 \cdot 3 \cdot 5$ **51.** $3 \cdot 5 \cdot 7$ **53.** $2 \cdot 2 \cdot 3 \cdot 3$

55. $2 \cdot 3 \cdot 5 \cdot 7$ **57.** What number is 25% of 350?   **59.** What percent of 24 is 16?   **61.** 46 is 75% of what number?

Problem Set 7.3

1. 70%   **3.** 84%   **5.** 45 mL   **7.** 18.2 acres for farming; 9.8 acres are not available for farming   **9.** 3,000 students

11. 400 students   **13.** 1,664 women students   **15.** 31.25%   **17.** 50%   **19.** 42 mL   **21.** $\dfrac{1}{5}$ **23.** $\dfrac{5}{12}$ **25.** $\dfrac{3}{4}$ **27.** $\dfrac{2}{3}$

29. 38.1%, to the nearest tenth of a percent   **31.** 66 hits   **33.** At least 19 hits

Problem Set 7.4

1. $52.50   **3.** $2.70; $47.70   **5.** $150; $156   **7.** 5%   **9.** $2,820   **11.** $200   **13.** 14%   **15.** $11.93   **17.** 4.5%   **19.** $3,995

21. $16,300   **23.** 860 items   **25.** $\frac{1}{2}$ **27.** $2\frac{2}{3}$ **29.** $1\frac{1}{2}$ **31.** $4\frac{1}{2}$ **33.** Sales tax = $3,180; luxury tax = $2,300   **35.** $2,300

37. You saved $1,600 on the sticker price and $150 in luxury tax. If you lived in a state with a 6% sales tax rate, you saved an additional 0.06($1600) = $96.

Problem Set 7.5

1. $24,610   **3.** $3,510   **5.** $13,200   **7.** 10%   **9.** 20%   **11. a.** 150% **b.** 540% **c.** 1.9 million   **13.** $45; $255   **15.** $381.60

17. $46,595.88   **19. a.** 51.9% **b.** 7.8%   **21.** 1   **23.** $\frac{3}{4}$ **25.** $1\frac{5}{12}$ **27.** 6   **29.** .500   **31.** .001   **33.** .333   **35.** .001

Problem Set 7.6

1. $2,160   **3.** $665   **5.** $8,560   **7.** $2,160   **9.** $5   **11.** $813.33   **13.** $5,618   **15.** $8,407.56   **17.** $974.59

19. a. $13,468.55 **b.** $13,488.50 **c.** $12,820.37 **d.** $12,833.59   **21.** $\frac{1}{2}$ **23.** $\frac{3}{8}$ **25.** $\frac{1}{12}$ **27.** $1\frac{1}{4}$ **29.** 5.7% alcohol

31. Percent increase in production costs: *Star Wars* 1 to 2, 63.6%; *Star Wars* 2 to 3, 80.6%; *Star Wars* 3 to 4, 254%

Chapter 7 Review

1. 0.35   **3.** 0.05   **5.** 95%   **7.** 49.5%   **9.** $\frac{3}{4}$ **11.** $1\frac{9}{20}$ **13.** 30%   **15.** $66\frac{2}{3}$%   **17.** 16.8   **19.** 80   **21.** $72   **23.** $477

25. $1.45   **27.** 55.4% increase   **29.** $42

Chapter 7 Test

1. 0.18   **2.** 0.04   **3.** 0.005   **4.** 45%   **5.** 70%   **6.** 135%   **7.** $\frac{13}{20}$ **8.** $1\frac{23}{50}$ **9.** $\frac{7}{200}$ **10.** 35%   **11.** 37.5%   **12.** 175%

13. 45   **14.** 45%   **15.** 80   **16.** 92%   **17.** $960   **18.** $40; 16% off   **19.** $220.50   **20.** $100   **21.** $2,520

22. 7.2% decrease in area

Chapters 1–7 Cumulative Review

1. 5,996   **3.** 4.559   **5.** $18x^2 - 9x - 35$ **7.** 42   **9.** 440,000   **11.** $6\frac{1}{3}$ **13.** 80%   **15.** $1\frac{6}{25}$ **17.** $4y\sqrt{2x}$ **19.** -7 **21.** 4

23. 56   **25.** 5   **27.** $4x + 1$ **29.** 82 in²   **31.** 16 mpg   **33.** 13 m   **35.** $21   **37.** $-$45   **39.** 240 glasses

Appendix 1

1. 4   **2.** 10   **3.** 9   **4.** 3   **5.** 5   **6.** 7   **7.** 11   **8.** 8   **9.** 17   **10.** 13   **11.** 10   **12.** 13   **13.** 9   **14.** 5   **15.** 1
16. 10   **17.** 8   **18.** 14   **19.** 13   **20.** 7   **21.** 6   **22.** 1   **23.** 11   **24.** 7   **25.** 12   **26.** 9   **27.** 7   **28.** 10   **29.** 6
30. 4   **31.** 8   **32.** 13   **33.** 6   **34.** 8   **35.** 9   **36.** 9   **37.** 10   **38.** 5   **39.** 14   **40.** 18   **41.** 0   **42.** 15   **43.** 11
44. 10   **45.** 10   **46.** 13   **47.** 12   **48.** 13   **49.** 9   **50.** 4   **51.** 12   **52.** 1   **53.** 11   **54.** 12   **55.** 15   **56.** 17   **57.** 3
58. 8   **59.** 3   **60.** 6   **61.** 12   **62.** 7   **63.** 2   **64.** 16   **65.** 11   **66.** 12   **67.** 11   **68.** 9   **69.** 16   **70.** 8   **71.** 15
72. 6   **73.** 5   **74.** 11   **75.** 16   **76.** 3   **77.** 9   **78.** 4   **79.** 11   **80.** 10   **81.** 8   **82.** 2   **83.** 10   **84.** 14   **85.** 8
86. 15   **87.** 5   **88.** 12   **89.** 14   **90.** 4   **91.** 6   **92.** 7   **93.** 8   **94.** 9   **95.** 9   **96.** 14   **97.** 7   **98.** 5   **99.** 7
100. 6

Appendix 2

1. 12   **2.** 20   **3.** 0   **4.** 0   **5.** 10   **6.** 7   **7.** 16   **8.** 14   **9.** 7   **10.** 45   **11.** 12   **12.** 15   **13.** 15   **14.** 20   **15.** 8
16. 21   **17.** 40   **18.** 0   **19.** 0   **20.** 0   **21.** 24   **22.** 16   **23.** 35   **24.** 0   **25.** 14   **26.** 6   **27.** 8   **28.** 28   **29.** 24
30. 48   **31.** 20   **32.** 21   **33.** 0   **34.** 32   **35.** 18   **36.** 30   **37.** 9   **38.** 16   **39.** 0   **40.** 12   **41.** 36   **42.** 0   **43.** 0

44. 54 **45.** 12 **46.** 24 **47.** 63 **48.** 9 **49.** 3 **50.** 81 **51.** 8 **52.** 56 **53.** 10 **54.** 56 **55.** 4 **56.** 0 **57.** 72

58. 6 **59.** 45 **60.** 49 **61.** 30 **62.** 0 **63.** 27 **64.** 72 **65.** 63 **66.** 24 **67.** 3 **68.** 5 **69.** 54 **70.** 2 **71.** 4

72. 4 **73.** 18 **74.** 5 **75.** 18 **76.** 54 **77.** 0 **78.** 0 **79.** 42 **80.** 0 **81.** 0 **82.** 27 **83.** 1 **84.** 48 **85.** 0

86. 28 **87.** 32 **88.** 0 **89.** 9 **90.** 40 **91.** 8 **92.** 24 **93.** 35 **94.** 2 **95.** 36 **96.** 6 **97.** 12 **98.** 6 **99.** 42

100. 0

Appendix 3

Problem Set A.3

1. $\dfrac{1}{32}$ **3.** $\dfrac{1}{100}$ **5.** $\dfrac{1}{x^3}$ **7.** $\dfrac{1}{16}$ **9.** $-\dfrac{1}{125}$ **11.** $\dfrac{1}{4}$ **13.** 1,000 **15.** $\dfrac{1}{x^7}$ **17.** 9 **19.** $\dfrac{1}{8}$ **21.** 1,000 **23.** $\dfrac{1}{x^{12}}$ **25.** 2

27. 4 **29.** $\dfrac{1}{4}$ **31.** x **33.** $\dfrac{1}{x}$ **35.** $\dfrac{1}{10^7}$ **37.** $\dfrac{1}{10^7}$ **39.** 1,000 **41.** x^{24} **43.** 4 **45.** 10,000 **47.** $15x^5$ **49.** $\dfrac{6}{x^5}$

51. $-\dfrac{12}{x^7}$ **53.** $42x^2$ **55.** x^5 **57.** x^2

Appendix 4

Problem Set A.4

1. 4.25×10^5 **3.** 6.78×10^6 **5.** 5.4×10^2 **7.** 1.1×10^4 **9.** 8.9×10^7 **11.** 38,400 **13.** 57,100,000 **15.** 600 **17.** 3,300

19. 89,130,000 **21.** 3.5×10^{-4} **23.** 7×10^{-4} **25.** 1.2×10^{-2} **27.** 6.035×10^{-2} **29.** 1.276×10^{-1} **31.** 0.00083

33. 0.0625 **35.** 0.000011 **37.** 0.3125 **39.** 0.005 **41.–49.**   **51.** 7.88×10^8

	STANDARD FORM		SCIENTIFIC NOTATION
41.	4,200,000	=	4.2×10^6
43.	275,000,000	=	2.75×10^8
45.	0.00496	=	4.96×10^{-3}
47.	0.00051	=	5.1×10^{-4}
49.	47,892,000	=	4.7892×10^7

53. $1,092,000 **55.**

GAME	YEAR	AD COST	COST IN SCIENTIFIC NOTATION
Super Bowl I	1967	$42,000	4.2×10^4
Super Bowl XXII	1985	$525,000	5.25×10^5
Super Bowl XXXIII	1999	$1,600,000	1.6×10^6

Appendix 5

Problem Set A.5

1. 6×10^{10} **3.** 1.5×10^{11} **5.** 6.84×10^{-2} **7.** 1.62×10^{10} **9.** 3.78×10^{-1} **11.** 2×10^3 **13.** 4×10^{-9} **15.** 5×10^{14}

17. 6×10^1 **19.** 2×10^{-8} **21.** 4×10^6 **23.** 4×10^{-4} **25.** 2.1×10^{-2} **27.** 3×10^{-5} **29.** 6.72×10^{-1}

Index

End-User License Agreement

This software and data on the enclosed Digital Video Companion ("DVC") is provided to the purchaser (the "User") of *Prealgebra, Fourth Edition* (the "Textbook") on the condition that the User agrees to the terms and conditions set forth in this End-User License Agreement. Read this End-User License Agreement carefully. You will be bound by the terms of this Agreement if you have opened the sealed DVC.

1. License

The author of the Textbook (the "Author") grants, and the User hereby accepts, a nonexclusive, nontransferable license to use the software, data, and documentation contained in the DVC on the terms and conditions set forth in this Agreement. The license granted hereunder authorizes the User to use the DVC in connection with the Textbook only, on a single personal computer system for User's own use. The DVC may not be loaned, given, sold or transferred, in any form or medium, to any other person, without the Author's prior written consent.

2. Distribution

The publisher and owner of the Textbook (the "Publisher") is distributing the DVC solely in connection with its sale of the Textbook.

3. Termination

The User may terminate this license at any time by destroying the CD-ROM, print-outs, and any files derived from the CD-ROM. All rights to use the materials in the DVC will terminate if the User fails to comply with any of the terms or conditions of this Agreement. Upon such termination, the User will destroy the CD-ROM, print-outs, and any files derived from the CD-ROM.

4. Copyright and Restrictions

All ownership rights in the DVC remains with the Author.

None of the content of the DVC, or any updates thereof, may be reproduced, transcribed, stored in a retrieval system, translated into any language or computer language, transmitted in any form or by any means (electronic, mechanical, photocopied, recorded, or otherwise), resold, or redistributed without the prior written consent of the Author, except that the User may reproduce limited excerpts of the data for his or her own use.

5. Protection and Security

The User shall ensure that no unauthorized person shall have access to DVC and that no unauthorized copy of any part of it is made in any form.

6. Limited Warranty

Author warrants that the media on which the DVC is distributed will conform with its specifications set forth in the documentation therefor.

THE FOREGOING WARRANTY IS IN LIEU OF ALL OTHER WARRANTIES, EXPRESSED OR IMPLIED. AUTHOR AND PUBLISHER SPECIFICALLY DISCLAIM ALL IMPLIED WARRANTIES OF MERCHANTABILITY AND FITNESS FOR A PARTICULAR PURPOSE WITH RESPECT TO SAID MATERIALS OR THE USE THEREOF. AUTHOR AND PUBLISHER SHALL NOT BE LIABLE UNDER ANY CLAIM, DEMAND, OR ACTION ARISING OUT OF, OR RELATING TO THIS AGREEMENT FOR DIRECT, SPECIAL, INCIDENTAL, OR CONSEQUENTIAL DAMAGES, INCLUDING, WITHOUT LIMITATION, LOSS OF PROFITS OR OTHER DAMAGES CAUSED BY THE INABILITY TO USE THE DVC, WHETHER OR NOT AUTHOR OR PUBLISHER HAS BEEN ADVISED OF THE POSSIBILITY OF SUCH CLAIM, DEMAND, OR ACTION. AUTHOR AND PUBLISHER MAKE NO WARRANTY THAT THE DVC IS COMPATIBLE OR OPERABLE WITH THE USER'S COMPUTER EQUIPMENT OR SOFTWARE, OR THAT THE DVC WILL PERFORM WITHOUT INTERRUPTION OR FREE OF ERRORS.

9. General

If any provision of this Agreement is determined to be invalid under any statute or rule of law, the same shall be deemed omitted and the remaining provisions shall continue in full force and effect.

This Agreement shall be deemed to be made in the State of New York and shall in all respects be interpreted, construed, and governed by and in accordance with the laws of the State of New York applicable to contracts executed and to be wholly performed therein.

This Agreement constitutes the entire agreement between the parties with respect to the subject matter hereof and supersedes all prior agreements, oral or written, and all other communications relating to the subject matter hereof. No amendment or modification of any provision of this Agreement will be effective unless set forth in a document that purports to amend this Agreement and that is accepted by both parties hereto.